T0323974

Essentials of the Finite Element Method

Essentials of the Finite Element Method

For Mechanical and Structural Engineers

Dimitrios G. Pavlou, PhD
Department of Mechanical and Structural Engineering and Materials Science,
University of Stavanger, Norway

AMSTERDAM • BOSTON • HEIDELBERG • LONDON
NEW YORK • OXFORD • PARIS • SAN DIEGO
SAN FRANCISCO • SINGAPORE • SYDNEY • TOKYO
Academic Press is an imprint of Elsevier

Academic Press is an imprint of Elsevier
125 London Wall, London, EC2Y 5AS, UK
525 B Street, Suite 1800, San Diego, CA 92101-4495, USA
225 Wyman Street,Waltham, MA 02451, USA
The Boulevard, Langford Lane, Kidlington, Oxford OX5 1GB, UK

Library of Congress Cataloging-in-Publication Data
A catalog record for this book is available from the Library of Congress

British Library Cataloguing-in-Publication Data
A catalogue record for this book is available from the British Library

ISBN: 978-0-12-802386-0

For information on all Academic Press publications
visit our website at elsevierdirect.com

Typeset by SPi Global, India

Printed and bound in the USA

ELSEVIER
Book Aid International
Working together to grow libraries in developing countries
www.elsevier.com • www.bookaid.org

To my children, Evangelia and Georgios Pavlou, and to my wife, Mina, for their love and patience.

Contents

Preface

In last decades, "finite elements" (FE) has become a standard course in study programs of mechanical and civil engineering specialties and tends to be unique to engineering design practice. Since FEM is a numerical method, its evolution concurs with the evolution of digital computing technology. Today, contemporary FEM software packages allow fast and accurate design of complex engineering problems pertaining in the fields of solid mechanics, heat transfer, fluid mechanics, and electrical engineering.

The aim of this book is to provide Bachelor (BSc) and Master (MSc) of Science students, as well as professional mechanical and civil engineers, with a complete and unified coverage of finite element analysis and to demonstrate how FEM can be programmed. Throughout the text, readers are shown step-by-step detailed FE analyses of integrated engineering problems. For this purpose, analytic mathematics is used for the development of stiffness matrices for widely used elements in mechanical and structural engineering practice. To help readers to understand how the boundary conditions can be taken into account in the procedure of the FE modeling, a special type of structural matrix equation incorporating both global stiffness matrix and submatrices containing the boundary conditions in a unified form is proposed.

After completing FEM courses, new engineers would be expected to have adequate knowledge to use a commercial FEM program in their first job. Among the targets of the book is to assist readers to understand the architecture of FEM software, as well as its limitations. Since computer coding to solve FEM problems is the required background for subsequent learning how to use commercial package software intelligently and critically, effort has been made to teach readers how to develop their own programs in order to understand the fundamental abilities of commercial packages. Therefore, the aims of the book are: (1) to provide the tools to help students and engineers to be software developers, and (2) to provide "how-to" knowledge in running a commercial FEM program. To achieve the first aim, the book uses a contemporary computer-aided learning platform, called CALFEM, which adopts the facilities of the well-known computational matrix laboratory MATLAB. For common types of structures (e.g., trusses, beams, frames, hybrid structures, structural dynamic problems, etc.), the book provides the logical steps for developing an FEM computer algorithm. In the offered examples, the analysis of the required software commands, the computer code, and the numerical/graphic results are also exhibited. To achieve the second aim, a self-learning on "how-to" run the widely used commercial FEM program ANSYS is provided. To this end, most of the examples treated with the CALFEM/MATLAB platform are used for step-by-step ANSYS learning. Therefore, the ANSYS windows for data entry are provided and the required commands are described through a combination of text and graphics.

This book introduces FE methodology with simple concepts (e.g., the method of direct equilibrium) and progresses to more complicated principles (such as variational methods), allowing a smooth transition of the reader to deeper knowledge of the method.

Dimitrios G. Pavlou, PhD
Professor
Department of Mechanical and Structural Engineering
and Materials Science, University of Stavanger, Norway

Acknowledgments

I would like to thank my students of the Department of Mechanical and Structural Engineering and Materials Science for their useful comments during the lectures.

I am grateful to Professor Ivar Langen and to Associate Professors Hirpa Lemu and Ove Mikkelsen for their support during the preparation of my FEM courses. Thanks also to Overingeniør Adugna Akessa for his help to implement the computer algorithms to the FEM lab.

I express my deepest appreciation to the Head of the Department, Professor Per Skjerpe, for the excellent working conditions allowing me to write this book.

Six reviewers provided immensely useful comments to improve the content of this book. For their time, I am deeply grateful.

Particular thanks go to Asimina Kechagia for her patience in reading this book and making valuable linguistic comments.

In particular, I wish to express my enormous appreciation to Joseph P. Hayton, Publisher of Elsevier, for giving me the opportunity to develop this book, and to Kattie Washington, Senior Editorial Project Manager, and to Chelsea Johnston, former Editorial Project Manager of Elsevier for their valuable help in issues regarding the publishing procedure.

CHAPTER

AN OVERVIEW OF THE FINITE ELEMENT METHOD

1

1.1 WHAT ARE FINITE ELEMENTS?

Since the differential equations describing the displacement field of a structure are difficult (or impossible) to solve by analytical methods, the domain of the structural problem can be divided into a large number of small subdomains, called finite elements (FE). The displacement field of each element is approximated by polynomials, which are interpolated with respect to prescribed points (nodes) located on the boundary (or within) the element. The polynomials are referred to as interpolation functions, where variational or weighted residual methods are applied to determine the unknown nodal values.

1.2 WHY FINITE ELEMENT METHOD IS VERY POPULAR?

The concept of the finite element method (FEM) was described in 1956, when Turner et al. used pin-jointed bars and triangular plates to calculate aircraft structures. However, as the method is based on the solution of systems of algebraic equations with large number of unknowns, in past few decades, FEM has become very popular due to the development of high-speed digital computers.

After 1980, new commercial software packages were developed, boosting the application of FEM to structural engineering, heat transfer, fluid mechanics, aerodynamics, and electrostatics.

Among the pioneers who founded and developed FEM are Przemieniecki, Zienkiewicz and Cheung, Gallagher, Argyris, etc.

1.3 MAIN ADVANTAGES OF FINITE ELEMENT METHOD

1. Analyzes problems with complex geometry.
2. Analyzes problems with complex loading (point loads, pressure, inertial forces, thermal loading, fluid-structure interactions, etc.).
3. Analyzes a wide variety of engineering problems (structural engineering, heat transfer, fluid mechanics, aerodynamics, and electrostatics).

1.4 MAIN DISADVANTAGES OF FINITE ELEMENT METHOD

1. FEM results are approximate. Their accuracy depends on the number of elements, the type of elements, the adopted assumptions, etc.

Essentials of the Finite Element Method. http://dx.doi.org/10.1016/B978-0-12-802386-0.00001-3

2. The accuracy of the results of FEM depends on the experience of the software user, for example, the use of the wrong type or distorted elements, insufficient supports to prevent all rigid body motions, and different units for the same quantity yields mistakes.
3. FEM has inherent errors (e.g., the geometry of the structure is approximate, the field deformation is assumed to be a polynomial over the element, the computer carries only a finite number of digits, the combination of elements with very large stiffness differences yields numerical difficulties).

1.5 WHAT IS STRUCTURAL MATRIX?

Structural matrix is a matrix correlating the forces and displacements in the nodal points of the elements. For a structural FE, the structural matrix contains the geometric and material behavior information that indicated the resistance of the element to deformation when subjected to loading. The primary characteristics of an FE are embodied in the element structural matrix. There are two types of structural matrices: stiffness matrices, and transfer matrices (Figure 1.1).

Taking into account the nomenclature of Figure 1.2, the stiffness and the transfer matrix for a simple beam are:

1.5.1 STIFFNESS MATRIX

$$\begin{bmatrix} V_a \\ M_a \\ V_b \\ M_b \end{bmatrix} = \begin{bmatrix} \frac{12EJ}{\ell^3} & -\frac{6EJ}{\ell^2} & -\frac{12EJ}{\ell^3} & -\frac{6EJ}{\ell^2} \\ -\frac{6EJ}{\ell^2} & \frac{4EJ}{\ell} & \frac{6EJ}{\ell^2} & \frac{2EJ}{\ell} \\ -\frac{12EJ}{\ell^3} & \frac{6EJ}{\ell^2} & \frac{12EJ}{\ell^3} & \frac{6EJ}{\ell^2} \\ -\frac{6EJ}{\ell^2} & \frac{2EJ}{\ell} & \frac{6EJ}{\ell^2} & \frac{4EJ}{\ell} \end{bmatrix} \begin{bmatrix} w_a \\ \theta_a \\ w_b \\ \theta_b \end{bmatrix}$$

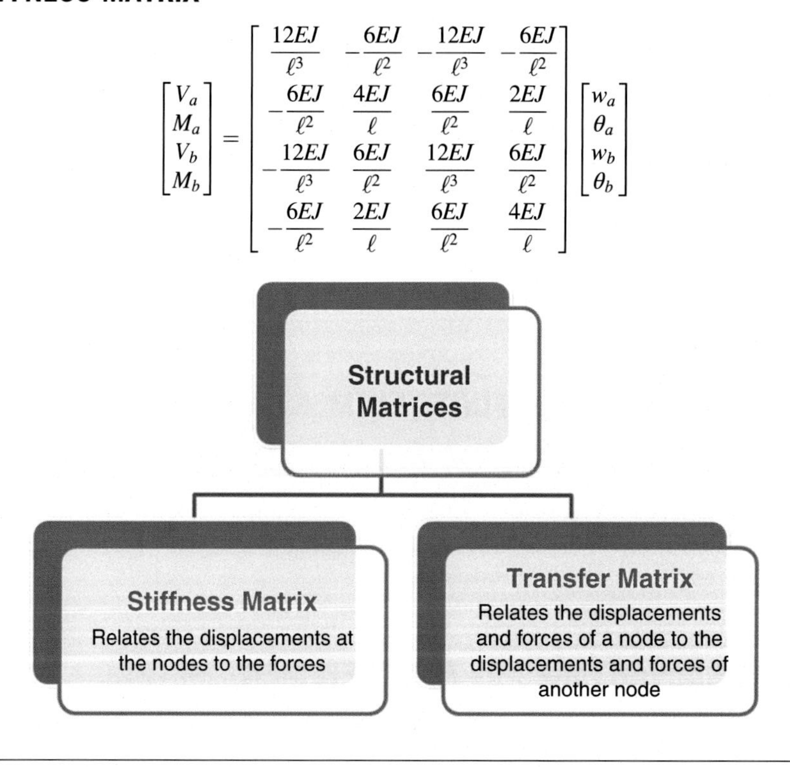
Structural Matrices
Stiffness Matrix
Relates the displacements at the nodes to the forces
Transfer Matrix
Relates the displacements and forces of a node to the displacements and forces of another node

FIGURE 1.1

Types of structural matrices.

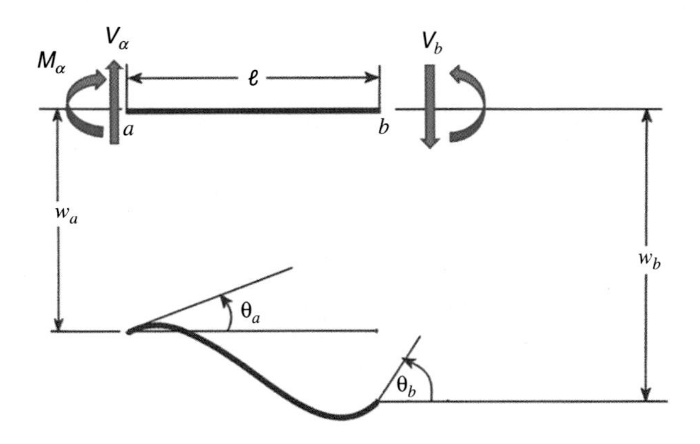

FIGURE 1.2

Nomenclature of the nodal forces and displacements at the ends of a beam.

1.5.2 TRANSFER MATRIX

$$\begin{bmatrix} w_a \\ \theta_a \\ V_b \\ M_b \end{bmatrix} = \begin{bmatrix} 1 & -\ell & -\ell^3/6EJ & -\ell^2/2EJ \\ 0 & 1 & \ell^2/2EJ & \ell/EJ \\ 0 & 0 & 1 & 0 \\ 0 & 0 & 0 & 1 \end{bmatrix} \begin{bmatrix} w_a \\ \theta_a \\ V_a \\ M_a \end{bmatrix}$$

1.6 WHAT ARE THE STEPS TO BE FOLLOWED FOR FINITE ELEMENT METHOD ANALYSIS OF STRUCTURE?

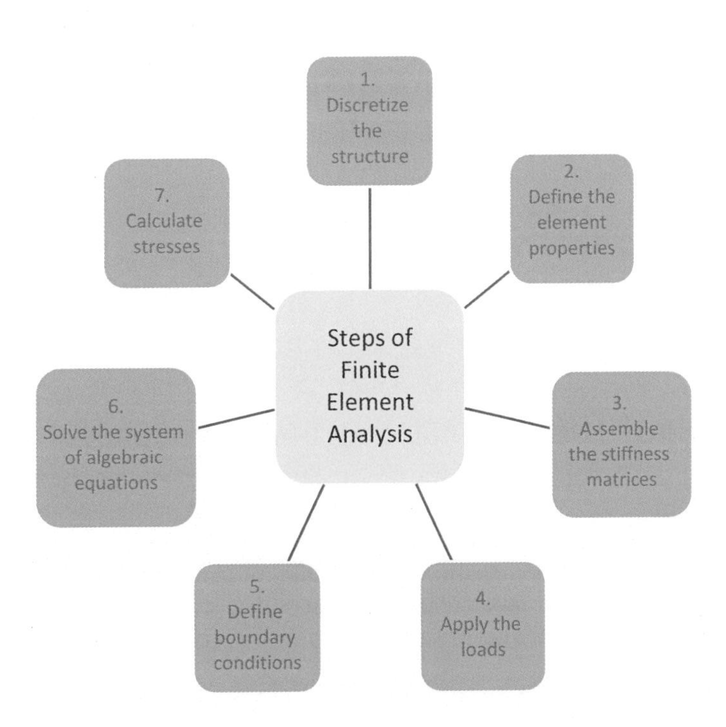

1.6.1 STEP 1. DISCRETIZE OR MODEL THE STRUCTURE

The structure is divided into FEs. This step is one of the most crucial in determining the solution accuracy of the problem.

1.6.2 STEP 2. DEFINE THE ELEMENT PROPERTIES

At this step, the user must define the element properties and select the types of FEs that are the most suitable to model the physical problem.

1.6.3 STEP 3. ASSEMBLE THE ELEMENT STRUCTURAL MATRICES

The structural matrix of an element consists of coefficients that can be derived, for example, from equilibrium. The structural matrix relates the nodal displacements to the applied forces at the nodes. Assembling of the element structural matrices implies application of equilibrium for the whole structure.

1.6.4 STEP 4. APPLY THE LOADS

At this step, externally applied concentrated or uniform forces, moments, or ground motions are provided.

1.6.5 STEP 5. DEFINE BOUNDARY CONDITIONS

At this step the support conditions must be provided, that is, several nodal displacements must be set to known values.

1.6.6 STEP 6. SOLVE THE SYSTEM OF LINEAR ALGEBRAIC EQUATIONS

The sequential application of the above steps leads to a system of simultaneous algebraic equations where the nodal displacements are usually the unknowns.

1.6.7 STEP 7. CALCULATE STRESSES

At the users discretion, the commercial programs can also calculate stresses, reactions, mode shapes, etc.

1.7 WHAT ABOUT THE AVAILABLE SOFTWARE PACKAGES?

Some of the important FEM packages that are available today include Ansys, Abaqus, Nastran, and Lusas. Their structure is based on pre-processor, solution process, post-processor.

Pre-processor stage: data preparation takes place, that is, selection of elements, selection of material properties, discretization of the structure, definition of boundary conditions, and definition of loadings. With these data, the computer algorithm creates the structural equations for every element. Since the data input takes place during this stage, the user interacts with the software only during the pre-processing step.

Solution process stage: computer software solves the system of algebraic equations derived at the pre-processor stage.

Post-processor stage: numerical results obtained by the previous stage are demonstrated graphically in order to represent the displacement, the strain, and the stress field.

1.8 PHYSICAL PRINCIPLES IN THE FINITE ELEMENT METHOD

As it is already mentioned, FEM is based on small subdomain elements. In order to derive the displacement field for the whole structure, a structural matrix and the associated element equation for each element should be derived first. This fundamental equation correlates the nodal displacements and the forces of each element according to the following abbreviated form

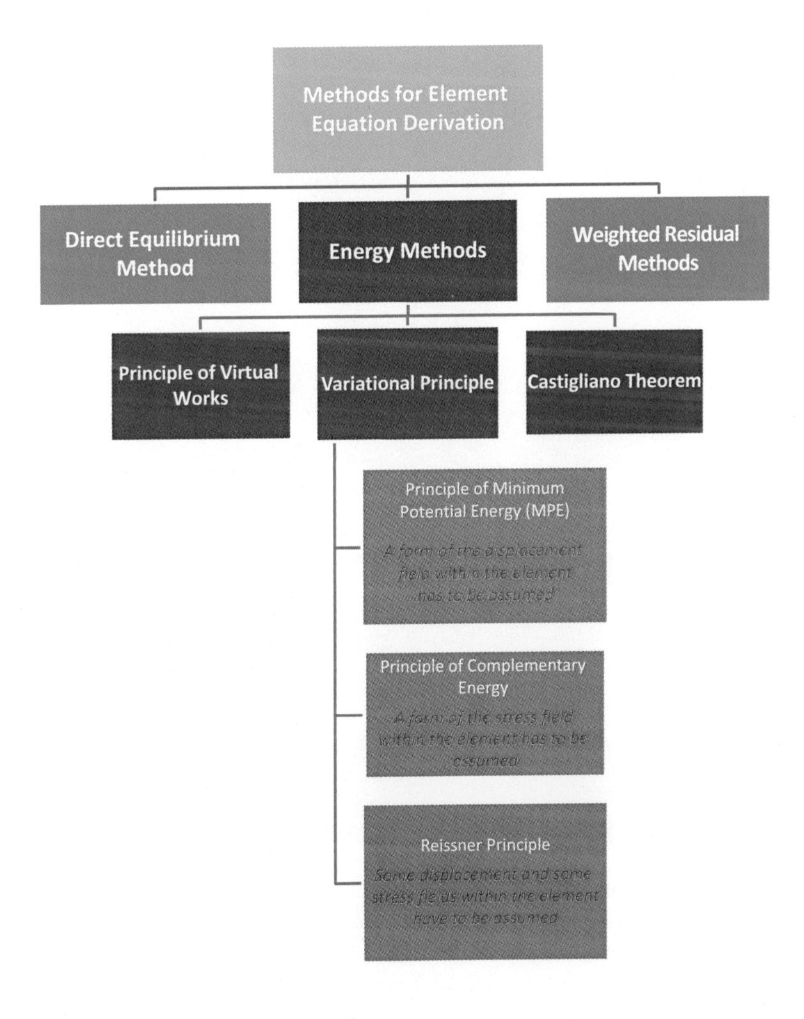

$$\{r\} = [k]\{d\}$$

where $[k]$ is the element stiffness matrix (or local matrix), $\{r\}$ the forces at the nodes, and $\{d\}$ the displacements at the nodes.

The derivation of the element stiffness matrix is very important for the accurate numerical simulation of the structure. To achieve this target, the methods for elements stiffness matrix derivation are classified in three categories: direct equilibrium method, energy methods, and weighted residual methods.

The characteristics of each of the above categories are shown in following figures.

Direct Equilibrium Method

It uses:
1. Force equilibrium at the nodes
2. Force-deformation relationships

Is suitable for:
Simple structural elements, i.e., springs, bars, trusses, beams, frames

Weighted Residual Methods

This method is used when functional is not available.
Galerking method is a representative method of this category.
It is considered as a pure mathematical method.
It is based on substantial mathematical treatment.

Principle of Virtual Works

Is suitable for any material behavior, linear or nonlinear

Variational Principle

Is suitable only for elastic materials

Castigliano's theorem

Is suitable only for elastic materials

In the following chapters, focus will be given on the direct equilibrium method for analyzing springs, bars, trusses, beams, and frames, and the minimum potential energy (MPE) for analyzing springs, bars, trusses, beams, frames, plane stress, and three-dimensional (3D) problems.

It should be noted that most of the FEM applications on structural engineering are based on the MPE method. This method uses the calculus of variations. Therefore, a functional should be initially identified. For solid mechanic problems, the total potential energy Π is the functional to be minimized. Since this functional is expressed in terms of nodal displacements $\{d\}$, the condition of its minimization $\partial\Pi/\partial\{d\}$ yields the required element equation $\{r\}=[k]\{d\}$.

1.9 FROM THE ELEMENT EQUATION TO THE STRUCTURE EQUATION

To obtain the FE solution, five steps should be followed:

1. the element equation $\{r\}=[k]\{d\}$ should be generated for each element;
2. all element equations should be assembled in order to produce the global system of equations of the structure $\{R\}=[K]\{d\}$;
3. the boundary conditions should be introduced in the above mathematical model $\{R\}=[K]\{d\}$;
4. the global system of equations along with the boundary conditions should be solved in order to obtain the displacements in the nodes of the domain; and
5. using the strain-displacement and stress-strain (Hooke's law) equations, the required field values, that is, the strain and stress distributions, should be obtained.

1.10 COMPUTER-AIDED LEARNING OF THE FINITE ELEMENT METHOD

The main target of the subsequent chapters regarding FEM learning is to support students and engineers for:

1. implementation of the theory of FEM on real engineering examples;
2. interactive learning on how to create computational models of physical problems;
3. developing computer algorithms for numerical analysis of engineering structures; and
4. encountering well-known FEM software packages.

To achieve the above targets, a series of numerical examples has been chosen for step-by-step learning of the FE modeling of engineering structures. The modeling is carried out in CALFEM/MATLAB and ANSYS environments. CALFEM (Computer Aided Learning of the Finite Element Method) is an interactive computational toolbox for teaching the FEM with the aid of the standard mathematical software MATLAB. The targets of the book are not limited to help only software users. Focus is given on the deep understanding of the essence of FEM and on integrated knowledge transfer of its physical and mathematical principles in order to provide support for software developing. Therefore, the main learning objective of teaching CALFEM modeling is to help readers to understand the architecture of an FEM computer algorithm and to develop their computer codes. Contact with ANSYS, a widely used FEM commercial software package, will support FEM software users.

1.10.1 INTRODUCTION TO CALFEM

The main advantage of using the CALFEM for computer code development is the in-built functions in MATLAB environment for (a) defining and assembly stiffness matrices, material properties matrices, boundary conditions matrices, topology matrices (i.e., how the elements are interconnected), etc.;

(b) solving the structure equation; and (c) graphical representation of the results. In fact, behind of each in-built function there is a subroutine doing standard computations, thus simplifying the development of the FE analysis code. According to the instructions manual (version 3.4), CALFEM is based on seven groups of functions, namely:

1. general purpose commands, for managing variables, workspace, output, etc.;
2. matrix functions, for matrix handling;
3. material functions, for material matrices derivation;
4. element functions, for composing the element matrices and element forces;
5. system functions, for solving systems of equations;
6. statement functions, for algorithm definitions; and
7. graphic functions, for deriving graphical representation of results.

Using CALFEM for the first time is not a difficult task. However, since its combination with the capabilities of MATLAB is almost endless, mastering MATLAB/CALFEM is time consuming. Therefore, in the present text, only commands required for FEM algorithms development for common types of structures are presented. Learning will be more effective through the interpretation of the integrated programs in the end of each chapter.

General purpose commands

"**clear** *name1 name2 name3*"	removes the variables *name1 name2 name3* from the workspace. *"clear"* removes all variables.
"**disp(A)**"	displays the matrix **A** on the screen.
"**format short**"	displays five-digit scaled fixed point (e.g., 2.1492).
"**format long**"	displays 15-digit scaled fixed point (e.g., 2.14929683187604).
"**format short e**"	displays five-digit floating point (e.g., 2.1492e+000).
"**format long e**"	displays 16-digit floating point (e.g., 2.14929683187604 e+000).
"**load** *filename*"	retrieves the variables from the binary file *filename.mat*.
"**quit**"	terminates the program without saving the workspace.
"**save** *filename*"	saves all variables residing in workspace in a file named *filename.mat*.
"..."	Continuation of an expression in the next line.
"*% text*"	this is a comment line, that is, the text to the right of the symbol % will not be executed.

Matrix functions

"[ ]"	the brackets form vectors and matrices. The columns are separated by space or comma, and the rows are separated by a semicolon. For example, the matrix $A=\begin{bmatrix} 3 & 7 \\ 2 & 9 \end{bmatrix}$ can be specified as **A**=[3 7;2 9] or **A**=[3,7;2,9] or as A=[3 7 2 9]. A matrix **M** can be constructed by submatrices as well. The command **M**=[**A**,**B**] constructs a matrix **M** from matrix **A** in column 1 and matrix **B** in column 2. The command **M**=[**A**;**B**] constructs a 6 × 1 column vector **M** with elements from matrix **A** in row 1 and matrix **B** in row 2.

"( )"	the parentheses after a matrix specify an element of the matrix. For example, the command **F(4,2)** returns the element located in the row 4 and column 2 of the matrix **F**.
+ - * / \	The symbols + - * / are used for addition, subtraction, multiplication, and division of matrices, respectively. The last symbol \ is used for deriving the solution of matrix systems of equations of the type **A*X = B**. By using the above symbol, the solution of this system is **X = B\A**
" **A′** "	**A′**: is the transpose of matrix **A**
"**det**(*A*)"	**det(A)** calculates the determinant of the matrix **A**
"**inv**(*A*)"	**inv(A)** inverses the square matrix **A**. Therefore, the solution of the system **A*X = B** is **X = inv(A) *B**
"**zeros**(*m,n*)"	generates an $m \times n$ matrix of zeros.
"**eye**(*n*)"	generates an $n \times n$ square identity matrix with all members along the main diagonal being equal to one (1) and the remainder being zero.

Material functions

"**hooke**(*ptype, E,v*)"	this command derives the material matrix for an elastic isotropic material. The parameter *E* denotes the modulus of elasticity and *v* the Poisson's ratio, and the variable *ptype* is used to define the type of analysis. Depending on the value of *ptype* the following material matrices can be derived:

ptype = *1* (plane stress):

$$hooke(1,E,v) = \frac{E}{1-v^2}\begin{bmatrix} 1 & v & 0 \\ v & 1 & 0 \\ 0 & 0 & \dfrac{1-v}{2} \end{bmatrix}$$

ptype = *2* (plane strain) and *ptype* = *3* (axisymmetry):

$$hooke(2,E,v) = hooke(3,E,v) = \frac{E}{(1+v)(1-2v)}\begin{bmatrix} 1-v & v & v & 0 \\ v & 1-v & v & 0 \\ v & v & 1-v & 0 \\ 0 & 0 & 0 & \dfrac{1-2v}{2} \end{bmatrix}$$

ptype = *4* (3D analysis):

$$hooke(4,E,v) = \frac{E}{(1+v)(1-2v)}\begin{bmatrix} 1-v & v & v & 0 & 0 & 0 \\ v & 1-v & v & 0 & 0 & 0 \\ v & v & 1-v & 0 & 0 & 0 \\ 0 & 0 & 0 & \dfrac{1-2v}{2} & 0 & 0 \\ 0 & 0 & 0 & 0 & \dfrac{1-2v}{2} & 0 \\ 0 & 0 & 0 & 0 & 0 & \dfrac{1-2v}{2} \end{bmatrix}$$

Element functions

The element functions are commands for derivation of the element stiffness matrix, and computation of the section forces for spring elements, bar elements, beam elements, solid elements, plate elements, and heat flow elements. In some element types (e.g., beams), in addition to the stiffness matrix, the element load vector can be derived. The syntax and the meaning of the main element functions needed for the exercises are as follows.

1.10.2 SPRING ELEMENTS

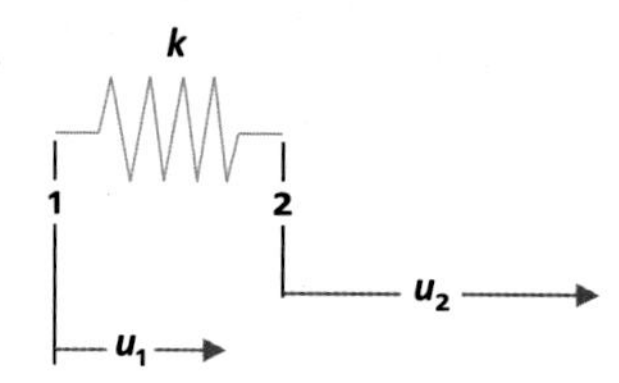

Input variable

ep =*[k]*	% supplies the spring stiffness k

Stiffness matrix

spring1e*(ep)*	% derives the stiffness matrix $\begin{bmatrix} k & -k \\ -k & k \end{bmatrix}$ for a spring element

Section force

spring1s *(ep,ed)*	% computes the axial force in a spring element. Apart from the already known input variable *ep*, the command **spring1s** needs the variable *ed*. This variable is a vector containing the element nodal displacements, and can be derived by the command **extract**. The command **extract** belongs to the category of the system functions and its syntax is presented in the next group of functions (system functions).

In addition to springs, the above commands can also be used for bar elements subjected to axial forces (one-dimensional (1D) analysis), as well as elements for 1D thermal conduction and 1D laminar flow in pipes. For such cases, the value of k should be replaced by the following parameters:

$$k = \begin{cases} EA/L & \text{for bar elements (1-D analysis)} \\ \lambda A/L & \text{for 1-D thermal conduction elements} \\ \pi D^4/(128\mu L) & \text{for 1-D flow elements (pipe elements)} \end{cases}$$

where E is the modulus of elasticity, A the cross-sectional area, L the element length, D the pipe diameter, and μ the fluid viscosity. For these type of elements, the command **spring1s** will compute the element axial force for bars, the element heat flow for 1D thermal conduction elements, and the element fluid flow for pipe elements.

1.10.3 BAR ELEMENTS FOR TWO-DIMENSIONAL ANALYSIS

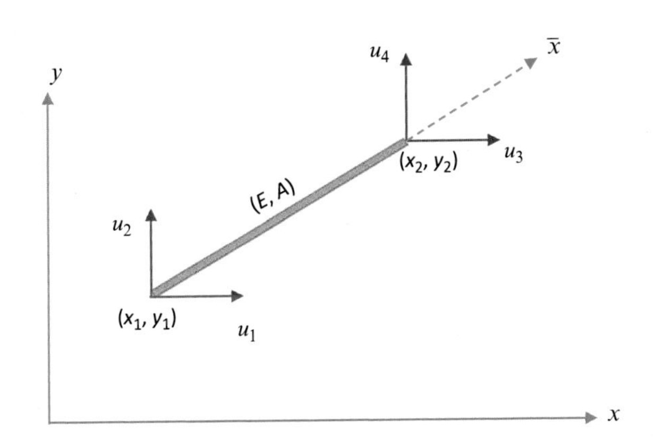

Input variables

ep=[*E A*]	% supplies the bar material and geometric properties
E:	% modulus of elasticity, and *A* cross-sectional area.
ex=[x_1 x_2]	% supplies the element coordinates in x-axis.
ey=[y_1 y_2]	% supplies the element coordinates in y-axis.

Stiffness matrix

bar2e(*ex,ey,ep*)	% derives the stiffness matrix of a bar element for two-dimensional (2D) analysis.

Section force

bar2s(*ex,ey, ep,ed*)	% computes the axial force of a bar element for 2D analysis. Apart from the already known input variable *ep*, the command **bar2s** needs the variable *ed*. This variable is a vector containing the element nodal displacements, and can be derived by the command **extract**. The command **extract** belongs to the category of the system functions and its syntax is presented in the next group of functions (system functions).

1.10.4 BAR ELEMENTS FOR THREE-DIMENSIONAL ANALYSIS

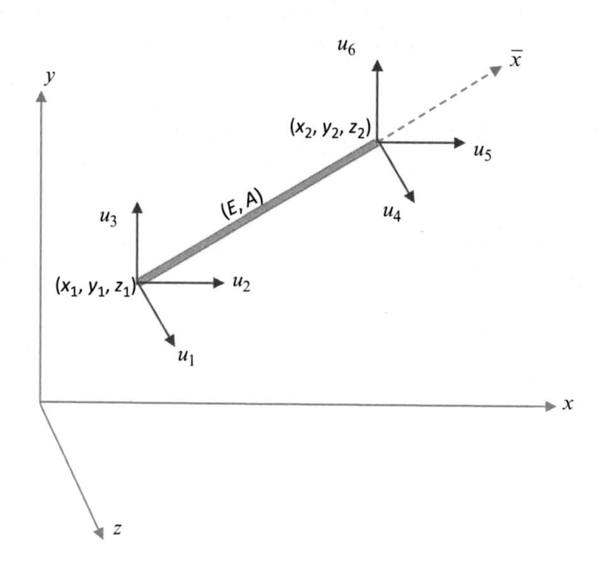

Input variables

ep =*[E A]*	% supplies the bar material and geometric properties. % *E*: modulus of elasticity, *A*: cross-section area.
ex=*[x_1 x_2]*	% supplies the element coordinates in *x*-axis.
ey=*[y_1 y_2]*	% supplies the element coordinates in *y*-axis.
ez=*[z_1 z_2]*	% supplies the element coordinates in *z*-axis.

Stiffness matrix

bar3e(*ex,ey,ez,ep*)	% derives the stiffness matrix of a bar element for 3D % analysis.

Section force

bar3s(*ex,ey,ez,* *ep,ed*)	% computes the axial force of a bar element for 3D analysis. Apart from the already known input variable *ep*, the command **bar3s** needs the variable *ed*. This variable is a vector containing the element nodal displacements, and can be derived by the command **extract**. The command **extract** belongs to the category of the system functions and its syntax is presented in the next group of functions (system functions).

1.10.5 BEAM ELEMENTS FOR TWO-DIMENSIONAL ANALYSIS

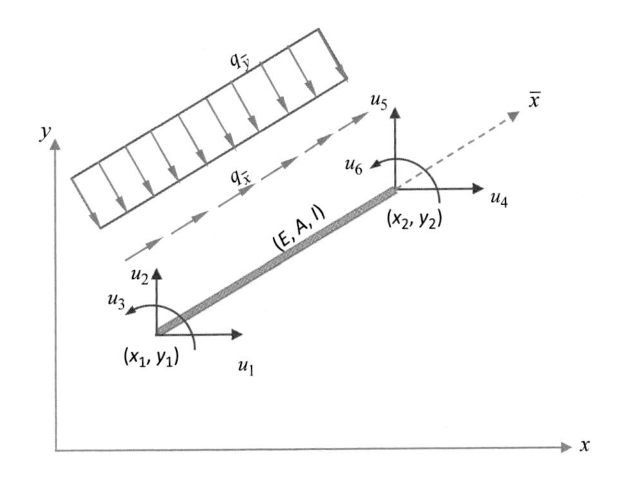

Input variables

ep=[E A I]	% supplies the bar material and geometric properties E: % modulus of elasticity, A: cross-section area, and I: moment % of inertia.
$ex=[x_1\ x_2]$	% supplies the element coordinates in x-axis
$ey=[y_1\ y_2]$	% supplies the element coordinates in y-axis
$eq=[q_{\bar{x}}\ q_{\bar{y}}]$	% uniformly distributed loads per unit length in the local % coordinate system $\bar{x}$, $\bar{y}$ of the element.

Stiffness matrix

beam2e(*ex,ey,ep,eq*)	% derives the stiffness matrix K_e and the load vector % f_e of a beam element for 2D analysis, for example, % $[K_e,f_e]$=**beam2e**(*ex,ey,ep,eq*).

Section force

beam2s(*ex,ey,ep,ed,eq,n*)

The above command computes the distribution of section forces *es=[N V M]* and the displacements *edi=[u v]* along the beam element for 2D analysis, for example, *[es,edi,eci]*=**beam2s**(*ex,* *ey,ep,ed,eq,n*). The parameter *n* is the number of evaluation points along the element for section forces and displacements computation. The variable *eci* expresses the local coordinate $\bar{x}$, that is, $eci = [\bar{x}]$. Apart from the already known input variables *ep*, *eq* the command **beam2s** needs the variable *ed*. This variable is a vector containing the element nodal displacements, and can be derived by the command **extract**. The command **extract** belongs to the category of the system functions and its syntax is presented in the next group of functions (system functions).

1.10.6 BEAM ELEMENTS FOR THREE-DIMENSIONAL ANALYSIS

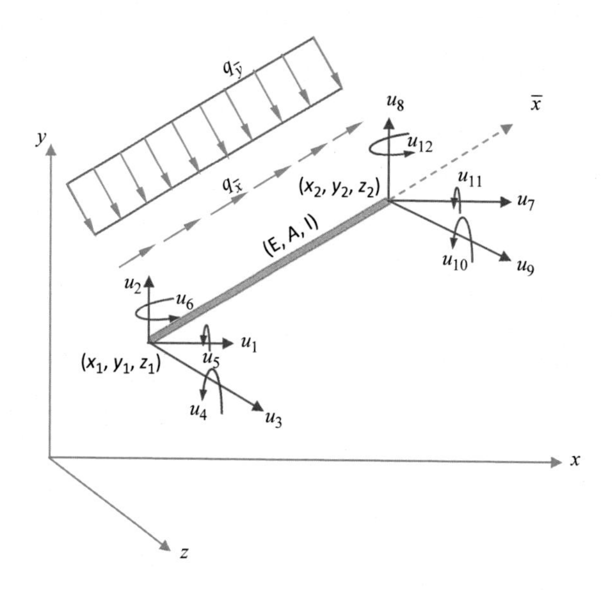

Input variables

$ep = [E\ G\ A\ I_y\ I_z\ K_\upsilon]$	% supplies the bar material and geometric properties: E is the modulus of elasticity, A is the cross-sectional area, I_y, I_z are the moments of inertia with respect to the local y and z axes of the element, and K_υ is the St Venant torsional stiffness.
$ex = [x_1\ x_2]$	% supplies the element coordinates in x-axis.
$ey = [y_1\ y_2]$	% supplies the element coordinates in y-axis.
$ez = [z_1\ z_2]$	% supplies the element coordinates in z-axis.
$eo = [x_{\bar{z}} y_{\bar{z}} z_{\bar{z}}]$	% supplies the direction of the local beam coordinate system.
$eq = [q_{\bar{x}} q_{\bar{y}} q_{\bar{z}} q_{\bar{\omega}}]$	% uniformly distributed loads per unit length in the local
	% coordinate system $\bar{x}, \bar{y}, \bar{z}$ of the element.
$q_{\bar{\omega}}$	% distributed torque along the local axis $\bar{x}$.

Stiffness matrix

beam3e(*ex,ey,ez,eo, ep,eq*)	% derives the stiffness matrix K_e and the load vector f_e of a beam element for 3-D analysis, for example, *[K_e,f_e]*=**beam3e**(*ex,ey,ez,eo,ep,eq*).

Section force

beam3s(*ex,ey,ez,eo,ep,ed,eq,n*).

The above command computes the distribution of section forces $es=[N\ V_{\bar{y}}\ V_{\bar{z}}\ T\ M_{\bar{y}}\ M_{\bar{z}}]$ and the displacements $edi=[u\ v\ w\ \phi]$ along the beam element for 3D analysis and put them in a matrix *[es, edi,eci]*, that is, *[es,edi,eci]*=**beam3s**(*ex,ey,ez,eo,ep,ed,eq,n*). The parameter *n* is the number of evaluation points along the element for section forces and displacements computation. The variable *eci* expresses the local coordinate $\bar{x}$, that is, $eci=[\bar{x}]$. Apart from the already known input variables *ep, eq, eo* the command **beam3s** needs the variable *ed*. This variable is a vector containing the element nodal displacements, and can be derived by the command **extract**. The command **extract** belongs to the category of the system functions and its syntax is presented in the next group of functions (system functions).

1.10.7 SYSTEM FUNCTIONS

[K,f]=**assem**(*edof,K, Ke,f,fe*)	% adds the element stiffness matrix *Ke* to the structure stiffness matrix *K* according to the topology matrix *edof*. The element load vector *fe* (derived from the command **beam2e** or **beam3e**) is also added to the global vector *f*.
ed=**extract**(*edof*,a)	% extracts element nodal displacements from the global solution vector a, according to the topology matrix *edof*.
*[*a*,r]*=**solveq**(*K,f,bc*)	% solves the matrix equation *K*a=*f*, where *K* is the global stiffness matrix, *f* is the global load vector, and *bc* is the matrix containing the boundary conditions. The results of the nodal displacements of all nodes are placed in the vector a, and the support reactions are placed in the vector *r*.

1.10.8 STATEMENT FUNCTIONS

if *logical* expr*ession* ⋮ *else* ⋮ *end*	% If the *logical expression* yields the value True, then the commands following ***if*** are executed, otherwise the commands following ***else*** are executed. The loop is closed by the command ***end***.
for i=a:b:c ⋮ *end*	% initiates a loop. The variable *i* takes the values from *a* to *c with step b*. The loop is closed by the command ***end***.

1.10.9 GRAPHIC FUNCTIONS

figure(*n*)	% creates graph windows. The parameter *n* specifies the order of the figure.
eldraw2(*Ex,Ey,plotpar*)	% displays the underformed 2D structure according to *Ex, Ey* coordinate matrices. The parameter *plotpar* specifies the line type, line color, and node marker according to the following nomenclature:
plotpar[*linetype linecolor nodemark*]	
linetype =	1 solid line
	2 dashed line
	3 dotted line
linecolor =	1 black
	2 blue
	3 magenta
	4 red
nodemark =	1 circle
	2 star
	0 no mark
[sfac] = **scalfact2**(*ex,ey,ed,n*)	% specifies a scale factor for drawing displacement or section force distributions contained in the matrix *ed*. Input data are the coordinate matrices *ex, ey* and the index *n*, which defines the ratio between the geometric quantity to be displayed and the element size. If *n* is not specified, *0.2* is the default value.
eldisp2(*Ex,Ey,Ed,plotpar,sfac*)	% makes the graphic of the deformed structure according to the displacement matrix *Ed*.
eldia2(*ex,ey,es,plotpar,sfac*)	% makes the graphic of the section force distributions according to the force results matrix *es*.
axis([*xmin xmax ymin ymax*])	% specifies the values at the ends of *x* and *y* axes.
xlabel('*text*') and **ylabel**('*text*')	% add legends to the corresponding axes.
title('*text*')	% add the legend '*text*' at the top of the graphic.

1.10.10 WORKING ENVIRONMENT IN ANSYS

ANSYS is one of the most widely used FEM software packages in academy and industry for solving structural, thermal, fluid, and electric problems as well as thermal–mechanical, thermal–electric, and thermal–structural–electric interaction problems.

The general steps of the FEM procedure are the following.

1. *Start-up*: the user should prepare the working environment by specifying the title and the job name of the file, the units of the physical problem, and the pre-processor.
2. *Pre-processing*: the user should specify the material data and the geometry of the structure, and then create the mesh of the mechanical model.
3. *Solution of the problem*: the user should specify the boundary conditions and apply the loads. Then order the program to solve the FE model.

4. *Post-processing*: the software presents the results. The user should select the type of graphs and the solution outputs to be demonstrated.

To perform FE analysis with ANSYS, the following windows are used for data entry.

1. **Utility Menu (UM)**: used for utility functions selection regarding file controls, graphic controls, etc.
2. **Input window (INPUT)**: user can type commands (e.g., to specify the units).
3. **ANSYS Toolbar (Toolbar)**: contains buttons for commonly used commands.
4. **Main Menu (MM)**: user can interact with the pre-processor, solution, post-processor, etc. in order to make the model of the structure. MM is the most important window.
5. **Graphics Window (GW)**: the working environment where ANSYS displays the FE model (key-points, nodes, elements) and the results (e.g., stress or strain distributions, temperature distributions, heat flux direction, etc.).
6. **Output Window (Output)**: provides text output from the program.

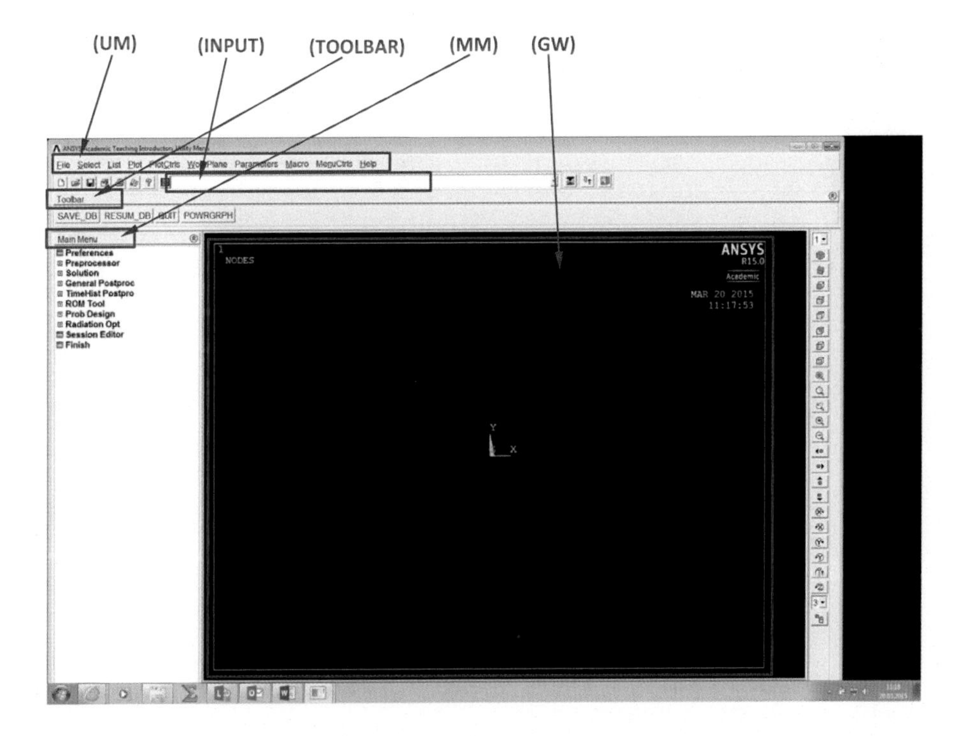

REFERENCES

[1] Austrell P-E, Dahlblom O, Lindemann J, Olsson A, Olsson K-G, Persson K, et al. CALFEM a finite element toolbox, Version 3.4., Division of Structural Mechanics, Lund University; 2004.

[2] Lawrence K. ANSYS Workbench Tutorial Release 14. SDC Publications; 2012.

[3] Cook RD, Malkus DS, Plesha ME, Witt RJ. Concepts and applications of finite element analysis. Hoboken: John Wiley & Sons; 2002.
[4] Spyrakos CC. Finite element modeling in engineering practice. Morgantown: West Virginia University Press; 1994.
[5] Alawadhi EM. Finite element simulations using ANSYS. Boca Raton: CRC Press; 2010.
[6] Bhatti MA. Fundamental finite element analysis and applications. Hoboken: John Wiley & Sons; 2005.
[7] Logan DL. A first course in the finite element method. Boston, MA: Gengage Learning; 2012.
[8] Oden JT, Becker EB, Carey GF. Finite elements: an introduction, volume I. New Jersey: Prentice Hall; 1981.
[9] Fish J, Belytschko T. A first course in finite elements. New York: Wiley; 2007.
[10] Zienkiewicz OC, Taylor RL, Fox DD. The finite element method for solid and structural mechanics. 7th ed. Oxford: Butterworth-Heinemann; 2013.
[11] Deif AS. Advanced matrix theory for scientists and engineers. 2nd ed. London: Abacus press; 1991.
[12] Hartmann F, Katz C. Structural analysis with finite elements. 2nd ed. Berlin: Springer; 2007.
[13] Rao SS. The finite element method in engineering. 5th ed. Burlington, MA: Butterworth-Heinemann; 2011.
[14] Przemieniecki JS. Theory of matrix structural analysis. New York: McGraw Hill; 1968.
[15] Zienkiewicz OC, Cheung YK. Finite elements in the solution of field problems. The Engineer 1965;507–10.
[16] Gallagher RH, Oden JT, Taylor C, Zienkiewicz OC, editors. Finite elements in fluids, vols. 1 and 2. London: Wiley; 1975.
[17] Argyris JH. Energy theorems and structural analysis. London: Butterworth; 1968.
[18] Young WC, Bodynas RG, Sadegh AM. Roarks formulas for stress and strain. 8th ed. New York: McGraw Hill; 2012.

CHAPTER

MATHEMATICAL BACKGROUND 2

2.1 VECTORS

2.1.1 DEFINITION OF VECTOR

It is well known from elementary physics that physical quantities can be represented by scalars or vectors. For a physical quantity described by a scalar, we just need a single value for its specification. Unlike scalar quantities, many other quantities (e.g., force, displacement, area, etc.), have direction in addition to magnitude. These quantities are called vectors and are represented graphically by directed line segments (Figure 2.1).

Any vector $\vec{a}$ is correlated with a system of coordinates x, y, z through its components a_x, a_y, a_z and can be expressed analytically by the following equation:

$$\vec{a}=a_x\,\vec{i}\,+a_y\,\vec{j}\,+a_z\,\vec{k} \tag{2.1}$$

where $\vec{i}$, $\vec{j}$, $\vec{k}$ are unit vectors directed along the x, y, z, axes, respectively.

Between vectors, there are two types of products; the scalar product and the vector product.

2.1.2 SCALAR PRODUCT

Let us consider the vectors $\vec{a}$ and $\vec{b}$ (Figure 2.2). If we denote by a, b their magnitudes, respectively, the scalar product $\vec{a}\cdot\vec{b}$ is defined by the following equation:

$$\vec{a}\cdot\vec{b}=a(b\cos\vartheta)=ab\cos\vartheta \tag{2.2}$$

Therefore, the scalar product of the vectors $\vec{a}$ and $\vec{b}$ is the magnitude of $\vec{a}$ multiplied by the magnitude of the projection of $\vec{b}$ on the direction of $\vec{a}$.

Taking into account Equation (2.2), it is obvious that

$$\vec{a}\cdot\vec{b}=\vec{b}\cdot\vec{a} \tag{2.3}$$

According to the definition of the scalar product, the following properties can be obtained:

$$\vec{i}\cdot\vec{i}=(1)(1)\cos(0)=1 \tag{2.4}$$

$$\vec{i}\cdot\vec{j}=(1)(1)\cos(\pi/2)=0 \tag{2.5}$$

Essentials of the Finite Element Method. http://dx.doi.org/10.1016/B978-0-12-802386-0.00002-5

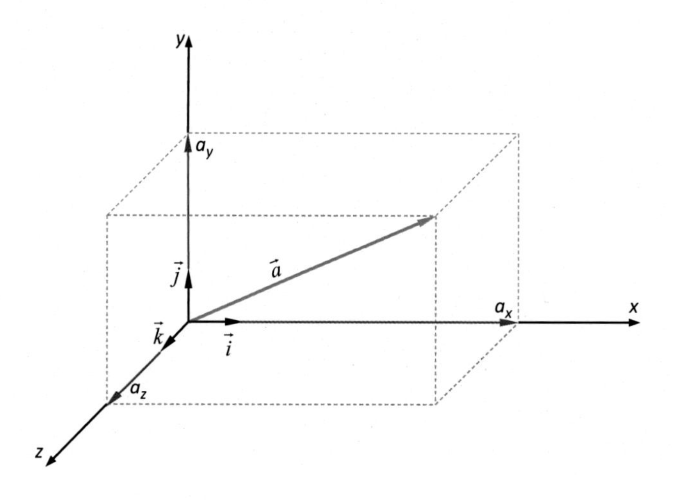

FIGURE 2.1

Graphical representation of a vector quantity $\vec{a}$.

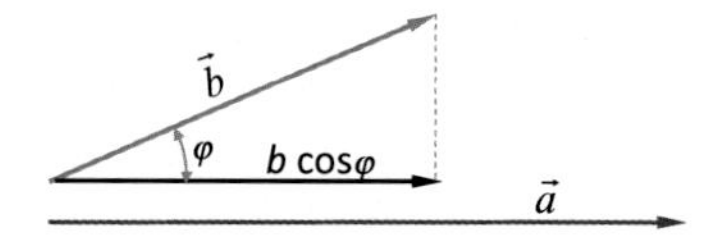

FIGURE 2.2

Scalar product of vectors $\vec{a}$ and $\vec{b}$.

$$\vec{i} \cdot \vec{k} = (1)(1)\cos(\pi/2) = 0 \tag{2.6}$$

and so on.

Therefore, the scalar product

$$\vec{a} \cdot \vec{b} = \left(a_x \vec{i} + a_y \vec{j} + a_z \vec{k}\right) \cdot \left(b_x \vec{i} + b_y \vec{j} + b_z \vec{k}\right) \tag{2.7}$$

yields

$$\begin{aligned} \vec{a} \cdot \vec{b} = a_x b_x \vec{i} \cdot \vec{i} + a_x b_y \vec{i} \cdot \vec{j} + a_x b_z \vec{i} \cdot \vec{k} \\ + a_y b_x \vec{j} \cdot \vec{i} + a_y b_y \vec{j} \cdot \vec{j} + a_y b_z \vec{j} \cdot \vec{k} \\ + a_z b_x \vec{k} \cdot \vec{i} + a_z b_y \vec{k} \cdot \vec{j} + a_z b_z \vec{k} \cdot \vec{k} \end{aligned} \tag{2.8}$$

or

$$\vec{a} \cdot \vec{b} = a_x b_x + a_y b_y + a_z b_z = [a_x \quad a_y \quad a_z] \cdot \begin{Bmatrix} b_x \\ b_y \\ b_z \end{Bmatrix} \tag{2.9}$$

2.1.3 VECTOR PRODUCT

The vector product of two vectors $\vec{a}$ and $\vec{b}$ (Figure 2.3) is defined by the following equation:

$$\vec{a} \times \vec{b} = \det \begin{bmatrix} \vec{i} & \vec{j} & \vec{k} \\ a_x & a_y & a_z \\ b_x & b_y & b_z \end{bmatrix} \tag{2.10}$$

Taking into account the above definition, the following properties can be obtained:

$$\vec{i} \times \vec{i} = \det \begin{bmatrix} \vec{i} & \vec{j} & \vec{k} \\ 1 & 0 & 0 \\ 1 & 0 & 0 \end{bmatrix} = 0 \tag{2.11}$$

$$\vec{i} \times \vec{j} = \det \begin{bmatrix} \vec{i} & \vec{j} & \vec{k} \\ 1 & 0 & 0 \\ 0 & 1 & 0 \end{bmatrix} = \vec{k} \tag{2.12}$$

$$\vec{j} \times \vec{i} = \det \begin{bmatrix} \vec{i} & \vec{j} & \vec{k} \\ 0 & 1 & 0 \\ 1 & 0 & 0 \end{bmatrix} = -\vec{k} \tag{2.13}$$

The resulting vector of the product $\vec{a} \times \vec{b}$ is directed perpendicular to the plane formed by the vectors $\vec{a}$ and $\vec{b}$, and its magnitude is

$$\left|\vec{a} \times \vec{b}\right| = a(b \sin\varphi) \tag{2.14}$$

Taking into account Figure 2.3, the area A of the triangle ABC is

$$A = \frac{1}{2} a(b \sin\varphi) \tag{2.15}$$

Therefore, combining Equations (2.14) and (2.15), the following result can be obtained:

$$A = \frac{1}{2}\left|\vec{a} \times \vec{b}\right| \tag{2.16}$$

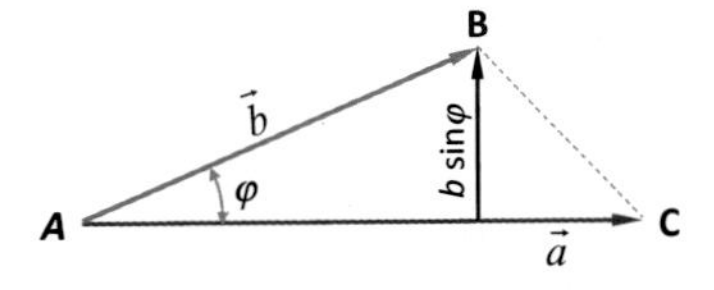

FIGURE 2.3

Vector product of vectors $\vec{a}$ and $\vec{b}$.

2.1.4 ROTATION OF COORDINATE SYSTEM

A vector $\vec{a}$ can be expressed with respect to two different coordinate systems, namely x-y-z and η-ξ-ζ (Figure 2.4). The vector's components with respect to two coordinate systems can be correlated. If we define the same vector $\vec{a}$ in the rotated coordinates η-ξ-ζ, the following expression can be written:

$$\vec{a}=a_\eta \vec{i'} + a_\xi \vec{j'} + a_\zeta \vec{k'} \tag{2.17}$$

where $\vec{i'}$, $\vec{j'}$, $\vec{k'}$ are the unit vectors along the η-ξ-ζ directions.

If we denote by $\varphi_x, \varphi_y, \varphi_z$ the angles of the vector $\vec{a}$ with the axes x, y, z, respectively, the following equations can be written:

$$\vec{i} \cdot \vec{a}=(1)\cdot(a \cos\varphi_x)=a_x \tag{2.18}$$

$$\vec{j} \cdot \vec{a}=(1)\cdot(a \cos\varphi_y)=a_y \tag{2.19}$$

$$\vec{k} \cdot \vec{a}=(1)\cdot(a \cos\varphi_z)=a_z \tag{2.20}$$

Using Equation (2.17), Equations (2.18)–(2.20) can now be written in the following form:

$$a_x=\vec{i} \cdot \vec{a}=a_n \vec{i} \cdot \vec{i'} + a_\xi \vec{i} \cdot \vec{j'} + a_\zeta \vec{i} \cdot \vec{k'} \tag{2.21}$$

$$a_y=\vec{j} \cdot \vec{a}=a_n \vec{j} \cdot \vec{i'} + a_\xi \vec{j} \cdot \vec{j'} + a_\zeta \vec{j} \cdot \vec{k'} \tag{2.22}$$

$$a_z=\vec{k} \cdot \vec{a}=a_n \vec{k} \cdot \vec{i'} + a_\xi \vec{k} \cdot \vec{j'} + a_\zeta \vec{k} \cdot \vec{k'} \tag{2.23}$$

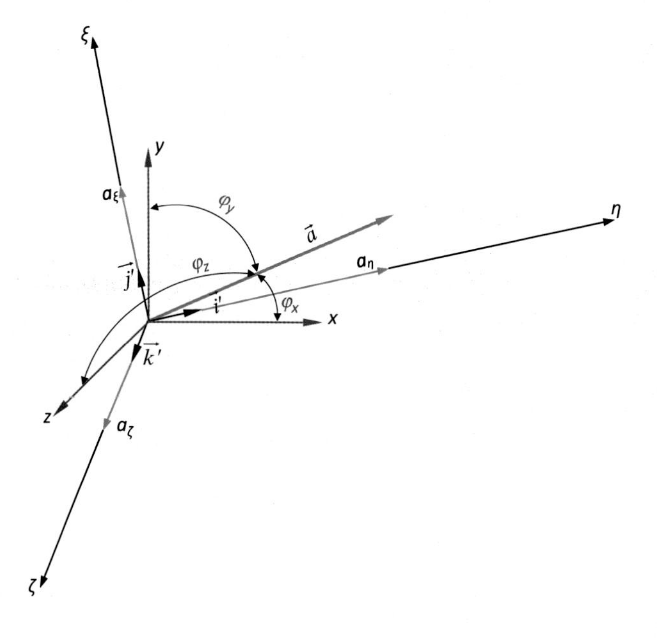

FIGURE 2.4

Representation of a vector $\vec{a}$ with respect to two different systems of coordinates x-y-z and η-ξ-ζ.

Taking into account Equations (2.4)–(2.6), the above equations yield:

$$\begin{Bmatrix} a_x \\ a_y \\ a_z \end{Bmatrix} = \begin{bmatrix} \cos(x,n) & \cos(x,\xi) & \cos(x,\zeta) \\ \cos(y,n) & \cos(y,\xi) & \cos(y,\zeta) \\ \cos(z,n) & \cos(z,\xi) & \cos(z,\zeta) \end{bmatrix} \cdot \begin{Bmatrix} a_n \\ a_\xi \\ a_\zeta \end{Bmatrix} \tag{2.24}$$

where

$$\cos(x,n) = \vec{i} \cdot \vec{i'} \tag{2.25}$$

$$\cos(x,\xi) = \vec{i} \cdot \vec{j'} \tag{2.26}$$

$$\cos(x,\zeta) = \vec{i} \cdot \vec{k'} \tag{2.27}$$

and so on are called directional cosines.

2.1.5 THE VECTOR DIFFERENTIAL OPERATOR (GRADIENT)

The vector differential operator $\vec{\nabla}$ is defined by the following equation:

$$\vec{\nabla} \equiv \frac{\partial}{\partial x}\vec{i} + \frac{\partial}{\partial y}\vec{j} + \frac{\partial}{\partial z}\vec{k} \tag{2.28}$$

This operator is used for defining the following Green's theorem.

2.1.6 GREEN'S THEOREM

Let us consider a scalar field function $\Phi = \Phi(x, y)$. Its gradient in a two-dimensional (2D) domain is the following vector:

$$\vec{\nabla}\,\Phi(x,y) \equiv \frac{\partial \Phi(x,y)}{\partial x}\vec{i} + \frac{\partial \Phi(x,y)}{\partial y}\vec{j} \tag{2.29}$$

If we denote by $\vec{n}$ the unit vector normal to the boundary S of the 2D domain Ω (Figure 2.5), where

$$\vec{n} = n_x\,\vec{i} + n_y\,\vec{j} \tag{2.30}$$

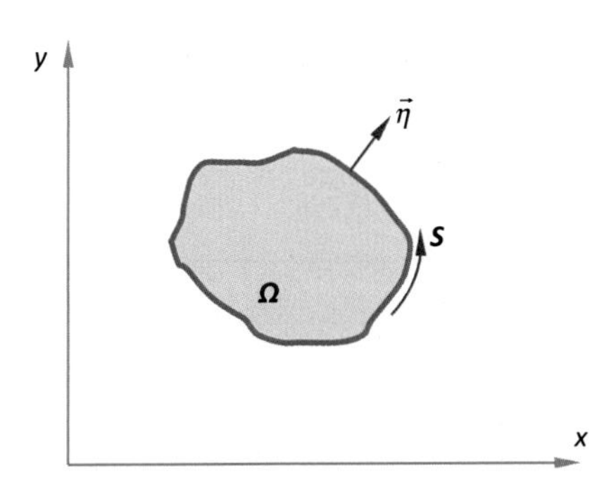

FIGURE 2.5

A two-dimensional domain of a scalar function.

and

$$|\vec{n}| = \sqrt{n_x^2 + n_y^2} = 1 \tag{2.31}$$

the following theorem can be used:

$$\iint \vec{\nabla}\, \Phi(x,y)\mathrm{d}x\mathrm{d}y = \oint_S \Phi(x,y)\, \vec{n}\, \mathrm{d}S \tag{2.32}$$

The above equation is known as Green's theorem and correlates an integral of a gradient of a function $\Phi = \Phi(x,y)$ over an area Ω with a contour integral over the boundary S.

2.2 COORDINATE SYSTEMS

In applied mechanics, the three common coordinate systems are rectangular (or Cartesian), cylindrical, and spherical. The aim of the present section is to describe these coordinate systems and to derive the formulae for the transformation of the vector components of each coordinate system with respect to the others.

2.2.1 RECTANGULAR (OR CARTESIAN) COORDINATE SYSTEM

The rectangular coordinate system has already been presented in Figure 2.1. The components of a vector $\vec{a}$ are specified in terms of three mutually perpendicular unit vectors $\vec{i}$, $\vec{j}$, $\vec{k}$. For convenience, we denote them as $\vec{x}$, $\vec{y}$, $\vec{z}$, as shown in Figure 2.6.

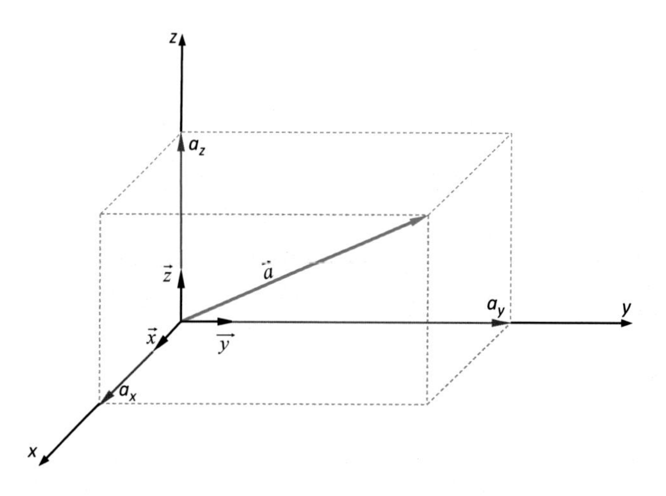

FIGURE 2.6

Rectangular (or Cartesian) coordinate system.

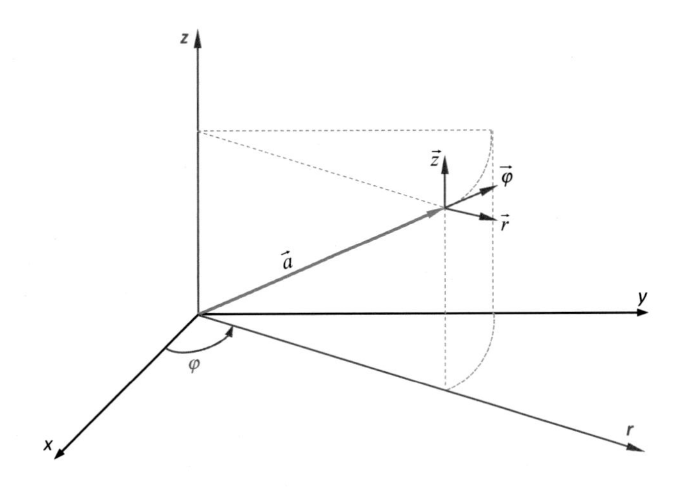

FIGURE 2.7

Cylindrical coordinate system.

2.2.2 CYLINDRICAL COORDINATE SYSTEM

In the cylindrical coordinate system, the components of a vector $\vec{a}$ are specified in terms of three mutually perpendicular unit vectors $\vec{r}$, $\vec{\varphi}$, $\vec{z}$, as shown in Figure 2.7. Therefore, its coordinates are r, φ, z.

2.2.3 SPHERICAL COORDINATE SYSTEM

In the spherical coordinate system, the coordinates are r, ϑ, φ. Therefore, the components of any vector $\vec{a}$ are specified in terms of the three mutually perpendicular unit vectors $\vec{r}$, $\vec{\vartheta}$, $\vec{\varphi}$, as shown in Figure 2.8.

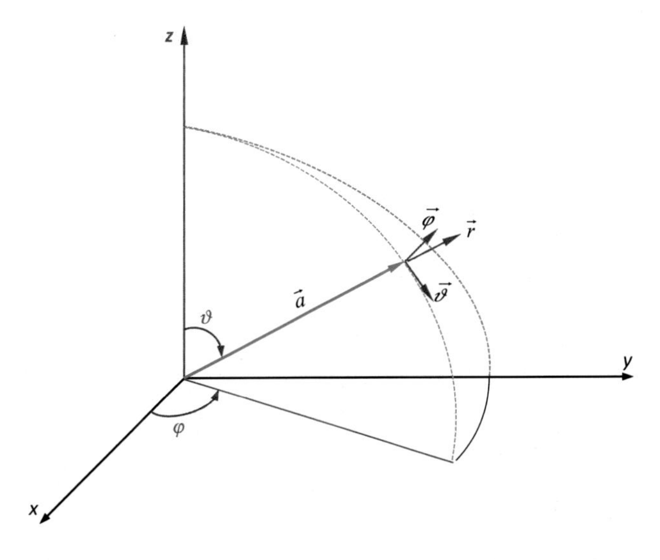

FIGURE 2.8

Spherical coordinate system.

2.2.4 COMPONENT TRANSFORMATION

The components of a vector with respect to a certain coordinate system must often be transformed to another coordinate system. The equations correlating the components of a vector in the three coordinate systems can be derived using the properties of vector analysis. Let us consider the vector $\vec{a}$ to be expressed in rectangular and spherical coordinate system as shown in Figure 2.9.

The vector $\vec{a}$ can be expressed with respect to the system x, y, z by the following equation:

$$\vec{a} = a_x \vec{x} + a_y \vec{y} + a_z \vec{z} \tag{2.33}$$

Since the scalar product $\vec{e} \cdot \vec{u}$ of two arbitrary vectors $\vec{e}, \vec{u}$ expresses the product of the magnitude of the vector $\vec{u}$ with the magnitude of the projection of $\vec{e}$ on the direction of $\vec{u}$, the following properties for the products $\vec{r} \cdot \vec{x}, \vec{r} \cdot \vec{y}, \vec{r} \cdot \vec{z}$ can be used:

$$\vec{r} \cdot \vec{x} = \sin\vartheta \cos\varphi \tag{2.34}$$

$$\vec{r} \cdot \vec{y} = \sin\vartheta \sin\varphi \tag{2.35}$$

$$\vec{r} \cdot \vec{z} = \cos\vartheta \tag{2.36}$$

and so on.

Therefore, if we multiply Equation (2.33) with $\vec{r}$, the following equation can be obtained:

$$\vec{r} \cdot \vec{a} = a_x \vec{r} \cdot \vec{x} + a_y \vec{r} \cdot \vec{y} + a_z \vec{r} \cdot \vec{z} \tag{2.37}$$

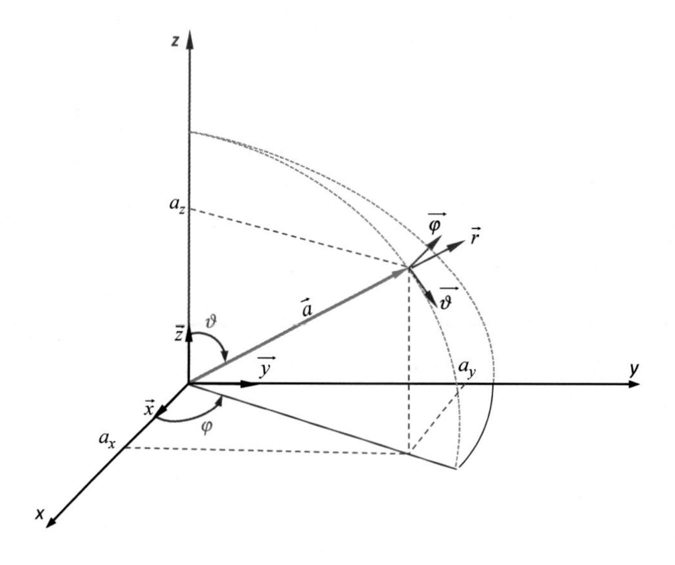

FIGURE 2.9

Transformation of a vector's components from a rectangular coordinate system to a spherical one.

or

$$\vec{r} \cdot \vec{a} = a_x \sin\vartheta \cos\varphi + a_y \sin\vartheta \sin\varphi + a_z \cos\vartheta \tag{2.38}$$

Since

$$\vec{r} \cdot \vec{a} = a_r \tag{2.39}$$

Equation (2.38) can now be written as:

$$a_r = \left[\sin\vartheta \cos\varphi \quad \sin\vartheta \sin\varphi \quad \cos\vartheta\right] \begin{Bmatrix} a_x \\ a_y \\ a_z \end{Bmatrix} \tag{2.40}$$

Following similar procedure (i.e., to multiply Equation (2.33) with ϑ and φ), the following equations can be derived:

$$a_\vartheta = \left[\cos\vartheta \cos\varphi \quad \cos\vartheta \sin\varphi \quad -\sin\vartheta\right] \begin{Bmatrix} a_x \\ a_y \\ a_z \end{Bmatrix} \tag{2.41}$$

and

$$a_\varphi = \left[-\sin\varphi \quad \cos\varphi \quad 0\right] \begin{Bmatrix} a_x \\ a_y \\ a_z \end{Bmatrix} \tag{2.42}$$

Combination of Equations (2.40)–(2.42) yields the following transformation of a vector's components from a rectangular coordinate system to a spherical one:

$$\begin{Bmatrix} a_r \\ a_\vartheta \\ a_\varphi \end{Bmatrix} = \begin{bmatrix} \sin\vartheta \cos\varphi & \sin\vartheta \sin\varphi & \cos\vartheta \\ \cos\vartheta \cos\varphi & \cos\vartheta \sin\varphi & -\sin\vartheta \\ -\sin\varphi & \cos\varphi & 0 \end{bmatrix} \begin{Bmatrix} a_x \\ a_y \\ a_z \end{Bmatrix} \tag{2.43}$$

Following the same concept, the transformation of a vector's components from a rectangular coordinate system to a cylindrical one can be performed by the following equation:

$$\begin{Bmatrix} a_r \\ a_\varphi \\ a_z \end{Bmatrix} = \begin{bmatrix} \cos\varphi & \sin\varphi & 0 \\ -\sin\varphi & \cos\varphi & 0 \\ 0 & 0 & 1 \end{bmatrix} \begin{Bmatrix} a_x \\ a_y \\ a_z \end{Bmatrix} \tag{2.44}$$

Inversion of the above equations yields the transformation of the components from a spherical and cylindrical coordinate system to a rectangular one, respectively:

$$\begin{Bmatrix} a_x \\ a_y \\ a_z \end{Bmatrix} = \begin{bmatrix} \sin\vartheta \cos\varphi & \cos\vartheta \cos\varphi & -\sin\varphi \\ \sin\vartheta \sin\varphi & \cos\vartheta \sin\varphi & \cos\varphi \\ \cos\vartheta & -\sin\vartheta & 0 \end{bmatrix} \begin{Bmatrix} a_r \\ a_\vartheta \\ a_\varphi \end{Bmatrix} \tag{2.45}$$

and

$$\begin{Bmatrix} a_x \\ a_y \\ a_z \end{Bmatrix} = \begin{bmatrix} \cos\varphi & -\sin\varphi & 0 \\ \sin\varphi & \cos\varphi & 0 \\ 0 & 0 & 1 \end{bmatrix} \begin{Bmatrix} a_r \\ a_\varphi \\ a_z \end{Bmatrix} \tag{2.46}$$

Finally, transformation of the components from a cylindrical coordinate system to a spherical one, and vice versa, can be performed from the following equations:

$$\begin{Bmatrix} a_r \\ a_\vartheta \\ a_\varphi \end{Bmatrix} = \begin{bmatrix} \sin\vartheta & 0 & \cos\vartheta \\ \cos\vartheta & 0 & -\sin\vartheta \\ 0 & 1 & 0 \end{bmatrix} \begin{Bmatrix} a_r \\ a_\varphi \\ a_z \end{Bmatrix} \tag{2.47}$$

and

$$\begin{Bmatrix} a_r \\ a_\varphi \\ a_z \end{Bmatrix} = \begin{bmatrix} \sin\vartheta & \cos\vartheta & 0 \\ 0 & 0 & 1 \\ \cos\vartheta & -\sin\vartheta & 0 \end{bmatrix} \begin{Bmatrix} a_r \\ a_\vartheta \\ a_\varphi \end{Bmatrix} \tag{2.48}$$

2.2.5 THE VECTOR DIFFERENTIAL OPERATOR (GRADIENT) IN CYLINDRICAL AND SPHERICAL COORDINATES

Taking into account the above results, the vector differential operator $\vec{\nabla}$ presented in Equation (2.28) for rectangular coordinates, can now be expressed in cylindrical and spherical coordinates:

$$\vec{\nabla} \equiv \frac{\partial}{\partial r}\vec{r} + \frac{1}{r}\frac{\partial}{\partial \varphi}\vec{\varphi} + \frac{\partial}{\partial z}\vec{z} \quad \text{for cylindrical coordinates} \tag{2.49}$$

and

$$\vec{\nabla} \equiv \frac{\partial}{\partial r}\vec{r} + \frac{1}{r}\frac{\partial}{\partial \vartheta}\vec{\vartheta} + \frac{1}{r\sin\vartheta}\frac{\partial}{\partial \varphi}\vec{\varphi} \quad \text{for spherical coordinates} \tag{2.50}$$

2.3 ELEMENTS OF MATRIX ALGEBRA

2.3.1 BASIC DEFINITIONS

A matrix $[A_{mn}]$ is an array of scalars consisting of m rows and n columns:

$$[A_{mn}] = \begin{bmatrix} a_{11} & a_{12} & \dots & a_{1n} \\ a_{21} & a_{22} & \dots & a_{2n} \\ \vdots & \vdots & & \vdots \\ a_{m1} & a_{m2} & \dots & a_{mn} \end{bmatrix} \tag{2.51}$$

The location of any element within the matrix is defined by the subscripts i, j. Therefore, the symbol a_{ij} means that this element is located at the i-th row and j-th column. A matrix $\{B_n\}$ containing only one column is called a vector:

$$\{B_n\} = \begin{Bmatrix} b_1 \\ b_2 \\ \vdots \\ b_n \end{Bmatrix} \tag{2.52}$$

In mechanical and structural engineering problems, we often encounter special cases of matrices such as:

1. *Diagonal matrix*, that is, square matrix containing non-zero elements only at the locations where $i \neq j$, for example,

$$[C_4] = \begin{bmatrix} 1 & 0 & 0 & 0 \\ 0 & -12 & 0 & 0 \\ 0 & 0 & 7 & 0 \\ 0 & 0 & 0 & 3 \end{bmatrix} \tag{2.53}$$

2. *Unit matrix*, that is, a diagonal matrix containing unit elements at the locations where $i = j$, for example,

$$[I_3] = \begin{bmatrix} 1 & 0 & 0 \\ 0 & 1 & 0 \\ 0 & 0 & 1 \end{bmatrix} \tag{2.54}$$

3. *Zero matrix*, that is, a matrix containing elements with zero value for all locations i,j, for example,

$$[O_{3,2}] = \begin{bmatrix} 0 & 0 \\ 0 & 0 \\ 0 & 0 \end{bmatrix} \tag{2.55}$$

4. *Symmetric matrix*, that is, a matrix where $a_{ij} = a_{ji}$ for all i,j, for example,

$$[F] = \begin{bmatrix} 1 & 2 & -6 \\ 2 & 8 & 9 \\ -6 & 9 & 11 \end{bmatrix} \tag{2.56}$$

2.3.2 BASIC OPERATIONS

Determinant of a square matrix

The determinant $\det[A_{nn}]$ of a square matrix $[A_{nn}]$ is a scalar quantity given by the following equation:

$$\det[A_{nn}] = \sum_{\varphi_1, \varphi_2, \ldots, \varphi_n} p(\varphi_1, \varphi_2, \ldots, \varphi_n) a_{1,\varphi_1} \cdot a_{2,\varphi_2} \cdot \cdots \cdot a_{n,\varphi_n} \tag{2.57}$$

where $p(\varphi_1, \varphi_2, \ldots, \varphi_n)$ is a permutation equal to ± 1. For example, for a 3×3 matrix, we have the following permutations:

$$p(1, 2, 3) = 1 \tag{2.58}$$

$$p(1, 3, 2) = -1 \tag{2.59}$$

$$p(3, 1, 2) = 1 \tag{2.60}$$

$$p(3, 2, 1) = -1 \tag{2.61}$$

$$p(2, 3, 1) = 1 \tag{2.62}$$

$$p(2, 1, 3) = -1 \tag{2.63}$$

According to the above definition, the determinant of a 3×3 matrix is

$$\det\begin{pmatrix} a_{11} & a_{12} & a_{13} \\ a_{21} & a_{22} & a_{23} \\ a_{31} & a_{32} & a_{33} \end{pmatrix} = a_{11}a_{22}a_{33} - a_{11}a_{23}a_{32} + a_{13}a_{21}a_{32} - a_{13}a_{22}a_{31} + a_{12}a_{23}a_{31} - a_{12}a_{21}a_{33} \tag{2.64}$$

EXAMPLE 2.1

Calculate the determinant of the following matrix:

$$A = \begin{pmatrix} 1 & 5 & 4 \\ 2 & 3 & 1 \\ 7 & 8 & 5 \end{pmatrix}$$

Solution

According to Equation (2.57), det[A] can be calculated as:

$$\det[A] = (1 \times 3 \times 5) - (1 \times 1 \times 8) + (4 \times 2 \times 8) - (4 \times 3 \times 7) + (5 \times 1 \times 7) - (2 \times 2 \times 5) = -28$$

Minor D_{ij} *of an element* a_{ij}

Minor D_{ij} of an element a_{ij} is the determinant of the submatrix obtained from the matrix [A] by deleting the i-th row and j-th column, for example, for a 3×3 matrix, the minor D_{12} is:

$$D_{12} = \det\begin{pmatrix} a_{11} & a_{12} & a_{13} \\ a_{21} & a_{22} & a_{23} \\ a_{31} & a_{32} & a_{33} \end{pmatrix} = \det\begin{pmatrix} a_{21} & a_{23} \\ a_{31} & a_{33} \end{pmatrix} \tag{2.65}$$

or

$$D_{12} = a_{21}a_{33} - a_{31}a_{23} \tag{2.66}$$

Cofactor A_{ij} *of an element* α_{ij}

The cofactor A_{ij} of an element a_{ij} is defined as:

$$A_{ij} = (-1)^{i+j} D_{ij} \tag{2.67}$$

For the above example of the 3×3 matrix, the cofactor A_{12} is

$$A_{12} = (-1)^{1+2}(a_{21}a_{33} - a_{31}a_{23}) = -a_{21}a_{33} + a_{31}a_{23} \tag{2.68}$$

The inverse of a matrix

The inverse of a square matrix [A], denoted by $[A]^{-1}$ is defined by the following equation:

$$[A]^{-1} = \frac{1}{\det[A]}\begin{pmatrix} A_{11} & A_{21} & \dots & A_{n1} \\ A_{12} & A_{22} & \dots & A_{n2} \\ \vdots & \vdots & & \vdots \\ A_{1n} & A_{2n} & \dots & A_{nn} \end{pmatrix} \tag{2.69}$$

where A_{ij} are the cofactors of the elements a_{ij}.

EXAMPLE 2.2

Calculate the inverse of the following matrix:

$$A = \begin{pmatrix} 1 & 5 & 4 \\ 2 & 3 & 1 \\ 7 & 8 & 5 \end{pmatrix}$$

Solution

The minors of the above matrix are:

$$D_{11} = \det\begin{pmatrix} 3 & 1 \\ 8 & 5 \end{pmatrix} = 7$$

$$D_{12} = \det\begin{pmatrix} 2 & 1 \\ 7 & 5 \end{pmatrix} = 3$$

$$D_{13} = \det\begin{pmatrix} 2 & 3 \\ 7 & 8 \end{pmatrix} = -5$$

$$D_{21} = \det\begin{pmatrix} 5 & 4 \\ 8 & 5 \end{pmatrix} = -7$$

$$D_{22} = \det\begin{pmatrix} 1 & 4 \\ 7 & 5 \end{pmatrix} = -23$$

$$D_{23} = \det\begin{pmatrix} 1 & 5 \\ 7 & 8 \end{pmatrix} = -27$$

$$D_{31} = \det\begin{pmatrix} 5 & 4 \\ 3 & 1 \end{pmatrix} = -7$$

$$D_{32} = \det\begin{pmatrix} 1 & 4 \\ 2 & 1 \end{pmatrix} = -7$$

$$D_{33} = \det\begin{pmatrix} 1 & 5 \\ 2 & 3 \end{pmatrix} = -7$$

Then, the cofactors of the matrix $[A]$ are

$$A_{11} = (-1)^{1+1}D_{11} = 7,\ A_{12} = (-1)^{1+2}D_{12} = -3,\ A_{13} = (-1)^{1+3}D_{13} = -5,$$
$$A_{21} = (-1)^{2+1}D_{21} = 7,\ A_{22} = (-1)^{2+2}D_{22} = -23,\ A_{23} = (-1)^{2+3}D_{23} = 27,$$
$$A_{31} = (-1)^{3+1}D_{31} = -7,\ A_{32} = (-1)^{3+2}D_{32} = 7,\ A_{33} = (-1)^{3+3}D_{33} = -7$$

Taking into account (from the previous example) that

$$\det[A] = -28$$

then, according to Equation (2.69) the matrix $[A]^{-1}$ is

$$[A]^{-1} = \frac{1}{-28}\begin{bmatrix} 7 & 7 & -7 \\ -3 & -23 & 7 \\ -5 & 27 & -7 \end{bmatrix}$$

Transpose of a matrix

The transpose of a matrix $[A]$, denoted by $[A]^{\mathrm{T}}$, is given by the following equation:

$$[A]^{\mathrm{T}} = \begin{pmatrix} a_{11} & a_{12} & \dots & a_{1n} \\ a_{21} & a_{22} & \dots & a_{2n} \\ \vdots & \vdots & & \vdots \\ a_{n1} & a_{n2} & \dots & a_{nn} \end{pmatrix}^{\mathrm{T}} = \begin{pmatrix} a_{11} & a_{21} & \dots & a_{n1} \\ a_{12} & a_{22} & \dots & a_{n2} \\ \vdots & \vdots & & \vdots \\ a_{1n} & a_{2n} & \dots & a_{nn} \end{pmatrix} \tag{2.70}$$

Addition/subtraction of two matrices

$$[A_{nm}] \pm [B_{nm}] = \begin{pmatrix} a_{11} \pm b_{11} & a_{12} \pm b_{12} & \dots & a_{1m} \pm b_{1m} \\ a_{21} \pm b_{21} & a_{22} \pm b_{22} & \dots & a_{2m} \pm b_{2m} \\ \vdots & \vdots & & \vdots \\ a_{n1} \pm b_{n1} & a_{n2} \pm b_{n2} & \dots & a_{nm} \pm b_{nm} \end{pmatrix} \tag{2.71}$$

Multiplication of two matrices

If a matrix $[A_{nm}]$ is multiplied by a scalar e, the matrix $e \cdot [A]$ is obtained by multiplying all elements of $[A]$ by the scalar e. For example,

$$e \cdot \begin{pmatrix} a_{11} & a_{12} & a_{13} \\ a_{21} & a_{22} & a_{23} \\ a_{31} & a_{32} & a_{33} \end{pmatrix} = \begin{pmatrix} e \cdot a_{11} & e \cdot a_{12} & e \cdot a_{13} \\ e \cdot a_{21} & e \cdot a_{22} & e \cdot a_{23} \\ e \cdot a_{31} & e \cdot a_{32} & e \cdot a_{33} \end{pmatrix}$$

If a matrix $[A_{nm}]$ is multiplied by another matrix $[B_{mk}]$, then the obtained matrix

$$[C_{nk}] = [A_{nm}] \cdot [B_{mk}] \tag{2.72}$$

is composed by elements given by the following equation:

$$c_{ij} = \sum_{\varphi=1}^{m} a_{i\varphi} b_{\varphi j} \tag{2.73}$$

Application of the above formula into multiplication of a matrix

$$[R] = [r_1 \;\; r_2 \;\; r_3] \tag{2.74}$$

with a vector

$$\{C\} = \begin{Bmatrix} c_1 \\ c_2 \\ c_3 \end{Bmatrix} \tag{2.75}$$

yields

$$[R]\cdot\{C\} = r_1c_1 + r_2c_2 + r_3c_3 \tag{2.76}$$

An alternative procedure for multiplication of the following matrices

$$[A_{23}] = \begin{pmatrix} a_{11} & a_{12} & a_{13} \\ a_{21} & a_{22} & a_{23} \end{pmatrix} \tag{2.77}$$

$$[B_{32}] = \begin{pmatrix} b_{11} & b_{12} \\ b_{21} & b_{22} \\ b_{31} & b_{32} \end{pmatrix} \tag{2.78}$$

is to express the above matrices as

$$[A_{23}] = \begin{pmatrix} [R_1] \\ [R_2] \end{pmatrix} \tag{2.79}$$

$$[B_{32}] = [\{C_1\} \ \ \{C_2\}] \tag{2.80}$$

where

$$[R_1] = [a_{11} \ \ a_{12} \ \ a_{13}] \tag{2.81}$$

$$[R_2] = [a_{21} \ \ a_{22} \ \ a_{23}] \tag{2.82}$$

$$\{C_1\} = \begin{Bmatrix} b_{11} \\ b_{21} \\ b_{31} \end{Bmatrix} \tag{2.83}$$

$$\{C_2\} = \begin{Bmatrix} b_{12} \\ b_{22} \\ b_{32} \end{Bmatrix} \tag{2.84}$$

Then,

$$[A_{23}]\cdot[A_{32}] = \begin{array}{c|c|c|} & \{C_1\} & \{C_2\} \\ & \downarrow & \downarrow \\ \hline [R_1] \ \rightarrow & [R_1]\{C_1\} & [R_1]\{C_2\} \\ \hline [R_2] \ \rightarrow & [R_2]\{C_1\} & [R_2]\{C_2\} \\ \hline \end{array} \tag{2.85}$$

Following the same procedure, multiplication of a matrix with a vector can be performed as follows:

$$\begin{pmatrix} a_{11} & a_{12} & a_{13} \\ a_{21} & a_{22} & a_{23} \\ a_{31} & a_{32} & a_{33} \end{pmatrix} \cdot \begin{Bmatrix} b_1 \\ b_2 \\ b_3 \end{Bmatrix} = \begin{matrix} [b_1 \ \ b_2 \ \ b_3] \\ \begin{pmatrix} a_{11} & a_{12} & a_{13} \\ a_{21} & a_{22} & a_{23} \\ a_{31} & a_{32} & a_{33} \end{pmatrix} \end{matrix} = \begin{Bmatrix} a_{11}b_1 + a_{12}b_2 + a_{13}b_3 \\ a_{21}b_1 + a_{22}b_2 + a_{23}b_3 \\ a_{31}b_1 + a_{32}b_2 + a_{33}b_3 \end{Bmatrix} \tag{2.86}$$

Properties of matrix operations

$$[A] + [B] = [B] + [A] \tag{2.87}$$

$$[A]\cdot[B] \neq [B]\cdot[A] \tag{2.88}$$

$$[A]\cdot([B]+[C]) = [A]\cdot[B]+[A]\cdot[C] \tag{2.89}$$

$$([A]+[B])\cdot[C] = [A]\cdot[C]+[B]\cdot[C] \tag{2.90}$$

$$([A]+[B])^{\mathrm{T}} = [A]^{\mathrm{T}}+[B]^{\mathrm{T}} \tag{2.91}$$

$$([A]\cdot[B])^{\mathrm{T}} = [B]^{\mathrm{T}}\cdot[A]^{\mathrm{T}} \tag{2.92}$$

$$([A]+[B])^{-1} = [A]^{-1}+[B]^{-1} \tag{2.93}$$

$$([A]\cdot[B])^{-1} = [B]^{-1}\cdot[A]^{-1} \tag{2.94}$$

2.4 VARIATIONAL FORMULATION OF ELASTICITY PROBLEMS

As it will be described in Chapters 7 and 9, the derivation of the finite element equation of elastic solids can be based on the principle of minimum potential energy (MPE). This principle belongs to a general category of energy methods that is known as the variational principle. The mathematical background behind MPE assumes that some classes of boundary value problems can be solved by the minimization of an integral functional (i.e., function of a function) whose necessary conditions for a minimum implies that the differential equation and the associated boundary conditions are satisfied. Therefore, the main task of the above procedure is the derivation of the integral functional.

Since the derivation of an integral functional is based on concepts from calculus of variations, this section explain the properties and the main concept of the variation of a function.

2.4.1 DEFINITION OF THE VARIATION OF A FUNCTION

Let us assume a continuous function $u(x)$ (e.g., the axial displacement of a bar due to axial loads). For simplicity, we have chosen a polynomial function

$$u(x) = a_0 + a_1x + a_2x^2 + \cdots + a_nx^n \tag{2.95}$$

For infinitesimal changes of the above coefficients, $a_0, a_1, a_2, \ldots, a_n$, namely, $\delta a_0, \delta a_1, \delta a_2, \ldots, \delta a_n$, a closely related function is written as follows:

$$\bar{u}(x) = (a_0+\delta a_0) + (a_1+\delta a_1)x + (a_2+\delta a_2)x^2 + \cdots + (a_n+\delta a_n)x^n \tag{2.96}$$

The variation $\delta u(x)$ of $u(x)$ is defined as the following difference

$$\delta u(x) = \bar{u}(x) - u(x) = \delta a_0 + \delta a_1x + \delta a_2x^2 + \cdots + \delta a_nx^n \tag{2.97}$$

Very often, apart from the variation of a function we need to calculate the variation of a functional (i.e., function of a function) $U(u(x))$:

$$\delta U = U(\bar{u}(x)) - U(u(x)) \tag{2.98}$$

The calculation of the variation of a functional can be simplified by using the following properties of variations.

2.4.2 PROPERTIES OF VARIATIONS

$$\delta\left(\frac{\mathrm{d}U}{\mathrm{d}x}\right)=\frac{\mathrm{d}(\delta U)}{\mathrm{d}x} \tag{2.99}$$

$$\delta\left(\int U\mathrm{d}x\right)=\int \delta U\mathrm{d}x \tag{2.100}$$

$$\delta(U+V)=\delta U+\delta V \tag{2.101}$$

$$\delta(UV)=(\delta U)V+U(\delta V) \tag{2.102}$$

$$\delta(U^n)=nU^{n-1}\delta U \tag{2.103}$$

$$f(x)\delta U(x)=\delta(f(x)U(x)) \tag{2.104}$$

$$U(x)\delta U(x)=\frac{1}{2}\delta(U^2(x)) \tag{2.105}$$

$$\frac{\mathrm{d}U}{\mathrm{d}x}\cdot\frac{\mathrm{d}(\delta U)}{\mathrm{d}x}=\frac{1}{2}\delta\left[\left(\frac{\mathrm{d}U}{\mathrm{d}x}\right)^2\right] \tag{2.106}$$

$$F\equiv \text{minimum}\Rightarrow \delta F=0 \tag{2.107}$$

2.4.3 DERIVATION OF THE FUNCTIONAL FROM THE BOUNDARY VALUE PROBLEM

Following mathematical manipulations, the functional to be minimized can be derived from the differential equation and the boundary conditions. This procedure is illustrated through the following example of a bar under axial loading.

Boundary value problem of a bar under axial loading

Let us consider a bar (Figure 2.10) with variable cross-sectional area $A=A(x)$. The bar is subjected to a distributed axial load $q(x)$ and a concentrated force F at its right end. The equilibrium equation of an infinitesimal part of the bar yields

$$N+\frac{\partial N}{\partial x}\mathrm{d}x+q(x)\mathrm{d}x-N=0 \tag{2.108}$$

or

$$\frac{\partial N}{\partial x}+q(x)=0 \tag{2.109}$$

where $N=N(x)$ is the axial force acting on a cross-section located at x.

According to the Hooke's law, the force $N(x)$ can be expressed as a function of strain $\varepsilon(x)$

$$\frac{N(x)}{A(x)}=E\varepsilon(x) \tag{2.110}$$

where E is the modulus of elasticity.

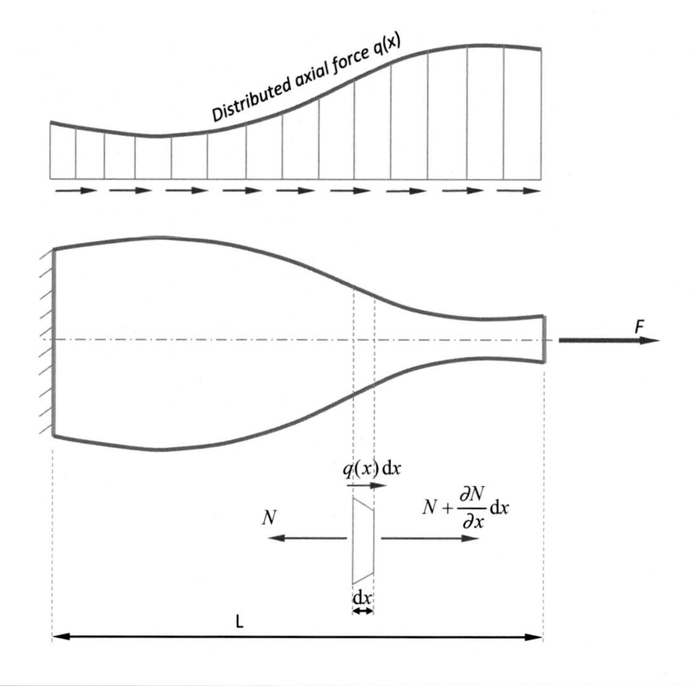

FIGURE 2.10

Bar under axial loads.

Since

$$\varepsilon(x) = \frac{\partial u(x)}{\partial x} \tag{2.111}$$

where $u(x)$ is the axial displacement, Equation (2.110) can be written as:

$$N(x) = EA(x)\frac{\partial u(x)}{\partial x} \tag{2.112}$$

Using the above equation, Equation (2.109) yields

$$\frac{\partial}{\partial x}\left[EA(x)\frac{\partial u(x)}{\partial x}\right] + q(x) = 0 \qquad 0 \leq x \leq L \tag{2.113}$$

Assuming that the bar is fixed at the left end and free at the right end, the following boundary conditions may associate the governing Equation (2.113):

$$u(0) = 0 \tag{2.114}$$

$$u'(0) = \varepsilon(0) = 0 \tag{2.115}$$

$$u'(L) = \varepsilon(L) = \frac{F}{E \cdot A(L)} \tag{2.116}$$

Total potential energy of a bar under axial loading

Total potential energy is defined by:

$$\Pi = U + W \tag{2.117}$$

where U is the strain energy of stresses or internal forces, and W is the energy possessed by the external loads (e.g., concentrated forces and surface tractions).

According to the above definition, the strain energy U and the work W are given by the following equations:

$$U = \frac{1}{2}\int_0^L \sigma\varepsilon A\mathrm{d}x \tag{2.118}$$

and

$$W = -Fu(L) - \int_0^L u(x)q(x)\mathrm{d}x \tag{2.119}$$

It should be noted that the external loads F and $q(x)$ are always acting at their full value (i.e., their work is independent of the elastic behavior of the bar). Their movement through the corresponding displacements $u(L)$ and $u(x)$ is doing work in amount $F \cdot u(L)$ and $u(x) \cdot q(x)\mathrm{d}x$, losing potential of equal amounts $-F \cdot u(L)$ and $-u(x) \cdot q(x)\mathrm{d}x$, respectively.

Since

$$\varepsilon = \frac{\partial u}{\partial x} \tag{2.120}$$

and

$$\sigma = E\varepsilon = E\frac{\partial u}{\partial x} \tag{2.121}$$

Equation (2.118) can be written as:

$$U = \frac{1}{2}\int_0^L EA(x)(u')^2\mathrm{d}x \tag{2.122}$$

Therefore, using Equations (2.119) and (2.122), Equation (2.117) yields

$$\Pi = -Fu(L) + \int_0^L \left[\frac{1}{2}EA(u')^2 - q \cdot u\right]\mathrm{d}x \tag{2.123}$$

Derivation of the total potential of a bar from the boundary value problem

The aim of this section is to explain the derivation of Equation (2.123) from Equation (2.113). To achieve this target, we have to manipulate Equation (2.113) in order to derive an expression of the form $\delta(\cdots) = 0$. To achieve this, mathematical manipulation starts by multiplying the differential Equation (2.113) by the variation $-\delta u$ and integrating over the solution domain (for simplicity, we assume that the cross-sectional area A is constant):

$$\int_0^L (-EAu'' - q)\delta u\mathrm{d}x = 0, \qquad 0 \le x \le L \tag{2.124}$$

Let us write the above equation as

$$I_1 + I_2 = 0 \tag{2.125}$$

where

$$I_1 = -\int_0^L EAu''\delta u\mathrm{d}x \tag{2.126}$$

and

$$I_2 = -\int_0^L q\delta u \mathrm{d}x \tag{2.127}$$

Taking into account the following well-known formula

$$\int_a^b w\mathrm{d}v = wv|_a^b - \int_a^b v\mathrm{d}w \tag{2.128}$$

and setting

$$a = 0 \tag{2.129}$$

$$b = L \tag{2.130}$$

$$w = \delta u \tag{2.131}$$

$$\mathrm{d}v = u''\mathrm{d}x \quad \text{or} \quad v = u' \tag{2.132}$$

the integral I_1 can be written as:

$$\frac{I_1}{EA} = -u'\delta u|_0^L + \int_0^L u'\mathrm{d}(\delta u) \tag{2.133}$$

or

$$\frac{I_1}{EA} = -u'(L)\delta u(L) + u'(0)\delta u(0) + \int_0^L u'\mathrm{d}(\delta u) \tag{2.134}$$

Taking into account Equation (2.99), that is,

$$\frac{\mathrm{d}(\delta u)}{\mathrm{d}x} = \delta\left(\frac{\mathrm{d}u}{\mathrm{d}x}\right) \tag{2.135}$$

the quantity $\mathrm{d}(\delta u)$ is

$$\mathrm{d}(\delta u) = \delta u'\mathrm{d}x \tag{2.136}$$

Then, Equation (2.134) yields

$$\frac{I_1}{EA} = -u'(L)\delta u(L) + u'(0)\delta u(0) + \int_0^L u'\delta u'\mathrm{d}x \tag{2.137}$$

We recall from the boundary conditions (Equation 2.115) that $u'(0) = 0$. In contrast, $u'(L) = \varepsilon(L)$. Then,

$$\frac{I_1}{EA} = -\varepsilon(L)\delta u(L) + \int_0^L u'\delta u'\mathrm{d}x \tag{2.138}$$

Taking into account Equation (2.105), the product $u'\delta u'$ can be written as:

$$u'\delta u' = \frac{1}{2}\delta(u')^2 \tag{2.139}$$

Therefore, Equation (2.138) can now take the form:

$$I_1 = -AE\varepsilon(L)\delta u(L) + AE\int_0^L \frac{1}{2}\delta(u')^2\mathrm{d}x \tag{2.140}$$

Using the boundary condition given in Equation (2.116), the product, $AE\varepsilon(L)$, is $AE\varepsilon(L)=F$.

Then, Equation (2.140) yields

$$I_1=-F\cdot\delta u(L)+AE\int_0^L\frac{1}{2}\delta(u')^2\mathrm{d}x \tag{2.141}$$

Using Equation (2.104), Equation (2.141) can be written as:

$$I_1=-\delta(F\cdot u(L))+\int_0^L\delta\left(AE\frac{1}{2}(u')^2\right)\mathrm{d}x \tag{2.142}$$

According to Equation (2.100), Equation (2.142) yields

$$I_1=-\delta(F\cdot u(L))+\delta\int_0^L AE\frac{1}{2}(u')^2\mathrm{d}x \tag{2.143}$$

Using Equation (2.101), Equation (2.143) can be formulated as:

$$I_1=\delta\left[-F\cdot u(L)+\int_0^L\frac{1}{2}AE(u')^2\mathrm{d}x\right] \tag{2.144}$$

Combining Equation (2.144) with Equations (2.125) and (2.127), the following formula can be obtained:

$$\delta\left[-F\cdot u(L)+\int_0^L\frac{1}{2}AE(u')^2\mathrm{d}x\right]-\int_0^L q\delta u\mathrm{d}x=0 \tag{2.145}$$

Using Equation (2.104), Equation (2.145) yields

$$\delta\left[-F\cdot u(L)+\int_0^L\frac{1}{2}AE(u')^2\mathrm{d}x\right]-\int_0^L\delta(qu)\cdot\mathrm{d}x=0 \tag{2.146}$$

According to Equation (2.100), Equation (2.146) can now be written as:

$$\delta\left[-F\cdot u(L)+\int_0^L\frac{1}{2}AE(u')^2\mathrm{d}x\right]-\delta\int_0^L q\cdot u\cdot\mathrm{d}x=0 \tag{2.147}$$

or

$$\delta\left\{-F\cdot u(L)+\int_0^L\left[\frac{1}{2}AE(u')^2-q\cdot u\right]\mathrm{d}x\right\}=0 \tag{2.148}$$

Equation (2.148) expresses the principle of MPE:

$$\delta\Pi=0 \tag{2.149}$$

Therefore,

$$\Pi=-F\cdot u(L)+\int_0^L\left[\frac{1}{2}AE(u')^2-q\cdot u\right]\mathrm{d}x \tag{2.150}$$

Equation (2.150) is same as Equation (2.123).

REFERENCES

[1] Deif AS. Advanced matrix theory for scientists and engineers. 2nd ed. London: Abacus press; 1991.
[2] Wunderlich W, Pilkey WD. Mechanics of structures: variational and computational methods. Boca Raton: CRC press; 2003.
[3] Hartmann F. Green's functions and finite elements. Heidelberg: Springer; 2013.
[4] Bhatti MA. Fundamental finite element analysis and applications. Hoboken: John Wiley & Sons; 2005.
[5] Logan DL. A first course in the finite element method. Boston, MA: Gengage Learning; 2012.
[6] Fish J, Belytschko T. A first course in finite elements. New York: Wiley; 2007.
[7] Boresi AP, Schmidt RJ. Advanced mechanics of materials. New York: John Wiley & Sons; 2003.

CHAPTER

LINEAR SPRING ELEMENTS

3

Linear springs are the simplest elements. In this chapter we use them to present the main concept of an element equation derivation, and to explain how the local stiffness matrices can be assembled in order to compose the global stiffness matrix of the structure, how the boundary conditions can be specified and adopted, and how the displacement field of the structural system can be derived. Even though linear springs are elements carrying axial forces in only one direction, the steps for the finite element (FE) formulation of the problem is similar even for more complicated cases.

3.1 THE ELEMENT EQUATION

As already mentioned, the first step for the FE formulation of a structure is the derivation of the element equation

$$\{r\} = [k]\{d\} \tag{3.1}$$

correlating the loads $\{r\}$ acting on the nodes of an element with the corresponding nodal displacements $\{d\}$. More specifically, the main target of this step is the derivation of the element's stiffness matrix $[k]$ correlating the above column matrices $\{r\}$ and $\{d\}$.

Figure 3.1a shows a system of linear springs before loading. As Figure 3.1b indicates, a consequence of the action of axial loads at the end of the spring members are axial displacements.

To derive the element equation, the method of direct equilibrium will be adopted.

3.1.1 THE MECHANICAL BEHAVIOR OF THE MATERIAL

It is well known that the equation describing the mechanical behavior of a spring is given by Hook's law:

$$F = k(d_{2x} - d_{1x}) \tag{3.2}$$

where k is the spring constant, $(d_{2x} - d_{1x})$ is the spring's elongation, and F is the axial force causing spring's elongation.

Essentials of the Finite Element Method. http://dx.doi.org/10.1016/B978-0-12-802386-0.00003-7

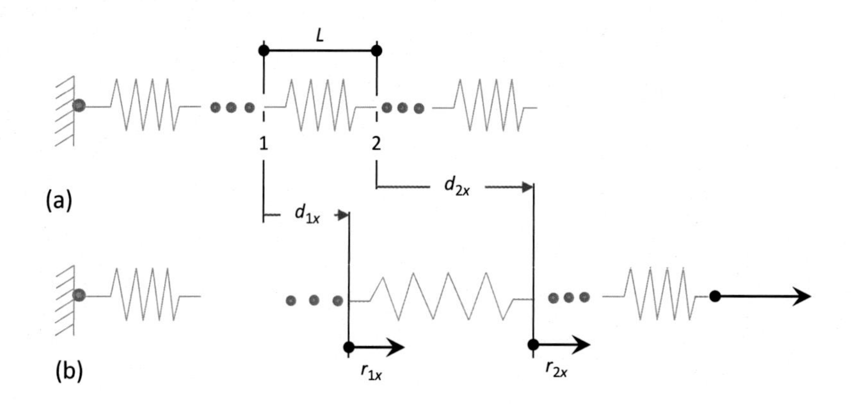

FIGURE 3.1

(a) System of springs before loading. (b) Deformation of a system of springs due to action of axial forces on its nodes.

3.1.2 THE PRINCIPLE OF DIRECT EQUILIBRIUM

In order to correlate the force F (that causes the spring's elongation) to nodal forces, the equilibrium equation will be applied in the two pieces of the spring 1-2 (see Figure 3.2).

Therefore:

$$r_{1x} + F = 0 \tag{3.3}$$

$$-F + r_{2x} = 0 \tag{3.4}$$

The Equations (3.3) and (3.4) can be written in the following matrix form:

$$\begin{Bmatrix} r_{1x} \\ r_{2x} \end{Bmatrix} = \begin{Bmatrix} -F \\ F \end{Bmatrix} \tag{3.5}$$

Using Equation (3.2), Equation (3.5) yields

$$\begin{Bmatrix} r_{1x} \\ r_{2x} \end{Bmatrix} = \begin{Bmatrix} -k(d_{2x} - d_{1x}) \\ k(d_{2x} - d_{1x}) \end{Bmatrix} \tag{3.6}$$

or

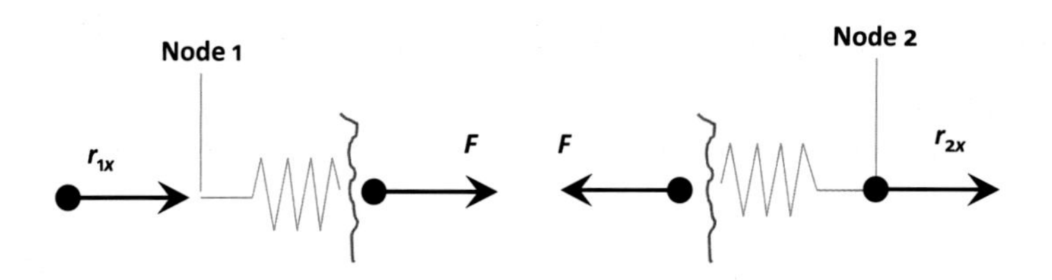

FIGURE 3.2

Equilibrium of the two pieces of the spring.

$$\begin{Bmatrix} r_{1x} \\ r_{2x} \end{Bmatrix} = \begin{bmatrix} k & -k \\ -k & k \end{bmatrix} \begin{Bmatrix} d_{1x} \\ d_{2x} \end{Bmatrix} \tag{3.7}$$

Therefore, Equation (3.7) is the element's equation, and the matrix

$$[k] = \begin{bmatrix} k & -k \\ -k & k \end{bmatrix} \tag{3.8}$$

is the stiffness matrix of the spring element.

3.2 THE STIFFNESS MATRIX OF A SYSTEM OF SPRINGS

The procedure for deriving the stiffness matrix for a system of springs will be demonstrated using the example of Figure 3.3.

The main steps to be followed are: (a) to derive the stiffness matrix for each spring element, (b) to expand the stiffness matrices to the degrees of freedom of the whole structure, and (c) to assembly the expanded stiffness matrices in order to derive the stiffness matrix of the structure.

3.2.1 DERIVATION OF ELEMENT MATRICES

Taking into account Equation (3.7), the following element equations for the corresponding elements can be derived:

Element 1, nodes 1 and 2

$$\begin{Bmatrix} r_{1x} \\ r_{2x} \end{Bmatrix} = \begin{bmatrix} k_1 & -k_1 \\ -k_1 & k_1 \end{bmatrix} \begin{Bmatrix} d_{1x} \\ d_{2x} \end{Bmatrix} \tag{3.9}$$

Element 2, nodes 2 and 3

$$\begin{Bmatrix} r_{2x} \\ r_{3x} \end{Bmatrix} = \begin{bmatrix} k_2 & -k_2 \\ -k_2 & k_2 \end{bmatrix} \begin{Bmatrix} d_{2x} \\ d_{3x} \end{Bmatrix} \tag{3.10}$$

Element 3, nodes 3 and 4

$$\begin{Bmatrix} r_{3x} \\ r_{4x} \end{Bmatrix} = \begin{bmatrix} k_3 & -k_3 \\ -k_3 & k_3 \end{bmatrix} \begin{Bmatrix} d_{3x} \\ d_{4x} \end{Bmatrix} \tag{3.11}$$

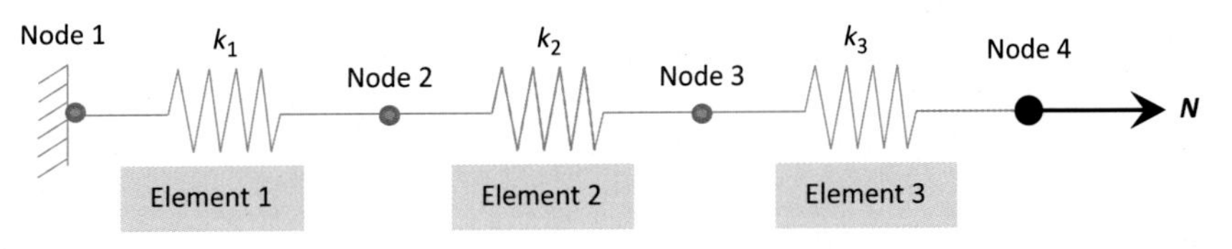

FIGURE 3.3

A simple structural system composed of three spring elements.

3.2.2 EXPANSION OF ELEMENT EQUATIONS TO THE DEGREES OF FREEDOM OF THE STRUCTURE

Since the structure contains four nodes, the above element equations (Equations 3.9–3.11) should be expanded to four degrees of freedom, that is:

Element 1, nodes 1 and 2

$$\begin{Bmatrix} r_{1x} \\ r_{2x} \\ r_{3x} \\ r_{4x} \end{Bmatrix} = \begin{bmatrix} k_1 & -k_1 & 0 & 0 \\ -k_1 & k_1 & 0 & 0 \\ 0 & 0 & 0 & 0 \\ 0 & 0 & 0 & 0 \end{bmatrix} \begin{Bmatrix} d_{1x} \\ d_{2x} \\ d_{3x} \\ d_{4x} \end{Bmatrix} \tag{3.12}$$

Element 2, nodes 2 and 3

$$\begin{Bmatrix} r_{1x} \\ r_{2x} \\ r_{3x} \\ r_{4x} \end{Bmatrix} = \begin{bmatrix} 0 & 0 & 0 & 0 \\ 0 & k_2 & -k_2 & 0 \\ 0 & -k_2 & k_2 & 0 \\ 0 & 0 & 0 & 0 \end{bmatrix} \begin{Bmatrix} d_{1x} \\ d_{2x} \\ d_{3x} \\ d_{4x} \end{Bmatrix} \tag{3.13}$$

Element 3, nodes 3 and 4

$$\begin{Bmatrix} r_{1x} \\ r_{2x} \\ r_{3x} \\ r_{4x} \end{Bmatrix} = \begin{bmatrix} 0 & 0 & 0 & 0 \\ 0 & 0 & 0 & 0 \\ 0 & 0 & k_3 & -k_3 \\ 0 & 0 & -k_3 & k_3 \end{bmatrix} \begin{Bmatrix} d_{1x} \\ d_{2x} \\ d_{3x} \\ d_{4x} \end{Bmatrix} \tag{3.14}$$

3.2.3 ASSEMBLY OF ELEMENT EQUATIONS

Superposition (addition) of the element Equations (3.12)–(3.14) yields:

$$\begin{Bmatrix} R_{1x} \\ R_{2x} \\ R_{3x} \\ R_{4x} \end{Bmatrix} = \begin{bmatrix} k_1 & -k_1 & 0 & 0 \\ -k_1 & k_1+k_2 & -k_2 & 0 \\ 0 & -k_2 & k_2+k_3 & -k_3 \\ 0 & 0 & -k_3 & k_3 \end{bmatrix} \begin{Bmatrix} d_{1x} \\ d_{2x} \\ d_{3x} \\ d_{4x} \end{Bmatrix} \tag{3.15}$$

It should be noted that the quantities R_{ix} $(i=1,2,3,4)$ are the resultants of the nodal forces.

3.2.4 DERIVATION OF THE FIELD VALUES

Equation (3.15) can be written in an abbreviated notation, that is, $\{R\}=[k]\{d\}$ or $[k]\{d\}-[I]\{R\}=\{O_{4\times1}\}$. An alternative formulation of the latter matrix equation is:

$$[[k] \ \ [-I]]\begin{Bmatrix} \{d\} \\ \{R\} \end{Bmatrix} = \{O_{4\times1}\} \tag{3.16}$$

where $[k]$ is the global stiffness matrix, $[I]$ is the 4×4 unit matrix, $\{O_{4x1}\}$ is a 4×1 vector containing zeros, and the vectors $\{d\}$, $\{R\}$ are $\{d\}=[d_{1x}\ d_{2x}\ d_{3x}\ d_{4x}]^{\mathrm{T}}$ and $\{R\}=[R_{1x}\ R_{2x}\ R_{3x}\ R_{4x}]^{\mathrm{T}}$. Therefore, Equation (3.16) expresses the following system of equations:

$$\begin{bmatrix} k_1 & -k_1 & 0 & 0 & -1 & 0 & 0 & 0 \\ -k_1 & k_1+k_2 & -k_2 & 0 & 0 & -1 & 0 & 0 \\ 0 & -k_2 & k_2+k_3 & -k_3 & 0 & 0 & -1 & 0 \\ 0 & 0 & -k_3 & k_3 & 0 & 0 & 0 & -1 \end{bmatrix} \begin{Bmatrix} d_{1x} \\ d_{2x} \\ d_{3x} \\ d_{4x} \\ R_{1x} \\ R_{2x} \\ R_{3x} \\ R_{4x} \end{Bmatrix} = \begin{Bmatrix} 0 \\ 0 \\ 0 \\ 0 \end{Bmatrix} \tag{3.17}$$

Boundary Conditions

The above matrix equation represents an algebraic system of four equations with eight unknowns. For the above system to be solved, four more equations should be added. The missing equations can be specified by the following boundary conditions:

$$d_{1x} = 0 \tag{3.18}$$

$$R_{2x} = 0 \tag{3.19}$$

$$R_{3x} = 0 \tag{3.20}$$

$$R_{4x} = N \tag{3.21}$$

The above boundary conditions can be formulated in a matrix form as follows:

$$\begin{bmatrix} 1 & 0 & 0 & 0 & 0 & 0 & 0 & 0 \\ 0 & 0 & 0 & 0 & 0 & 1 & 0 & 0 \\ 0 & 0 & 0 & 0 & 0 & 0 & 1 & 0 \\ 0 & 0 & 0 & 0 & 0 & 0 & 0 & 1 \end{bmatrix} \begin{Bmatrix} d_{1x} \\ d_{2x} \\ d_{3x} \\ d_{4x} \\ R_{1x} \\ R_{2x} \\ R_{3x} \\ R_{4x} \end{Bmatrix} = \begin{Bmatrix} 0 \\ 0 \\ 0 \\ N \end{Bmatrix} \tag{3.22}$$

Final Solution

Therefore, superposition of Equations (3.17) and (3.22) yields an algebraic system of eight equations with eight unknowns, providing the nodal displacements and forces:

$$\begin{bmatrix} k1 & -k1 & 0 & 0 & -1 & 0 & 0 & 0 \\ -k1 & k1+k2 & -k2 & 0 & 0 & -1 & 0 & 0 \\ 0 & -k2 & k2+k3 & -k3 & 0 & 0 & -1 & 0 \\ 0 & 0 & -k3 & k3 & 0 & 0 & 0 & -1 \\ 1 & 0 & 0 & 0 & 0 & 0 & 0 & 0 \\ 0 & 0 & 0 & 0 & 0 & 1 & 0 & 0 \\ 0 & 0 & 0 & 0 & 0 & 0 & 1 & 0 \\ 0 & 0 & 0 & 0 & 0 & 0 & 0 & 1 \end{bmatrix} \begin{bmatrix} d_{1x} \\ d_{2x} \\ d_{3x} \\ d_{4x} \\ R_{1x} \\ R_{2x} \\ R_{3x} \\ R_{4x} \end{bmatrix} = \begin{bmatrix} 0 \\ 0 \\ 0 \\ 0 \\ 0 \\ 0 \\ 0 \\ N \end{bmatrix} \tag{3.23}$$

Knowing the nodal displacements d_{1x}, d_{2x}, d_{3x}, and d_{4x}, the internal forces r_{1x}, r_{2x}, r_{3x}, and r_{4x} can now be obtained using Equations (3.9)–(3.11).

EXAMPLE 3.1

Determine the field values (nodal displacements and internal forces) for the following structural system.

Data

$k = 40{,}000$ N/m, $F = 5000$ N

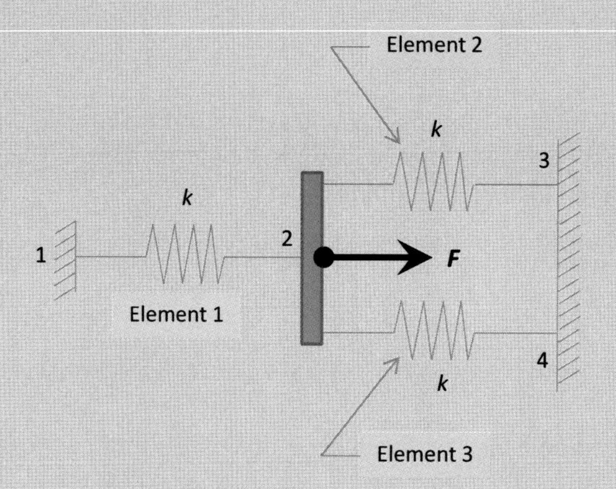

Step 1: Derivation of Element Equations

$$\begin{Bmatrix} r_{1x} \\ r_{2x} \end{Bmatrix} = \begin{bmatrix} k & -k \\ -k & k \end{bmatrix} \begin{Bmatrix} d_{1x} \\ d_{2x} \end{Bmatrix} \quad \text{for element 1} \tag{3.24}$$

$$\begin{Bmatrix} r_{2x} \\ r_{3x} \end{Bmatrix} = \begin{bmatrix} k & -k \\ -k & k \end{bmatrix} \begin{Bmatrix} d_{2x} \\ d_{3x} \end{Bmatrix} \quad \text{for element 2} \tag{3.25}$$

$$\begin{Bmatrix} r_{2x} \\ r_{4x} \end{Bmatrix} = \begin{bmatrix} k & -k \\ -k & k \end{bmatrix} \begin{Bmatrix} d_{2x} \\ d_{4x} \end{Bmatrix} \quad \text{for element 3} \tag{3.26}$$

Step 2: Expansion of the Element Equations to the Degrees of Freedom of the Structure

Element 1, nodes 1 and 2

$$\begin{Bmatrix} r_{1x} \\ r_{2x} \\ r_{3x} \\ r_{4x} \end{Bmatrix} = \begin{bmatrix} k & -k & 0 & 0 \\ -k & k & 0 & 0 \\ 0 & 0 & 0 & 0 \\ 0 & 0 & 0 & 0 \end{bmatrix} \begin{Bmatrix} d_{1x} \\ d_{2x} \\ d_{3x} \\ d_{4x} \end{Bmatrix} \tag{3.27}$$

Element 2, nodes 2 and 3

$$\begin{Bmatrix} r_{1x} \\ r_{2x} \\ r_{3x} \\ r_{4x} \end{Bmatrix} = \begin{bmatrix} 0 & 0 & 0 & 0 \\ 0 & k & -k & 0 \\ 0 & -k & k & 0 \\ 0 & 0 & 0 & 0 \end{bmatrix} \begin{Bmatrix} d_{1x} \\ d_{2x} \\ d_{3x} \\ d_{4x} \end{Bmatrix} \tag{3.28}$$

Element 3, nodes 2 and 4

$$\begin{Bmatrix} r_{1x} \\ r_{2x} \\ r_{3x} \\ r_{4x} \end{Bmatrix} = \begin{bmatrix} 0 & 0 & 0 & 0 \\ 0 & k & 0 & -k \\ 0 & 0 & 0 & 0 \\ 0 & -k & 0 & k \end{bmatrix} \begin{Bmatrix} d_{1x} \\ d_{2x} \\ d_{3x} \\ d_{4x} \end{Bmatrix} \tag{3.29}$$

Step 3: Superposition of Elements Equation and Derivation of the Global Matrix

Addition of Equations (3.27)–(3.29) yields

$$\begin{Bmatrix} R_{1x} \\ R_{2x} \\ R_{3x} \\ R_{4x} \end{Bmatrix} = \begin{bmatrix} k & -k & 0 & 0 \\ -k & 3k & -k & -k \\ 0 & -k & k & 0 \\ 0 & -k & 0 & k \end{bmatrix} \begin{Bmatrix} d_{1x} \\ d_{2x} \\ d_{3x} \\ d_{4x} \end{Bmatrix} \tag{3.30}$$

Equation (3.30) can be written in the following abbreviated notation:

$$\{R\} = [k]\{d\} \tag{3.31}$$

or

$$[k]\{d\} - [I]\{R\} = \{O_{4\times 1}\} \tag{3.32}$$

or

$$[[k] \; [-I]]\begin{Bmatrix} \{d\} \\ \{R\} \end{Bmatrix} = \{O_{4\times 1}\} \tag{3.33}$$

where $[k]$ is the global stiffness matrix, $[I]$ is the 4×4 unit matrix, $\{O_{4\times 1}\}$ is a 4×1 vector containing zeros, and the vectors $\{d\}$, $\{R\}$ are

$$\{d\} = \begin{Bmatrix} d_{1x} \\ d_{2x} \\ d_{3x} \\ d_{4x} \end{Bmatrix} \tag{3.34}$$

$$\{R\} = \begin{Bmatrix} R_{1x} \\ R_{2x} \\ R_{3x} \\ R_{4x} \end{Bmatrix} \tag{3.35}$$

Continued

EXAMPLE 3.1—CONT'D

Step 4: Boundary Conditions

(a) Boundary conditions for nodal displacements

Due to the supports on the nodes 1, 3, and 4, the boundary conditions regarding displacements are:

$$\left.\begin{array}{l} d_{1x}=0 \\ d_{3x}=0 \\ d_{4x}=0 \end{array}\right\} \tag{3.36}$$

The above boundary conditions can be written in the following matrix form:

$$\begin{bmatrix} 1 & 0 & 0 & 0 \\ 0 & 0 & 1 & 0 \\ 0 & 0 & 0 & 1 \\ 0 & 0 & 0 & 0 \end{bmatrix} \begin{Bmatrix} d_{1x} \\ d_{2x} \\ d_{3x} \\ d_{4x} \end{Bmatrix} = \begin{Bmatrix} 0 \\ 0 \\ 0 \\ 0 \end{Bmatrix} \tag{3.37}$$

or in an abbreviated notation

$$[BCd]\{d\}=\{DO\} \tag{3.38}$$

or

$$[[BCd] \;\; [O_{4x4}]]\begin{Bmatrix} \{d\} \\ \{R\} \end{Bmatrix}=\{DO\} \tag{3.39}$$

where

$$[BCd]=\begin{bmatrix} 1 & 0 & 0 & 0 \\ 0 & 0 & 1 & 0 \\ 0 & 0 & 0 & 1 \\ 0 & 0 & 0 & 0 \end{bmatrix} \tag{3.40}$$

$$[O_{4x4}]=\begin{bmatrix} 0 & 0 & 0 & 0 \\ 0 & 0 & 0 & 0 \\ 0 & 0 & 0 & 0 \\ 0 & 0 & 0 & 0 \end{bmatrix} \tag{3.41}$$

$$\{DO\}=[0 \;\; 0 \;\; 0 \;\; 0]^{\mathrm{T}} \tag{3.42}$$

(b) Boundary conditions for nodal forces

Taking into account the force acting on node 2, the following boundary condition can be used:

$$R_{2x}=F \tag{3.43}$$

The above boundary condition can be expressed in the following matrix form:

$$\begin{bmatrix} 0 & 0 & 0 & 0 \\ 0 & 0 & 0 & 0 \\ 0 & 0 & 0 & 0 \\ 0 & 1 & 0 & 0 \end{bmatrix} \begin{Bmatrix} R_{1x} \\ R_{2x} \\ R_{3x} \\ R_{4x} \end{Bmatrix} = \begin{Bmatrix} 0 \\ 0 \\ 0 \\ F \end{Bmatrix} \tag{3.44}$$

or in an abbreviated notation

$$[BCR]\{R\} = \{RO\} \tag{3.45}$$

or

$$[[O_{4\times4}]\ [BCR]]\begin{Bmatrix}\{d\}\\\{R\}\end{Bmatrix} = \{RO\} \tag{3.46}$$

where

$$[BCR] = \begin{bmatrix}0 & 0 & 0 & 0\\0 & 0 & 0 & 0\\0 & 0 & 0 & 0\\0 & 1 & 0 & 0\end{bmatrix} \tag{3.47}$$

$$\{RO\} = [0\ \ 0\ \ 0\ \ F]^{\mathrm{T}} \tag{3.48}$$

It should be noted that since the first three rows are occupied by the boundary conditions for displacements, the boundary condition for the forces should be placed in the fourth line.

(c) Matrix equation of boundary conditions

Adding Equations (3.39) and (3.46), the following matrix equation of boundary conditions can be obtained:

$$[[BCd]\ [BCR]]\begin{Bmatrix}\{d\}\\\{R\}\end{Bmatrix} = \{DO\} + \{RO\} \tag{3.49}$$

Step 5: Equation of Structure

Combining Equations (3.33) and (3.49), the following equation of the structure, containing both stiffness matrix and boundary conditions, can be obtained:

$$\begin{bmatrix}[k] & [-I]\\ [BCd] & [BCR]\end{bmatrix}\begin{Bmatrix}\{d\}\\\{R\}\end{Bmatrix} = \begin{Bmatrix}\{O_{4\times1}\}\\\{DO\}+\{RO\}\end{Bmatrix} \tag{3.50}$$

The above system expresses the following set of algebraic equations:

$$\left[\begin{array}{cccc|cccc} k & -k & 0 & 0 & -1 & 0 & 0 & 0\\ -k & 3k & -k & -k & 0 & -1 & 0 & 0\\ 0 & -k & k & 0 & 0 & 0 & -1 & 0\\ 0 & -k & 0 & k & 0 & 0 & 0 & -1\\ \hline 1 & 0 & 0 & 0 & 0 & 0 & 0 & 0\\ 0 & 0 & 1 & 0 & 0 & 0 & 0 & 0\\ 0 & 0 & 0 & 1 & 0 & 0 & 0 & 0\\ 0 & 0 & 0 & 0 & 0 & 1 & 0 & 0\end{array}\right]\begin{Bmatrix}d_{1x}\\d_{2x}\\d_{3x}\\d_{4x}\\R_{1x}\\R_{2x}\\R_{3x}\\R_{4x}\end{Bmatrix} = \begin{Bmatrix}0\\0\\0\\0\\0\\0\\0\\F\end{Bmatrix} \tag{3.51}$$

Continued

EXAMPLE 3.1—CONT'D

Step 6: Solution of the Algebraic System

Taking into account that $k=40{,}000$ N/m and $F=5000$ N, the following results can be obtained:

$$\begin{Bmatrix} d_{1x} \\ d_{2x} \\ d_{3x} \\ d_{4x} \\ R_{1x} \\ R_{2x} \\ R_{3x} \\ R_{4x} \end{Bmatrix} = \begin{Bmatrix} 0 \\ 0.041 \\ 0 \\ 0 \\ -1666.67 \\ 5000 \\ -1666.67 \\ -1666.67 \end{Bmatrix} \tag{3.52}$$

Step 7: Internal Forces

Using the above values for the nodal displacements, the internal forces can now be obtained by Equations (3.27)–(3.29):

Element 1, nodes 1 and 2

$$\begin{Bmatrix} r_{1x} \\ r_{2x} \\ r_{3x} \\ r_{4x} \end{Bmatrix} = \begin{Bmatrix} -1666.67 \\ 1666.67 \\ 0 \\ 0 \end{Bmatrix} \tag{3.53}$$

Element 2, nodes 2 and 3

$$\begin{Bmatrix} r_{1x} \\ r_{2x} \\ r_{3x} \\ r_{4x} \end{Bmatrix} = \begin{Bmatrix} 0 \\ 1666.67 \\ -1666.67 \\ 0 \end{Bmatrix} \tag{3.54}$$

Element 3, nodes 2 and 4

$$\begin{Bmatrix} r_{1x} \\ r_{2x} \\ r_{3x} \\ r_{4x} \end{Bmatrix} = \begin{Bmatrix} 0 \\ 1666.67 \\ 0 \\ -1666.67 \end{Bmatrix} \tag{3.55}$$

EXAMPLE 3.2: SOLUTION OF A SIMPLE STRUCTURE COMPOSED OF SPRINGS BY CALFEM/MATLAB

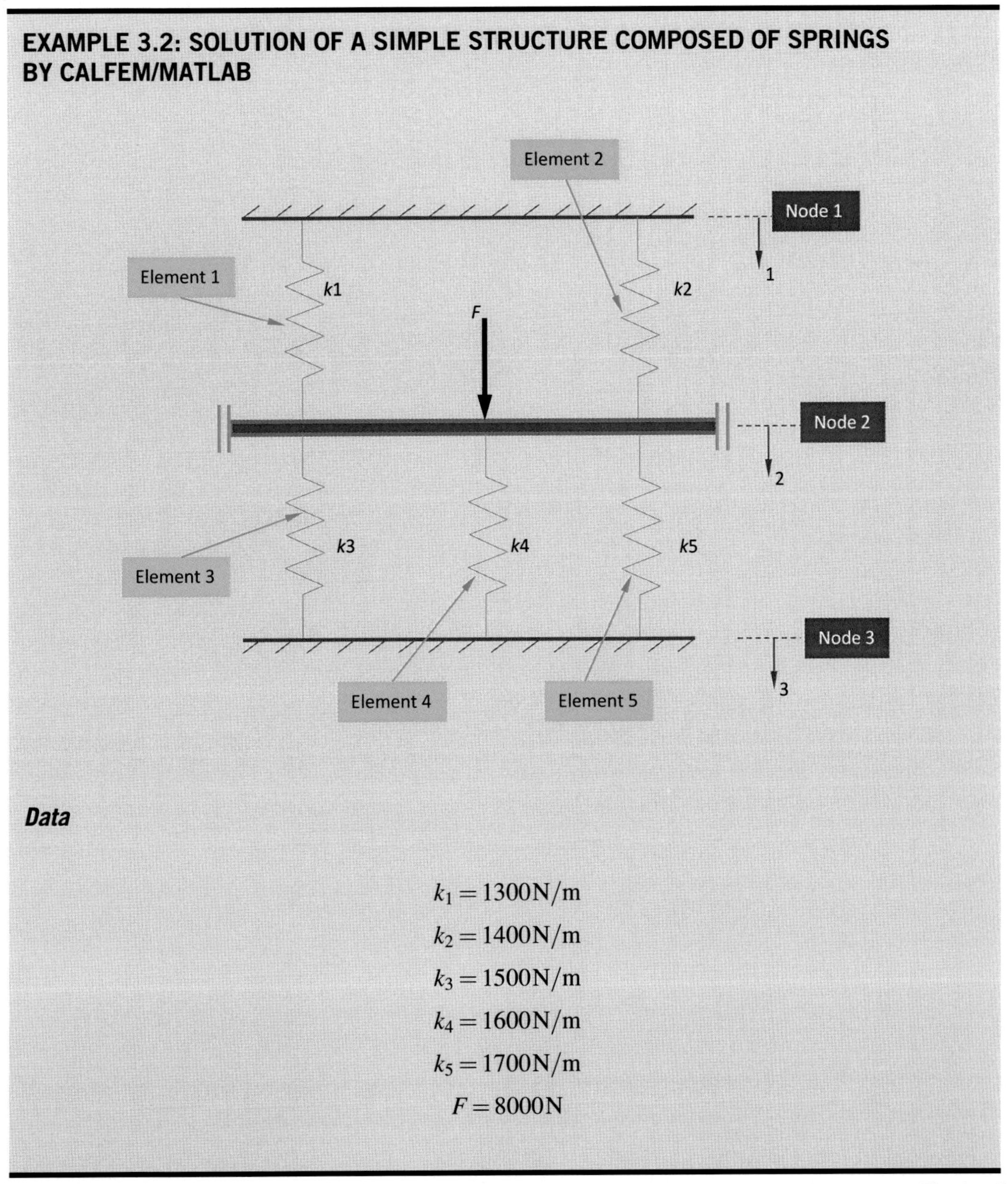

Data

$$k_1 = 1300\text{N/m}$$
$$k_2 = 1400\text{N/m}$$
$$k_3 = 1500\text{N/m}$$
$$k_4 = 1600\text{N/m}$$
$$k_5 = 1700\text{N/m}$$
$$F = 8000\text{N}$$

Continued

EXAMPLE 3.2: SOLUTION OF A SIMPLE STRUCTURE COMPOSED OF SPRINGS BY CALFEM/MATLAB—CONT'D

Step 1: Topology Matrix

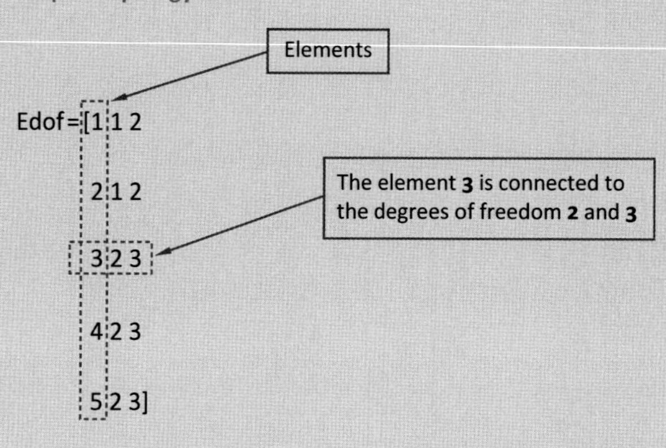

Step 2: Loads

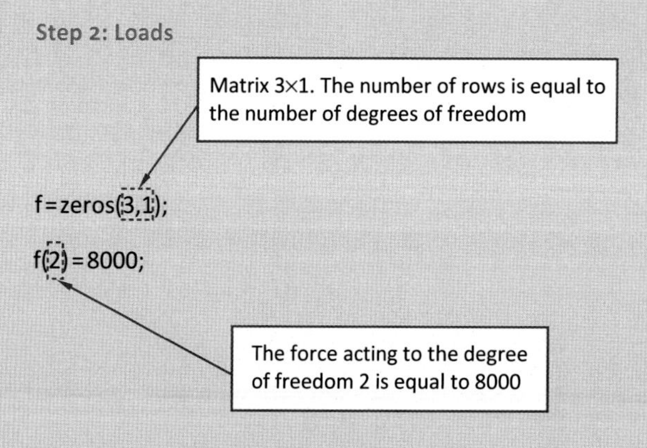

Step 3: Element Matrices

```
k1=13000;ep1=k1;Ke1=spring1e(ep1);
k2=14000;ep2=k2;Ke2=spring1e(ep2);
k3=15000;ep3=k3;Ke3=spring1e(ep3);
k4=16000;ep4=k4;Ke4=spring1e(ep4);
k5=17000;ep5=k5;Ke5=spring1e(ep5);
```

Step 4: Assembly of the Element Stiffness Matrices

```
K=zeros(3,3);
K=assem(Edof(1,:),K,Ke1);
K=assem(Edof(2,:),K,Ke2);
K=assem(Edof(3,:),K,Ke3);
K=assem(Edof(4,:),K,Ke4);
K=assem(Edof(5,:),K,Ke5);
```

Step 5: Boundary Conditions

Step 5: Boundary conditions

On the node 1 the displacement is 0
bc=[10;30];
On the node 3 the displacement is 0

Step 6: Solution of the System of Equations

Step 6: Solution of the system of equations

Output of the solution is the vector **a** containing the displacements on each degree of freedom and the vector **r** containing the reactions
[a,r]=solveq(K,f,bc)

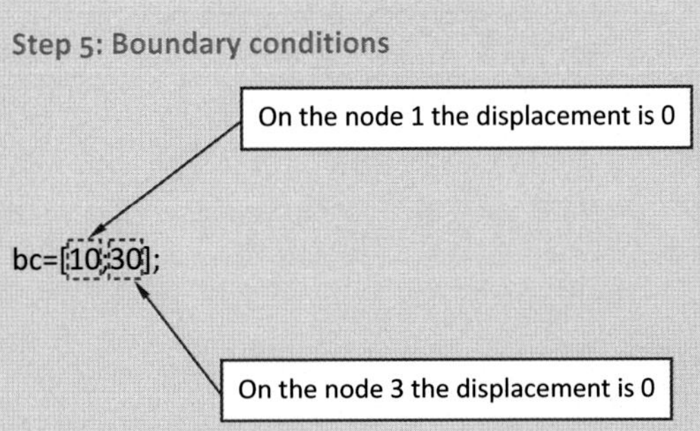

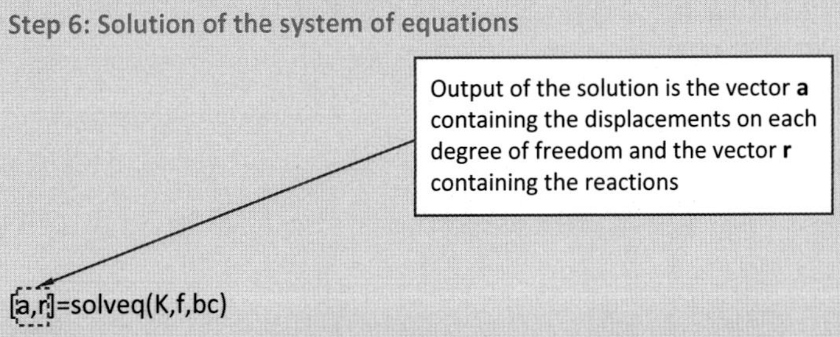

Results of the Step 6

```
a =
      0
 0.1067
      0
r =
 1.0e+03 *
 -2.8800
 0.0000
 -5.1200
```

Continued

EXAMPLE 3.2: SOLUTION OF A SIMPLE STRUCTURE COMPOSED OF SPRINGS BY CALFEM/MATLAB—CONT'D

Step 7: Computation of the Nodal Displacements of Each Element

```
ed1=extract(Edof(1,:),a)
ed2=extract(Edof(2,:),a)
ed3=extract(Edof(3,:),a)
ed4=extract(Edof(4,:),a)
ed5=extract(Edof(5,:),a)
```

Results of Step 7

```
ed1 =
      0   0.1067
ed2 =
      0   0.1067
ed3 =
  0.1067      0
ed4 =
  0.1067      0
ed5 =
  0.1067      0
```

Step 8: Computation of the Spring Forces

```
es1=spring1s(ep1,ed1)
es2=spring1s(ep2,ed2)
es3=spring1s(ep3,ed3)
es4=spring1s(ep4,ed4)
es5=spring1s(ep5,ed5)
```

Results of Step 8

```
es1 =
  1.3867e+03
es2 =
  1.4933e+03
es3 =
  -1600
es4 =
  -1.7067e+03
es5 =
  -1.8133e+03
```

The CALFEM/MATLAB Computer Code

```
>>Edof=[1 1 2;2 1 2;3 2 3;4 2 3;5 2 3];
    f=zeros(3,1);f(2)=8000;
```

```
k1=13000;ep1=k1;Ke1=spring1e(ep1);
k2=14000;ep2=k2;Ke2=spring1e(ep2);
k3=15000;ep3=k3;Ke3=spring1e(ep3);
k4=16000;ep4=k4;Ke4=spring1e(ep4);
k5=17000;ep5=k5;Ke5=spring1e(ep5);
K=zeros(3,3);
K=assem(Edof(1,:),K,Ke1);
K=assem(Edof(2,:),K,Ke2);
K=assem(Edof(3,:),K,Ke3);
K=assem(Edof(4,:),K,Ke4);
K=assem(Edof(5,:),K,Ke5);
bc=[1 0;3 0];
[a,r]=solveq(K,f,bc)
ed1=extract(Edof(1,:),a)
ed2=extract(Edof(2,:),a)
ed3=extract(Edof(3,:),a)
ed4=extract(Edof(4,:),a)
ed5=extract(Edof(5,:),a)
es1=spring1s(ep1,ed1)
es2=spring1s(ep2,ed2)
es3=spring1s(ep3,ed3)
es4=spring1s(ep4,ed4)
es5=spring1s(ep5,ed5)
```

REFERENCES

[1] Austrell P-E, Dahlblom O, Lindemann J, Olsson A, Olsson K-G, Persson K, et al. CALFEM a finite element toolbox, Version 3.4., Division of Structural Mechanics, Lund University; 2004.
[2] Cook RD, Malkus DS, Plesha ME, Witt RJ. Concepts and applications of finite element analysis. Hoboken: John Wiley & Sons; 2002.
[3] Bhatti MA. Fundamental finite element analysis and applications. Hoboken: John Wiley & Sons; 2005.
[4] Logan DL. A first course in the finite element method. Boston, MA: Gengage Learning; 2012.
[5] Oden JT, Becker EB, Carey GF. Finite elements: an introduction, volume I. New Jersey: Prentice Hall; 1981.
[6] Fish J, Belytschko T. A first course in finite elements. New York: Wiley; 2007.
[7] Zienkiewicz OC, Taylor RL, Fox DD. The finite element method for solid and structural mechanics. 7th ed. Oxford: Butterworth-Heinemann; 2013.
[8] Deif AS. Advanced matrix theory for scientists and engineers. 2nd ed. London: Abacus Press; 1991.

CHAPTER

BAR ELEMENTS AND HYDRAULIC NETWORKS

4

The mechanical behavior of bars is similar to the mechanical behavior of springs. They carry only axial (tensile or compressive) forces, and the corresponding displacements take place along their axis. It should be noted that compressive loads acting on long bars might cause buckling. However, analysis of buckling phenomena is beyond the target of this book.

4.1 DISPLACEMENT INTERPOLATION FUNCTIONS

Since bar elements are the simplest structural elements, a procedure based on interpolation functions and the principle of direct equilibrium is going to be adopted. The parameters characterizing a bar element are:

1. geometric parameters: length L, and cross-sectional area A;
2. mechanical properties: modulus of elasticity E.

The aim of this step is the derivation of the element equation $\{r\} = [k]\{d\}$. Figure 4.1 shows the deformation of a bar element 1-2 due to axial loads acting on the nodes of the bar system.

4.1.1 FUNCTIONAL FORM OF DISPLACEMENT DISTRIBUTION

Since a bar is a 1D linear element, a linear function for the distribution of displacements can be adopted:

$$d(x) = a_1 + a_2 x \tag{4.1}$$

where $d(x)$ is the displacement at any point x of the bar element 1-2, and α_1, α_2 are unknown constants. Equation (4.1) can be written in matrix form:

$$d(x) = [1 \;\; x]\begin{Bmatrix} \alpha_1 \\ \alpha_2 \end{Bmatrix} \tag{4.2}$$

Taking into account the conditions at the ends of the bar 1-2

$$d(0) = d_{1x} \tag{4.3}$$

$$d(L) = d_{2x} \tag{4.4}$$

Essentials of the Finite Element Method. http://dx.doi.org/10.1016/B978-0-12-802386-0.00004-9

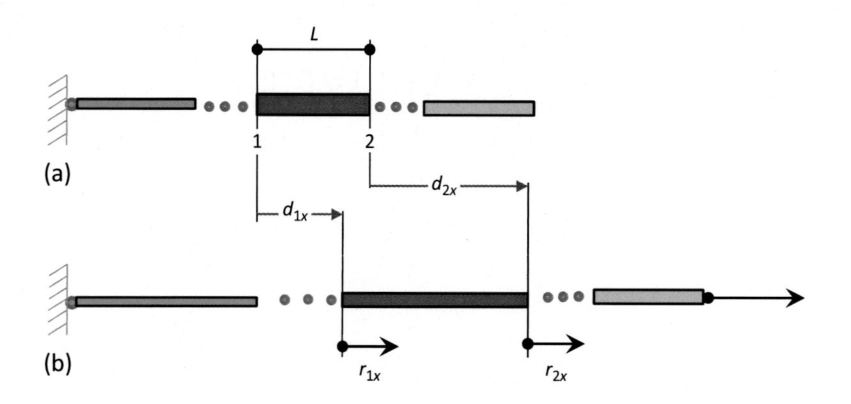

FIGURE 4.1

(a) System of bars before loading. (b) Deformation of a system of bars due to action of axial forces on its nodes.

Equation (4.3), Equation (4.4) can now be written as follows:

$$d_{1x} = [1 \;\; 0]\begin{Bmatrix} \alpha_1 \\ \alpha_2 \end{Bmatrix} \tag{4.5}$$

$$d_{2x} = [1 \;\; L]\begin{Bmatrix} \alpha_1 \\ \alpha_2 \end{Bmatrix} \tag{4.6}$$

or

$$\begin{Bmatrix} d_{1x} \\ d_{2x} \end{Bmatrix} = \begin{bmatrix} 1 & 0 \\ 1 & L \end{bmatrix}\begin{Bmatrix} \alpha_1 \\ \alpha_2 \end{Bmatrix} \tag{4.7}$$

Equation (4.7) yields

$$\begin{Bmatrix} \alpha_1 \\ \alpha_2 \end{Bmatrix} = \begin{bmatrix} 1 & 0 \\ 1 & L \end{bmatrix}^{-1}\begin{Bmatrix} d_{1x} \\ d_{2x} \end{Bmatrix} \tag{4.8}$$

or

$$\begin{Bmatrix} \alpha_1 \\ \alpha_2 \end{Bmatrix} = \begin{bmatrix} 1 & 0 \\ -1/L & 1/L \end{bmatrix}\begin{Bmatrix} d_{1x} \\ d_{2x} \end{Bmatrix} \tag{4.9}$$

Using Equation (4.9), Equation (4.2) can be formulated as:

$$d(x) = [1 \;\; x]\begin{bmatrix} 1 & 0 \\ -1/L & 1/L \end{bmatrix}\begin{Bmatrix} d_{1x} \\ d_{2x} \end{Bmatrix} \tag{4.10}$$

or

$$d(x) = [N_1 \;\; N_2]\begin{Bmatrix} d_{1x} \\ d_{2x} \end{Bmatrix} \tag{4.11}$$

where

$$N_1 = 1 - \frac{x}{L} \tag{4.12}$$

$$N_2 = \frac{x}{L} \tag{4.13}$$

4.1.2 DERIVATION OF THE ELEMENT EQUATION

It is well known from elementary mechanics of solids that the strain $\varepsilon(x)$ is defined as the first derivative (with respect to x) of the displacement $d(x)$. Therefore:

$$\varepsilon(x) = \frac{d}{dx}[N_1 \;\; N_2]\begin{Bmatrix} d_{1x} \\ d_{2x} \end{Bmatrix} \tag{4.14}$$

or

$$\varepsilon(x) = \begin{bmatrix} -\frac{1}{L} & \frac{1}{L} \end{bmatrix}\begin{Bmatrix} d_{1x} \\ d_{2x} \end{Bmatrix} \tag{4.15}$$

Then, the Hooke's law

$$\sigma(x) = E\varepsilon(x) \tag{4.16}$$

yields

$$\sigma(x) = \begin{bmatrix} -\frac{E}{L} & \frac{E}{L} \end{bmatrix}\begin{Bmatrix} d_{1x} \\ d_{2x} \end{Bmatrix} \tag{4.17}$$

or

$$\frac{F(x)}{A} = \begin{bmatrix} -\frac{E}{L} & \frac{E}{L} \end{bmatrix}\begin{Bmatrix} d_{1x} \\ d_{2x} \end{Bmatrix} \tag{4.18}$$

or

$$F(x) = \begin{bmatrix} -\frac{EA}{L} & \frac{EA}{L} \end{bmatrix}\begin{Bmatrix} d_{1x} \\ d_{2x} \end{Bmatrix} \tag{4.19}$$

where $F(x)$ is the axial force at any arbitrary point x.

Taking into account the sign convention of Figure 4.2, the following conditions can be written:

$$x = 0: \;\; F(x) = -r_{1x} \tag{4.20}$$

$$x = L: \;\; F(x) = +r_{2x} \tag{4.21}$$

Taking into account Equation (4.19), Equations (4.20) and (4.21) yield

$$-r_{1x} = \begin{bmatrix} -\frac{EA}{L} & \frac{EA}{L} \end{bmatrix}\begin{Bmatrix} d_{1x} \\ d_{2x} \end{Bmatrix} \tag{4.22}$$

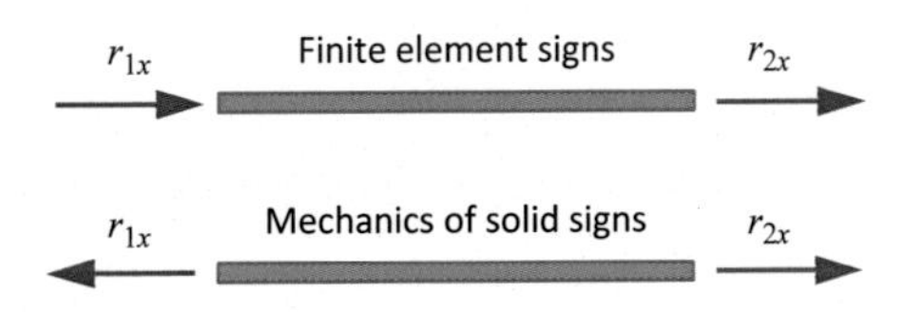

FIGURE 4.2

Sign convention for the nodal forces.

$$r_{2x} = \left[-\frac{EA}{L} \ \frac{EA}{L} \right] \begin{Bmatrix} d_{1x} \\ d_{2x} \end{Bmatrix} \tag{4.23}$$

Combining the above equations, the following matrix equation can be obtained:

$$\begin{Bmatrix} r_{1x} \\ r_{2x} \end{Bmatrix} = \begin{bmatrix} EA/L & -EA/L \\ -EA/L & EA/L \end{bmatrix} \begin{Bmatrix} d_{1x} \\ d_{2x} \end{Bmatrix} \tag{4.24}$$

Equation (4.24) is called the "element equation" and can be written in the following abbreviated form:

$$\{r\} = [k]\{d\} \tag{4.25}$$

where the 2×2 matrix $[k]$ is called "stiffness matrix."

4.2 ALTERNATIVE PROCEDURE BASED ON THE PRINCIPLE OF DIRECT EQUILIBRIUM

Since bars are very simple elements, the element equation can be derived without using interpolation functions. However, this is not possible for more complicated elements (e.g., plane stress or brick elements). The procedure followed for element equation derivation of springs can also be followed for element equation derivation of bars.

4.2.1 THE MECHANICAL BEHAVIOR OF THE MATERIAL

As it is well known from the mechanics of solids, the mechanical behavior of a bar is governed by Hooke's law

$$\sigma = E\varepsilon \tag{4.26}$$

where σ is the applied normal stress and ε is the corresponding strain.

Since the stress is defined as the axial force F over the cross-sectional area A

$$\sigma = \frac{F}{A} \tag{4.27}$$

and the strain is the ratio of the axial deformation $(d_{2x} - d_{1x})$ with the initial length L of the bar

$$\varepsilon = \frac{(d_{2x} - d_{1x})}{L} \tag{4.28}$$

Hooke's law can be written in the following form:

$$F = \frac{AE}{L}(d_{2x} - d_{1x}) \tag{4.29}$$

Comparing Equation (4.29) with Equation (3.2), it can be concluded that a bar can be simulated by a spring with elastic constant

$$k = \frac{AE}{L} \tag{4.30}$$

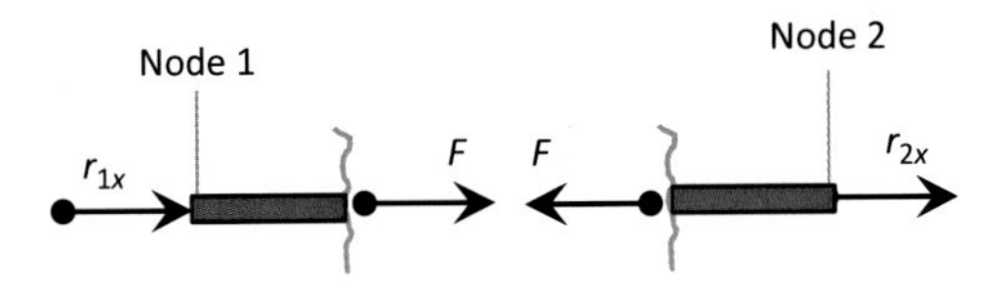

FIGURE 4.3

Equilibrium of the two pieces of the bar.

4.2.2 THE PRINCIPLE OF DIRECT EQUILIBRIUM

In order to correlate the force F (that causes the bar's deformation) to nodal forces, the equilibrium equation will be applied in the two pieces of the bar element 1-2 (see Figure 4.3).

Therefore:

$$r_{1x} + F = 0 \tag{4.31}$$

$$-F + r_{2x} = 0 \tag{4.32}$$

The Equations (4.31) and (4.32) can be written in the following matrix form:

$$\begin{Bmatrix} r_{1x} \\ r_{2x} \end{Bmatrix} = \begin{Bmatrix} -F \\ F \end{Bmatrix} \tag{4.33}$$

Using Equation (4.29), Equation (4.33) yields

$$\begin{Bmatrix} r_{1x} \\ r_{2x} \end{Bmatrix} = \begin{Bmatrix} -(EA/L)\ (d_{2x} - d_{1x}) \\ (EA/L)\ (d_{2x} - d_{1x}) \end{Bmatrix} \tag{4.34}$$

or

$$\begin{Bmatrix} r_{1x} \\ r_{2x} \end{Bmatrix} = \begin{bmatrix} (EA/L) & -(EA/L) \\ -(EA/L) & (EA/L) \end{bmatrix} \begin{Bmatrix} d_{1x} \\ d_{2x} \end{Bmatrix} \tag{4.35}$$

Therefore, Equation (4.35) is the element's equation, and the matrix

$$[k] = \begin{bmatrix} (EA/L) & -(EA/L) \\ -(EA/L) & (EA/L) \end{bmatrix} \tag{4.36}$$

is the stiffness matrix of the bar element.

4.3 FINITE ELEMENT METHOD MODELING OF A SYSTEM OF BARS

The procedure for deriving the stiffness matrix for a system of bars is demonstrated in the example of Figure 4.4.

The main steps to be followed are: (a) to derive the stiffness matrix for each bar element, (b) to expand the stiffness matrices to the degrees of freedom of the whole structure, and (c) to assembly the expanded stiffness matrices in order to derive the stiffness matrix of the structure.

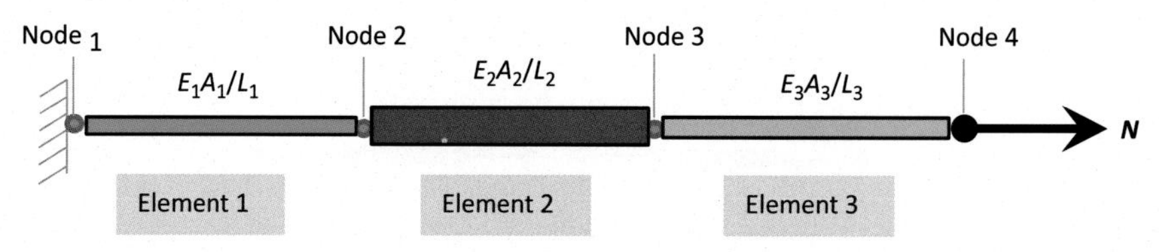

FIGURE 4.4

A simple structural system composed of three bar elements.

4.3.1 DERIVATION OF ELEMENT MATRICES

Taking into account Equation (4.35), the following element equations for the corresponding elements-can be derived:

Element 1, nodes 1 and 2

$$\begin{Bmatrix} r_{1x} \\ r_{2x} \end{Bmatrix} = \begin{bmatrix} (E_1A_1/L_1) & -(E_1A_1/L_1) \\ -(E_1A_1/L_1) & (E_1A_1/L_1) \end{bmatrix} \begin{Bmatrix} d_{1x} \\ d_{2x} \end{Bmatrix} \tag{4.37}$$

Element 2, nodes 2 and 3

$$\begin{Bmatrix} r_{2x} \\ r_{3x} \end{Bmatrix} = \begin{bmatrix} (E_2A_2/L_2) & -(E_2A_2/L_2) \\ -(E_2A_2/L_2) & (E_2A_2/L_2) \end{bmatrix} \begin{Bmatrix} d_{2x} \\ d_{3x} \end{Bmatrix} \tag{4.38}$$

Element 3, nodes 3 and 4

$$\begin{Bmatrix} r_{3x} \\ r_{4x} \end{Bmatrix} = \begin{bmatrix} (E_3A_3/L_3) & -(E_3A_3/L_3) \\ -(E_3A_3/L_{31}) & (E_3A_3/L_3) \end{bmatrix} \begin{Bmatrix} d_{3x} \\ d_{4x} \end{Bmatrix} \tag{4.39}$$

4.3.2 EXPANSION OF ELEMENT EQUATIONS TO THE DEGREES OF FREEDOM OF THE STRUCTURE

Since the structure contains four nodes, element equations (4.37)–(4.39) should be expanded to four-degrees of freedom, that is:

Element 1, nodes 1 and 2

$$\begin{Bmatrix} r_{1x} \\ r_{2x} \\ r_{3x} \\ r_{4x} \end{Bmatrix} = \begin{bmatrix} (E_1A_1/L_1) & -(E_1A_1/L_1) & 0 & 0 \\ -(E_1A_1/L_1) & (E_1A_1/L_1) & 0 & 0 \\ 0 & 0 & 0 & 0 \\ 0 & 0 & 0 & 0 \end{bmatrix} \begin{Bmatrix} d_{1x} \\ d_{2x} \\ d_{3x} \\ d_{4x} \end{Bmatrix} \tag{4.40}$$

Element 2, nodes 2 and 3

$$\begin{Bmatrix} r_{1x} \\ r_{2x} \\ r_{3x} \\ r_{4x} \end{Bmatrix} = \begin{bmatrix} 0 & 0 & 0 & 0 \\ 0 & (E_2A_2/L_2) & -(E_2A_2/L_2) & 0 \\ 0 & -(E_2A_2/L_2) & (E_2A_2/L_2) & 0 \\ 0 & 0 & 0 & 0 \end{bmatrix} \begin{Bmatrix} d_{1x} \\ d_{2x} \\ d_{3x} \\ d_{4x} \end{Bmatrix} \tag{4.41}$$

Element 3, nodes 3 and 4

$$\begin{Bmatrix} r_{1x} \\ r_{2x} \\ r_{3x} \\ r_{4x} \end{Bmatrix} = \begin{bmatrix} 0 & 0 & 0 & 0 \\ 0 & 0 & 0 & 0 \\ 0 & 0 & (E_3A_3/L_3) & -(E_3A_3/L_3) \\ 0 & 0 & -(E_3A_3/L_3) & (E_3A_3/L_3) \end{bmatrix} \begin{Bmatrix} d_{1x} \\ d_{2x} \\ d_{3x} \\ d_{4x} \end{Bmatrix} \tag{4.42}$$

4.3.3 ASSEMBLY OF ELEMENT EQUATIONS

Superposition (addition) of element equations (4.40)–(4.42) yields:

$$\begin{Bmatrix} R_{1x} \\ R_{2x} \\ R_{3x} \\ R_{4x} \end{Bmatrix} = \begin{bmatrix} (E_1A_1/L_1) & -(E_1A_1/L_1) & 0 & 0 \\ -(E_1A_1/L_1) & (E_1A_1/L_1)+(E_2A_2/L_2) & -(E_2A_2/L_2) & 0 \\ 0 & -(E_2A_2/L_2) & (E_2A_2/L_2)+(E_3A_3/L_3) & -(E_3A_3/L_3) \\ 0 & 0 & -(E_3A_3/L_3) & (E_3A_3/L_3) \end{bmatrix} \begin{Bmatrix} d_{1x} \\ d_{2x} \\ d_{3x} \\ d_{4x} \end{Bmatrix} \tag{4.43}$$

It should be noted that the quantities R_{ix} $(i=1,2,3,4)$ are the resultants of the nodal forces.

4.3.4 DERIVATION OF THE FIELD VALUES

An alternative formulation of Equation (4.43) is the following:

$$\begin{bmatrix} (E_1A_1/L_1) & -(E_1A_1/L_1) & 0 & 0 & -1 & 0 & 0 & 0 \\ -(E_1A_1/L_1) & (E_1A_1/L_1)+(E_2A_2/L_2) & -(E_2A_2/L_2) & 0 & 0 & -1 & 0 & 0 \\ 0 & -(E_2A_2/L_2) & (E_2A_2/L_2)+(E_3A_3/L_3) & -(E_3A_3/L_3) & 0 & 0 & -1 & 0 \\ 0 & 0 & -(E_3A_3/L_3) & (E_3A_3/L_3) & 0 & 0 & 0 & -1 \end{bmatrix} \begin{Bmatrix} d_{1x} \\ d_{2x} \\ d_{3x} \\ d_{4x} \\ R_{1x} \\ R_{2x} \\ R_{3x} \\ R_{4x} \end{Bmatrix} = \begin{Bmatrix} 0 \\ 0 \\ 0 \\ 0 \end{Bmatrix} \tag{4.44}$$

Boundary conditions

The matrix equation (4.44) represents an algebraic system of four equations with eight unknowns. In order for the above system to be solvable, four more equations should be added. The missing equations can be specified by the following boundary conditions:

$$d_{1x}=0 \tag{4.45}$$

$$R_{2x}=0 \tag{4.46}$$

$$R_{3x}=0 \tag{4.47}$$

$$R_{4x}=N \tag{4.48}$$

The boundary conditions of Equations (4.45)–(4.48) can be expanded in a 4×8 matrix as follows:

$$\begin{bmatrix} 1 & 0 & 0 & 0 & 0 & 0 & 0 & 0 \\ 0 & 0 & 0 & 0 & 0 & 1 & 0 & 0 \\ 0 & 0 & 0 & 0 & 0 & 0 & 1 & 0 \\ 0 & 0 & 0 & 0 & 0 & 0 & 0 & 1 \end{bmatrix} \begin{Bmatrix} d_{1x} \\ d_{2x} \\ d_{3x} \\ d_{4x} \\ R_{1x} \\ R_{2x} \\ R_{3x} \\ R_{4x} \end{Bmatrix} = \begin{Bmatrix} 0 \\ 0 \\ 0 \\ N \end{Bmatrix} \tag{4.49}$$

Final solution

Therefore, superposition of Equations (4.44) and (4.49) yields an algebraic system of eight equations with eight unknowns, providing the nodal displacements and forces:

$$\begin{bmatrix} E_1A_1/L_1 & -E_1A_1/L_1 & 0 & 0 & -1 & 0 & 0 & 0 \\ -E_1A_1/L_1 & E_1A_1/L_1+E_2A_2/L_2 & -E_2A_2/L_2 & 0 & 0 & -1 & 0 & 0 \\ 0 & -E_2A_2/L_2 & E_2A_2/L_2+E_3A_3/L_3 & -E_3A_3/L_3 & 0 & 0 & -1 & 0 \\ 0 & 0 & -E_3A_3/L_3 & E_3A_3/L_3 & 0 & 0 & 0 & -1 \\ 1 & 0 & 0 & 0 & 0 & 0 & 0 & 0 \\ 0 & 0 & 0 & 0 & 0 & 1 & 0 & 0 \\ 0 & 0 & 0 & 0 & 0 & 0 & 1 & 0 \\ 0 & 0 & 0 & 0 & 0 & 0 & 0 & 1 \end{bmatrix} \begin{Bmatrix} d_{1x} \\ d_{2x} \\ d_{3x} \\ d_{4x} \\ R_{1x} \\ R_{2x} \\ R_{3x} \\ R_{4x} \end{Bmatrix} = \begin{Bmatrix} 0 \\ 0 \\ 0 \\ 0 \\ 0 \\ 0 \\ 0 \\ N \end{Bmatrix} \tag{4.50}$$

Knowing the nodal displacements d_{1x}, d_{2x}, d_{3x}, d_{4x}, the element forces F_1, F_2, F_3, F_4 for each element can be calculated from the following equations:

$$F_1 = \frac{AE}{L}(d_{2x} - d_{1x}) \quad \text{for the element 1} \tag{5.51}$$

$$F_2 = \frac{AE}{L}(d_{3x} - d_{2x}) \quad \text{for the element 2} \tag{5.52}$$

$$F_3 = \frac{AE}{L}(d_{4x} - d_{3x}) \quad \text{for the element 3} \tag{5.53}$$

EXAMPLE 4.1

Determine the field values (nodal displacements and internal forces) for the following structural system.

Data

$$F = 50{,}000\text{N}$$

$$A_a = 1.5 \times 10^{-4}\text{m}^2$$

$$A_b = 1.0 \times 10^{-4}\text{m}^2$$

$$E_a = 203 \times 10^9 \,\text{Pa}$$

$$E_b = 69 \times 10^9 \,\text{Pa}$$

$$L_a = L_b = 1.0\text{m}$$

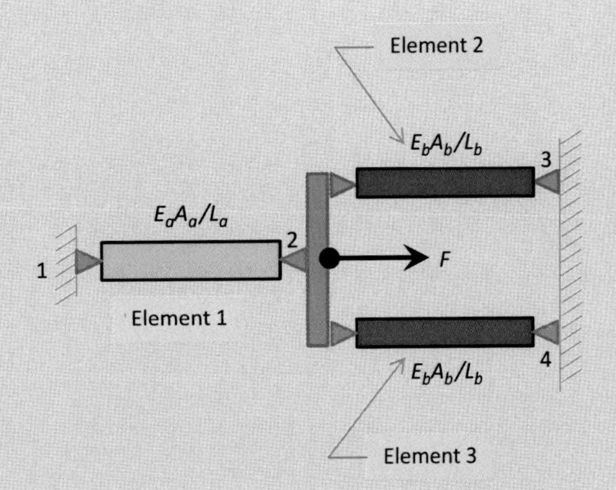

Element 1, nodes 1 and 2

$$\begin{Bmatrix} r_{1x} \\ r_{2x} \\ r_{3x} \\ r_{4x} \end{Bmatrix} = \begin{bmatrix} E_aA_a/L_a & -E_aA_a/L_a & 0 & 0 \\ -E_aA_a/L_a & E_aA_a/L_a & 0 & 0 \\ 0 & 0 & 0 & 0 \\ 0 & 0 & 0 & 0 \end{bmatrix} \begin{Bmatrix} d_{1x} \\ d_{2x} \\ d_{3x} \\ d_{4x} \end{Bmatrix}$$

Element 2, nodes 2 and 3

$$\begin{Bmatrix} r_{1x} \\ r_{2x} \\ r_{3x} \\ r_{4x} \end{Bmatrix} = \begin{bmatrix} 0 & 0 & 0 & 0 \\ 0 & E_bA_b/L_b & -E_bA_b/L_b & 0 \\ 0 & -E_bA_b/L_b & E_bA_b/L_b & 0 \\ 0 & 0 & 0 & 0 \end{bmatrix} \begin{Bmatrix} d_{1x} \\ d_{2x} \\ d_{3x} \\ d_{4x} \end{Bmatrix}$$

Element 3, nodes 2 and 4

$$\begin{Bmatrix} r_{1x} \\ r_{2x} \\ r_{3x} \\ r_{4x} \end{Bmatrix} = \begin{bmatrix} 0 & 0 & 0 & 0 \\ 0 & E_bA_b/L_b & 0 & -E_bA_b/L_b \\ 0 & 0 & 0 & 0 \\ 0 & -E_bA_b/L_b & 0 & E_bA_b/L_b \end{bmatrix} \begin{Bmatrix} d_{1x} \\ d_{2x} \\ d_{3x} \\ d_{4x} \end{Bmatrix}$$

Continued

EXAMPLE 4.1—CONT'D

Superposition of element equations and derivation of the global matrix

$$\begin{Bmatrix} R_{1x} \\ R_{2x} \\ R_{3x} \\ R_{4x} \end{Bmatrix} = \begin{bmatrix} E_aA_a/L_a & -E_aA_a/L_a & 0 & 0 \\ -E_aA_a/L_a & E_aA_a/L_a+2E_bA_b/L_b & -E_bA_b/L_b & -E_bA_b/L_b \\ 0 & -E_bA_b/L_b & E_bA_b/L_b & 0 \\ 0 & -E_bA_b/L_b & 0 & E_bA_b/L_b \end{bmatrix} \begin{Bmatrix} d_{1x} \\ d_{2x} \\ d_{3x} \\ d_{4x} \end{Bmatrix}$$

Matrix model containing the boundary conditions

$$\begin{bmatrix} E_aA_a/L_a & -E_aA_a/L_a & 0 & 0 & -1 & 0 & 0 & 0 \\ -E_aA_a/L_a & E_aA_a/L_a+2E_bA_b/L_b & -E_bA_b/L_b & -E_bA_b/L_b & 0 & -1 & 0 & 0 \\ 0 & -E_bA_b/L_b & E_bA_b/L_b & 0 & 0 & 0 & -1 & 0 \\ 0 & -E_bA_b/L_b & 0 & E_bA_b/L_b & 0 & 0 & 0 & -1 \\ 1 & 0 & 0 & 0 & 0 & 0 & 0 & 0 \\ 0 & 0 & 1 & 0 & 0 & 0 & 0 & 0 \\ 0 & 0 & 0 & 1 & 0 & 0 & 0 & 0 \\ 0 & 0 & 0 & 0 & 0 & 1 & 0 & 0 \end{bmatrix} \begin{Bmatrix} d_{1x} \\ d_{2x} \\ d_{3x} \\ d_{4x} \\ R_{1x} \\ R_{2x} \\ R_{3x} \\ R_{4x} \end{Bmatrix} = \begin{Bmatrix} 0 \\ 0 \\ 0 \\ 0 \\ 0 \\ 0 \\ 0 \\ F \end{Bmatrix}$$

Numerical results

Taking into account the numerical values of the data, the above matrix equation yields the following results:

1. Displacement field

$$\begin{Bmatrix} d_{1x} \\ d_{2x} \\ d_{3x} \\ d_{4x} \\ R_{1x} \\ R_{2x} \\ R_{3x} \\ R_{4x} \end{Bmatrix} = \begin{Bmatrix} 0 \\ 0.00098 \\ 0 \\ 0 \\ -29,794 \\ 50,000 \\ -10,127 \\ -10,127 \end{Bmatrix}$$

2. Internal forces

Using the above values for nodal displacements, the internal forces can now be obtained using Equations (4.40)–(4.42):

Element 1, nodes 1 and 2

$$\begin{Bmatrix} r_{1x} \\ r_{2x} \\ r_{3x} \\ r_{4x} \end{Bmatrix} = \begin{Bmatrix} -29,794 \\ 29,794 \\ 0 \\ 0 \end{Bmatrix}$$

Element 2, nodes 2 and 3

$$\begin{Bmatrix} r_{1x} \\ r_{2x} \\ r_{3x} \\ r_{4x} \end{Bmatrix} = \begin{Bmatrix} 0 \\ 10,127 \\ -10,127 \\ 0 \end{Bmatrix}$$

Element 3, nodes 2 and 4

$$\begin{Bmatrix} r_{1x} \\ r_{2x} \\ r_{3x} \\ r_{4x} \end{Bmatrix} = \begin{Bmatrix} 0 \\ 10,127 \\ 0 \\ -10,127 \end{Bmatrix}$$

4.4 FINITE ELEMENTS METHOD MODELING OF A PIPING NETWORK

Apart from bars, the spring element equation can also be implemented to pipes. Pipes with a constant cross-sectional area are used for fluid transportation. The nodal variables describing the pipe operation are the values $p_1,\ p_2$ of the pressure at the nodes 1 and 2, respectively, and the values $q_1,\ q_2$ of the corresponding volumetric flows (Figure 4.5). The SI units of $q,\ p$ are meters cubed per second and Pascals, respectively.

During the flow, there is a pressure drop along the length L of the pipe due to the friction of the fluid with the interior wall surface. Under the condition of laminar, incompressible, and steady state flow, the equation correlating the nodal pressures to the nodal flows is similar to the element equation of a spring:

$$\begin{Bmatrix} q_1 \\ q_2 \end{Bmatrix} = \frac{\pi D^4}{128\mu L} \begin{bmatrix} 1 & -1 \\ -1 & 1 \end{bmatrix} \begin{Bmatrix} p_1 \\ p_2 \end{Bmatrix} \tag{4.54}$$

where D is the internal diameter of the pipe, L is its length, and μ is the viscosity of the fluid. Therefore, if we use the following nomenclature

$$k = \frac{\pi D^4}{128\mu L} \tag{4.55}$$

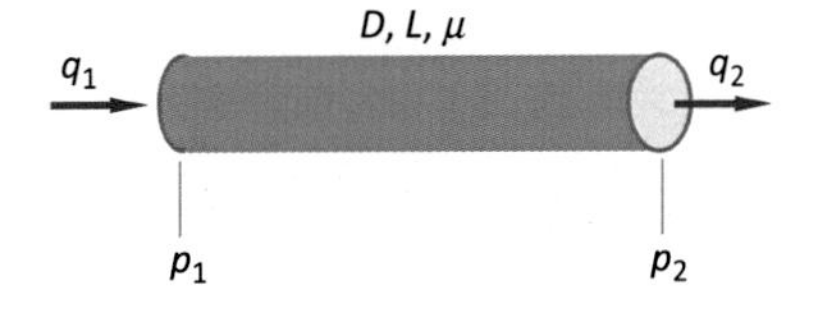

FIGURE 4.5

Pipe element.

$$q = r \tag{4.56}$$

$$p = d \tag{4.57}$$

Equation (4.54) is same as that of the spring element Equation (3.7).

Let us now discuss how the finite element method can be implemented to analysis of a simple piping network. Assuming the piping system of Figure 4.6, the local element equations of the five pipe elements are:

Pipe element 1, nodes 1-2

$$\begin{Bmatrix} q_1 \\ q_2 \end{Bmatrix} = \frac{\pi D_1^4}{128\mu L_1} \begin{bmatrix} 1 & -1 \\ -1 & 1 \end{bmatrix} \begin{Bmatrix} p_1 \\ p_2 \end{Bmatrix} \tag{4.58}$$

Pipe element 2, nodes 1-3

$$\begin{Bmatrix} q_1 \\ q_3 \end{Bmatrix} = \frac{\pi D_2^4}{128\mu L_2} \begin{bmatrix} 1 & -1 \\ -1 & 1 \end{bmatrix} \begin{Bmatrix} p_1 \\ p_3 \end{Bmatrix} \tag{4.59}$$

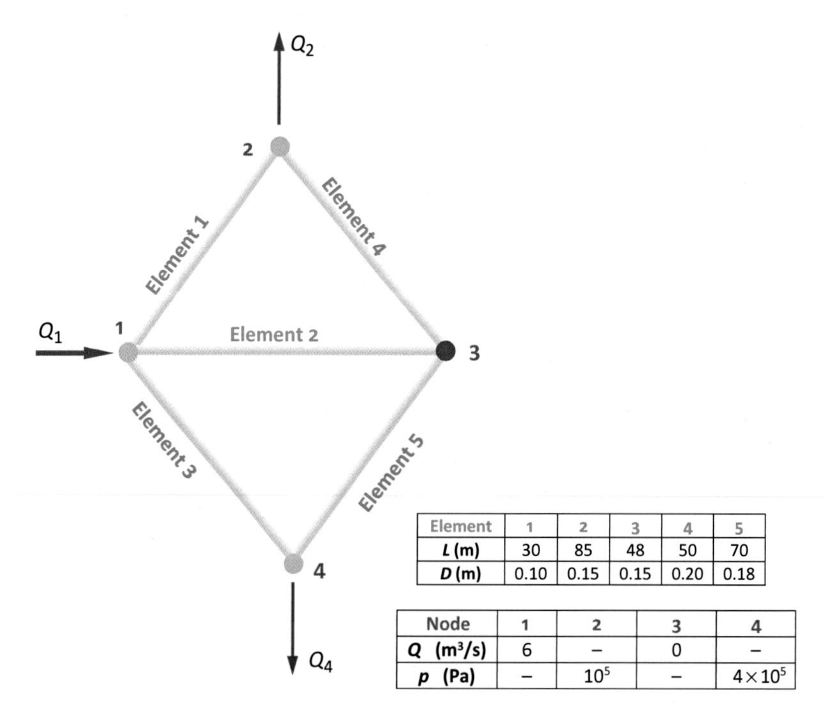

Element	1	2	3	4	5
L (m)	30	85	48	50	70
D (m)	0.10	0.15	0.15	0.20	0.18

Node	1	2	3	4
Q (m³/s)	6	–	0	–
p (Pa)	–	10^5	–	4×10^5

FIGURE 4.6

A simple pipe network system.

Pipe element 3, nodes 1-4

$$\begin{Bmatrix} q_1 \\ q_4 \end{Bmatrix} = \frac{\pi D_3^4}{128\mu L_3} \begin{bmatrix} 1 & -1 \\ -1 & 1 \end{bmatrix} \begin{Bmatrix} p_1 \\ p_4 \end{Bmatrix} \tag{4.60}$$

Pipe element 4, nodes 2-3

$$\begin{Bmatrix} q_2 \\ q_3 \end{Bmatrix} = \frac{\pi D_4^4}{128\mu L_4} \begin{bmatrix} 1 & -1 \\ -1 & 1 \end{bmatrix} \begin{Bmatrix} p_2 \\ p_3 \end{Bmatrix} \tag{4.61}$$

Pipe element 5, nodes 3-4

$$\begin{Bmatrix} q_3 \\ q_4 \end{Bmatrix} = \frac{\pi D_5^4}{128\mu L_5} \begin{bmatrix} 1 & -1 \\ -1 & 1 \end{bmatrix} \begin{Bmatrix} p_3 \\ p_4 \end{Bmatrix} \tag{4.62}$$

The local element equations (4.58)–(4.62) should now be expanded to the degrees of freedom of the whole structure:

Pipe element 1, nodes 1-2

$$\begin{Bmatrix} q_1 \\ q_2 \\ q_3 \\ q_4 \end{Bmatrix} = \frac{\pi D_1^4}{128\mu L_1} \begin{bmatrix} 1 & -1 & 0 & 0 \\ -1 & 1 & 0 & 0 \\ 0 & 0 & 0 & 0 \\ 0 & 0 & 0 & 0 \end{bmatrix} \begin{Bmatrix} p_1 \\ p_2 \\ p_3 \\ p_4 \end{Bmatrix} \tag{4.63}$$

Pipe element 2, nodes 1-3

$$\begin{Bmatrix} q_1 \\ q_2 \\ q_3 \\ q_4 \end{Bmatrix} = \frac{\pi D_2^4}{128\mu L_2} \begin{bmatrix} 1 & 0 & -1 & 0 \\ 0 & 0 & 0 & 0 \\ -1 & 0 & 1 & 0 \\ 0 & 0 & 0 & 0 \end{bmatrix} \begin{Bmatrix} p_1 \\ p_2 \\ p_3 \\ p_4 \end{Bmatrix} \tag{4.64}$$

Pipe element 3, nodes 1-4

$$\begin{Bmatrix} q_1 \\ q_2 \\ q_3 \\ q_4 \end{Bmatrix} = \frac{\pi D_3^4}{128\mu L_3} \begin{bmatrix} 1 & 0 & 0 & -1 \\ 0 & 0 & 0 & 0 \\ 0 & 0 & 0 & 0 \\ -1 & 0 & 0 & 1 \end{bmatrix} \begin{Bmatrix} p_1 \\ p_2 \\ p_3 \\ p_4 \end{Bmatrix} \tag{4.65}$$

Pipe element 4, nodes 2-3

$$\begin{Bmatrix} q_1 \\ q_2 \\ q_3 \\ q_4 \end{Bmatrix} = \frac{\pi D_4^4}{128\mu L_4} \begin{bmatrix} 0 & 0 & 0 & 0 \\ 0 & 1 & -1 & 0 \\ 0 & -1 & 1 & 0 \\ 0 & 0 & 0 & 0 \end{bmatrix} \begin{Bmatrix} p_1 \\ p_2 \\ p_3 \\ p_4 \end{Bmatrix} \tag{4.66}$$

Pipe element 5, nodes 3-4

$$\begin{Bmatrix} q_1 \\ q_2 \\ q_3 \\ q_4 \end{Bmatrix} = \frac{\pi D_5^4}{128\mu L_5} \begin{bmatrix} 0 & 0 & 0 & 0 \\ 0 & 0 & 0 & 0 \\ 0 & 0 & 1 & -1 \\ 0 & 0 & -1 & 1 \end{bmatrix} \begin{Bmatrix} p_1 \\ p_2 \\ p_3 \\ p_4 \end{Bmatrix} \tag{4.67}$$

Superposition of Equations (4.63)–(4.67) yields the following global equation of the pipe network:

$$\begin{Bmatrix} Q_1 \\ Q_2 \\ Q_3 \\ Q_4 \end{Bmatrix} = \frac{\pi}{128\mu} \begin{bmatrix} \frac{D_1^4}{L_1}+\frac{D_2^4}{L_2}+\frac{D_3^4}{L_3} & -\frac{D_1^4}{L_1} & -\frac{D_2^4}{L_2} & -\frac{D_3^4}{L_3} \\ -\frac{D_1^4}{L_1} & \frac{D_1^4}{L_1}+\frac{D_4^4}{L_4} & -\frac{D_4^4}{L_4} & 0 \\ -\frac{D_2^4}{L_2} & -\frac{D_4^4}{L_4} & \frac{D_2^4}{L_2}+\frac{D_4^4}{L_4}+\frac{D_5^4}{L_5} & -\frac{D_5^4}{L_5} \\ -\frac{D_3^4}{L_3} & 0 & -\frac{D_5^4}{L_5} & \frac{D_3^4}{L_3}+\frac{D_5^4}{L_5} \end{bmatrix} \begin{Bmatrix} p_1 \\ p_2 \\ p_3 \\ p_4 \end{Bmatrix} \tag{4.68}$$

Matrix Equation (4.68) can be written in the following format:

$$\{Q\} = [k]\{p\} \tag{4.69}$$

or

$$[k]\{p\} - [I]\{Q\} = \{O_{4\times 1}\} \tag{4.70}$$

or

$$[[k] \ \ [-I]] \begin{Bmatrix} \{p\} \\ \{Q\} \end{Bmatrix} = \{O_{4\times 1}\} \tag{4.71}$$

The submatrix $[I]$ is the unit matrix with size 4×4, the submatrix $\{O_{4\times 1}\}$ is the zero matrix with size 4×1, and the submatrices $\{p\}$, $\{Q\}$ are

$$\{p\} = [p_1 \ p_2 \ p_3 \ p_4]^{\mathrm{T}} \tag{4.72}$$

$$\{Q\} = [Q_1 \ Q_2 \ Q_3 \ Q_4]^{\mathrm{T}} \tag{4.73}$$

The boundary conditions of the pipe network are the following:

$$Q_1 = 6\,\mathrm{m^3/s} \tag{4.74}$$

$$Q_3 = 0\,\mathrm{m^3/s} \tag{4.75}$$

$$p_2 = 10^5\,\mathrm{Pa} \tag{4.76}$$

$$p_4 = 4 \times 10^5\,\mathrm{Pa} \tag{4.77}$$

The boundary conditions in Equations (4.74) –(4.77) can be written in the following format:

$$\underbrace{\begin{bmatrix} 0 & 1 & 0 & 0 \\ 0 & 0 & 0 & 1 \\ 0 & 0 & 0 & 0 \\ 0 & 0 & 0 & 0 \end{bmatrix}}_{[BCd]} \underbrace{\begin{Bmatrix} p_1 \\ p_2 \\ p_3 \\ p_4 \end{Bmatrix}}_{\{p\}} = \underbrace{\begin{Bmatrix} 10^5 \\ 4\times 10^5 \\ 0 \\ 0 \end{Bmatrix}}_{\{DO\}} \tag{4.78}$$

and

$$\underbrace{\begin{bmatrix} 0 & 0 & 0 & 0 \\ 0 & 0 & 0 & 0 \\ 1 & 0 & 0 & 0 \\ 0 & 0 & 1 & 0 \end{bmatrix}}_{[BCR]} \underbrace{\begin{Bmatrix} Q_1 \\ Q_2 \\ Q_3 \\ Q_4 \end{Bmatrix}}_{[Q]} = \underbrace{\begin{Bmatrix} 0 \\ 0 \\ 6 \\ 0 \end{Bmatrix}}_{\{RO\}} \tag{4.79}$$

Combination of Equations (4.78) and (4.79) yields:

$$\left[\begin{array}{cccc|cccc} 0 & 1 & 0 & 0 & 0 & 0 & 0 & 0 \\ 0 & 0 & 0 & 1 & 0 & 0 & 0 & 0 \\ 0 & 0 & 0 & 0 & 1 & 0 & 0 & 0 \\ 0 & 0 & 0 & 0 & 0 & 0 & 1 & 0 \end{array}\right] \begin{Bmatrix} P_1 \\ P_2 \\ P_3 \\ P_4 \\ Q_1 \\ Q_2 \\ Q_3 \\ Q_4 \end{Bmatrix} = \begin{Bmatrix} 10^5 \\ 4\times 10^5 \\ 6 \\ 0 \end{Bmatrix} \tag{4.80}$$

or in an abbreviated notation

$$[[BCd]\ \ [BCR]] \begin{Bmatrix} \{p\} \\ \{Q\} \end{Bmatrix} = \{\{DO\} + \{RO\}\} \tag{4.81}$$

Equations (4.71) and (4.81) can now be combined, yielding:

$$\begin{bmatrix} [k] & [-I] \\ [BCd] & [BCR] \end{bmatrix} \begin{Bmatrix} \{p\} \\ \{Q\} \end{Bmatrix} = \begin{Bmatrix} \{O_{4\times 1}\} \\ \{DO\} + \{RO\} \end{Bmatrix} \tag{4.82}$$

Matrix equation (4.82) is a complete algebraic system. The numerical values of the matrices containing the coefficients of the variables {p}, {Q}, and the constant terms are the following:

$$\begin{bmatrix} [k] & [-I] \\ [BCd] & [BCR] \end{bmatrix} = \left(\begin{array}{cccc|cccc} 0.00048587 & -0.00008165 & -0.00014589 & -0.00025834 & -1 & 0 & 0 & 0 \\ -0.00008165 & 0.00086548 & -0.00078383 & 0. & 0 & -1 & 0 & 0 \\ -0.00014589 & -0.00078383 & 0.00129705 & -0.00036734 & 0 & 0 & -1 & 0 \\ -0.00025834 & 0. & -0.00036733 & 0.00062568 & 0 & 0 & 0 & -1 \\ \hline 0 & 1 & 0 & 0 & 0 & 0 & 0 & 0 \\ 0 & 0 & 0 & 1 & 0 & 0 & 0 & 0 \\ 0 & 0 & 0 & 0 & 1 & 0 & 0 & 0 \\ 0 & 0 & 0 & 0 & 0 & 0 & 1 & 0 \end{array}\right)$$

$$\begin{Bmatrix} \{O_{4\times 1}\} \\ \{DO\}+\{RO\} \end{Bmatrix} = \left(\begin{array}{c} 0 \\ 0 \\ 0 \\ 0 \\ \hline 100{,}000 \\ 400{,}000 \\ 6 \\ 0 \end{array}\right)$$

Then, the solution of Equation (4.82) provides the following results:

$$\{p\} = \begin{Bmatrix} p_1 \\ p_2 \\ p_3 \\ p_4 \end{Bmatrix} = \begin{Bmatrix} 304{,}268 \\ 100{,}000 \\ 207{,}938 \\ 400{,}000 \end{Bmatrix}$$

$$\{Q\} = \begin{Bmatrix} Q_1 \\ Q_2 \\ Q_3 \\ Q_4 \end{Bmatrix} = \begin{Bmatrix} 6 \\ -101 \\ 0 \\ 95 \end{Bmatrix}$$

EXAMPLE 4.2: ANALYSIS OF A HYDRAULIC NETWORK BY CALFEM/MATLAB

The hydraulic network shown in the following figure is composed of 25 pipes. At nodes 1 and 5, there are flow sources with values $Q_1 = 0.45\ m^3/s$ and $Q_5 = 0.60\ m^3/s$, respectively. At nodes 2, 11, 12, 13, and 14, flow is zero ($Q_2 = Q_{11} = Q_{12} = Q_{13} = Q_{14} = 0$). At the remainder of the nodes, the pressure is prescribed according to Table 4.1. Table 4.2 provides the values for the length and diameter of each pipe element. The flowing liquid is water with viscosity $\mu = 1.002 \times 10^{-3}$ Pa.s. Determine the flow outlet at the nodes 3, 4, 6, 7, 8, 9, and 10.

Table 4.1 Data for nodal pressures

Node	3	4	6	7	8	9	10
Pressure (kPa)	37	35	38	35	30	34	33

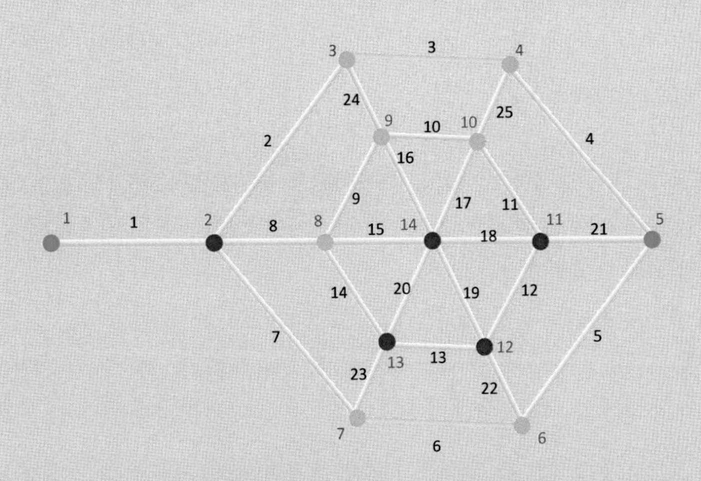

Table 4.2 Data for pipe element lengths and diameters

Pipe	1	2	3	4	5	6	7	8	9	10
L (km)	1.9	1.2	0.8	1.0	0.9	0.9	1.1	0.6	0.5	0.55
D (cm)	16	10	10	10	10	10	10	10	10	10

Pipe	11	12	13	14	15	16	17	18	19	20	21	22	23	24	25
L (km)	0.5	0.6	0.61	0.5	0.6	0.55	0.6	0.4	0.5	0.5	0.35	0.38	0.37	0.39	0.40
D (cm)	10	10	10	10	10	10	10	10	10	10	10	0.10	0.10	0.10	0.10

Data

mi = 1.002e-3;

L1 = 1900;L2 = 1200;L3 = 800;L4 = 1000;L5 = 900;L6 = 900;L7 = 1100;L8 = 600;L9 = 500; L10 = 550;L11 = 500;L12 = 600;L13 = 610;L14 = 500;L15 = 600;L16 = 550;L17 = 600;L18 = 400;L19 = 500;L20 = 500;L21 = 350;L22 = 380;L23 = 370;L24 = 390;L25 = 400;

D1 = 0.16;D2 = 0.10;D3 = 0.10;D4 = 0.10;D5 = 0.10;D6 = 0.10;D7 = 0.10;D8 = 0.10;D9 = 0.10; D10 = 0.10;D11 = 0.10;D12 = 0.10;D13 = 0.10;D14 = 0.10;D15 = 0.10;D16 = 0.10;D17 = 0.10; D18 = 0.10;D19 = 0.10;D20 = 0.10;D21 = 0.10;D22 = 0.10;D23 = 0.10;D24 = 0.10;D25 = 0.10;

Continued

EXAMPLE 4.2: ANALYSIS OF A HYDRAULIC NETWORK BY CALFEM/MATLAB—CONT'D

Step 1: Topology Matrix

```
Edof = [1 1 2;2 2 3;3 3 4;4 4 5;5 5 6;6 6 7;7 7 2;8 2 8;9 8 9;10 9 10;11 10 11;12 11 12;13 12 13;14 13 8;15 8 14;16 9 14;17 10 14;18 11 14;19 12 14;20 13 14;21 5 11;22 6 12;23 7 13;24 3 9;25 4 10];
```

Step 2: Nodal Flows (Loads)

```
f = zeros(14,1);f(1) = 0.45;f(2) = 0;f(5) = 0.60;f(11) = 0;f(12) = 0;f(13) = 0;f(14) = 0;
```

Step 3: Element Matrices

```
k1 = pi*(D1^4)/(128*mi*L1);ep1 = k1;Ke1 = spring1e(ep1);
k2 = pi*(D2^4)/(128*mi*L2);ep2 = k2;Ke2 = spring1e(ep2);
k3 = pi*(D3^4)/(128*mi*L3);ep3 = k3;Ke3 = spring1e(ep3);
k4 = pi*(D4^4)/(128*mi*L4);ep4 = k4;Ke4 = spring1e(ep4);
k5 = pi*(D5^4)/(128*mi*L5);ep5 = k5;Ke5 = spring1e(ep5);
k6 = pi*(D6^4)/(128*mi*L6);ep6 = k6;Ke6 = spring1e(ep6);
k7 = pi*(D7^4)/(128*mi*L7);ep7 = k7;Ke7 = spring1e(ep7);
k8 = pi*(D8^4)/(128*mi*L8);ep8 = k8;Ke8 = spring1e(ep8);
k9 = pi*(D9^4)/(128*mi*L9);ep9 = k9;Ke9 = spring1e(ep9);
k10 = pi*(D10^4)/(128*mi*L10);ep10 = k10;Ke10 = spring1e(ep10);
k11 = pi*(D11^4)/(128*mi*L11);ep11 = k11;Ke11 = spring1e(ep11);
k12 = pi*(D12^4)/(128*mi*L12);ep12 = k12;Ke12 = spring1e(ep12);
k13 = pi*(D13^4)/(128*mi*L13);ep13 = k13;Ke13 = spring1e(ep13);
k14 = pi*(D14^4)/(128*mi*L14);ep14 = k14;Ke14 = spring1e(ep14);
k15 = pi*(D15^4)/(128*mi*L15);ep15 = k15;Ke15 = spring1e(ep15);
k16 = pi*(D16^4)/(128*mi*L16);ep16 = k16;Ke16 = spring1e(ep16);
k17 = pi*(D17^4)/(128*mi*L17);ep17 = k17;Ke17 = spring1e(ep17);
k18 = pi*(D18^4)/(128*mi*L18);ep18 = k18;Ke18 = spring1e(ep18);
k19 = pi*(D19^4)/(128*mi*L19);ep19 = k19;Ke19 = spring1e(ep19);
k20 = pi*(D20^4)/(128*mi*L20);ep20 = k20;Ke20 = spring1e(ep20);
k21 = pi*(D21^4)/(128*mi*L21);ep21 = k21;Ke21 = spring1e(ep21);
k22 = pi*(D22^4)/(128*mi*L22);ep22 = k22;Ke22 = spring1e(ep22);
k23 = pi*(D23^4)/(128*mi*L23);ep23 = k23;Ke23 = spring1e(ep23);
```

```
k24 = pi*(D24^4)/(128*mi*L24);ep24 = k24;Ke24 = spring1e(ep24);
k25 = pi*(D25^4)/(128*mi*L25);ep25 = k25;Ke25 = spring1e(ep25);
```

Step 4: Assembly of the Element Matrices

```
K = zeros(14,14);
K = assem(Edof(1,:),K,Ke1);
K = assem(Edof(2,:),K,Ke2);
K = assem(Edof(3,:),K,Ke3);
K = assem(Edof(4,:),K,Ke4);
K = assem(Edof(5,:),K,Ke5);
K = assem(Edof(6,:),K,Ke6);
K = assem(Edof(7,:),K,Ke7);
K = assem(Edof(8,:),K,Ke8);
K = assem(Edof(9,:),K,Ke9);
K = assem(Edof(10,:),K,Ke10);
K = assem(Edof(11,:),K,Ke11);
K = assem(Edof(12,:),K,Ke12);
K = assem(Edof(13,:),K,Ke13);
K = assem(Edof(14,:),K,Ke14);
K = assem(Edof(15,:),K,Ke15);
K = assem(Edof(16,:),K,Ke16);
K = assem(Edof(17,:),K,Ke17);
K = assem(Edof(18,:),K,Ke18);
K = assem(Edof(19,:),K,Ke19);
K = assem(Edof(20,:),K,Ke20);
K = assem(Edof(21,:),K,Ke21);
K = assem(Edof(22,:),K,Ke22);
K = assem(Edof(23,:),K,Ke23);
K = assem(Edof(24,:),K,Ke24);
K = assem(Edof(25,:),K,Ke25);
```

Continued

EXAMPLE 4.2: ANALYSIS OF A HYDRAULIC NETWORK BY CALFEM/MATLAB—CONT'D

Step 5: Boundary Conditions (Nodal Pressure Values)

```
bc = [3 37000;4 35000;6 38000;7 35000;8 30000;9 34000;10 33000];
```

Step 6: Solution of the System of Equations

```
[a,r] = solveq(K,f,bc)
```

Results of the Step 6 (Nodal Flow Values)

```
r =

   0.0000
   0.0000
  -0.0770
  -0.1475
  -0.0000
  -0.1816
  -0.1333
  -0.3247
  -0.0219
  -0.1640
  -0.0000
        0
  -0.0000
  -0.0000
```

Step 8: Computation of the Flow Within Each Pipe Element

```
ed1 = extract(Edof(1,:),a);
ed2 = extract(Edof(2,:),a);
ed3 = extract(Edof(3,:),a);
ed4 = extract(Edof(4,:),a);
ed5 = extract(Edof(5,:),a);
ed6 = extract(Edof(6,:),a);
ed7 = extract(Edof(7,:),a);
```

```
ed8 = extract(Edof(8,:),a);
ed9 = extract(Edof(9,:),a);
ed10 = extract(Edof(10,:),a);
ed11 = extract(Edof(11,:),a);
ed12 = extract(Edof(12,:),a);
ed13 = extract(Edof(13,:),a);
ed14 = extract(Edof(14,:),a);
ed15 = extract(Edof(15,:),a);
ed16 = extract(Edof(16,:),a);
ed17 = extract(Edof(17,:),a);
ed18 = extract(Edof(18,:),a);
ed19 = extract(Edof(19,:),a);
ed20 = extract(Edof(20,:),a);
ed21 = extract(Edof(21,:),a);
ed22 = extract(Edof(22,:),a);
ed23 = extract(Edof(23,:),a);
ed24 = extract(Edof(24,:),a);
ed25 = extract(Edof(25,:),a);
es1 = spring1s(ep1,ed1)
es2 = spring1s(ep2,ed2)
es3 = spring1s(ep3,ed3)
es4 = spring1s(ep4,ed4)
es5 = spring1s(ep5,ed5)
es6 = spring1s(ep6,ed6)
es7 = spring1s(ep7,ed7)
es8 = spring1s(ep8,ed8)
es9 = spring1s(ep9,ed9)
es10 = spring1s(ep10,ed10)
es11 = spring1s(ep11,ed11)
```

Continued

EXAMPLE 4.2: ANALYSIS OF A HYDRAULIC NETWORK BY CALFEM/MATLAB—CONT'D

```
es12 = spring1s(ep12,ed12)
es13 = spring1s(ep13,ed13)
es14 = spring1s(ep14,ed14)
es15 = spring1s(ep15,ed15)
es16 = spring1s(ep16,ed16)
es17 = spring1s(ep17,ed17)
es18 = spring1s(ep18,ed18)
es19 = spring1s(ep19,ed19)
es20 = spring1s(ep20,ed20)
es21 = spring1s(ep21,ed21)
es22 = spring1s(ep22,ed22)
es23 = spring1s(ep23,ed23)
es24 = spring1s(ep24,ed24)
es25 = spring1s(ep25,ed25)
```

Results of the Step 8 (Pipe Element Flow Values)

es1 = -0.4500

es2 = -0.1019

es3 = -0.0061

es4 = 0.1536

es5 = -0.1625

es6 = -0.0082

es7 = 0.1156

es8 = -0.2324

es9 = 0.0196

es10 = -0.0045

es11 = 0.1183

es12 = -0.0610

es13 = -0.0232

```
es14 = -0.0315
es15 = 0.0412
es16 = 0.0271
es17 = 0.0289
es18 = -0.1045
es19 = -0.0105
es20 = 0.0179
es21 = -0.2839
es22 = 0.0272
es23 = 0.0095
es24 = -0.0188
es25 = -0.0122
```

REFERENCES

[1] Austrell P-E, Dahlblom O, Lindemann J, Olsson A, Olsson K-G, Persson K, et al. CALFEM a finite element toolbox, Version 3.4., Division of Structural Mechanics, Lund University; 2004.

[2] Cook RD, Malkus DS, Plesha ME, Witt RJ. Concepts and applications of finite element analysis. Hoboken: John Wiley & Sons; 2002.

[3] Bhatti MA. Fundamental finite element analysis and applications. Hoboken: John Wiley & Sons; 2005.

[4] Logan DL. A first course in the finite element method. Boston MA: Gengage Learning; 2012.

[5] Oden JT, Becker EB, Carey GF. Finite elements: an introduction, volume I. New Jersey: Prentice Hall; 1981.

[6] Fish J, Belytschko T. A first course in finite elements. New York: Wiley; 2007.

[7] Zienkiewicz OC, Taylor RL, Fox DD. The finite element method for solid and structural mechanics. 7th ed. Oxford: Butterworth-Heinemann; 2013.

[8] Deif AS. Advanced matrix theory for scientists and engineers. 2nd ed. London: Abacus Press; 1991.

CHAPTER

TRUSSES

5

Offshore structures, free-form structures, bridges, roofs, crane booms, antenna towers, etc. are examples of common structural systems consisting of bars of various cross-sections. Structures consisting of bars are called trusses (e.g., Figure 5.1). The members of the trusses are pin joined and carry only axial forces (tensile or compressive). The external forces are acting only on the nodes of the truss.

The basic element to analyze any plane truss is a two-node bar element arbitrary oriented in the x-y plane (Figure 5.2).

5.1 THE ELEMENT EQUATION FOR PLANE TRUSS MEMBERS

In any truss member, two coordinate systems exist: the local system $\bar{x}, \bar{y}$, which is aligned to the member, and the global system x, y, which is the system of coordinates of the whole structure and is common for all truss members. In the previous chapter, we analyzed bar structures where the local and the global coordinate systems coincided. Moreover, in this treatment, we assumed that the direction of the external forces coincided with the axes of the bar elements.

In the present chapter, we analyze plane trusses where the nodal forces have two directions x and y. Therefore, the corresponding displacements of forces r_{jx}, r_{jy} also have two directions, that is, d_{jx}, d_{jy}. The basis of the analysis is the expansion of the bar element equation to the global coordinate system. Taking into account the nomenclature of the coordinate systems demonstrated in Figure 5.2, we recall the bar element equation:

$$\begin{Bmatrix} r_{1\bar{x}} \\ r_{2\bar{x}} \end{Bmatrix} = \begin{bmatrix} (EA/L) & -(EA/L) \\ -(EA/L) & (EA/L) \end{bmatrix} \begin{Bmatrix} d_{1\bar{x}} \\ d_{2\bar{x}} \end{Bmatrix} \tag{5.1}$$

However, it is known from mathematics that any pair of displacements d_{ix}, d_{iy} of a node i in the global x, y system causes a displacement $d_{i\bar{x}}$ along the axis $\bar{x}$ given by:

$$d_{i\bar{x}} = (\cos\vartheta)d_{ix} + (\sin\vartheta)d_{iy} \tag{5.2}$$

Application of the above equation to nodes 1 and 2 (see Figure 5.2) yields

$$d_{1\bar{x}} = Cd_{1x} + Sd_{1y} \tag{5.3}$$

$$d_{2\bar{x}} = Cd_{2x} + Sd_{2y} \tag{5.4}$$

where $C = \cos\vartheta$ and $S = \sin\vartheta$.

Essentials of the Finite Element Method. http://dx.doi.org/10.1016/B978-0-12-802386-0.00005-0

FIGURE 5.1

Examples of structures consisting of bars (trusses).

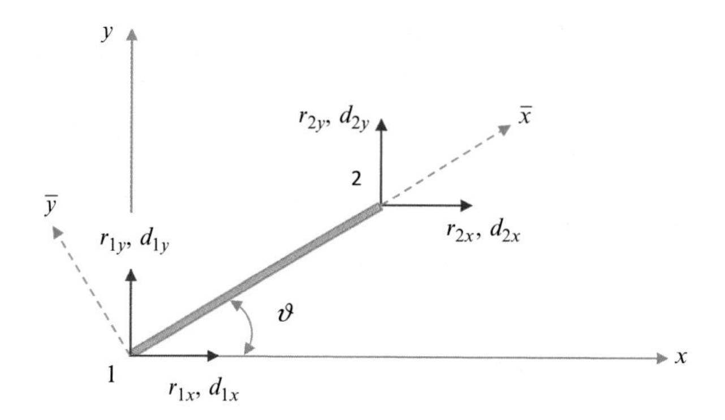

FIGURE 5.2

A two-node truss member of a plane truss.

Equations (5.3) and (5.4) can be expressed in matrix form as:

$$\begin{Bmatrix} d_{1\bar{x}} \\ d_{2\bar{x}} \end{Bmatrix} = \begin{bmatrix} C & S & 0 & 0 \\ 0 & 0 & C & S \end{bmatrix} \begin{Bmatrix} d_{1x} \\ d_{1y} \\ d_{2x} \\ d_{2y} \end{Bmatrix} \tag{5.5}$$

Similarly, the resultant $r_{i\bar{x}}$ of a pair of nodal forces r_{ix}, r_{iy} acting on a node i, can be expressed by the following equation:

$$\begin{Bmatrix} r_{1\bar{x}} \\ r_{2\bar{x}} \end{Bmatrix} = \begin{bmatrix} C & S & 0 & 0 \\ 0 & 0 & C & S \end{bmatrix} \begin{Bmatrix} r_{1x} \\ r_{1y} \\ r_{2x} \\ r_{2y} \end{Bmatrix} \tag{5.6}$$

Taking into account Equations (5.5) and (5.6), Equation (5.1) can now be written:

$$\begin{Bmatrix} r_{1x} \\ r_{1y} \\ r_{2x} \\ r_{2y} \end{Bmatrix} = \frac{AE}{L} \begin{bmatrix} C^2 & CS & -C^2 & -CS \\ CS & S^2 & -CS & -S^2 \\ -C^2 & -CS & C^2 & CS \\ -CS & -S^2 & CS & S^2 \end{bmatrix} \begin{Bmatrix} d_{1x} \\ d_{1y} \\ d_{2x} \\ d_{2y} \end{Bmatrix} \tag{5.7}$$

Again, Equation (5.7) can be written in the following abbreviated form:

$$\{r\} = [k]\{d\} \tag{5.8}$$

where

$$[k] = \frac{AE}{L} \begin{bmatrix} C^2 & CS & -C^2 & -CS \\ CS & S^2 & -CS & -S^2 \\ -C^2 & -CS & C^2 & CS \\ -CS & -S^2 & CS & S^2 \end{bmatrix} \tag{5.9}$$

5.2 THE ELEMENT EQUATION FOR 3D TRUSSES

Let us consider now a bar within the 3D system x, y, z (Figure 5.3).

Following a similar procedure, we are going to derive the element equation of a member of a 3D truss.

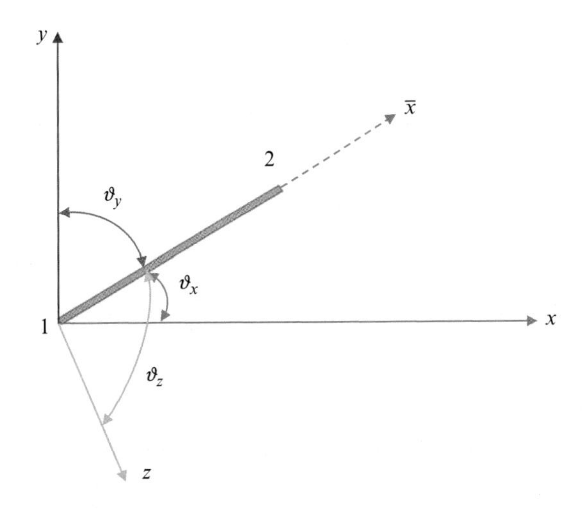

FIGURE 5.3

A two-node truss member of a three-dimensional truss.

It is well known from mathematics that any set of displacements d_{ix}, d_{iy}, d_{iz} of a node i in the x-y-z system (Figure 5.3) causes a displacement $d_{i\bar{x}}$ along the axis $\bar{x}$ given by:

$$d_{i\bar{x}} = (\cos\vartheta_x)\, d_{ix} + (\cos\vartheta_y)\, d_{iy} + (\cos\vartheta_z)\, d_{iz} \tag{5.10}$$

where

$$\cos\vartheta_x = \frac{x_2 - x_1}{L} \tag{5.11}$$

$$\cos\vartheta_y = \frac{y_2 - y_1}{L} \tag{5.12}$$

$$\cos\vartheta_z = \frac{z_2 - z_1}{L} \tag{5.13}$$

In the above equations, (x_1, y_1, z_1) and (x_2, y_2, z_2) are the coordinates of nodes 1 and 2 of the bar, and L is the length of the bar given by:

$$L = \sqrt{(x_2 - x_1)^2 + (y_2 - y_1)^2 + (z_2 - z_1)^2} \tag{5.14}$$

Application of Equation (5.10) to nodes 1 and 2 yields

$$d_{1\bar{x}} = C_x d_{1x} + C_y d_{1y} + C_z d_{1z} \tag{5.15}$$

$$d_{2\bar{x}} = C_x d_{2x} + C_y d_{2y} + C_z d_{2z} \tag{5.16}$$

where

$$C_x = \cos\vartheta_x, \quad C_y = \cos\vartheta_y, \quad C_z = \cos\vartheta_z$$

Equations (5.15) and (5.16) can now be written in the following matrix form:

$$\begin{Bmatrix} d_{1\bar{x}} \\ d_{2\bar{x}} \end{Bmatrix} = \begin{bmatrix} C_x & C_y & C_z & 0 & 0 & 0 \\ 0 & 0 & 0 & C_x & C_y & C_z \end{bmatrix} \begin{Bmatrix} d_{1x} \\ d_{1y} \\ d_{1z} \\ d_{2x} \\ d_{2y} \\ d_{2z} \end{Bmatrix} \tag{5.17}$$

In the same way, the resultant $r_{i\bar{x}}$ of a set of nodal forces r_{ix}, r_{iy}, r_{iz} acting on a node i can be expressed by the following equation:

$$\begin{Bmatrix} r_{1\bar{x}} \\ r_{2\bar{x}} \end{Bmatrix} = \begin{bmatrix} C_x & C_y & C_z & 0 & 0 & 0 \\ 0 & 0 & 0 & C_x & C_y & C_z \end{bmatrix} \begin{Bmatrix} r_{1x} \\ r_{1y} \\ r_{1z} \\ r_{2x} \\ r_{2y} \\ r_{2z} \end{Bmatrix} \tag{5.18}$$

Taking into account Equations (5.17) and (5.18), Equation (5.1) yields

$$\begin{Bmatrix} r_{1x} \\ r_{1y} \\ r_{1z} \\ r_{2x} \\ r_{2y} \\ r_{2z} \end{Bmatrix} = \frac{AE}{L} \begin{bmatrix} C_x^2 & C_xC_y & C_xC_z & -C_x^2 & -C_xC_y & -C_xC_z \\ & C_y^2 & C_yC_z & -C_xC_y & -C_y^2 & -C_yC_z \\ & & C_z^2 & -C_xC_z & -C_yC_z & -C_z^2 \\ & & & C_x^2 & C_xC_y & C_xC_z \\ & \text{symmetric} & & & C_y^2 & C_yC_z \\ & & & & & C_z^2 \end{bmatrix} \begin{Bmatrix} d_{1x} \\ d_{1y} \\ d_{1z} \\ d_{2x} \\ d_{2y} \\ d_{2z} \end{Bmatrix} \tag{5.19}$$

5.3 CALCULATION OF THE BAR'S AXIAL FORCES (INTERNAL FORCES)

Knowing the nodal displacements d_{1x}, d_{1y} and d_{2x}, d_{2y} of a 2D truss (or the nodal displacements d_{1x}, d_{1y}, d_{1z} and d_{2x}, d_{2y}, d_{2z} of a 3D truss), the axial force $F = -r_{1\bar{x}} = r_{2\bar{x}}$ of any truss member (internal force) can be calculated using Equation (5.1). Taking into account Equations (5.5) and (5.17) correlating the axial displacements $d_{1\bar{x}}, d_{2\bar{x}}$ with the global displacements d_{1x}, d_{1y} and d_{2x}, d_{2y} for a 2D problem (or the global displacements d_{1x}, d_{1y}, d_{1z} and d_{2x}, d_{2y}, d_{2z} for a 3D problem), Equation (5.1) yields

$$\begin{Bmatrix} r_{1\bar{x}} \\ r_{2\bar{x}} \end{Bmatrix} = \begin{bmatrix} EA/L & -EA/L \\ -EA/L & EA/L \end{bmatrix} \begin{bmatrix} C & S & 0 & 0 \\ 0 & 0 & C & S \end{bmatrix} \begin{Bmatrix} d_{1x} \\ d_{1y} \\ d_{2x} \\ d_{2y} \end{Bmatrix} \quad \text{for 2D problems} \tag{5.20}$$

$$\begin{Bmatrix} r_{1\bar{x}} \\ r_{2\bar{x}} \end{Bmatrix} = \begin{bmatrix} EA/L & -EA/L \\ -EA/L & EA/L \end{bmatrix} \begin{bmatrix} C_x & C_y & C_z & 0 & 0 & 0 \\ 0 & 0 & 0 & C_x & C_y & C_z \end{bmatrix} \begin{Bmatrix} d_{1x} \\ d_{1y} \\ d_{1z} \\ d_{2x} \\ d_{2y} \\ d_{2z} \end{Bmatrix} \quad \text{for 3D problems} \tag{5.21}$$

EXAMPLE 5.1

Calculate the displacement field, the reactions of the supports and the internal forces of the following plane truss (Table 5.1). The values of the forces are $V_1 = 10$ kN, $V_2 = 5$ kN. All members are made from steel bars with cross-sectional area $A = 10^{-3}$ m^2 and modulus of elasticity $E = 200$ GPa.

Table 5.1 Coordinates of Nodes

Node	1	2	3	4	5
x (m)	0	0	0	3	6
y (m)	0	3	6	3	0

Continued

EXAMPLE 5.1—CONT'D

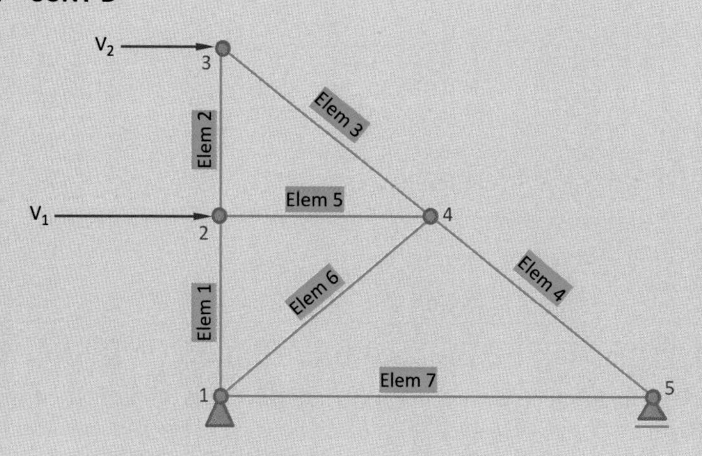

Step 1: Derivation of the Local Stiffness Matrices

We are going to derive the element stiffness matrices using Equation (5.7). To achieve this target, we initially need the values *L*, *C*, and *S* of each element.

Element 1, nodes 1 and 2

$$L_1 = \sqrt{(x_2 - x_1)^2 + (y_2 - y_1)^2} = \sqrt{(0-0)^2 + (3-0)^2} = 3$$

$$\theta_1 = \arctan\frac{y_2 - y_1}{x_2 - x_1} = \arctan\frac{3-0}{0-0} = \arctan(\infty) = 90°$$

$$C_1 = \cos\theta_1 = \cos 90 = 0$$

$$S_1 = \sin\theta_1 = \sin 90 = 1$$

Using the above values, Equation (5.7) yields

$$\begin{Bmatrix} r_{1x} \\ r_{1y} \\ r_{2x} \\ r_{2y} \end{Bmatrix} = \frac{(10^{-3})\times(2\times10^{11})}{3} \begin{bmatrix} 0^2 & 0\times1 & -0^2 & -0\times1 \\ 0\times1 & 1^2 & -0\times1 & -1^2 \\ -0^2 & -0\times1 & 0^2 & 0\times1 \\ -0\times1 & -1^2 & 0\times1 & 1^2 \end{bmatrix} \begin{Bmatrix} d_{1x} \\ d_{1y} \\ d_{2x} \\ d_{2y} \end{Bmatrix}$$

or

$$\begin{Bmatrix} r_{1x} \\ r_{1y} \\ r_{2x} \\ r_{2y} \end{Bmatrix} = 10^6 \times \begin{pmatrix} 0. & 0. & 0. & 0. \\ 0. & 66.666 & 0. & -66.666 \\ 0. & 0. & 0. & 0. \\ 0. & -66.666 & 0. & 66.666 \end{pmatrix} \begin{Bmatrix} d_{1x} \\ d_{1y} \\ d_{2x} \\ d_{2y} \end{Bmatrix}$$

Element 2, nodes 2 and 3

$$L_2 = \sqrt{(x_3 - x_2)^2 + (y_3 - y_2)^2} = \sqrt{(0-0)^2 + (6-3)^2} = 3$$

$$\theta_2 = \arctan\frac{y_3 - y_2}{x_3 - x_2} = \arctan\frac{6-3}{0-0} = \arctan(\infty) = 90°$$

$$C_2 = \cos\theta_2 = \cos 90 = 0$$

$$S_2 = \sin\theta_2 = \sin 90 = 1$$

Using the above values, Equation (5.7) yields

$$\begin{Bmatrix} r_{2x} \\ r_{2y} \\ r_{3x} \\ r_{3y} \end{Bmatrix} = 10^6 \times \begin{pmatrix} 0. & 0. & 0. & 0. \\ 0. & 66.666 & 0. & -66.666 \\ 0. & 0. & 0. & 0. \\ 0. & -66.666 & 0. & 66.666 \end{pmatrix} \begin{Bmatrix} d_{2x} \\ d_{2y} \\ d_{3x} \\ d_{3y} \end{Bmatrix}$$

Element 3, nodes 3 and 4

$$L_3 = \sqrt{(x_4 - x_3)^2 + (y_4 - y_3)^2} = \sqrt{(3-0)^2 + (3-6)^2} = \sqrt{18}$$

$$\theta_3 = \arctan\frac{y_4 - y_3}{x_4 - x_3} = \arctan\frac{3-6}{3-0} = \arctan(-1) = -45°$$

$$C_3 = \cos\theta_3 = \cos(-45) = \sqrt{2}/2$$

$$S_3 = \sin\theta_3 = \sin(-45) = -\sqrt{2}/2$$

Using the above values, Equation (5.7) yields

$$\begin{Bmatrix} r_{3x} \\ r_{3y} \\ r_{4x} \\ r_{4y} \end{Bmatrix} = 10^6 \times \begin{pmatrix} 23.570 & -23.570 & -23.570 & 23.570 \\ -23.570 & 23.570 & 23.570 & -23.570 \\ -23.570 & 23.570 & 23.570 & -23.570 \\ 23.570 & -23.570 & -23.570 & 23.570 \end{pmatrix} \begin{Bmatrix} d_{3x} \\ d_{3y} \\ d_{4x} \\ d_{4y} \end{Bmatrix}$$

Element 4, nodes 4 and 5

$$L_4 = \sqrt{(x_5 - x_4)^2 + (y_5 - y_4)^2} = \sqrt{(6-3)^2 + (0-3)^2} = \sqrt{18}$$

$$\theta_4 = \arctan\frac{y_5 - y_4}{x_5 - x_4} = \arctan\frac{0-3}{6-3} = \arctan(-1) = -45°$$

$$C_4 = \cos\theta_4 = \cos(-45) = \sqrt{2}/2$$

$$S_4 = \sin\theta_4 = \sin(-45) = -\sqrt{2}/2$$

Continued

EXAMPLE 5.1—CONT'D

Using the above values, Equation (5.7) yields

$$\begin{Bmatrix} r_{4x} \\ r_{4y} \\ r_{5x} \\ r_{5y} \end{Bmatrix} = 10^6 \times \begin{pmatrix} 23.570 & -23.570 & -23.570 & 23.570 \\ -23.570 & 23.570 & 23.570 & -23.570 \\ -23.570 & 23.570 & 23.570 & -23.570 \\ 23.570 & -23.570 & -23.570 & 23.570 \end{pmatrix} \begin{Bmatrix} d_{4x} \\ d_{4y} \\ d_{5x} \\ d_{5y} \end{Bmatrix}$$

Element 5, nodes 2 and 4

$$L_5 = \sqrt{(x_4 - x_2)^2 + (y_4 - y_2)^2} = \sqrt{(3-0)^2 + (3-3)^2} = 3$$

$$\theta_5 = \arctan\frac{y_4 - y_2}{x_4 - x_2} = \arctan\frac{3-3}{3-0} = \arctan(0) = 0°$$

$$C_5 = \cos\theta_5 = \cos(0) = 1$$

$$S_5 = \sin\theta_5 = \sin(0) = 0$$

Using the above values, Equation (5.7) yields

$$\begin{Bmatrix} r_{2x} \\ r_{2y} \\ r_{4x} \\ r_{4y} \end{Bmatrix} = 10^6 \times \begin{pmatrix} 66.666 & 0. & -66.666 & 0. \\ 0. & 0. & 0. & 0. \\ -66.666 & 0. & 66.666 & 0. \\ 0. & 0. & 0. & 0. \end{pmatrix} \begin{Bmatrix} d_{2x} \\ d_{2y} \\ d_{4x} \\ d_{4y} \end{Bmatrix}$$

Element 6, nodes 1 and 4

$$L_6 = \sqrt{(x_4 - x_1)^2 + (y_4 - y_1)^2} = \sqrt{(3-0)^2 + (3-0)^2} = \sqrt{18}$$

$$\theta_6 = \arctan\frac{y_4 - y_1}{x_4 - x_1} = \arctan\frac{3-0}{3-0} = \arctan(1) = 45°$$

$$C_6 = \cos\theta_6 = \cos(45) = \sqrt{2}/2$$

$$S_6 = \sin\theta_6 = \sin(45) = \sqrt{2}/2$$

Using the above values, Equation (5.7) yields

$$\begin{Bmatrix} r_{1x} \\ r_{1y} \\ r_{4x} \\ r_{4y} \end{Bmatrix} = 10^6 \times \begin{pmatrix} 23.570 & 23.570 & -23.570 & -23.570 \\ 23.570 & 23.570 & -23.570 & -23.570 \\ -23.570 & -23.570 & 23.570 & 23.570 \\ -23.570 & -23.570 & 23.570 & 23.570 \end{pmatrix} \begin{Bmatrix} d_{1x} \\ d_{1y} \\ d_{4x} \\ d_{4y} \end{Bmatrix}$$

Element 7, nodes 1 and 5

$$L_7=\sqrt{(x_1-x_5)^2+(y_1-y_5)^2}=\sqrt{(6-0)^2+(0-0)^2}=6$$

$$\theta_7=\arctan\frac{y_1-y_5}{x_1-x_5}=\arctan\frac{0-0}{6-0}=\arctan(0)=0°$$

$$C_7=\cos\theta_7=\cos(0)=1$$

$$S_7=\sin\theta_7=\sin(0)=0$$

Using the above values, Equation (5.7) yields

$$\begin{Bmatrix} r_{1x}\\ r_{1y}\\ r_{5x}\\ r_{5y}\end{Bmatrix}=10^6\times\begin{pmatrix} 33.333 & 0. & -33.333 & 0.\\ 0. & 0. & 0. & 0.\\ -33.333 & 0. & 33.333 & 0.\\ 0. & 0. & 0. & 0.\end{pmatrix}\begin{Bmatrix} d_{1x}\\ d_{1y}\\ d_{5x}\\ d_{5y}\end{Bmatrix}$$

Step 2: Expansion of the Local Element Equations to the Degrees of Freedom of the Structure

Element 1, nodes 1 and 2

$$\begin{Bmatrix} r_{1x}\\ r_{1y}\\ r_{2x}\\ r_{2y}\\ r_{3x}\\ r_{3y}\\ r_{4x}\\ r_{4y}\\ r_{5x}\\ r_{5y}\end{Bmatrix}=10^6\times\begin{bmatrix} 0 & 0 & 0 & 0 & 0 & 0 & 0 & 0 & 0 & 0\\ 0 & 66.666 & 0 & -66.666 & 0 & 0 & 0 & 0 & 0 & 0\\ 0 & 0 & 0 & 0 & 0 & 0 & 0 & 0 & 0 & 0\\ 0 & -66.666 & 0 & 66.666 & 0 & 0 & 0 & 0 & 0 & 0\\ 0 & 0 & 0 & 0 & 0 & 0 & 0 & 0 & 0 & 0\\ 0 & 0 & 0 & 0 & 0 & 0 & 0 & 0 & 0 & 0\\ 0 & 0 & 0 & 0 & 0 & 0 & 0 & 0 & 0 & 0\\ 0 & 0 & 0 & 0 & 0 & 0 & 0 & 0 & 0 & 0\\ 0 & 0 & 0 & 0 & 0 & 0 & 0 & 0 & 0 & 0\\ 0 & 0 & 0 & 0 & 0 & 0 & 0 & 0 & 0 & 0\end{bmatrix}\begin{Bmatrix} d_{1x}\\ d_{1y}\\ d_{2x}\\ d_{2y}\\ d_{3x}\\ d_{3y}\\ d_{4x}\\ d_{4y}\\ d_{5x}\\ d_{5y}\end{Bmatrix}$$

Element 2, nodes 2 and 3

$$\begin{Bmatrix} r_{1x}\\ r_{1y}\\ r_{2x}\\ r_{2y}\\ r_{3x}\\ r_{3y}\\ r_{4x}\\ r_{4y}\\ r_{5x}\\ r_{5y}\end{Bmatrix}=10^6\times\begin{bmatrix} 0 & 0 & 0 & 0 & 0 & 0 & 0 & 0 & 0 & 0\\ 0 & 0 & 0 & 0 & 0 & 0 & 0 & 0 & 0 & 0\\ 0 & 0 & 0 & 0 & 0 & 0 & 0 & 0 & 0 & 0\\ 0 & 0 & 0 & 66.666 & 0 & -66.666 & 0 & 0 & 0 & 0\\ 0 & 0 & 0 & 0 & 0 & 0 & 0 & 0 & 0 & 0\\ 0 & 0 & 0 & -66.666 & 0 & 66.666 & 0 & 0 & 0 & 0\\ 0 & 0 & 0 & 0 & 0 & 0 & 0 & 0 & 0 & 0\\ 0 & 0 & 0 & 0 & 0 & 0 & 0 & 0 & 0 & 0\\ 0 & 0 & 0 & 0 & 0 & 0 & 0 & 0 & 0 & 0\\ 0 & 0 & 0 & 0 & 0 & 0 & 0 & 0 & 0 & 0\end{bmatrix}\begin{Bmatrix} d_{1x}\\ d_{1y}\\ d_{2x}\\ d_{2y}\\ d_{3x}\\ d_{3y}\\ d_{4x}\\ d_{4y}\\ d_{5x}\\ d_{5y}\end{Bmatrix}$$

Continued

EXAMPLE 5.1—CONT'D

Element 3, nodes 3 and 4

$$\begin{Bmatrix} r_{1x} \\ r_{1y} \\ r_{2x} \\ r_{2y} \\ r_{3x} \\ r_{3y} \\ r_{4x} \\ r_{4y} \\ r_{5x} \\ r_{5y} \end{Bmatrix} = 10^6 \times \begin{bmatrix} 0 & 0 & 0 & 0 & 0 & 0 & 0 & 0 & 0 & 0 \\ 0 & 0 & 0 & 0 & 0 & 0 & 0 & 0 & 0 & 0 \\ 0 & 0 & 0 & 0 & 0 & 0 & 0 & 0 & 0 & 0 \\ 0 & 0 & 0 & 0 & 0 & 0 & 0 & 0 & 0 & 0 \\ 0 & 0 & 0 & 0 & 23.570 & -23.570 & -23.570 & 23.570 & 0 & 0 \\ 0 & 0 & 0 & 0 & -23.570 & 23.570 & 23.570 & -23.570 & 0 & 0 \\ 0 & 0 & 0 & 0 & -23.570 & 23.570 & 23.570 & -23.570 & 0 & 0 \\ 0 & 0 & 0 & 0 & 23.570 & -23.570 & -23.570 & 23.570 & 0 & 0 \\ 0 & 0 & 0 & 0 & 0 & 0 & 0 & 0 & 0 & 0 \\ 0 & 0 & 0 & 0 & 0 & 0 & 0 & 0 & 0 & 0 \end{bmatrix} \begin{Bmatrix} d_{1x} \\ d_{1y} \\ d_{2x} \\ d_{2y} \\ d_{3x} \\ d_{3y} \\ d_{4x} \\ d_{4y} \\ d_{5x} \\ d_{5y} \end{Bmatrix}$$

Element 4, nodes 4 and 5

$$\begin{Bmatrix} r_{1x} \\ r_{1y} \\ r_{2x} \\ r_{2y} \\ r_{3x} \\ r_{3y} \\ r_{4x} \\ r_{4y} \\ r_{5x} \\ r_{5y} \end{Bmatrix} = 10^6 \times \begin{bmatrix} 0 & 0 & 0 & 0 & 0 & 0 & 0 & 0 & 0 & 0 \\ 0 & 0 & 0 & 0 & 0 & 0 & 0 & 0 & 0 & 0 \\ 0 & 0 & 0 & 0 & 0 & 0 & 0 & 0 & 0 & 0 \\ 0 & 0 & 0 & 0 & 0 & 0 & 0 & 0 & 0 & 0 \\ 0 & 0 & 0 & 0 & 0 & 0 & 0 & 0 & 0 & 0 \\ 0 & 0 & 0 & 0 & 0 & 0 & 0 & 0 & 0 & 0 \\ 0 & 0 & 0 & 0 & 0 & 0 & 23.570 & -23.570 & -23.570 & 23.570 \\ 0 & 0 & 0 & 0 & 0 & 0 & -23.570 & 23.570 & 23.570 & -23.570 \\ 0 & 0 & 0 & 0 & 0 & 0 & -23.570 & 23.570 & 23.570 & -23.570 \\ 0 & 0 & 0 & 0 & 0 & 0 & 23.570 & -23.570 & -23.570 & 23.570 \end{bmatrix} \begin{Bmatrix} d_{1x} \\ d_{1y} \\ d_{2x} \\ d_{2y} \\ d_{3x} \\ d_{3y} \\ d_{4x} \\ d_{4y} \\ d_{5x} \\ d_{5y} \end{Bmatrix}$$

Element 5, nodes 2 and 4

$$\begin{Bmatrix} r_{1x} \\ r_{1y} \\ r_{2x} \\ r_{2y} \\ r_{3x} \\ r_{3y} \\ r_{4x} \\ r_{4y} \\ r_{5x} \\ r_{5y} \end{Bmatrix} = 10^6 \times \begin{bmatrix} 0 & 0 & 0 & 0 & 0 & 0 & 0 & 0 & 0 & 0 \\ 0 & 0 & 0 & 0 & 0 & 0 & 0 & 0 & 0 & 0 \\ 0 & 0 & 66.666 & 0 & 0 & 0 & -66.666 & 0 & 0 & 0 \\ 0 & 0 & 0 & 0 & 0 & 0 & 0 & 0 & 0 & 0 \\ 0 & 0 & 0 & 0 & 0 & 0 & 0 & 0 & 0 & 0 \\ 0 & 0 & 0 & 0 & 0 & 0 & 0 & 0 & 0 & 0 \\ 0 & 0 & -66.666 & 0 & 0 & 0 & 66.666 & 0 & 0 & 0 \\ 0 & 0 & 0 & 0 & 0 & 0 & 0 & 0 & 0 & 0 \\ 0 & 0 & 0 & 0 & 0 & 0 & 0 & 0 & 0 & 0 \\ 0 & 0 & 0 & 0 & 0 & 0 & 0 & 0 & 0 & 0 \end{bmatrix} \begin{Bmatrix} d_{1x} \\ d_{1y} \\ d_{2x} \\ d_{2y} \\ d_{3x} \\ d_{3y} \\ d_{4x} \\ d_{4y} \\ d_{5x} \\ d_{5y} \end{Bmatrix}$$

Element 6, nodes 1 and 4

$$\begin{Bmatrix} r_{1x} \\ r_{1y} \\ r_{2x} \\ r_{2y} \\ r_{3x} \\ r_{3y} \\ r_{4x} \\ r_{4y} \\ r_{5x} \\ r_{5y} \end{Bmatrix} = 10^6 \times \begin{bmatrix} 23.570 & 23.570 & 0 & 0 & 0 & 0 & -23.570 & -23.570 & 0 & 0 \\ 23.570 & 23.570 & 0 & 0 & 0 & 0 & -23.570 & -23.570 & 0 & 0 \\ 0 & 0 & 0 & 0 & 0 & 0 & 0 & 0 & 0 & 0 \\ 0 & 0 & 0 & 0 & 0 & 0 & 0 & 0 & 0 & 0 \\ 0 & 0 & 0 & 0 & 0 & 0 & 0 & 0 & 0 & 0 \\ 0 & 0 & 0 & 0 & 0 & 0 & 0 & 0 & 0 & 0 \\ -23.570 & -23.570 & 0 & 0 & 0 & 0 & 23.570 & 23.570 & 0 & 0 \\ -23.570 & -23.570 & 0 & 0 & 0 & 0 & 23.570 & 23.570 & 0 & 0 \\ 0 & 0 & 0 & 0 & 0 & 0 & 0 & 0 & 0 & 0 \\ 0 & 0 & 0 & 0 & 0 & 0 & 0 & 0 & 0 & 0 \end{bmatrix} \begin{Bmatrix} d_{1x} \\ d_{1y} \\ d_{2x} \\ d_{2y} \\ d_{3x} \\ d_{3y} \\ d_{4x} \\ d_{4y} \\ d_{5x} \\ d_{5y} \end{Bmatrix}$$

Element 7, nodes 1 and 5

$$\begin{Bmatrix} r_{1x} \\ r_{1y} \\ r_{2x} \\ r_{2y} \\ r_{3x} \\ r_{3y} \\ r_{4x} \\ r_{4y} \\ r_{5x} \\ r_{5y} \end{Bmatrix} = 10^6 \times \begin{bmatrix} 33.333 & 0 & 0 & 0 & 0 & 0 & 0 & 0 & -33.333 & 0 \\ 0 & 0 & 0 & 0 & 0 & 0 & 0 & 0 & 0 & 0 \\ 0 & 0 & 0 & 0 & 0 & 0 & 0 & 0 & 0 & 0 \\ 0 & 0 & 0 & 0 & 0 & 0 & 0 & 0 & 0 & 0 \\ 0 & 0 & 0 & 0 & 0 & 0 & 0 & 0 & 0 & 0 \\ 0 & 0 & 0 & 0 & 0 & 0 & 0 & 0 & 0 & 0 \\ 0 & 0 & 0 & 0 & 0 & 0 & 0 & 0 & 0 & 0 \\ 0 & 0 & 0 & 0 & 0 & 0 & 0 & 0 & 0 & 0 \\ -33.333 & 0 & 0 & 0 & 0 & 0 & 0 & 0 & 33.333 & 0 \\ 0 & 0 & 0 & 0 & 0 & 0 & 0 & 0 & 0 & 0 \end{bmatrix} \begin{Bmatrix} d_{1x} \\ d_{1y} \\ d_{2x} \\ d_{2y} \\ d_{3x} \\ d_{3y} \\ d_{4x} \\ d_{4y} \\ d_{5x} \\ d_{5y} \end{Bmatrix}$$

Step 3: Assembly of the Local Element Equations to Compose the Structure Equation

Superposition (addition) of the extended element matrices yields the following structure equation:

$$\begin{Bmatrix} R_{1x} \\ R_{1y} \\ R_{2x} \\ R_{2y} \\ R_{3x} \\ R_{3y} \\ R_{4x} \\ R_{4y} \\ R_{5x} \\ R_{5y} \end{Bmatrix} = 10^6 \times \begin{bmatrix} 59.903 & 23.570 & 0 & 0 & 0 & 0 & -23.570 & -23.570 & -33.333 & 0 \\ 23.570 & 90.236 & 0 & -66.666 & 0 & 0 & -23.570 & -23.570 & 0 & 0 \\ 0 & 0 & 66.666 & 0 & 0 & 0 & -66.666 & 0 & 0 & 0 \\ 0 & -66.666 & 0 & 133.333 & 0 & -66.666 & 0 & 0 & 0 & 0 \\ 0 & 0 & 0 & 23.570 & -23.570 & -23.570 & 23.570 & 0 & 0 & 0 \\ 0 & 0 & 0 & -90.236 & 23.570 & 90.236 & -23.570 & 0 & 0 & 0 \\ -23.570 & -23.570 & -66.666 & -23.570 & 23.570 & 23.570 & 90.236 & 0 & -23.570 & 23.570 \\ -23.570 & -23.570 & 0 & 23.570 & -23.570 & -23.570 & 23.570 & 47.140 & 23.570 & -23.570 \\ -33.333 & 0 & 0 & 0 & 0 & 0 & -23.570 & 23.570 & 56.903 & -23.570 \\ 0 & 0 & 0 & 0 & 0 & 0 & 23.570 & -23.570 & -23.570 & 23.570 \end{bmatrix} \begin{Bmatrix} d_{1x} \\ d_{1y} \\ d_{2x} \\ d_{2y} \\ d_{3x} \\ d_{3y} \\ d_{4x} \\ d_{4y} \\ d_{5x} \\ d_{5y} \end{Bmatrix}$$

Continued

EXAMPLE 5.1—CONT'D

The above equation can be written in the following matrix form:

$$[K]\{d\} - [I]\{R\} = \{O_{10\times1}\}$$

or

$$\left[[K] \quad -[I]\right]\begin{Bmatrix} \{d\} \\ \{R\} \end{Bmatrix} = \{O_{10\times1}\} \quad (e1)$$

Step 4: Boundary Conditions

We are looking for 10 boundary conditions:

(a) Boundary Conditions Regarding Nodal Displacements

Taking into account the physical model of the structure, the boundary conditions regarding the displacements at the nodes of the structure are

$$d_{1x} = 0$$
$$d_{1y} = 0$$
$$d_{5y} = 0$$

The above boundary conditions can be written in the following matrix format:

$$\begin{bmatrix} 1 & 0 & 0 & 0 & 0 & 0 & 0 & 0 & 0 & 0 \\ 0 & 1 & 0 & 0 & 0 & 0 & 0 & 0 & 0 & 0 \\ 0 & 0 & 0 & 0 & 0 & 0 & 0 & 0 & 0 & 1 \\ 0 & 0 & 0 & 0 & 0 & 0 & 0 & 0 & 0 & 0 \\ 0 & 0 & 0 & 0 & 0 & 0 & 0 & 0 & 0 & 0 \\ 0 & 0 & 0 & 0 & 0 & 0 & 0 & 0 & 0 & 0 \\ 0 & 0 & 0 & 0 & 0 & 0 & 0 & 0 & 0 & 0 \\ 0 & 0 & 0 & 0 & 0 & 0 & 0 & 0 & 0 & 0 \\ 0 & 0 & 0 & 0 & 0 & 0 & 0 & 0 & 0 & 0 \\ 0 & 0 & 0 & 0 & 0 & 0 & 0 & 0 & 0 & 0 \end{bmatrix} \begin{Bmatrix} d_{1x} \\ d_{1y} \\ d_{2x} \\ d_{2y} \\ d_{3x} \\ d_{3y} \\ d_{4x} \\ d_{4y} \\ d_{5x} \\ d_{5y} \end{Bmatrix} = \begin{Bmatrix} 0 \\ 0 \\ 0 \\ 0 \\ 0 \\ 0 \\ 0 \\ 0 \\ 0 \\ 0 \end{Bmatrix}$$

or in an abbreviated form

$$[BCd]\{d\} = \{DO\} \quad (e2)$$

(b) Boundary Conditions Regarding Nodal Forces

Taking into account the physical model, the boundary conditions regarding the forces at the nodes are

$$R_{2x}=V_1$$
$$R_{2y}=0$$
$$R_{3x}=V_2$$
$$R_{3y}=0$$
$$R_{4x}=0$$
$$R_{4y}=0$$
$$R_{5x}=0$$

The above boundary conditions can be written in a matrix format. It should be noted that the first three lines of the following matrix are occupied from the three boundary conditions regarding the nodal displacements. Therefore, the seven boundary conditions regarding the nodal forces should be placed in the remainder of the lines (4th to 10th lines):

$$\begin{bmatrix} 0 & 0 & 0 & 0 & 0 & 0 & 0 & 0 & 0 & 0 \\ 0 & 0 & 0 & 0 & 0 & 0 & 0 & 0 & 0 & 0 \\ 0 & 0 & 0 & 0 & 0 & 0 & 0 & 0 & 0 & 0 \\ 0 & 0 & 1 & 0 & 0 & 0 & 0 & 0 & 0 & 0 \\ 0 & 0 & 0 & 1 & 0 & 0 & 0 & 0 & 0 & 0 \\ 0 & 0 & 0 & 0 & 1 & 0 & 0 & 0 & 0 & 0 \\ 0 & 0 & 0 & 0 & 0 & 1 & 0 & 0 & 0 & 0 \\ 0 & 0 & 0 & 0 & 0 & 0 & 1 & 0 & 0 & 0 \\ 0 & 0 & 0 & 0 & 0 & 0 & 0 & 1 & 0 & 0 \\ 0 & 0 & 0 & 0 & 0 & 0 & 0 & 0 & 1 & 0 \end{bmatrix} \begin{Bmatrix} R_{1x} \\ R_{1y} \\ R_{2x} \\ R_{2y} \\ R_{3x} \\ R_{3y} \\ R_{4x} \\ R_{4y} \\ R_{5x} \\ R_{5y} \end{Bmatrix} = \begin{Bmatrix} 0 \\ 0 \\ 0 \\ V_1 \\ 0 \\ V_2 \\ 0 \\ 0 \\ 0 \\ 0 \end{Bmatrix}$$

The above equation can be written in the following abbreviated form:

$$[BCR]\{R\}=\{RO\} \quad (e3)$$

Combining Equations (e1)–(e3), the following 20×20 matrix equation can be composed:

$$\begin{bmatrix} [K] & [-I] \\ [BCd] & [BCR] \end{bmatrix} \begin{Bmatrix} \{d\} \\ \{R\} \end{Bmatrix} = \begin{Bmatrix} \{O\}_{10x1} \\ \{DO+RO\}_{10x1} \end{Bmatrix}$$

The solution of the above equation provides the following values of the nodal displacements and forces $\{d\},\{R\}$, respectively:

Continued

EXAMPLE 5.1—CONT'D

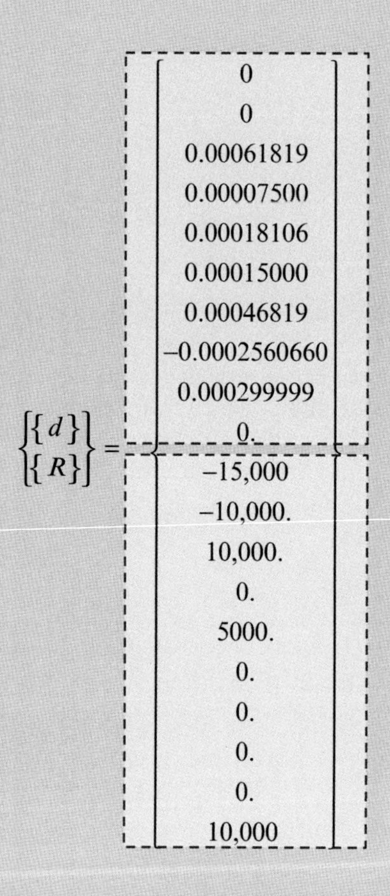

$$\begin{Bmatrix}\{d\}\\\{R\}\end{Bmatrix}=\begin{bmatrix}0\\0\\0.00061819\\0.00007500\\0.00018106\\0.00015000\\0.00046819\\-0.0002560660\\0.000299999\\0.\\-15{,}000\\-10{,}000.\\10{,}000.\\0.\\5000.\\0.\\0.\\0.\\0.\\10{,}000\end{bmatrix}$$

Step 5: Internal Forces

Since we have calculated the nodal displacements, the internal forces for each element can now be calculated by Equation (5.20):

Element 1

$$\begin{Bmatrix}r_{1\bar{x}}\\r_{2\bar{x}}\end{Bmatrix}=\begin{Bmatrix}-F\\F\end{Bmatrix}=\begin{bmatrix}EA/L_1 & -EA/L_1\\-EA/L_1 & EA/L_1\end{bmatrix}\begin{bmatrix}C_1 & S_1 & 0 & 0\\0 & 0 & C_1 & S_1\end{bmatrix}\begin{Bmatrix}d_{1x}\\d_{1y}\\d_{2x}\\d_{2y}\end{Bmatrix}$$

or

$$\begin{Bmatrix}-F\\F\end{Bmatrix}=\frac{(10^{-3})\times(2\times10^{11})}{3}\begin{bmatrix}1 & -1\\-1 & 1\end{bmatrix}\begin{bmatrix}0 & 1 & 0 & 0\\0 & 0 & 0 & 1\end{bmatrix}\begin{Bmatrix}0\\0\\0.00061819\\0.00007500\end{Bmatrix}=\begin{pmatrix}-5000\\5000\end{pmatrix}$$

Element 2

$$\begin{Bmatrix} r_{2\bar{x}} \\ r_{3\bar{x}} \end{Bmatrix} = \begin{Bmatrix} -F \\ F \end{Bmatrix} = \begin{bmatrix} EA/L_2 & -EA/L_2 \\ -EA/L_2 & EA/L_2 \end{bmatrix} \begin{bmatrix} C_2 & S_2 & 0 & 0 \\ 0 & 0 & C_2 & S_2 \end{bmatrix} \begin{Bmatrix} d_{2x} \\ d_{2y} \\ d_{3x} \\ d_{3y} \end{Bmatrix}$$

or

$$\begin{Bmatrix} -F \\ F \end{Bmatrix} = \frac{(10^{-3}) \times (2 \times 10^{11})}{3} \begin{bmatrix} 1 & -1 \\ -1 & 1 \end{bmatrix} \begin{bmatrix} 0 & 1 & 0 & 0 \\ 0 & 0 & 0 & 1 \end{bmatrix} \begin{Bmatrix} 0.00061819 \\ 0.00007500 \\ 0.00018106 \\ 0.00015000 \end{Bmatrix} = \begin{pmatrix} -5000 \\ 5000 \end{pmatrix}$$

Element 3

$$\begin{Bmatrix} r_{3\bar{x}} \\ r_{4\bar{x}} \end{Bmatrix} = \begin{Bmatrix} -F \\ F \end{Bmatrix} = \begin{bmatrix} EA/L_3 & -EA/L_3 \\ -EA/L_3 & EA/L_3 \end{bmatrix} \begin{bmatrix} C_3 & S_3 & 0 & 0 \\ 0 & 0 & C_3 & S_3 \end{bmatrix} \begin{Bmatrix} d_{3x} \\ d_{3y} \\ d_{4x} \\ d_{4y} \end{Bmatrix}$$

or

$$\begin{Bmatrix} -F \\ F \end{Bmatrix} = \frac{(10^{-3}) \times (2 \times 10^{11})}{\sqrt{18}} \begin{bmatrix} 1 & -1 \\ -1 & 1 \end{bmatrix} \begin{bmatrix} \sqrt{2}/2 & -\sqrt{2}/2 & 0 & 0 \\ 0 & 0 & \sqrt{2}/2 & -\sqrt{2}/2 \end{bmatrix} \begin{Bmatrix} 0.00018106 \\ 0.00015000 \\ 0.00046819 \\ -0.00025606 \end{Bmatrix}$$

$$= \begin{pmatrix} 7071 \\ -7071 \end{pmatrix}$$

Element 4

$$\begin{Bmatrix} r_{4\bar{x}} \\ r_{5\bar{x}} \end{Bmatrix} = \begin{Bmatrix} -F \\ F \end{Bmatrix} = \begin{bmatrix} EA/L_4 & -EA/L_4 \\ -EA/L_4 & EA/L_4 \end{bmatrix} \begin{bmatrix} C_4 & S_4 & 0 & 0 \\ 0 & 0 & C_4 & S_4 \end{bmatrix} \begin{Bmatrix} d_{4x} \\ d_{4y} \\ d_{5x} \\ d_{5y} \end{Bmatrix}$$

or

$$\begin{Bmatrix} -F \\ F \end{Bmatrix} = \frac{(10^{-3}) \times (2 \times 10^{11})}{\sqrt{18}} \begin{bmatrix} 1 & -1 \\ -1 & 1 \end{bmatrix} \begin{bmatrix} \sqrt{2}/2 & -\sqrt{2}/2 & 0 & 0 \\ 0 & 0 & \sqrt{2}/2 & -\sqrt{2}/2 \end{bmatrix} \begin{Bmatrix} 0.00046819 \\ -0.00025606 \\ 0.00029999 \\ 0.00000000 \end{Bmatrix}$$

$$= \begin{pmatrix} 14,142 \\ -14,142 \end{pmatrix}$$

Continued

EXAMPLE 5.1—CONT'D

Element 5

$$\begin{Bmatrix} r_{2\bar{x}} \\ r_{4\bar{x}} \end{Bmatrix} = \begin{Bmatrix} -F \\ F \end{Bmatrix} = \begin{bmatrix} EA/L_5 & -EA/L_5 \\ -EA/L_5 & EA/L_5 \end{bmatrix} \begin{bmatrix} C_5 & S_5 & 0 & 0 \\ 0 & 0 & C_5 & S_5 \end{bmatrix} \begin{Bmatrix} d_{2x} \\ d_{2y} \\ d_{4x} \\ d_{4y} \end{Bmatrix}$$

or

$$\begin{Bmatrix} -F \\ F \end{Bmatrix} = \frac{(10^{-3}) \times (2 \times 10^{11})}{3} \begin{bmatrix} 1 & -1 \\ -1 & 1 \end{bmatrix} \begin{bmatrix} 1 & 0 & 0 & 0 \\ 0 & 0 & 1 & 0 \end{bmatrix} \begin{Bmatrix} 0.000618198 \\ 0.000075000 \\ 0.00046819 \\ -0.000256066 \end{Bmatrix} = \begin{pmatrix} 10{,}000 \\ -10{,}000 \end{pmatrix}$$

Element 6

$$\begin{Bmatrix} r_{1\bar{x}} \\ r_{4\bar{x}} \end{Bmatrix} = \begin{Bmatrix} -F \\ F \end{Bmatrix} = \begin{bmatrix} EA/L_6 & -EA/L_6 \\ -EA/L_6 & EA/L_6 \end{bmatrix} \begin{bmatrix} C_6 & S_6 & 0 & 0 \\ 0 & 0 & C_6 & S_6 \end{bmatrix} \begin{Bmatrix} d_{1x} \\ d_{1y} \\ d_{4x} \\ d_{4y} \end{Bmatrix}$$

or

$$\begin{Bmatrix} -F \\ F \end{Bmatrix} = \frac{(10^{-3}) \times (2 \times 10^{11})}{\sqrt{18}} \begin{bmatrix} 1 & -1 \\ -1 & 1 \end{bmatrix} \begin{bmatrix} \sqrt{2}/2 & \sqrt{2}/2 & 0 & 0 \\ 0 & 0 & \sqrt{2}/2 & \sqrt{2}/2 \end{bmatrix} \begin{Bmatrix} 0.000000000 \\ 0.000000000 \\ 0.00046819 \\ -0.000256066 \end{Bmatrix}$$

$$= \begin{pmatrix} -7070.8 \\ 7070.8 \end{pmatrix}$$

Element 7

$$\begin{Bmatrix} r_{1\bar{x}} \\ r_{5\bar{x}} \end{Bmatrix} = \begin{Bmatrix} -F \\ F \end{Bmatrix} = \begin{bmatrix} EA/L_7 & -EA/L_7 \\ -EA/L_7 & EA/L_7 \end{bmatrix} \begin{bmatrix} C_7 & S_7 & 0 & 0 \\ 0 & 0 & C_7 & S_7 \end{bmatrix} \begin{Bmatrix} d_{1x} \\ d_{1y} \\ d_{5x} \\ d_{5y} \end{Bmatrix}$$

or

$$\begin{Bmatrix} -F \\ F \end{Bmatrix} = \frac{(10^{-3}) \times (2 \times 10^{11})}{6} \begin{bmatrix} 1 & -1 \\ -1 & 1 \end{bmatrix} \begin{bmatrix} 1 & 0 & 0 & 0 \\ 0 & 0 & 1 & 0 \end{bmatrix} \begin{Bmatrix} 0.000000000 \\ 0.000000000 \\ 0.00029999 \\ 0.00000000 \end{Bmatrix} = \begin{pmatrix} -10{,}000 \\ 10{,}000 \end{pmatrix}$$

EXAMPLE 5.2

Calculate the displacement field and the reactions of the supports of the following plane truss (Table 5.2). The values of the forces are $F_x = 6$ kN and $F_y = 8$ kN. All members are made from steel bars with cross-sectional area $A = 10^{-3}$ m^2 and modulus of elasticity $E = 200$ GPa.

Table 5.2 Coordinates of Nodes

Node	1	2	3	4
x (m)	0	3	0	0
y (m)	0	0	3	0
z (m)	0	0	0	4

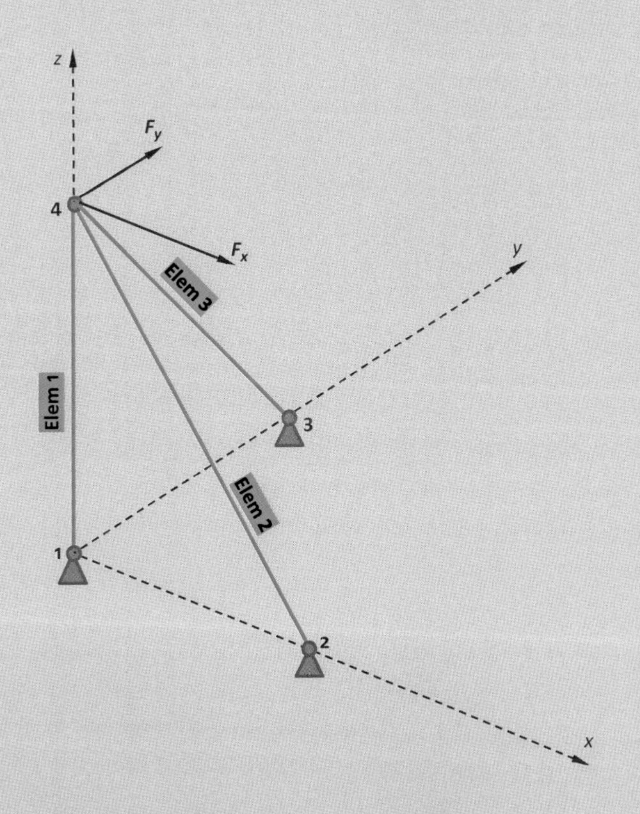

Step 1: Calculation of the Direction Cosines

Element 1, nodes 1 and 4

In order to calculate the direction cosines we have first to derive the vectors corresponding to each element. For the element 1 (nodes 1 and 4) the vector $\vec{e}_1 = \vec{a}_{(1-4)}$ can be derived as follows:

Continued

EXAMPLE 5.2—CONT'D

$$\vec{e}_1 = \vec{a}_{(1-4)} = (x_4 - x_1)\vec{i} + (y_4 - y_1)\vec{j} + (z_4 - z_1)\vec{k}$$
$$= (0-0)\vec{i} + (0-0)\vec{j} + (4-0)\vec{k} = 4\vec{k}$$

Then, the direction cosines for the element 1 can be calculated using the scalar products $\vec{e}_1 \cdot \vec{i}, \vec{e}_1 \cdot \vec{j}, \vec{e}_1 \cdot \vec{k}$:

$$\vec{e}_1 \cdot \vec{i} = |\vec{e}_1| \cdot |\vec{i}| \cdot C_x$$

or

$$4\vec{k} \cdot \vec{i} = |4| \cdot |1| \cdot C_x$$

Since $\vec{k} \cdot \vec{i} = 0$ the above equation yields $C_x = 0$.
Following same procedure, the direction cosines C_y, C_z can be calculated as follows:

$$\vec{e}_1 \cdot \vec{j} = |\vec{e}_1| \cdot |\vec{j}| \cdot C_y \Rightarrow 4\vec{k} \cdot \vec{j} = |4| \cdot |1| \cdot C_y \Rightarrow C_y = 0$$

and

$$\vec{e}_1 \cdot \vec{k} = |\vec{e}_1| \cdot |\vec{k}| \cdot C_z \Rightarrow 4\vec{k} \cdot \vec{k} = |4| \cdot |1| \cdot C_z \Rightarrow C_z = 1$$

Element 2, nodes 2 and 4

$$\vec{e}_2 = \vec{a}_{(2-4)} = (x_4 - x_2)\vec{i} + (y_4 - y_2)\vec{j} + (z_4 - z_2)\vec{k}$$
$$= (0-3)\vec{i} + (0-0)\vec{j} + (4-0)\vec{k} = -3\vec{i} + 4\vec{k}$$

Then,

$$\vec{e}_2 \cdot \vec{i} = |\vec{e}_2| \cdot |\vec{i}| \cdot C_x \Rightarrow \left(-3\vec{i} + 4\vec{k}\right) \cdot \vec{i} = \left|\sqrt{3^2 + 4^2}\right| \cdot |1| \cdot C_x \Rightarrow C_x = -\frac{3}{5}$$

$$\vec{e}_2 \cdot \vec{j} = |\vec{e}_2| \cdot |\vec{j}| \cdot C_y \Rightarrow \left(-3\vec{i} + 4\vec{k}\right) \cdot \vec{j} = \left|\sqrt{3^2 + 4^2}\right| \cdot |1| \cdot C_y \Rightarrow C_y = 0$$

$$\vec{e}_2 \cdot \vec{k} = |\vec{e}_2| \cdot |\vec{k}| \cdot C_z \Rightarrow \left(-3\vec{i} + 4\vec{k}\right) \cdot \vec{k} = \left|\sqrt{3^2 + 4^2}\right| \cdot |1| \cdot C_z \Rightarrow C_z = \frac{4}{5}$$

Element 3, nodes 3 and 4

$$\vec{e}_3 = \vec{a}_{(3-4)} = (x_4 - x_3)\vec{i} + (y_4 - y_3)\vec{j} + (z_4 - z_3)\vec{k}$$
$$= (0-0)\vec{i} + (0-3)\vec{j} + (4-0)\vec{k} = -3\vec{j} + 4\vec{k}$$

Then,

$$\vec{e}_3 \cdot \vec{i} = |\vec{e}_3| \cdot |\vec{i}| \cdot C_x \Rightarrow \left(-3\vec{j} + 4\vec{k}\right) \cdot \vec{i} = \left|\sqrt{3^2+4^2}\right| \cdot |1| \cdot C_x \Rightarrow C_x = 0$$

$$\vec{e}_3 \cdot \vec{j} = |\vec{e}_3| \cdot |\vec{j}| \cdot C_y \Rightarrow \left(-3\vec{j} + 4\vec{k}\right) \cdot \vec{j} = \left|\sqrt{3^2+4^2}\right| \cdot |1| \cdot C_y \Rightarrow C_y = \frac{-3}{5}$$

$$\vec{e}_3 \cdot \vec{k} = |\vec{e}_3| \cdot |\vec{k}| \cdot C_z \Rightarrow \left(-3\vec{j} + 4\vec{k}\right) \cdot \vec{k} = \left|\sqrt{3^2+4^2}\right| \cdot |1| \cdot C_z \Rightarrow C_z = \frac{4}{5}$$

Step 2: Derivation of the Local Stiffness Matrices

We derive the element stiffness matrices using Equation (5.19). To achieve this target, we initially need the values of the length L_i and the directions cosine C_{xi}, C_{yi}, C_{zi} of each element $i = 1,2,3$, respectively.

Element 1, nodes 1 and 4

$$L_1 = \sqrt{(x_4 - x_1)^2 + (y_4 - y_1)^2 + (z_4 - z_1)^2} = \sqrt{(0-0)^2 + (0-0)^2 + (4-0)^2} = 4$$

$$C_{1x} = 0$$

$$C_{1y} = 0$$

$$C_{1z} = 0$$

Using the above values, Equation (5.19) yields

$$\begin{Bmatrix} r_{1x} \\ r_{1y} \\ r_{1z} \\ r_{4x} \\ r_{4y} \\ r_{4z} \end{Bmatrix} = 10^7 \times \begin{pmatrix} 0. & 0. & 0. & 0. & 0. & 0. \\ 0. & 0. & 0. & 0. & 0. & 0. \\ 0. & 0. & 5. & 0. & 0. & -5. \\ 0. & 0. & 0. & 0. & 0. & 0. \\ 0. & 0. & 0. & 0. & 0. & 0. \\ 0. & 0. & -5. & 0. & 0. & 5. \end{pmatrix} \cdot \begin{Bmatrix} d_{1x} \\ d_{1y} \\ d_{1z} \\ d_{4x} \\ d_{4y} \\ d_{4z} \end{Bmatrix}$$

Element 2, nodes 2 and 4

$$L_2 = \sqrt{(x_4 - x_2)^2 + (y_4 - y_2)^2 + (z_4 - z_2)^2} = \sqrt{(0-3)^2 + (0-0)^2 + (4-0)^2} = 5$$

$$C_{2x} = -\frac{3}{5}$$

$$C_{2y} = 0$$

$$C_{2z} = \frac{4}{5}$$

Continued

EXAMPLE 5.2—CONT'D

Using the above values, Equation (5.19) yields

$$\begin{Bmatrix} r_{2x} \\ r_{2y} \\ r_{2z} \\ r_{4x} \\ r_{4y} \\ r_{4z} \end{Bmatrix} = 10^7 \times \begin{pmatrix} 1.44 & 0. & -1.92 & -1.44 & 0. & 1.92 \\ 0. & 0. & 0. & 0. & 0. & 0. \\ -1.92 & 0. & 2.56 & 1.92 & 0. & -2.56 \\ 1.44 & 0. & 1.92 & 1.44 & 0. & -1.92 \\ 0. & 0. & 0. & 0. & 0. & 0. \\ 1.92 & 0. & -2.56 & -1.92 & 0. & 2.56 \end{pmatrix} \cdot \begin{Bmatrix} d_{2x} \\ d_{2y} \\ d_{2z} \\ d_{4x} \\ d_{4y} \\ d_{4z} \end{Bmatrix}$$

Element 3, nodes 3 and 4

$$L_3 = \sqrt{(x_4 - x_3)^2 + (y_4 - y_3)^2 + (z_4 - z_3)^2} = \sqrt{(0-0)^2 + (0-3)^2 + (4-0)^2} = 5$$

$$C_{3x} = 0$$

$$C_{3y} = -\frac{3}{5}$$

$$C_{3z} = \frac{4}{5}$$

Using the above values, Equation (5.19) yields

$$\begin{Bmatrix} r_{3x} \\ r_{3y} \\ r_{3z} \\ r_{4x} \\ r_{4y} \\ r_{4z} \end{Bmatrix} = 10^7 \times \begin{pmatrix} 0. & 0. & 0. & 0. & 0. & 0. \\ 0. & 1.44 & -1.92 & 0. & -1.44 & 1.92 \\ 0. & -1.92 & 2.56 & 0. & 1.92 & -2.56 \\ 0. & 0. & 0. & 0. & 0. & 0. \\ 0. & -1.44 & 1.92 & 0. & 1.44 & -1.92 \\ 0. & 1.92 & -2.56 & 0. & -1.92 & 2.56 \end{pmatrix} \cdot \begin{Bmatrix} d_{3x} \\ d_{3y} \\ d_{3z} \\ d_{4x} \\ d_{4y} \\ d_{4z} \end{Bmatrix}$$

Step 3: Expansion of the Local Element Equations to the Degrees of Freedom of the Structure

Element 1, nodes 1 and 4

$$\begin{Bmatrix} r_{1x} \\ r_{1y} \\ r_{1z} \\ r_{2x} \\ r_{2y} \\ r_{2z} \\ r_{3x} \\ r_{3y} \\ r_{3z} \\ r_{4x} \\ r_{4y} \\ r_{4z} \end{Bmatrix} = (10^7) \cdot \begin{bmatrix} 0 & 0 & 0 & 0 & 0 & 0 & 0 & 0 & 0 & 0 & 0 & 0 \\ 0 & 0 & 0 & 0 & 0 & 0 & 0 & 0 & 0 & 0 & 0 & 0 \\ 0 & 0 & 5 & 0 & 0 & 0 & 0 & 0 & 0 & 0 & 0 & -5 \\ 0 & 0 & 0 & 0 & 0 & 0 & 0 & 0 & 0 & 0 & 0 & 0 \\ 0 & 0 & 0 & 0 & 0 & 0 & 0 & 0 & 0 & 0 & 0 & 0 \\ 0 & 0 & 0 & 0 & 0 & 0 & 0 & 0 & 0 & 0 & 0 & 0 \\ 0 & 0 & 0 & 0 & 0 & 0 & 0 & 0 & 0 & 0 & 0 & 0 \\ 0 & 0 & 0 & 0 & 0 & 0 & 0 & 0 & 0 & 0 & 0 & 0 \\ 0 & 0 & 0 & 0 & 0 & 0 & 0 & 0 & 0 & 0 & 0 & 0 \\ 0 & 0 & 0 & 0 & 0 & 0 & 0 & 0 & 0 & 0 & 0 & 0 \\ 0 & 0 & 0 & 0 & 0 & 0 & 0 & 0 & 0 & 0 & 0 & 0 \\ 0 & 0 & -5 & 0 & 0 & 0 & 0 & 0 & 0 & 0 & 0 & 5 \end{bmatrix} \cdot \begin{Bmatrix} d_{1x} \\ d_{1y} \\ d_{1z} \\ d_{2x} \\ d_{2y} \\ d_{2z} \\ d_{3x} \\ d_{3y} \\ d_{3z} \\ d_{4x} \\ d_{4y} \\ d_{4z} \end{Bmatrix}$$

Element 2, nodes 2 and 4

$$
\begin{Bmatrix} r_{1x} \\ r_{1y} \\ r_{1z} \\ r_{2x} \\ r_{2y} \\ r_{2z} \\ r_{3x} \\ r_{3y} \\ r_{3z} \\ r_{4x} \\ r_{4y} \\ r_{4z} \end{Bmatrix} = (10^7) \cdot \begin{bmatrix}
0 & 0 & 0 & 0 & 0 & 0 & 0 & 0 & 0 & 0 & 0 & 0 \\
0 & 0 & 0 & 0 & 0 & 0 & 0 & 0 & 0 & 0 & 0 & 0 \\
0 & 0 & 0 & 0 & 0 & 0 & 0 & 0 & 0 & 0 & 0 & 0 \\
0 & 0 & 0 & 1.44 & 0 & -1.92 & 0 & 0 & 0 & -1.44 & 0. & 1.92 \\
0 & 0 & 0 & 0 & 0 & 0 & 0 & 0 & 0 & 0. & 0. & 0. \\
0 & 0 & 0 & -1.92 & 0 & 2.56 & 0 & 0 & 0 & 1.92 & 0. & -2.56 \\
0 & 0 & 0 & 0 & 0 & 0 & 0 & 0 & 0 & 0 & 0 & 0 \\
0 & 0 & 0 & 0 & 0 & 0 & 0 & 0 & 0 & 0 & 0 & 0 \\
0 & 0 & 0 & 0 & 0 & 0 & 0 & 0 & 0 & 0 & 0 & 0 \\
0 & 0 & 0 & -1.44 & 0. & 1.92 & 0 & 0 & 0 & 1.44 & 0. & -1.92 \\
0 & 0 & 0 & 0. & 0. & 0. & 0 & 0 & 0 & 0. & 0. & 0. \\
0 & 0 & 0 & 1.92 & 0. & -2.56 & 0 & 0 & 0 & -1.92 & 0. & 2.56
\end{bmatrix} \cdot \begin{Bmatrix} d_{1x} \\ d_{1y} \\ d_{1z} \\ d_{2x} \\ d_{2y} \\ d_{2z} \\ d_{3x} \\ d_{3y} \\ d_{3z} \\ d_{4x} \\ d_{4y} \\ d_{4z} \end{Bmatrix}
$$

Element 3, nodes 3 and 4

$$
\begin{Bmatrix} r_{1x} \\ r_{1y} \\ r_{1z} \\ r_{2x} \\ r_{2y} \\ r_{2z} \\ r_{3x} \\ r_{3y} \\ r_{3z} \\ r_{4x} \\ r_{4y} \\ r_{4z} \end{Bmatrix} = \left(10^7\right) \cdot \begin{bmatrix}
0 & 0 & 0 & 0 & 0 & 0 & 0 & 0 & 0 & 0 & 0 & 0 \\
0 & 0 & 0 & 0 & 0 & 0 & 0 & 0 & 0 & 0 & 0 & 0 \\
0 & 0 & 0 & 0 & 0 & 0 & 0 & 0 & 0 & 0 & 0 & 0 \\
0 & 0 & 0 & 0 & 0 & 0 & 0 & 0 & 0 & 0 & 0 & 0 \\
0 & 0 & 0 & 0 & 0 & 0 & 0 & 0 & 0 & 0 & 0 & 0 \\
0 & 0 & 0 & 0 & 0 & 0 & 0 & 0 & 0 & 0 & 0 & 0 \\
0 & 0 & 0 & 0 & 0 & 0 & 0 & 0 & 0 & 0 & 0 & 0 \\
0 & 0 & 0 & 0 & 0 & 0 & 0 & 1.44 & -1.92 & 0 & -1.44 & 1.92 \\
0 & 0 & 0 & 0 & 0 & 0 & 0 & -1.92 & 2.56 & 0 & 1.92 & -2.56 \\
0 & 0 & 0 & 0 & 0 & 0 & 0 & 0 & 0 & 0 & 0 & 0 \\
0 & 0 & 0 & 0 & 0 & 0 & 0 & -1.44 & 1.92 & 0 & 1.44 & -1.92 \\
0 & 0 & 0 & 0 & 0 & 0 & 0 & 1.92 & -2.56 & 0 & -1.92 & 2.56
\end{bmatrix} \cdot \begin{Bmatrix} d_{1x} \\ d_{1y} \\ d_{1z} \\ d_{2x} \\ d_{2y} \\ d_{2z} \\ d_{3x} \\ d_{3y} \\ d_{3z} \\ d_{4x} \\ d_{4y} \\ d_{4z} \end{Bmatrix}
$$

Step 4: Assembly of the Local Element Equations to Compose the Structure Equation

Superposition (addition) of the expanded element matrices yields the following structure equation:

$$
\begin{Bmatrix} R_{1x} \\ R_{1y} \\ R_{1z} \\ R_{2x} \\ R_{2y} \\ R_{2z} \\ R_{3x} \\ R_{3y} \\ R_{3z} \\ R_{4x} \\ R_{4y} \\ R_{4z} \end{Bmatrix} = \left(10^7\right) \cdot \begin{bmatrix}
0 & 0 & 0 & 0 & 0 & 0 & 0 & 0 & 0 & 0 & 0 & 0 \\
0 & 0 & 0 & 0 & 0 & 0 & 0 & 0 & 0 & 0 & 0 & 0 \\
0 & 0 & 5 & 0 & 0 & 0 & 0 & 0 & 0 & 0 & 0 & -5 \\
0 & 0 & 0 & 1.44 & 0 & -1.92 & 0 & 0 & 0 & -1.44 & 0 & 1.92 \\
0 & 0 & 0 & 0 & 0 & 0 & 0 & 0 & 0 & 0 & 0 & 0 \\
0 & 0 & 0 & -1.92 & 0 & 2.56 & 0 & 0 & 0 & 1.92 & 0 & -2.56 \\
0 & 0 & 0 & 0 & 0 & 0 & 0 & 0 & 0 & 0 & 0 & 0 \\
0 & 0 & 0 & 0 & 0 & 0 & 0 & 1.44 & -1.92 & 0 & -1.44 & 1.92 \\
0 & 0 & 0 & 0 & 0 & 0 & 0 & -1.92 & 2.56 & 0 & 1.92 & -2.56 \\
0 & 0 & 0 & -1.44 & 0 & 1.92 & 0 & 0 & 0 & 1.44 & 0 & -1.92 \\
0 & 0 & 0 & 0 & 0 & 0 & 0 & -1.44 & 1.92 & 0 & 1.44 & -1.92 \\
0 & 0 & -5 & 1.92 & 0 & -2.56 & 0 & 1.92 & -2.56 & -1.92 & -1.92 & 10.12
\end{bmatrix} \cdot \begin{Bmatrix} d_{1x} \\ d_{1y} \\ d_{1z} \\ d_{2x} \\ d_{2y} \\ d_{2z} \\ d_{3x} \\ d_{3y} \\ d_{3z} \\ d_{4x} \\ d_{4y} \\ d_{4z} \end{Bmatrix}
$$

Continued

EXAMPLE 5.2—CONT'D

The above equation can be written in the following matrix form:

$$[K]\{d\} - [I]\{R\} = \{O_{12\times1}\}$$

or

$$\begin{bmatrix}[K] & -[I]\end{bmatrix}\begin{Bmatrix}\{d\}\\ \{R\}\end{Bmatrix} = \{O_{12\times1}\} \quad \text{(el)}$$

Step 5: Boundary Conditions

We are looking for 12 boundary conditions:

(a) Boundary Conditions Regarding Nodal Displacements

Taking into account the physical model shown in the figure, the boundary conditions regarding the displacements at the nodes of the structure are

$$d_{1x} = 0$$
$$d_{1y} = 0$$
$$d_{1z} = 0$$
$$d_{2x} = 0$$
$$d_{2y} = 0$$
$$d_{2z} = 0$$
$$d_{3x} = 0$$
$$d_{3y} = 0$$
$$d_{3z} = 0$$

The above boundary conditions can be written in the following matrix format:

$$\begin{bmatrix}
1 & 0 & 0 & 0 & 0 & 0 & 0 & 0 & 0 & 0 & 0 & 0\\
0 & 1 & 0 & 0 & 0 & 0 & 0 & 0 & 0 & 0 & 0 & 0\\
0 & 0 & 1 & 0 & 0 & 0 & 0 & 0 & 0 & 0 & 0 & 0\\
0 & 0 & 0 & 1 & 0 & 0 & 0 & 0 & 0 & 0 & 0 & 0\\
0 & 0 & 0 & 0 & 1 & 0 & 0 & 0 & 0 & 0 & 0 & 0\\
0 & 0 & 0 & 0 & 0 & 1 & 0 & 0 & 0 & 0 & 0 & 0\\
0 & 0 & 0 & 0 & 0 & 0 & 1 & 0 & 0 & 0 & 0 & 0\\
0 & 0 & 0 & 0 & 0 & 0 & 0 & 1 & 0 & 0 & 0 & 0\\
0 & 0 & 0 & 0 & 0 & 0 & 0 & 0 & 1 & 0 & 0 & 0\\
0 & 0 & 0 & 0 & 0 & 0 & 0 & 0 & 0 & 0 & 0 & 0\\
0 & 0 & 0 & 0 & 0 & 0 & 0 & 0 & 0 & 0 & 0 & 0\\
0 & 0 & 0 & 0 & 0 & 0 & 0 & 0 & 0 & 0 & 0 & 0
\end{bmatrix} \cdot \begin{Bmatrix} d_{1x}\\ d_{1y}\\ d_{1z}\\ d_{2x}\\ d_{2y}\\ d_{2z}\\ d_{3x}\\ d_{3y}\\ d_{3z}\\ d_{4x}\\ d_{4y}\\ d_{4z}\end{Bmatrix} = \begin{Bmatrix}0\\0\\0\\0\\0\\0\\0\\0\\0\\0\\0\\0\end{Bmatrix}$$

or in an abbreviated form

$$[BCd]\{d\} = \{DO\}$$

or

$$\begin{bmatrix}[BCd] & [O]\end{bmatrix}\begin{Bmatrix}\{d\} \\ \{R\}\end{Bmatrix} = \{DO\} \tag{e2}$$

(b) Boundary Conditions Regarding Nodal Forces

Taking into account the physical model, the boundary conditions regarding the forces at the nodes are

$$R_{4x} = F_x$$
$$R_{4y} = F_y$$
$$R_{4z} = 0$$

The above boundary conditions can be written in a matrix format. It should be noted that the first nine lines of the following matrix are occupied from the nine boundary conditions regarding the nodal displacements. Therefore, the three boundary conditions regarding the nodal forces should be placed in the remainder of the lines (10th to 12th lines):

$$\begin{bmatrix} 0&0&0&0&0&0&0&0&0&0&0&0 \\ 0&0&0&0&0&0&0&0&0&0&0&0 \\ 0&0&0&0&0&0&0&0&0&0&0&0 \\ 0&0&0&0&0&0&0&0&0&0&0&0 \\ 0&0&0&0&0&0&0&0&0&0&0&0 \\ 0&0&0&0&0&0&0&0&0&0&0&0 \\ 0&0&0&0&0&0&0&0&0&0&0&0 \\ 0&0&0&0&0&0&0&0&0&0&0&0 \\ 0&0&0&0&0&0&0&0&0&0&0&0 \\ 0&0&0&0&0&0&0&0&0&1&0&0 \\ 0&0&0&0&0&0&0&0&0&0&1&0 \\ 0&0&0&0&0&0&0&0&0&0&0&1 \end{bmatrix} \cdot \begin{Bmatrix} R_{1x} \\ R_{1y} \\ R_{1z} \\ R_{2x} \\ R_{2y} \\ R_{2z} \\ R_{3x} \\ R_{3y} \\ R_{3z} \\ R_{4x} \\ R_{4y} \\ R_{4z} \end{Bmatrix} = \begin{Bmatrix} 0 \\ 0 \\ 0 \\ 0 \\ 0 \\ 0 \\ 0 \\ 0 \\ 0 \\ F_x \\ F_y \\ 0 \end{Bmatrix}$$

The above equation can be written in the following abbreviated form:

$$[BCR]\{R\} = \{RO\}$$

$$\begin{bmatrix}[O] & [BCR]\end{bmatrix}\begin{Bmatrix}\{d\} \\ \{R\}\end{Bmatrix} = \{RO\} \tag{e3}$$

Continued

EXAMPLE 5.2—CONT'D

Summation of Equations (e2) and (e3) yields

$$\begin{bmatrix}[BCd] & [BCR]\end{bmatrix}\begin{Bmatrix}\{d\}\\ \{R\}\end{Bmatrix}=\{DO+RO\} \qquad \text{(e3)}$$

Combining Equations (e1) and (e4), the following 24×24 matrix equation can be composed

$$\begin{bmatrix}[K] & [-I]\\ [BCd] & [BCR]\end{bmatrix}\begin{Bmatrix}\{d\}\\ \{R\}\end{Bmatrix}=\begin{Bmatrix}\{O\}_{12\times1}\\ \{DO+RO\}_{12\times1}\end{Bmatrix}$$

The solution of the above equation provides the following values of the nodal displacements and forces $\{d\}, \{R\}$, respectively:

$$\begin{bmatrix} d_{1x}\\ d_{1y}\\ d_{1z}\\ d_{2x}\\ d_{2y}\\ d_{2z}\\ d_{3x}\\ d_{3y}\\ d_{3z}\\ d_{4x}\\ d_{4y}\\ d_{4z}\\ R_{1x}\\ R_{1y}\\ R_{1z}\\ R_{2x}\\ R_{2y}\\ R_{2z}\\ R_{3x}\\ R_{3y}\\ R_{3z}\\ R_{4x}\\ R_{4y}\\ R_{4z} \end{bmatrix} = \begin{pmatrix} 0.\\ 0.\\ 0.\\ 1.084202172485504\times10^{-19}\\ 0.\\ 8.131516293641287\times10^{-20}\\ 0.\\ 4.063184973664609\times10^{-19}\\ 1.626303258728257\times10^{-19}\\ 0.0009144444444444447\\ 0.0010533333333333336\\ 0.0003733333333333354\\ 0.\\ 0.\\ -18{,}666.666666666675\\ -6000.\\ 0.\\ 8000.000000000004\\ 0.\\ -8000.\\ 10{,}666.66666666667\\ 6000.\\ 8000.\\ 0. \end{pmatrix}$$

If the above results satisfy the following equilibrium equations, it can be concluded that the above solution is correct:

$$\sum_{i=1}^{4} R_{ix} = 0$$
$$\sum_{i=1}^{4} R_{iy} = 0$$
$$\sum_{i=1}^{4} R_{iz} = 0$$

Indeed:

$$R_{1x} + R_{2x} + R_{3x} + R_{4x} = 0 - 6000 + 0 + 6000 = 0$$
$$R_{1y} + R_{2y} + R_{3y} + R_{4y} = 0 + 0 + -8000 + 8000 = 0$$
$$R_{1z} + R_{2z} + R_{3z} + R_{4z} = -18,666.7 + 8000 + 10,666.7 + 0 = 0$$

EXAMPLE 5.3

Calculate the expanded stiffness matrices of the following plane truss (Table 5.3). All members are made from steel bars with cross-sectional area $A = 10^{-3}$ m^2 and modulus of elasticity $E = 200$ GPa.

Table 5.3 Coordinates of Nodes

Node	1	2	3	4	5	6	7	8	9	10
x (m)	0	1	2	3	4	8	9	10	11	12
y (m)	0	10	20	30	40	40	30	20	10	0

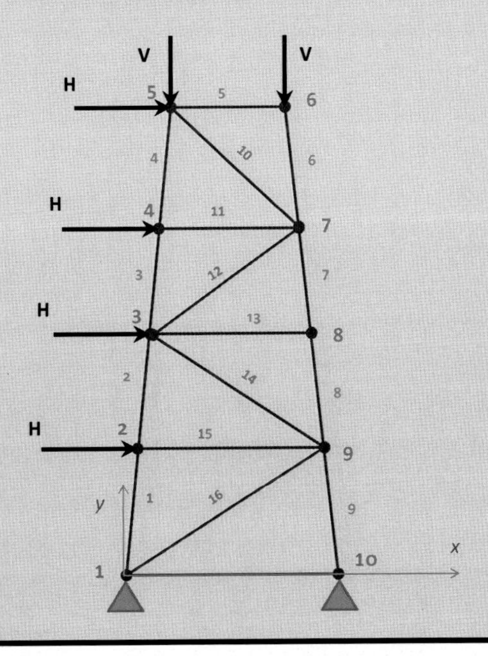

Continued

EXAMPLE 5.3—CONT'D

Stiffness Matrix for Element 1

$$
\left(\begin{array}{cccccccccccccccccccc}
197,029.7518619619 & 1,970,297.518619618 & -197,029.7518619619 & -1,970,297.518619618 & 0. & 0. & 0. & 0. & 0. & 0. & 0. & 0. & 0. & 0. & 0. & 0. & 0. & 0. & 0. & 0. \\
1,970,297.518619618 & 1.970297518619619\times10^{7} & -1,970,297.518619618 & -1.970297518619619\times10^{7} & 0. & 0. & 0. & 0. & 0. & 0. & 0. & 0. & 0. & 0. & 0. & 0. & 0. & 0. & 0. & 0. \\
-197,029.7518619619 & -1,970,297.518619618 & 197,029.7518619619 & 1,970,297.518619618 & 0. & 0. & 0. & 0. & 0. & 0. & 0. & 0. & 0. & 0. & 0. & 0. & 0. & 0. & 0. & 0. \\
-1,970,297.518619618 & -1.970297518619619\times10^{7} & 1,970,297.518619618 & 1.970297518619619\times10^{7} & 0. & 0. & 0. & 0. & 0. & 0. & 0. & 0. & 0. & 0. & 0. & 0. & 0. & 0. & 0. & 0. \\
0. & 0. & 0. & 0. & 0. & 0. & 0. & 0. & 0. & 0. & 0. & 0. & 0. & 0. & 0. & 0. & 0. & 0. & 0. & 0. \\
0. & 0. & 0. & 0. & 0. & 0. & 0. & 0. & 0. & 0. & 0. & 0. & 0. & 0. & 0. & 0. & 0. & 0. & 0. & 0. \\
0. & 0. & 0. & 0. & 0. & 0. & 0. & 0. & 0. & 0. & 0. & 0. & 0. & 0. & 0. & 0. & 0. & 0. & 0. & 0. \\
0. & 0. & 0. & 0. & 0. & 0. & 0. & 0. & 0. & 0. & 0. & 0. & 0. & 0. & 0. & 0. & 0. & 0. & 0. & 0. \\
0. & 0. & 0. & 0. & 0. & 0. & 0. & 0. & 0. & 0. & 0. & 0. & 0. & 0. & 0. & 0. & 0. & 0. & 0. & 0. \\
0. & 0. & 0. & 0. & 0. & 0. & 0. & 0. & 0. & 0. & 0. & 0. & 0. & 0. & 0. & 0. & 0. & 0. & 0. & 0. \\
0. & 0. & 0. & 0. & 0. & 0. & 0. & 0. & 0. & 0. & 0. & 0. & 0. & 0. & 0. & 0. & 0. & 0. & 0. & 0. \\
0. & 0. & 0. & 0. & 0. & 0. & 0. & 0. & 0. & 0. & 0. & 0. & 0. & 0. & 0. & 0. & 0. & 0. & 0. & 0. \\
0. & 0. & 0. & 0. & 0. & 0. & 0. & 0. & 0. & 0. & 0. & 0. & 0. & 0. & 0. & 0. & 0. & 0. & 0. & 0. \\
0. & 0. & 0. & 0. & 0. & 0. & 0. & 0. & 0. & 0. & 0. & 0. & 0. & 0. & 0. & 0. & 0. & 0. & 0. & 0. \\
0. & 0. & 0. & 0. & 0. & 0. & 0. & 0. & 0. & 0. & 0. & 0. & 0. & 0. & 0. & 0. & 0. & 0. & 0. & 0. \\
0. & 0. & 0. & 0. & 0. & 0. & 0. & 0. & 0. & 0. & 0. & 0. & 0. & 0. & 0. & 0. & 0. & 0. & 0. & 0. \\
0. & 0. & 0. & 0. & 0. & 0. & 0. & 0. & 0. & 0. & 0. & 0. & 0. & 0. & 0. & 0. & 0. & 0. & 0. & 0. \\
0. & 0. & 0. & 0. & 0. & 0. & 0. & 0. & 0. & 0. & 0. & 0. & 0. & 0. & 0. & 0. & 0. & 0. & 0. & 0. \\
0. & 0. & 0. & 0. & 0. & 0. & 0. & 0. & 0. & 0. & 0. & 0. & 0. & 0. & 0. & 0. & 0. & 0. & 0. & 0. \\
0. & 0. & 0. & 0. & 0. & 0. & 0. & 0. & 0. & 0. & 0. & 0. & 0. & 0. & 0. & 0. & 0. & 0. & 0. & 0.
\end{array}\right)
$$

Stiffness Matrix for Element 2

$$
\left(\begin{array}{cccccccccccccccccccc}
0. & 0. & 0. & 0. & 0. & 0. & 0. & 0. & 0. & 0. & 0. & 0. & 0. & 0. & 0. & 0. & 0. & 0. & 0. & 0. \\
0. & 0. & 0. & 0. & 0. & 0. & 0. & 0. & 0. & 0. & 0. & 0. & 0. & 0. & 0. & 0. & 0. & 0. & 0. & 0. \\
0. & 0. & 197,029.7518619619 & 1,970,297.518619618 & -197,029.7518619619 & -1,970,297.518619618 & 0. & 0. & 0. & 0. & 0. & 0. & 0. & 0. & 0. & 0. & 0. & 0. & 0. & 0. \\
0. & 0. & 1,970,297.518619618 & 1.970297518619619\times10^{7} & -1,970,297.518619618 & -1.970297518619619\times10^{7} & 0. & 0. & 0. & 0. & 0. & 0. & 0. & 0. & 0. & 0. & 0. & 0. & 0. & 0. \\
0. & 0. & -197,029.7518619619 & -1,970,297.518619618 & 197,029.7518619619 & 1,970,297.518619618 & 0. & 0. & 0. & 0. & 0. & 0. & 0. & 0. & 0. & 0. & 0. & 0. & 0. & 0. \\
0. & 0. & -1,970,297.518619618 & -1.970297518619619\times10^{7} & 1,970,297.518619618 & 1.970297518619619\times10^{7} & 0. & 0. & 0. & 0. & 0. & 0. & 0. & 0. & 0. & 0. & 0. & 0. & 0. & 0. \\
0. & 0. & 0. & 0. & 0. & 0. & 0. & 0. & 0. & 0. & 0. & 0. & 0. & 0. & 0. & 0. & 0. & 0. & 0. & 0. \\
0. & 0. & 0. & 0. & 0. & 0. & 0. & 0. & 0. & 0. & 0. & 0. & 0. & 0. & 0. & 0. & 0. & 0. & 0. & 0. \\
0. & 0. & 0. & 0. & 0. & 0. & 0. & 0. & 0. & 0. & 0. & 0. & 0. & 0. & 0. & 0. & 0. & 0. & 0. & 0. \\
0. & 0. & 0. & 0. & 0. & 0. & 0. & 0. & 0. & 0. & 0. & 0. & 0. & 0. & 0. & 0. & 0. & 0. & 0. & 0. \\
0. & 0. & 0. & 0. & 0. & 0. & 0. & 0. & 0. & 0. & 0. & 0. & 0. & 0. & 0. & 0. & 0. & 0. & 0. & 0. \\
0. & 0. & 0. & 0. & 0. & 0. & 0. & 0. & 0. & 0. & 0. & 0. & 0. & 0. & 0. & 0. & 0. & 0. & 0. & 0. \\
0. & 0. & 0. & 0. & 0. & 0. & 0. & 0. & 0. & 0. & 0. & 0. & 0. & 0. & 0. & 0. & 0. & 0. & 0. & 0. \\
0. & 0. & 0. & 0. & 0. & 0. & 0. & 0. & 0. & 0. & 0. & 0. & 0. & 0. & 0. & 0. & 0. & 0. & 0. & 0. \\
0. & 0. & 0. & 0. & 0. & 0. & 0. & 0. & 0. & 0. & 0. & 0. & 0. & 0. & 0. & 0. & 0. & 0. & 0. & 0. \\
0. & 0. & 0. & 0. & 0. & 0. & 0. & 0. & 0. & 0. & 0. & 0. & 0. & 0. & 0. & 0. & 0. & 0. & 0. & 0. \\
0. & 0. & 0. & 0. & 0. & 0. & 0. & 0. & 0. & 0. & 0. & 0. & 0. & 0. & 0. & 0. & 0. & 0. & 0. & 0. \\
0. & 0. & 0. & 0. & 0. & 0. & 0. & 0. & 0. & 0. & 0. & 0. & 0. & 0. & 0. & 0. & 0. & 0. & 0. & 0. \\
0. & 0. & 0. & 0. & 0. & 0. & 0. & 0. & 0. & 0. & 0. & 0. & 0. & 0. & 0. & 0. & 0. & 0. & 0. & 0. \\
0. & 0. & 0. & 0. & 0. & 0. & 0. & 0. & 0. & 0. & 0. & 0. & 0. & 0. & 0. & 0. & 0. & 0. & 0. & 0.
\end{array}\right)
$$

Stiffness Matrix for Element 3

0.	0.	0.	0.	0.	0.	0.	0.	0.	0.	0.	0.	0.	0.	0.	0.	0.	0.	0.	0.
0.	0.	0.	0.	0.	0.	0.	0.	0.	0.	0.	0.	0.	0.	0.	0.	0.	0.	0.	0.
0.	0.	0.	0.	0.	0.	0.	0.	0.	0.	0.	0.	0.	0.	0.	0.	0.	0.	0.	0.
0.	0.	0.	0.	0.	0.	0.	0.	0.	0.	0.	0.	0.	0.	0.	0.	0.	0.	0.	0.
0.	0.	0.	0.	197,029.7518619619	1,970,297.518619618	−197,029.7518619619	−1,970,297.518619618	0.	0.	0.	0.	0.	0.	0.	0.	0.	0.	0.	0.
0.	0.	0.	0.	1,970,297.518619618	$1.970297518619619 \times 10^7$	−1,970,297.518619618	$-1.970297518619619 \times 10^7$	0.	0.	0.	0.	0.	0.	0.	0.	0.	0.	0.	0.
0.	0.	0.	0.	−197,029.7518619619	−1,970,297.518619618	197,029.7518619619	1,970,297.518619618	0.	0.	0.	0.	0.	0.	0.	0.	0.	0.	0.	0.
0.	0.	0.	0.	−1,970,297.518619618	$-1.970297518619619 \times 10^7$	1,970,297.518619618	$1.970297518619619 \times 10^7$	0.	0.	0.	0.	0.	0.	0.	0.	0.	0.	0.	0.
0.	0.	0.	0.	0.	0.	0.	0.	0.	0.	0.	0.	0.	0.	0.	0.	0.	0.	0.	0.
0.	0.	0.	0.	0.	0.	0.	0.	0.	0.	0.	0.	0.	0.	0.	0.	0.	0.	0.	0.
0.	0.	0.	0.	0.	0.	0.	0.	0.	0.	0.	0.	0.	0.	0.	0.	0.	0.	0.	0.
0.	0.	0.	0.	0.	0.	0.	0.	0.	0.	0.	0.	0.	0.	0.	0.	0.	0.	0.	0.
0.	0.	0.	0.	0.	0.	0.	0.	0.	0.	0.	0.	0.	0.	0.	0.	0.	0.	0.	0.
0.	0.	0.	0.	0.	0.	0.	0.	0.	0.	0.	0.	0.	0.	0.	0.	0.	0.	0.	0.
0.	0.	0.	0.	0.	0.	0.	0.	0.	0.	0.	0.	0.	0.	0.	0.	0.	0.	0.	0.
0.	0.	0.	0.	0.	0.	0.	0.	0.	0.	0.	0.	0.	0.	0.	0.	0.	0.	0.	0.
0.	0.	0.	0.	0.	0.	0.	0.	0.	0.	0.	0.	0.	0.	0.	0.	0.	0.	0.	0.
0.	0.	0.	0.	0.	0.	0.	0.	0.	0.	0.	0.	0.	0.	0.	0.	0.	0.	0.	0.
0.	0.	0.	0.	0.	0.	0.	0.	0.	0.	0.	0.	0.	0.	0.	0.	0.	0.	0.	0.
0.	0.	0.	0.	0.	0.	0.	0.	0.	0.	0.	0.	0.	0.	0.	0.	0.	0.	0.	0.

Stiffness Matrix for Element 4

0.	0.	0.	0.	0.	0.	0.	0.	0.	0.	0.	0.	0.	0.	0.	0.	0.	0.	0.	0.
0.	0.	0.	0.	0.	0.	0.	0.	0.	0.	0.	0.	0.	0.	0.	0.	0.	0.	0.	0.
0.	0.	0.	0.	0.	0.	0.	0.	0.	0.	0.	0.	0.	0.	0.	0.	0.	0.	0.	0.
0.	0.	0.	0.	0.	0.	0.	0.	0.	0.	0.	0.	0.	0.	0.	0.	0.	0.	0.	0.
0.	0.	0.	0.	0.	0.	0.	0.	0.	0.	0.	0.	0.	0.	0.	0.	0.	0.	0.	0.
0.	0.	0.	0.	0.	0.	0.	0.	0.	0.	0.	0.	0.	0.	0.	0.	0.	0.	0.	0.
0.	0.	0.	0.	0.	0.	197,029.7518619619	1,970,297.518619618	−197,029.7518619619	−1,970,297.518619618	0.	0.	0.	0.	0.	0.	0.	0.	0.	0.
0.	0.	0.	0.	0.	0.	1,970,297.518619618	$1.970297518619619 \times 10^7$	−1,970,297.518619618	$-1.970297518619619 \times 10^7$	0.	0.	0.	0.	0.	0.	0.	0.	0.	0.
0.	0.	0.	0.	0.	0.	−197,029.7518619619	−1,970,297.518619618	197,029.7518619619	1,970,297.518619618	0.	0.	0.	0.	0.	0.	0.	0.	0.	0.
0.	0.	0.	0.	0.	0.	−1,970,297.518619618	$-1.970297518619619 \times 10^7$	1,970,297.518619618	$1.970297518619619 \times 10^7$	0.	0.	0.	0.	0.	0.	0.	0.	0.	0.
0.	0.	0.	0.	0.	0.	0.	0.	0.	0.	0.	0.	0.	0.	0.	0.	0.	0.	0.	0.
0.	0.	0.	0.	0.	0.	0.	0.	0.	0.	0.	0.	0.	0.	0.	0.	0.	0.	0.	0.
0.	0.	0.	0.	0.	0.	0.	0.	0.	0.	0.	0.	0.	0.	0.	0.	0.	0.	0.	0.
0.	0.	0.	0.	0.	0.	0.	0.	0.	0.	0.	0.	0.	0.	0.	0.	0.	0.	0.	0.
0.	0.	0.	0.	0.	0.	0.	0.	0.	0.	0.	0.	0.	0.	0.	0.	0.	0.	0.	0.
0.	0.	0.	0.	0.	0.	0.	0.	0.	0.	0.	0.	0.	0.	0.	0.	0.	0.	0.	0.
0.	0.	0.	0.	0.	0.	0.	0.	0.	0.	0.	0.	0.	0.	0.	0.	0.	0.	0.	0.
0.	0.	0.	0.	0.	0.	0.	0.	0.	0.	0.	0.	0.	0.	0.	0.	0.	0.	0.	0.
0.	0.	0.	0.	0.	0.	0.	0.	0.	0.	0.	0.	0.	0.	0.	0.	0.	0.	0.	0.
0.	0.	0.	0.	0.	0.	0.	0.	0.	0.	0.	0.	0.	0.	0.	0.	0.	0.	0.	0.

Continued

EXAMPLE 5.3—CONT'D

Stiffness Matrix for Element 5

$$
\begin{pmatrix}
0. & 0. & 0. & 0. & 0. & 0. & 0. & 0. & 0. & 0. & 0. & 0. & 0. & 0. & 0. & 0. & 0. & 0. & 0. & 0. \\
0. & 0. & 0. & 0. & 0. & 0. & 0. & 0. & 0. & 0. & 0. & 0. & 0. & 0. & 0. & 0. & 0. & 0. & 0. & 0. \\
0. & 0. & 0. & 0. & 0. & 0. & 0. & 0. & 0. & 0. & 0. & 0. & 0. & 0. & 0. & 0. & 0. & 0. & 0. & 0. \\
0. & 0. & 0. & 0. & 0. & 0. & 0. & 0. & 0. & 0. & 0. & 0. & 0. & 0. & 0. & 0. & 0. & 0. & 0. & 0. \\
0. & 0. & 0. & 0. & 0. & 0. & 0. & 0. & 0. & 0. & 0. & 0. & 0. & 0. & 0. & 0. & 0. & 0. & 0. & 0. \\
0. & 0. & 0. & 0. & 0. & 0. & 0. & 0. & 0. & 0. & 0. & 0. & 0. & 0. & 0. & 0. & 0. & 0. & 0. & 0. \\
0. & 0. & 0. & 0. & 0. & 0. & 0. & 0. & 0. & 0. & 0. & 0. & 0. & 0. & 0. & 0. & 0. & 0. & 0. & 0. \\
0. & 0. & 0. & 0. & 0. & 0. & 0. & 0. & 0. & 0. & 0. & 0. & 0. & 0. & 0. & 0. & 0. & 0. & 0. & 0. \\
0. & 0. & 0. & 0. & 0. & 0. & 0. & 0. & 495,037.25155317935 & 4,950,372.515531793 & -495,037.25155317935 & -4,950,372.515531793 & 0. & 0. & 0. & 0. & 0. & 0. & 0. & 0. \\
0. & 0. & 0. & 0. & 0. & 0. & 0. & 0. & 4,950,372.515531793 & 4.950372515531793 \times 10^7 & -4,950,372.515531793 & -4.950372515531793 \times 10^7 & 0. & 0. & 0. & 0. & 0. & 0. & 0. & 0. \\
0. & 0. & 0. & 0. & 0. & 0. & 0. & 0. & -495,037.25155317935 & -4,950,372.515531793 & 495,037.25155317935 & 4,950,372.515531793 & 0. & 0. & 0. & 0. & 0. & 0. & 0. & 0. \\
0. & 0. & 0. & 0. & 0. & 0. & 0. & 0. & -4,950,372.515531793 & -4.950372515531793 \times 10^7 & 4,950,372.515531793 & 4.950372515531793 \times 10^7 & 0. & 0. & 0. & 0. & 0. & 0. & 0. & 0. \\
0. & 0. & 0. & 0. & 0. & 0. & 0. & 0. & 0. & 0. & 0. & 0. & 0. & 0. & 0. & 0. & 0. & 0. & 0. & 0. \\
0. & 0. & 0. & 0. & 0. & 0. & 0. & 0. & 0. & 0. & 0. & 0. & 0. & 0. & 0. & 0. & 0. & 0. & 0. & 0. \\
0. & 0. & 0. & 0. & 0. & 0. & 0. & 0. & 0. & 0. & 0. & 0. & 0. & 0. & 0. & 0. & 0. & 0. & 0. & 0. \\
0. & 0. & 0. & 0. & 0. & 0. & 0. & 0. & 0. & 0. & 0. & 0. & 0. & 0. & 0. & 0. & 0. & 0. & 0. & 0. \\
0. & 0. & 0. & 0. & 0. & 0. & 0. & 0. & 0. & 0. & 0. & 0. & 0. & 0. & 0. & 0. & 0. & 0. & 0. & 0. \\
0. & 0. & 0. & 0. & 0. & 0. & 0. & 0. & 0. & 0. & 0. & 0. & 0. & 0. & 0. & 0. & 0. & 0. & 0. & 0. \\
0. & 0. & 0. & 0. & 0. & 0. & 0. & 0. & 0. & 0. & 0. & 0. & 0. & 0. & 0. & 0. & 0. & 0. & 0. & 0. \\
0. & 0. & 0. & 0. & 0. & 0. & 0. & 0. & 0. & 0. & 0. & 0. & 0. & 0. & 0. & 0. & 0. & 0. & 0. & 0.
\end{pmatrix}
$$

Stiffness Matrix for Element 6

$$
\begin{pmatrix}
0. & 0. & 0. & 0. & 0. & 0. & 0. & 0. & 0. & 0. & 0. & 0. & 0. & 0. & 0. & 0. & 0. & 0. & 0. & 0. \\
0. & 0. & 0. & 0. & 0. & 0. & 0. & 0. & 0. & 0. & 0. & 0. & 0. & 0. & 0. & 0. & 0. & 0. & 0. & 0. \\
0. & 0. & 0. & 0. & 0. & 0. & 0. & 0. & 0. & 0. & 0. & 0. & 0. & 0. & 0. & 0. & 0. & 0. & 0. & 0. \\
0. & 0. & 0. & 0. & 0. & 0. & 0. & 0. & 0. & 0. & 0. & 0. & 0. & 0. & 0. & 0. & 0. & 0. & 0. & 0. \\
0. & 0. & 0. & 0. & 0. & 0. & 0. & 0. & 0. & 0. & 0. & 0. & 0. & 0. & 0. & 0. & 0. & 0. & 0. & 0. \\
0. & 0. & 0. & 0. & 0. & 0. & 0. & 0. & 0. & 0. & 0. & 0. & 0. & 0. & 0. & 0. & 0. & 0. & 0. & 0. \\
0. & 0. & 0. & 0. & 0. & 0. & 0. & 0. & 0. & 0. & 0. & 0. & 0. & 0. & 0. & 0. & 0. & 0. & 0. & 0. \\
0. & 0. & 0. & 0. & 0. & 0. & 0. & 0. & 0. & 0. & 0. & 0. & 0. & 0. & 0. & 0. & 0. & 0. & 0. & 0. \\
0. & 0. & 0. & 0. & 0. & 0. & 0. & 0. & 0. & 0. & 0. & 0. & 0. & 0. & 0. & 0. & 0. & 0. & 0. & 0. \\
0. & 0. & 0. & 0. & 0. & 0. & 0. & 0. & 0. & 0. & 0. & 0. & 0. & 0. & 0. & 0. & 0. & 0. & 0. & 0. \\
0. & 0. & 0. & 0. & 0. & 0. & 0. & 0. & 0. & 0. & 197,029.7518619619 & -1,970,297.518619618 & -197,029.7518619619 & 1,970,297.518619618 & 0. & 0. & 0. & 0. & 0. & 0. \\
0. & 0. & 0. & 0. & 0. & 0. & 0. & 0. & 0. & 0. & -1,970,297.518619618 & 1.970297518619619 \times 10^7 & 1,970,297.518619618 & -1.970297518619619 \times 10^7 & 0. & 0. & 0. & 0. & 0. & 0. \\
0. & 0. & 0. & 0. & 0. & 0. & 0. & 0. & 0. & 0. & -197,029.7518619619 & 1,970,297.518619618 & 197,029.7518619619 & -1,970,297.518619618 & 0. & 0. & 0. & 0. & 0. & 0. \\
0. & 0. & 0. & 0. & 0. & 0. & 0. & 0. & 0. & 0. & 1,970,297.518619618 & -1.970297518619619 \times 10^7 & -1,970,297.518619618 & 1.970297518619619 \times 10^7 & 0. & 0. & 0. & 0. & 0. & 0. \\
0. & 0. & 0. & 0. & 0. & 0. & 0. & 0. & 0. & 0. & 0. & 0. & 0. & 0. & 0. & 0. & 0. & 0. & 0. & 0. \\
0. & 0. & 0. & 0. & 0. & 0. & 0. & 0. & 0. & 0. & 0. & 0. & 0. & 0. & 0. & 0. & 0. & 0. & 0. & 0. \\
0. & 0. & 0. & 0. & 0. & 0. & 0. & 0. & 0. & 0. & 0. & 0. & 0. & 0. & 0. & 0. & 0. & 0. & 0. & 0. \\
0. & 0. & 0. & 0. & 0. & 0. & 0. & 0. & 0. & 0. & 0. & 0. & 0. & 0. & 0. & 0. & 0. & 0. & 0. & 0. \\
0. & 0. & 0. & 0. & 0. & 0. & 0. & 0. & 0. & 0. & 0. & 0. & 0. & 0. & 0. & 0. & 0. & 0. & 0. & 0. \\
0. & 0. & 0. & 0. & 0. & 0. & 0. & 0. & 0. & 0. & 0. & 0. & 0. & 0. & 0. & 0. & 0. & 0. & 0. & 0.
\end{pmatrix}
$$

Stiffness Matrix for Element 7

0.	0.	0.	0.	0.	0.	0.	0.	0.	0.	0.	0.	0.	0.	0.	0.	0.	0.	0.	0.
0.	0.	0.	0.	0.	0.	0.	0.	0.	0.	0.	0.	0.	0.	0.	0.	0.	0.	0.	0.
0.	0.	0.	0.	0.	0.	0.	0.	0.	0.	0.	0.	0.	0.	0.	0.	0.	0.	0.	0.
0.	0.	0.	0.	0.	0.	0.	0.	0.	0.	0.	0.	0.	0.	0.	0.	0.	0.	0.	0.
0.	0.	0.	0.	0.	0.	0.	0.	0.	0.	0.	0.	0.	0.	0.	0.	0.	0.	0.	0.
0.	0.	0.	0.	0.	0.	0.	0.	0.	0.	0.	0.	0.	0.	0.	0.	0.	0.	0.	0.
0.	0.	0.	0.	0.	0.	0.	0.	0.	0.	0.	0.	0.	0.	0.	0.	0.	0.	0.	0.
0.	0.	0.	0.	0.	0.	0.	0.	0.	0.	0.	0.	0.	0.	0.	0.	0.	0.	0.	0.
0.	0.	0.	0.	0.	0.	0.	0.	0.	0.	0.	0.	0.	0.	0.	0.	0.	0.	0.	0.
0.	0.	0.	0.	0.	0.	0.	0.	0.	0.	0.	0.	0.	0.	0.	0.	0.	0.	0.	0.
0.	0.	0.	0.	0.	0.	0.	0.	0.	0.	0.	0.	0.	0.	0.	0.	0.	0.	0.	0.
0.	0.	0.	0.	0.	0.	0.	0.	0.	0.	0.	0.	0.	0.	0.	0.	0.	0.	0.	0.
0.	0.	0.	0.	0.	0.	0.	0.	0.	0.	0.	0.	197,029.7518619619	−1,970,297.518619618	−197,029.7518619619	1,970,297.518619618	0.	0.	0.	0.
0.	0.	0.	0.	0.	0.	0.	0.	0.	0.	0.	0.	−1,970,297.518619618	$1.970297518619619 \times 10^7$	1,970,297.518619618	$-1.970297518619619 \times 10^7$	0.	0.	0.	0.
0.	0.	0.	0.	0.	0.	0.	0.	0.	0.	0.	0.	−197,029.7518619619	1,970,297.518619618	197,029.7518619619	−1,970,297.518619618	0.	0.	0.	0.
0.	0.	0.	0.	0.	0.	0.	0.	0.	0.	0.	0.	1,970,297.518619618	$-1.970297518619619 \times 10^7$	−1,970,297.518619618	$1.970297518619619 \times 10^7$	0.	0.	0.	0.
0.	0.	0.	0.	0.	0.	0.	0.	0.	0.	0.	0.	0.	0.	0.	0.	0.	0.	0.	0.
0.	0.	0.	0.	0.	0.	0.	0.	0.	0.	0.	0.	0.	0.	0.	0.	0.	0.	0.	0.
0.	0.	0.	0.	0.	0.	0.	0.	0.	0.	0.	0.	0.	0.	0.	0.	0.	0.	0.	0.
0.	0.	0.	0.	0.	0.	0.	0.	0.	0.	0.	0.	0.	0.	0.	0.	0.	0.	0.	0.

Stiffness Matrix for Element 8

0.	0.	0.	0.	0.	0.	0.	0.	0.	0.	0.	0.	0.	0.	0.	0.	0.	0.	0.	0.
0.	0.	0.	0.	0.	0.	0.	0.	0.	0.	0.	0.	0.	0.	0.	0.	0.	0.	0.	0.
0.	0.	0.	0.	0.	0.	0.	0.	0.	0.	0.	0.	0.	0.	0.	0.	0.	0.	0.	0.
0.	0.	0.	0.	0.	0.	0.	0.	0.	0.	0.	0.	0.	0.	0.	0.	0.	0.	0.	0.
0.	0.	0.	0.	0.	0.	0.	0.	0.	0.	0.	0.	0.	0.	0.	0.	0.	0.	0.	0.
0.	0.	0.	0.	0.	0.	0.	0.	0.	0.	0.	0.	0.	0.	0.	0.	0.	0.	0.	0.
0.	0.	0.	0.	0.	0.	0.	0.	0.	0.	0.	0.	0.	0.	0.	0.	0.	0.	0.	0.
0.	0.	0.	0.	0.	0.	0.	0.	0.	0.	0.	0.	0.	0.	0.	0.	0.	0.	0.	0.
0.	0.	0.	0.	0.	0.	0.	0.	0.	0.	0.	0.	0.	0.	0.	0.	0.	0.	0.	0.
0.	0.	0.	0.	0.	0.	0.	0.	0.	0.	0.	0.	0.	0.	0.	0.	0.	0.	0.	0.
0.	0.	0.	0.	0.	0.	0.	0.	0.	0.	0.	0.	0.	0.	0.	0.	0.	0.	0.	0.
0.	0.	0.	0.	0.	0.	0.	0.	0.	0.	0.	0.	0.	0.	0.	0.	0.	0.	0.	0.
0.	0.	0.	0.	0.	0.	0.	0.	0.	0.	0.	0.	0.	0.	0.	0.	0.	0.	0.	0.
0.	0.	0.	0.	0.	0.	0.	0.	0.	0.	0.	0.	0.	0.	0.	0.	0.	0.	0.	0.
0.	0.	0.	0.	0.	0.	0.	0.	0.	0.	0.	0.	0.	0.	197,029.7518619619	−1,970,297.518619618	−197,029.7518619619	1,970,297.518619618	0.	0.
0.	0.	0.	0.	0.	0.	0.	0.	0.	0.	0.	0.	0.	0.	−1,970,297.518619618	$1.970297518619619 \times 10^7$	1,970,297.518619618	$-1.970297518619619 \times 10^7$	0.	0.
0.	0.	0.	0.	0.	0.	0.	0.	0.	0.	0.	0.	0.	0.	−197,029.7518619619	1,970,297.518619618	197,029.7518619619	−1,970,297.518619618	0.	0.
0.	0.	0.	0.	0.	0.	0.	0.	0.	0.	0.	0.	0.	0.	1,970,297.518619618	$-1.970297518619619 \times 10^7$	−1,970,297.518619618	$1.970297518619619 \times 10^7$	0.	0.
0.	0.	0.	0.	0.	0.	0.	0.	0.	0.	0.	0.	0.	0.	0.	0.	0.	0.	0.	0.
0.	0.	0.	0.	0.	0.	0.	0.	0.	0.	0.	0.	0.	0.	0.	0.	0.	0.	0.	0.

Continued

EXAMPLE 5.3—CONT'D

Stiffness Matrix for Element 9

0.	0.	0.	0.	0.	0.	0.	0.	0.	0.	0.	0.	0.	0.	0.	0.	0.	0.	0.	0.
0.	0.	0.	0.	0.	0.	0.	0.	0.	0.	0.	0.	0.	0.	0.	0.	0.	0.	0.	0.
0.	0.	0.	0.	0.	0.	0.	0.	0.	0.	0.	0.	0.	0.	0.	0.	0.	0.	0.	0.
0.	0.	0.	0.	0.	0.	0.	0.	0.	0.	0.	0.	0.	0.	0.	0.	0.	0.	0.	0.
0.	0.	0.	0.	0.	0.	0.	0.	0.	0.	0.	0.	0.	0.	0.	0.	0.	0.	0.	0.
0.	0.	0.	0.	0.	0.	0.	0.	0.	0.	0.	0.	0.	0.	0.	0.	0.	0.	0.	0.
0.	0.	0.	0.	0.	0.	0.	0.	0.	0.	0.	0.	0.	0.	0.	0.	0.	0.	0.	0.
0.	0.	0.	0.	0.	0.	0.	0.	0.	0.	0.	0.	0.	0.	0.	0.	0.	0.	0.	0.
0.	0.	0.	0.	0.	0.	0.	0.	0.	0.	0.	0.	0.	0.	0.	0.	0.	0.	0.	0.
0.	0.	0.	0.	0.	0.	0.	0.	0.	0.	0.	0.	0.	0.	0.	0.	0.	0.	0.	0.
0.	0.	0.	0.	0.	0.	0.	0.	0.	0.	0.	0.	0.	0.	0.	0.	0.	0.	0.	0.
0.	0.	0.	0.	0.	0.	0.	0.	0.	0.	0.	0.	0.	0.	0.	0.	0.	0.	0.	0.
0.	0.	0.	0.	0.	0.	0.	0.	0.	0.	0.	0.	0.	0.	0.	0.	0.	0.	0.	0.
0.	0.	0.	0.	0.	0.	0.	0.	0.	0.	0.	0.	0.	0.	0.	0.	0.	0.	0.	0.
0.	0.	0.	0.	0.	0.	0.	0.	0.	0.	0.	0.	0.	0.	0.	0.	0.	0.	0.	0.
0.	0.	0.	0.	0.	0.	0.	0.	0.	0.	0.	0.	0.	0.	0.	0.	0.	0.	0.	0.
0.	0.	0.	0.	0.	0.	0.	0.	0.	0.	0.	0.	0.	0.	0.	0.	197,029.7518619619	−1,970,297.518619618	−197,029.7518619619	1,970,297.518619618
0.	0.	0.	0.	0.	0.	0.	0.	0.	0.	0.	0.	0.	0.	0.	0.	−1,970,297.518619618	$1.970297518619619 \times 10^7$	1,970,297.518619618	$-1.970297518619619 \times 10^7$
0.	0.	0.	0.	0.	0.	0.	0.	0.	0.	0.	0.	0.	0.	0.	0.	−197,029.7518619619	1,970,297.518619618	197,029.7518619619	−1,970,297.518619618
0.	0.	0.	0.	0.	0.	0.	0.	0.	0.	0.	0.	0.	0.	0.	0.	1,970,297.518619618	$-1.970297518619619 \times 10^7$	−1,970,297.518619618	$1.970297518619619 \times 10^7$

Stiffness Matrix for Element 10

0.	0.	0.	0.	0.	0.	0.	0.	0.	0.	0.	0.	0.	0.	0.	0.	0.	0.	0.	0.
0.	0.	0.	0.	0.	0.	0.	0.	0.	0.	0.	0.	0.	0.	0.	0.	0.	0.	0.	0.
0.	0.	0.	0.	0.	0.	0.	0.	0.	0.	0.	0.	0.	0.	0.	0.	0.	0.	0.	0.
0.	0.	0.	0.	0.	0.	0.	0.	0.	0.	0.	0.	0.	0.	0.	0.	0.	0.	0.	0.
0.	0.	0.	0.	0.	0.	0.	0.	0.	0.	0.	0.	0.	0.	0.	0.	0.	0.	0.	0.
0.	0.	0.	0.	0.	0.	0.	0.	0.	0.	0.	0.	0.	0.	0.	0.	0.	0.	0.	0.
0.	0.	0.	0.	0.	0.	0.	0.	0.	0.	0.	0.	0.	0.	0.	0.	0.	0.	0.	0.
0.	0.	0.	0.	0.	0.	0.	0.	0.	0.	0.	0.	0.	0.	0.	0.	0.	0.	0.	0.
0.	0.	0.	0.	0.	0.	0.	0.	3,578,035.075838516	−7,156,070.151677032	0.	0.	−3,578,035.075838516	7,156,070.151677032	0.	0.	0.	0.	0.	0.
0.	0.	0.	0.	0.	0.	0.	0.	−7,156,070.151677032	$1.431214030335406 \times 10^7$	0.	0.	7,156,070.151677032	$-1.431214030335406 \times 10^7$	0.	0.	0.	0.	0.	0.
0.	0.	0.	0.	0.	0.	0.	0.	0.	0.	0.	0.	0.	0.	0.	0.	0.	0.	0.	0.
0.	0.	0.	0.	0.	0.	0.	0.	0.	0.	0.	0.	0.	0.	0.	0.	0.	0.	0.	0.
0.	0.	0.	0.	0.	0.	0.	0.	−3,578,035.075838516	7,156,070.151677032	0.	0.	3,578,035.075838516	−7,156,070.151677032	0.	0.	0.	0.	0.	0.
0.	0.	0.	0.	0.	0.	0.	0.	7,156,070.151677032	$-1.431214030335406 \times 10^7$	0.	0.	−7,156,070.151677032	$1.431214030335406 \times 10^7$	0.	0.	0.	0.	0.	0.
0.	0.	0.	0.	0.	0.	0.	0.	0.	0.	0.	0.	0.	0.	0.	0.	0.	0.	0.	0.
0.	0.	0.	0.	0.	0.	0.	0.	0.	0.	0.	0.	0.	0.	0.	0.	0.	0.	0.	0.
0.	0.	0.	0.	0.	0.	0.	0.	0.	0.	0.	0.	0.	0.	0.	0.	0.	0.	0.	0.
0.	0.	0.	0.	0.	0.	0.	0.	0.	0.	0.	0.	0.	0.	0.	0.	0.	0.	0.	0.
0.	0.	0.	0.	0.	0.	0.	0.	0.	0.	0.	0.	0.	0.	0.	0.	0.	0.	0.	0.
0.	0.	0.	0.	0.	0.	0.	0.	0.	0.	0.	0.	0.	0.	0.	0.	0.	0.	0.	0.

Stiffness Matrix for Element 11

$$
\left(\begin{array}{cccccccccccccccccccc}
0. & 0. & 0. & 0. & 0. & 0. & 0. & 0. & 0. & 0. & 0. & 0. & 0. & 0. & 0. & 0. & 0. & 0. & 0. & 0. \\
0. & 0. & 0. & 0. & 0. & 0. & 0. & 0. & 0. & 0. & 0. & 0. & 0. & 0. & 0. & 0. & 0. & 0. & 0. & 0. \\
0. & 0. & 0. & 0. & 0. & 0. & 0. & 0. & 0. & 0. & 0. & 0. & 0. & 0. & 0. & 0. & 0. & 0. & 0. & 0. \\
0. & 0. & 0. & 0. & 0. & 0. & 0. & 0. & 0. & 0. & 0. & 0. & 0. & 0. & 0. & 0. & 0. & 0. & 0. & 0. \\
0. & 0. & 0. & 0. & 0. & 0. & 0. & 0. & 0. & 0. & 0. & 0. & 0. & 0. & 0. & 0. & 0. & 0. & 0. & 0. \\
0. & 0. & 0. & 0. & 0. & 0. & 0. & 0. & 0. & 0. & 0. & 0. & 0. & 0. & 0. & 0. & 0. & 0. & 0. & 0. \\
0. & 0. & 0. & 0. & 0. & 0. & 3.333333333333333 \times 10^{7} & 0. & 0. & 0. & 0. & 0. & -3.333333333333333 \times 10^{7} & 0. & 0. & 0. & 0. & 0. & 0. & 0. \\
0. & 0. & 0. & 0. & 0. & 0. & 0. & 0. & 0. & 0. & 0. & 0. & 0. & 0. & 0. & 0. & 0. & 0. & 0. & 0. \\
0. & 0. & 0. & 0. & 0. & 0. & 0. & 0. & 0. & 0. & 0. & 0. & 0. & 0. & 0. & 0. & 0. & 0. & 0. & 0. \\
0. & 0. & 0. & 0. & 0. & 0. & 0. & 0. & 0. & 0. & 0. & 0. & 0. & 0. & 0. & 0. & 0. & 0. & 0. & 0. \\
0. & 0. & 0. & 0. & 0. & 0. & 0. & 0. & 0. & 0. & 0. & 0. & 0. & 0. & 0. & 0. & 0. & 0. & 0. & 0. \\
0. & 0. & 0. & 0. & 0. & 0. & 0. & 0. & 0. & 0. & 0. & 0. & 0. & 0. & 0. & 0. & 0. & 0. & 0. & 0. \\
0. & 0. & 0. & 0. & 0. & 0. & -3.333333333333333 \times 10^{7} & 0. & 0. & 0. & 0. & 0. & 3.333333333333333 \times 10^{7} & 0. & 0. & 0. & 0. & 0. & 0. & 0. \\
0. & 0. & 0. & 0. & 0. & 0. & 0. & 0. & 0. & 0. & 0. & 0. & 0. & 0. & 0. & 0. & 0. & 0. & 0. & 0. \\
0. & 0. & 0. & 0. & 0. & 0. & 0. & 0. & 0. & 0. & 0. & 0. & 0. & 0. & 0. & 0. & 0. & 0. & 0. & 0. \\
0. & 0. & 0. & 0. & 0. & 0. & 0. & 0. & 0. & 0. & 0. & 0. & 0. & 0. & 0. & 0. & 0. & 0. & 0. & 0. \\
0. & 0. & 0. & 0. & 0. & 0. & 0. & 0. & 0. & 0. & 0. & 0. & 0. & 0. & 0. & 0. & 0. & 0. & 0. & 0. \\
0. & 0. & 0. & 0. & 0. & 0. & 0. & 0. & 0. & 0. & 0. & 0. & 0. & 0. & 0. & 0. & 0. & 0. & 0. & 0. \\
0. & 0. & 0. & 0. & 0. & 0. & 0. & 0. & 0. & 0. & 0. & 0. & 0. & 0. & 0. & 0. & 0. & 0. & 0. & 0. \\
0. & 0. & 0. & 0. & 0. & 0. & 0. & 0. & 0. & 0. & 0. & 0. & 0. & 0. & 0. & 0. & 0. & 0. & 0. & 0.
\end{array}\right)
$$

Stiffness Matrix for Element 12

$$
\left(\begin{array}{cccccccccccccccccccc}
0. & 0. & 0. & 0. & 0. & 0. & 0. & 0. & 0. & 0. & 0. & 0. & 0. & 0. & 0. & 0. & 0. & 0. & 0. & 0. \\
0. & 0. & 0. & 0. & 0. & 0. & 0. & 0. & 0. & 0. & 0. & 0. & 0. & 0. & 0. & 0. & 0. & 0. & 0. & 0. \\
0. & 0. & 0. & 0. & 0. & 0. & 0. & 0. & 0. & 0. & 0. & 0. & 0. & 0. & 0. & 0. & 0. & 0. & 0. & 0. \\
0. & 0. & 0. & 0. & 0. & 0. & 0. & 0. & 0. & 0. & 0. & 0. & 0. & 0. & 0. & 0. & 0. & 0. & 0. & 0. \\
0. & 0. & 0. & 0. & 5{,}396{,}927.49613404 & 7{,}709{,}896.423048628 & 0. & 0. & 0. & 0. & 0. & 0. & -5{,}396{,}927.49613404 & -7{,}709{,}896.423048628 & 0. & 0. & 0. & 0. & 0. & 0. \\
0. & 0. & 0. & 0. & 7{,}709{,}896.423048628 & 1.101413774721232 \times 10^{7} & 0. & 0. & 0. & 0. & 0. & 0. & -7{,}709{,}896.423048628 & -1.101413774721232 \times 10^{7} & 0. & 0. & 0. & 0. & 0. & 0. \\
0. & 0. & 0. & 0. & 0. & 0. & 0. & 0. & 0. & 0. & 0. & 0. & 0. & 0. & 0. & 0. & 0. & 0. & 0. & 0. \\
0. & 0. & 0. & 0. & 0. & 0. & 0. & 0. & 0. & 0. & 0. & 0. & 0. & 0. & 0. & 0. & 0. & 0. & 0. & 0. \\
0. & 0. & 0. & 0. & 0. & 0. & 0. & 0. & 0. & 0. & 0. & 0. & 0. & 0. & 0. & 0. & 0. & 0. & 0. & 0. \\
0. & 0. & 0. & 0. & 0. & 0. & 0. & 0. & 0. & 0. & 0. & 0. & 0. & 0. & 0. & 0. & 0. & 0. & 0. & 0. \\
0. & 0. & 0. & 0. & 0. & 0. & 0. & 0. & 0. & 0. & 0. & 0. & 0. & 0. & 0. & 0. & 0. & 0. & 0. & 0. \\
0. & 0. & 0. & 0. & 0. & 0. & 0. & 0. & 0. & 0. & 0. & 0. & 0. & 0. & 0. & 0. & 0. & 0. & 0. & 0. \\
0. & 0. & 0. & 0. & -5{,}396{,}927.49613404 & -7{,}709{,}896.423048628 & 0. & 0. & 0. & 0. & 0. & 0. & 5{,}396{,}927.49613404 & 7{,}709{,}896.423048628 & 0. & 0. & 0. & 0. & 0. & 0. \\
0. & 0. & 0. & 0. & -7{,}709{,}896.423048628 & -1.101413774721232 \times 10^{7} & 0. & 0. & 0. & 0. & 0. & 0. & 7{,}709{,}896.423048628 & 1.101413774721232 \times 10^{7} & 0. & 0. & 0. & 0. & 0. & 0. \\
0. & 0. & 0. & 0. & 0. & 0. & 0. & 0. & 0. & 0. & 0. & 0. & 0. & 0. & 0. & 0. & 0. & 0. & 0. & 0. \\
0. & 0. & 0. & 0. & 0. & 0. & 0. & 0. & 0. & 0. & 0. & 0. & 0. & 0. & 0. & 0. & 0. & 0. & 0. & 0. \\
0. & 0. & 0. & 0. & 0. & 0. & 0. & 0. & 0. & 0. & 0. & 0. & 0. & 0. & 0. & 0. & 0. & 0. & 0. & 0. \\
0. & 0. & 0. & 0. & 0. & 0. & 0. & 0. & 0. & 0. & 0. & 0. & 0. & 0. & 0. & 0. & 0. & 0. & 0. & 0. \\
0. & 0. & 0. & 0. & 0. & 0. & 0. & 0. & 0. & 0. & 0. & 0. & 0. & 0. & 0. & 0. & 0. & 0. & 0. & 0. \\
0. & 0. & 0. & 0. & 0. & 0. & 0. & 0. & 0. & 0. & 0. & 0. & 0. & 0. & 0. & 0. & 0. & 0. & 0. & 0.
\end{array}\right)
$$

Continued

EXAMPLE 5.3—CONT'D

Stiffness Matrix for Element 13

$$
\begin{pmatrix}
0. & 0. & 0. & 0. & 0. & 0. & 0. & 0. & 0. & 0. & 0. & 0. & 0. & 0. & 0. & 0. & 0. & 0. & 0. & 0. \\
0. & 0. & 0. & 0. & 0. & 0. & 0. & 0. & 0. & 0. & 0. & 0. & 0. & 0. & 0. & 0. & 0. & 0. & 0. & 0. \\
0. & 0. & 0. & 0. & 0. & 0. & 0. & 0. & 0. & 0. & 0. & 0. & 0. & 0. & 0. & 0. & 0. & 0. & 0. & 0. \\
0. & 0. & 0. & 0. & 0. & 0. & 0. & 0. & 0. & 0. & 0. & 0. & 0. & 0. & 0. & 0. & 0. & 0. & 0. & 0. \\
0. & 0. & 0. & 0. & 2.5\times10^{7} & 0. & 0. & 0. & 0. & 0. & 0. & 0. & 0. & 0. & -2.5\times10^{7} & 0. & 0. & 0. & 0. & 0. \\
0. & 0. & 0. & 0. & 0. & 0. & 0. & 0. & 0. & 0. & 0. & 0. & 0. & 0. & 0. & 0. & 0. & 0. & 0. & 0. \\
0. & 0. & 0. & 0. & 0. & 0. & 0. & 0. & 0. & 0. & 0. & 0. & 0. & 0. & 0. & 0. & 0. & 0. & 0. & 0. \\
0. & 0. & 0. & 0. & 0. & 0. & 0. & 0. & 0. & 0. & 0. & 0. & 0. & 0. & 0. & 0. & 0. & 0. & 0. & 0. \\
0. & 0. & 0. & 0. & 0. & 0. & 0. & 0. & 0. & 0. & 0. & 0. & 0. & 0. & 0. & 0. & 0. & 0. & 0. & 0. \\
0. & 0. & 0. & 0. & 0. & 0. & 0. & 0. & 0. & 0. & 0. & 0. & 0. & 0. & 0. & 0. & 0. & 0. & 0. & 0. \\
0. & 0. & 0. & 0. & 0. & 0. & 0. & 0. & 0. & 0. & 0. & 0. & 0. & 0. & 0. & 0. & 0. & 0. & 0. & 0. \\
0. & 0. & 0. & 0. & 0. & 0. & 0. & 0. & 0. & 0. & 0. & 0. & 0. & 0. & 0. & 0. & 0. & 0. & 0. & 0. \\
0. & 0. & 0. & 0. & 0. & 0. & 0. & 0. & 0. & 0. & 0. & 0. & 0. & 0. & 0. & 0. & 0. & 0. & 0. & 0. \\
0. & 0. & 0. & 0. & 0. & 0. & 0. & 0. & 0. & 0. & 0. & 0. & 0. & 0. & 0. & 0. & 0. & 0. & 0. & 0. \\
0. & 0. & 0. & 0. & -2.5\times10^{7} & 0. & 0. & 0. & 0. & 0. & 0. & 0. & 0. & 0. & 2.5\times10^{7} & 0. & 0. & 0. & 0. & 0. \\
0. & 0. & 0. & 0. & 0. & 0. & 0. & 0. & 0. & 0. & 0. & 0. & 0. & 0. & 0. & 0. & 0. & 0. & 0. & 0. \\
0. & 0. & 0. & 0. & 0. & 0. & 0. & 0. & 0. & 0. & 0. & 0. & 0. & 0. & 0. & 0. & 0. & 0. & 0. & 0. \\
0. & 0. & 0. & 0. & 0. & 0. & 0. & 0. & 0. & 0. & 0. & 0. & 0. & 0. & 0. & 0. & 0. & 0. & 0. & 0. \\
0. & 0. & 0. & 0. & 0. & 0. & 0. & 0. & 0. & 0. & 0. & 0. & 0. & 0. & 0. & 0. & 0. & 0. & 0. & 0. \\
0. & 0. & 0. & 0. & 0. & 0. & 0. & 0. & 0. & 0. & 0. & 0. & 0. & 0. & 0. & 0. & 0. & 0. & 0. & 0.
\end{pmatrix}
$$

Stiffness Matrix for Element 14

$$
\begin{pmatrix}
0. & 0. & 0. & 0. & 0. & 0. & 0. & 0. & 0. & 0. & 0. & 0. & 0. & 0. & 0. & 0. & 0. & 0. & 0. & 0. \\
0. & 0. & 0. & 0. & 0. & 0. & 0. & 0. & 0. & 0. & 0. & 0. & 0. & 0. & 0. & 0. & 0. & 0. & 0. & 0. \\
0. & 0. & 0. & 0. & 0. & 0. & 0. & 0. & 0. & 0. & 0. & 0. & 0. & 0. & 0. & 0. & 0. & 0. & 0. & 0. \\
0. & 0. & 0. & 0. & 0. & 0. & 0. & 0. & 0. & 0. & 0. & 0. & 0. & 0. & 0. & 0. & 0. & 0. & 0. & 0. \\
0. & 0. & 0. & 0. & 1.486988847583643\times10^{7} & 0. & 0. & 0. & 0. & 0. & 0. & 0. & 0. & 0. & 0. & 0. & -1.486988847583643\times10^{7} & 0. & 0. & 0. \\
0. & 0. & 0. & 0. & 0. & 0. & 0. & 0. & 0. & 0. & 0. & 0. & 0. & 0. & 0. & 0. & 0. & 0. & 0. & 0. \\
0. & 0. & 0. & 0. & 0. & 0. & 0. & 0. & 0. & 0. & 0. & 0. & 0. & 0. & 0. & 0. & 0. & 0. & 0. & 0. \\
0. & 0. & 0. & 0. & 0. & 0. & 0. & 0. & 0. & 0. & 0. & 0. & 0. & 0. & 0. & 0. & 0. & 0. & 0. & 0. \\
0. & 0. & 0. & 0. & 0. & 0. & 0. & 0. & 0. & 0. & 0. & 0. & 0. & 0. & 0. & 0. & 0. & 0. & 0. & 0. \\
0. & 0. & 0. & 0. & 0. & 0. & 0. & 0. & 0. & 0. & 0. & 0. & 0. & 0. & 0. & 0. & 0. & 0. & 0. & 0. \\
0. & 0. & 0. & 0. & 0. & 0. & 0. & 0. & 0. & 0. & 0. & 0. & 0. & 0. & 0. & 0. & 0. & 0. & 0. & 0. \\
0. & 0. & 0. & 0. & 0. & 0. & 0. & 0. & 0. & 0. & 0. & 0. & 0. & 0. & 0. & 0. & 0. & 0. & 0. & 0. \\
0. & 0. & 0. & 0. & 0. & 0. & 0. & 0. & 0. & 0. & 0. & 0. & 0. & 0. & 0. & 0. & 0. & 0. & 0. & 0. \\
0. & 0. & 0. & 0. & 0. & 0. & 0. & 0. & 0. & 0. & 0. & 0. & 0. & 0. & 0. & 0. & 0. & 0. & 0. & 0. \\
0. & 0. & 0. & 0. & 0. & 0. & 0. & 0. & 0. & 0. & 0. & 0. & 0. & 0. & 0. & 0. & 0. & 0. & 0. & 0. \\
0. & 0. & 0. & 0. & 0. & 0. & 0. & 0. & 0. & 0. & 0. & 0. & 0. & 0. & 0. & 0. & 0. & 0. & 0. & 0. \\
0. & 0. & 0. & 0. & -1.486988847583643\times10^{7} & 0. & 0. & 0. & 0. & 0. & 0. & 0. & 0. & 0. & 0. & 0. & 1.486988847583643\times10^{7} & 0. & 0. & 0. \\
0. & 0. & 0. & 0. & 0. & 0. & 0. & 0. & 0. & 0. & 0. & 0. & 0. & 0. & 0. & 0. & 0. & 0. & 0. & 0. \\
0. & 0. & 0. & 0. & 0. & 0. & 0. & 0. & 0. & 0. & 0. & 0. & 0. & 0. & 0. & 0. & 0. & 0. & 0. & 0. \\
0. & 0. & 0. & 0. & 0. & 0. & 0. & 0. & 0. & 0. & 0. & 0. & 0. & 0. & 0. & 0. & 0. & 0. & 0. & 0.
\end{pmatrix}
$$

Stiffness Matrix for Element 15

$$
\begin{pmatrix}
0. & 0. & 0. & 0. & 0. & 0. & 0. & 0. & 0. & 0. & 0. & 0. & 0. & 0. & 0. & 0. & 0. & 0. & 0. & 0. \\
0. & 0. & 0. & 0. & 0. & 0. & 0. & 0. & 0. & 0. & 0. & 0. & 0. & 0. & 0. & 0. & 0. & 0. & 0. & 0. \\
0. & 0. & 2.\times 10^7 & 0. & 0. & 0. & 0. & 0. & 0. & 0. & 0. & 0. & 0. & 0. & 0. & 0. & -2.\times 10^7 & 0. & 0. & 0. \\
0. & 0. & 0. & 0. & 0. & 0. & 0. & 0. & 0. & 0. & 0. & 0. & 0. & 0. & 0. & 0. & 0. & 0. & 0. & 0. \\
0. & 0. & 0. & 0. & 0. & 0. & 0. & 0. & 0. & 0. & 0. & 0. & 0. & 0. & 0. & 0. & 0. & 0. & 0. & 0. \\
0. & 0. & 0. & 0. & 0. & 0. & 0. & 0. & 0. & 0. & 0. & 0. & 0. & 0. & 0. & 0. & 0. & 0. & 0. & 0. \\
0. & 0. & 0. & 0. & 0. & 0. & 0. & 0. & 0. & 0. & 0. & 0. & 0. & 0. & 0. & 0. & 0. & 0. & 0. & 0. \\
0. & 0. & 0. & 0. & 0. & 0. & 0. & 0. & 0. & 0. & 0. & 0. & 0. & 0. & 0. & 0. & 0. & 0. & 0. & 0. \\
0. & 0. & 0. & 0. & 0. & 0. & 0. & 0. & 0. & 0. & 0. & 0. & 0. & 0. & 0. & 0. & 0. & 0. & 0. & 0. \\
0. & 0. & 0. & 0. & 0. & 0. & 0. & 0. & 0. & 0. & 0. & 0. & 0. & 0. & 0. & 0. & 0. & 0. & 0. & 0. \\
0. & 0. & 0. & 0. & 0. & 0. & 0. & 0. & 0. & 0. & 0. & 0. & 0. & 0. & 0. & 0. & 0. & 0. & 0. & 0. \\
0. & 0. & 0. & 0. & 0. & 0. & 0. & 0. & 0. & 0. & 0. & 0. & 0. & 0. & 0. & 0. & 0. & 0. & 0. & 0. \\
0. & 0. & 0. & 0. & 0. & 0. & 0. & 0. & 0. & 0. & 0. & 0. & 0. & 0. & 0. & 0. & 0. & 0. & 0. & 0. \\
0. & 0. & 0. & 0. & 0. & 0. & 0. & 0. & 0. & 0. & 0. & 0. & 0. & 0. & 0. & 0. & 0. & 0. & 0. & 0. \\
0. & 0. & 0. & 0. & 0. & 0. & 0. & 0. & 0. & 0. & 0. & 0. & 0. & 0. & 0. & 0. & 0. & 0. & 0. & 0. \\
0. & 0. & 0. & 0. & 0. & 0. & 0. & 0. & 0. & 0. & 0. & 0. & 0. & 0. & 0. & 0. & 0. & 0. & 0. & 0. \\
0. & 0. & -2.\times 10^7 & 0. & 0. & 0. & 0. & 0. & 0. & 0. & 0. & 0. & 0. & 0. & 0. & 0. & 2.\times 10^7 & 0. & 0. & 0. \\
0. & 0. & 0. & 0. & 0. & 0. & 0. & 0. & 0. & 0. & 0. & 0. & 0. & 0. & 0. & 0. & 0. & 0. & 0. & 0. \\
0. & 0. & 0. & 0. & 0. & 0. & 0. & 0. & 0. & 0. & 0. & 0. & 0. & 0. & 0. & 0. & 0. & 0. & 0. & 0. \\
0. & 0. & 0. & 0. & 0. & 0. & 0. & 0. & 0. & 0. & 0. & 0. & 0. & 0. & 0. & 0. & 0. & 0. & 0. & 0.
\end{pmatrix}
$$

Continued

EXAMPLE 5.3—CONT'D

Stiffness Matrix for Element 16

7,404,805.630241144	6,731,641.482037405	0.	0.	0.	0.	0.	0.	0.	0.	0.	0.	0.	0.	0.	0.	−7,404,805.630241144	−6,731,641.482037405	0.	0.
6,731,641.482037405	6,119,674.074579459	0.	0.	0.	0.	0.	0.	0.	0.	0.	0.	0.	0.	0.	0.	−6,731,641.482037405	−6,119,674.074579459	0.	0.
0.	0.	0.	0.	0.	0.	0.	0.	0.	0.	0.	0.	0.	0.	0.	0.	0.	0.	0.	0.
0.	0.	0.	0.	0.	0.	0.	0.	0.	0.	0.	0.	0.	0.	0.	0.	0.	0.	0.	0.
0.	0.	0.	0.	0.	0.	0.	0.	0.	0.	0.	0.	0.	0.	0.	0.	0.	0.	0.	0.
0.	0.	0.	0.	0.	0.	0.	0.	0.	0.	0.	0.	0.	0.	0.	0.	0.	0.	0.	0.
0.	0.	0.	0.	0.	0.	0.	0.	0.	0.	0.	0.	0.	0.	0.	0.	0.	0.	0.	0.
0.	0.	0.	0.	0.	0.	0.	0.	0.	0.	0.	0.	0.	0.	0.	0.	0.	0.	0.	0.
0.	0.	0.	0.	0.	0.	0.	0.	0.	0.	0.	0.	0.	0.	0.	0.	0.	0.	0.	0.
0.	0.	0.	0.	0.	0.	0.	0.	0.	0.	0.	0.	0.	0.	0.	0.	0.	0.	0.	0.
0.	0.	0.	0.	0.	0.	0.	0.	0.	0.	0.	0.	0.	0.	0.	0.	0.	0.	0.	0.
0.	0.	0.	0.	0.	0.	0.	0.	0.	0.	0.	0.	0.	0.	0.	0.	0.	0.	0.	0.
0.	0.	0.	0.	0.	0.	0.	0.	0.	0.	0.	0.	0.	0.	0.	0.	0.	0.	0.	0.
0.	0.	0.	0.	0.	0.	0.	0.	0.	0.	0.	0.	0.	0.	0.	0.	0.	0.	0.	0.
0.	0.	0.	0.	0.	0.	0.	0.	0.	0.	0.	0.	0.	0.	0.	0.	0.	0.	0.	0.
0.	0.	0.	0.	0.	0.	0.	0.	0.	0.	0.	0.	0.	0.	0.	0.	0.	0.	0.	0.
−7,404,805.630241144	−6,731,641.482037405	0.	0.	0.	0.	0.	0.	0.	0.	0.	0.	0.	0.	0.	0.	7,404,805.630241144	6,731,641.482037405	0.	0.
−6,731,641.482037405	−6,119,674.074579459	0.	0.	0.	0.	0.	0.	0.	0.	0.	0.	0.	0.	0.	0.	6,731,641.482037405	6,119,674.074579459	0.	0.
0.	0.	0.	0.	0.	0.	0.	0.	0.	0.	0.	0.	0.	0.	0.	0.	0.	0.	0.	0.
0.	0.	0.	0.	0.	0.	0.	0.	0.	0.	0.	0.	0.	0.	0.	0.	0.	0.	0.	0.

EXAMPLE 5.4: CALFEM/MATLAB IMPLEMENTATION

Calculate the displacement field, the reactions of the supports and the internal forces of the following plane truss (Table 5.4). The values of the forces are $H=10$ kN and $V=30$ kN. All members are made from steel bars with cross-sectional area $A=10^{-3}$ m^2 and modulus of elasticity $E=200$ GPa.

Table 5.4 Coordinates of Nodes

Node	1	2	3	4	5	6	7	8	9	10
x (m)	0	1	2	3	4	8	9	10	11	12
y (m)	0	10	20	30	40	40	30	20	10	0

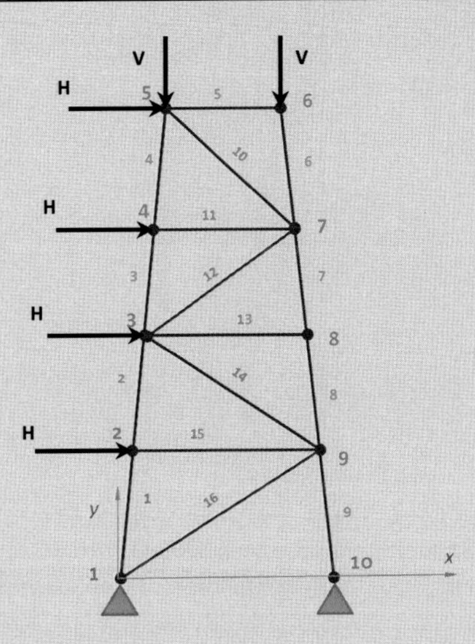

Step 1: Degrees of Freedom

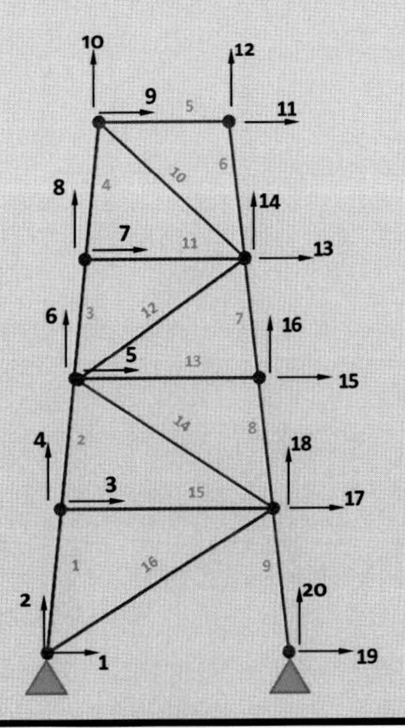

Continued

EXAMPLE 5.4: CALFEM/MATLAB IMPLEMENTATION—CONT'D

We can derive everything we need from the nodal displacements and reactions of the supports. The nodal displacements are included in a matrix a and the reactions of the supports in a matrix of nodal forces r, where

$$a = \{u_1 \; u_2 \; u_3 \; u_4 \; u_5 \; u_6 \; \cdots \; u_{19} \; u_{20}\}^T$$
$$r = \{F_1 \; F_2 \; F_3 \; F_4 \; F_5 \; F_6 \; \cdots \; F_{19} \; F_{20}\}^T$$

In the present example, the reactions of the supports are the components F_1, F_2, F_{19}, and F_{20} of the above matrix r.

The required matrices a and r can be derived by the solution of the following system:

```
[a,r] = solveq(K,f,bc)
```

Therefore, in order to obtain the solution, we have to derive the following matrices:

```
K: stiffness matrix of the structure
f: load vector
bc: boundary conditions
```

Let us start from the simpler matrices **f** and **bc**.

Step 2: Load Vector {f}

First, we should define the cells of the vector f, initially filled with zeros. Since the structure has 20 degrees of freedom, the vector **f** should contain 20 cells, that is,

```
f=zeros(20,1);
```

Then, in the above vector, we must now enter the external loads of the structure. The forces are acting on the degrees of freedom **3**, **5**, **7**, **9**, **10**, **12**, that is:

```
f(3)=10,000;
f(5)=10,000;
f(7)=10,000;
f(9)=10,000;
f(10)=−30,000;
f(12)=−30,000;
```

Step 3: Boundary Conditions {bc}

From the physical point of view, the displacements in the degrees of freedom **1**, **2**, **19**, **20** are zero. These boundary conditions can be incorporated in the matric **bc** as follows:

```
bc=[1 0;2 0;19 0;20 0];
```

In the above command, every row has the following meaning:

1 0; *in the degree of freedom 1 the displacement is 0*
2 0; *in the degree of freedom 2 the displacement is 0*
19 0; *in the degree of freedom 19 the displacement is 0*
20 0; *in the degree of freedom 20 the displacement is 0*

Step 4: Derivation of Global Stiffness Matrix [K]

Nodal Coordinates

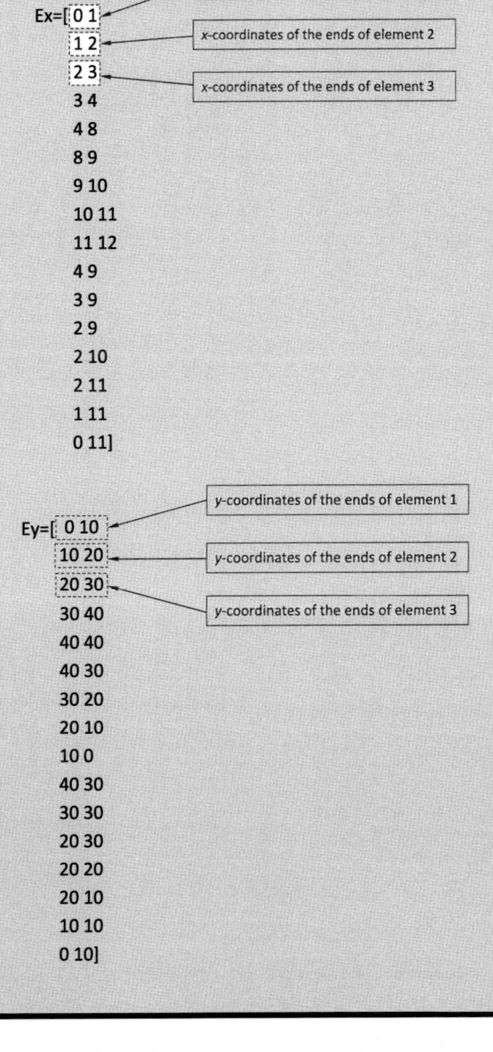

Continued

EXAMPLE 5.4: CALFEM/MATLAB IMPLEMENTATION—CONT'D

Material Properties

Since the material for all elements is same, we can derive only one vector **ep**, that is,

```
ep = [2e11 1e-3]
```

Topology Matrix

Until now, the algorithm "*knows*" the coordinates of each element (through the matrices **Ex** and **Ey**) but it does not "*know*" how the elements are interconnected. The following matrix incorporates the assembly information of the structure:

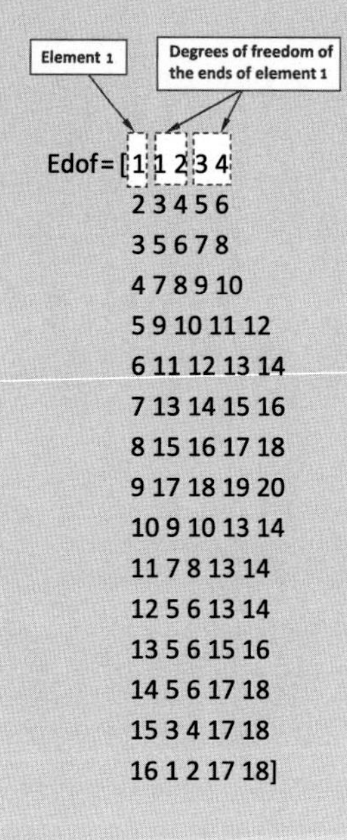

Element 1
Degrees of freedom of the ends of element 1
Edof= [1 1 2 3 4
2 3 4 5 6
3 5 6 7 8
4 7 8 9 10
5 9 10 11 12
6 11 12 13 14
7 13 14 15 16
8 15 16 17 18
9 17 18 19 20
10 9 10 13 14
11 7 8 13 14
12 5 6 13 14
13 5 6 15 16
14 5 6 17 18
15 3 4 17 18
16 1 2 17 18]

Assembly of the Individual Element Matrices

Let us derive the initial form of the **K** matrix containing zeros, that is,

```
K =zeros(20)
```

Now we can assembly the element stiffness matrices and derive the global stiffness matrix **K** by the following do-loop:

```
for i =1:16
Ke = bar2e(Ex(i,:),Ey(i,:),ep);
K = assem(Edof(i,:),K,Ke);
end;
```

Computation of the Displacement Field [a] and the Reactions [r]

```
[a,r] = solveq(K,f,bc)
```

Results

```
a =

     0
     0
0.0085
0.0007
0.0179
0.0013
0.0357
-0.0012
0.0493
-0.0032
0.0493
-0.0092
0.0354
-0.0091
0.0179
-0.0074
0.0080
-0.0050
     0
     0
```

```
r =

1.0e+05 *

-0.2867
-0.5333
     0
-0.0000
0.0000
0.0000
-0.0000
     0
0.0000
0.0000
-0.0000
-0.0000
-0.0000
0.0000
-0.0000
0.0000
0.0000
-0.0000
-0.1133
1.1333
```

Computation of the Element Axial Forces N_i

We can avoid writing individual commands for computation of the axial forces $\mathbf{N_1}$, $\mathbf{N_2}$, …, $\mathbf{N_{16}}$. These outputs can be generated by the following loop:

```
ed = extract(Edof,a);
for i = 1:16
N(i,:) = bar2s(Ex(i,:),Ey(i,:),ep,ed(i,:));
end
N
```

Continued

EXAMPLE 5.4: CALFEM/MATLAB IMPLEMENTATION—CONT'D

Results

```
N =

  1.0e+05 *

   0.3015
   0.3015
  -0.1340
  -0.1340
  -0.0300
  -0.3015
  -0.6784
  -0.6784
  -1.1390
  -0.1863
  -0.1000
   0.2543
  -0.0000
  -0.3027
  -0.1000
   0.3469
```

Graphical Demonstration of the Undeformed Structure

```
eldraw2(Ex,Ey,[1 2 1],Edof(:,1));
```

Graphical Demonstration of the Deformed Structure

```
[sfac] = scalfact2(Ex,Ey,ed,0.1);
   eldisp2(Ex,Ey,ed,[2 1 1],sfac);
```

Results

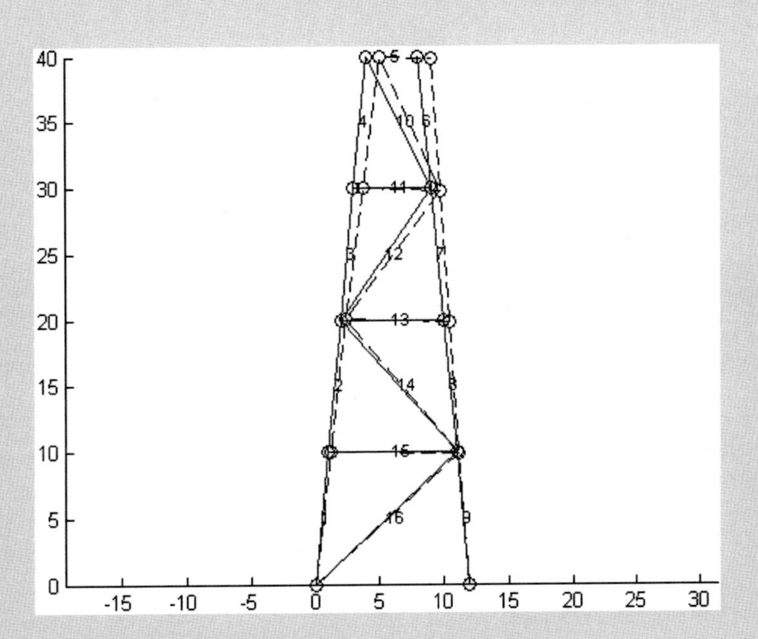

The CALFEM/MATLAB Computer Code

```
f = zeros(20,1);
  f(3) = 10000;
  f(5) = 10000;
  f(7) = 10000;
  f(9) = 10000;
  f(10) = -30000;
  f(12) = -30000;
  bc = [1 0;2 0;19 0;20 0];

  Ex = [0 1;1 2;2 3;3 4;4 8;8 9;9 10;10 11;11 12;4 9;3 9;2 9;2 10;2 11;1 11;0 11];
  Ey = [0 10;10 20;20 30;30 40;40 40;40 30;30 20;20 10;10 0;40 30;30 30;20 30;20 20;20 10;10
10;0 10];

  ep = [2e11 1e-3];

  Edof = [1 1 2 3 4;2 3 4 5 6;3 5 6 7 8;4 7 8 9 10;5 9 10 11 12;6 11 12 13 14;7 13 14 15 16;8 15 16 17
18;9 17 18 19 20;10 9 10 13 14;11 7 8 13 14;12 5 6 13 14;13 5 6 15 16;14 5 6 17 18;15 3 4 17 18;16 1 2
17 18];

  K = zeros(20);
  for i = 1:16
  Ke = bar2e(Ex(i,:),Ey(i,:),ep);
```

Continued

EXAMPLE 5.4: CALFEM/MATLAB IMPLEMENTATION—CONT'D

```
K = assem(Edof(i,:),K,Ke);
end;

[a,r] = solveq(K,f,bc)

ed = extract(Edof,a);
for i = 1:16
N(i,:) = bar2s(Ex(i,:),Ey(i,:),ep,ed(i,:));
end
N

eldraw2(Ex,Ey,[1 2 1],Edof(:,1));
[sfac] = scalfact2(Ex,Ey,ed,0.1);
eldisp2(Ex,Ey,ed,[2 1 1],sfac);
```

EXAMPLE 5.5: ANSYS IMPLEMENTATION

Derive the deformed mode of the following plane truss (Table 5.5). The values of the forces are $H = 10$ kN and $V = 30$ kN. All members are made from steel bars with cross-sectional area $A = 10^{-3}$ m^2 and modulus of elasticity $E = 200$ GPa.

Table 5.5 Coordinates of Nodes

Node	1	2	3	4	5	6	7	8	9	10
x (m)	0	1	2	3	4	8	9	10	11	12
y (m)	0	10	20	30	40	40	30	20	10	0

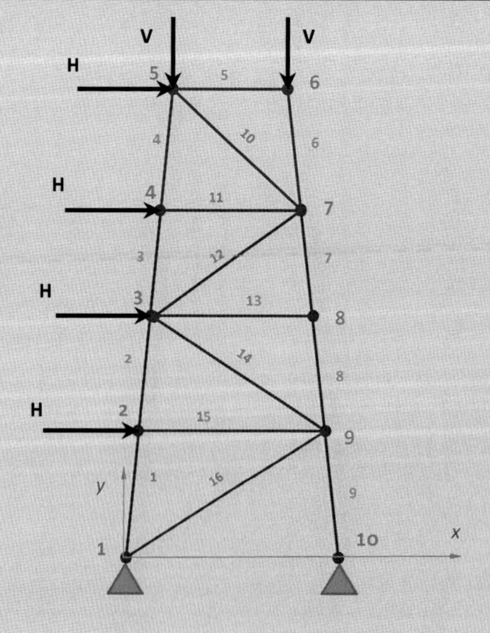

<u>UM</u>: **File → Change Directory**

Using this command we select the directory to save all files generated for this exercise. Let us choose Directory **F**, and file **ANSYS TUTORIALS**.

<u>INPUT</u>: **/UNITS, SI**

This command defines the units to SI.

<u>UM</u>: **File → Change Job Name**

With this command, we specify the job name. Let us choose **Tutorial 5**.

<u>UM</u>: **File → Change Title**

We can choose again **Tutorial 5**.

<u>MM</u>: Preferences

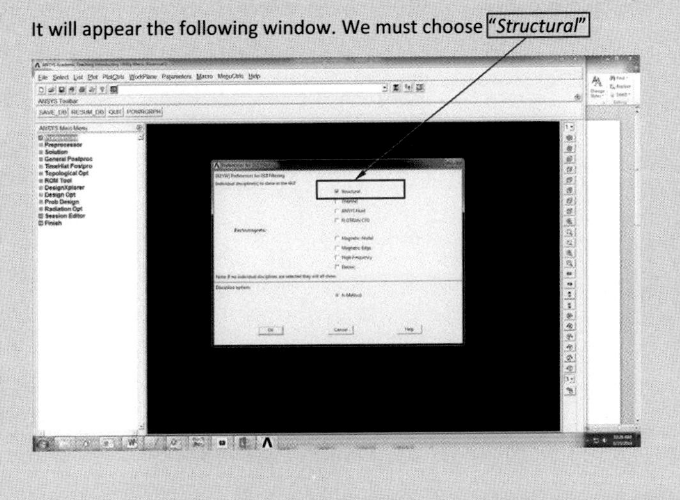

It will appear the following window. We must choose "Structural"

Continued

EXAMPLE 5.5: ANSYS IMPLEMENTATION—CONT'D

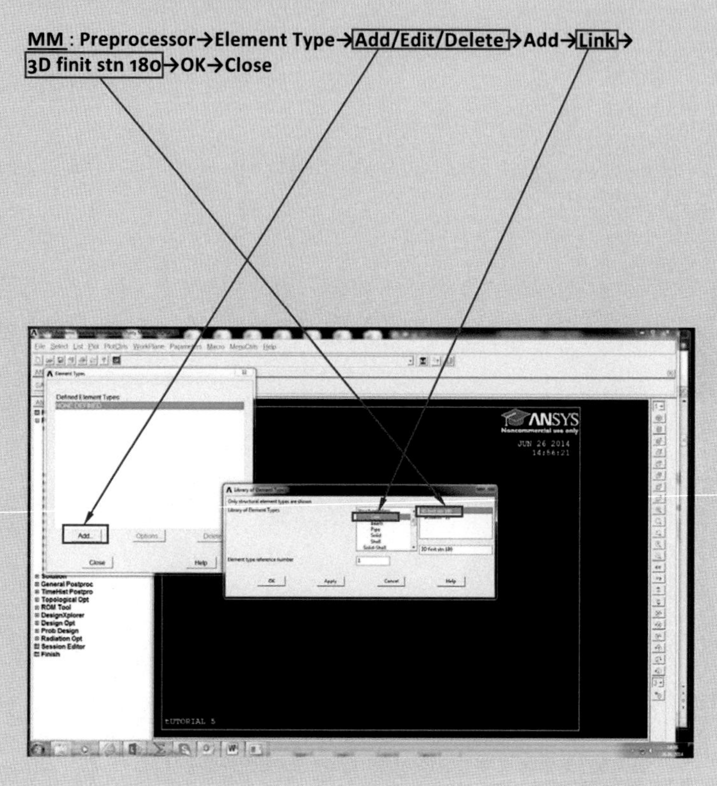

With the following commands, we will now specify the cross-sectional area.

MM: Preprocessor → Real Constants → Add/Edit/Delete

We click "**Add,**" and then "**OK**" to choose the element type "**Link180.**" We must now fill the value "**10e-3**" in the box of the "**cross section area,**" and press "**OK.**"

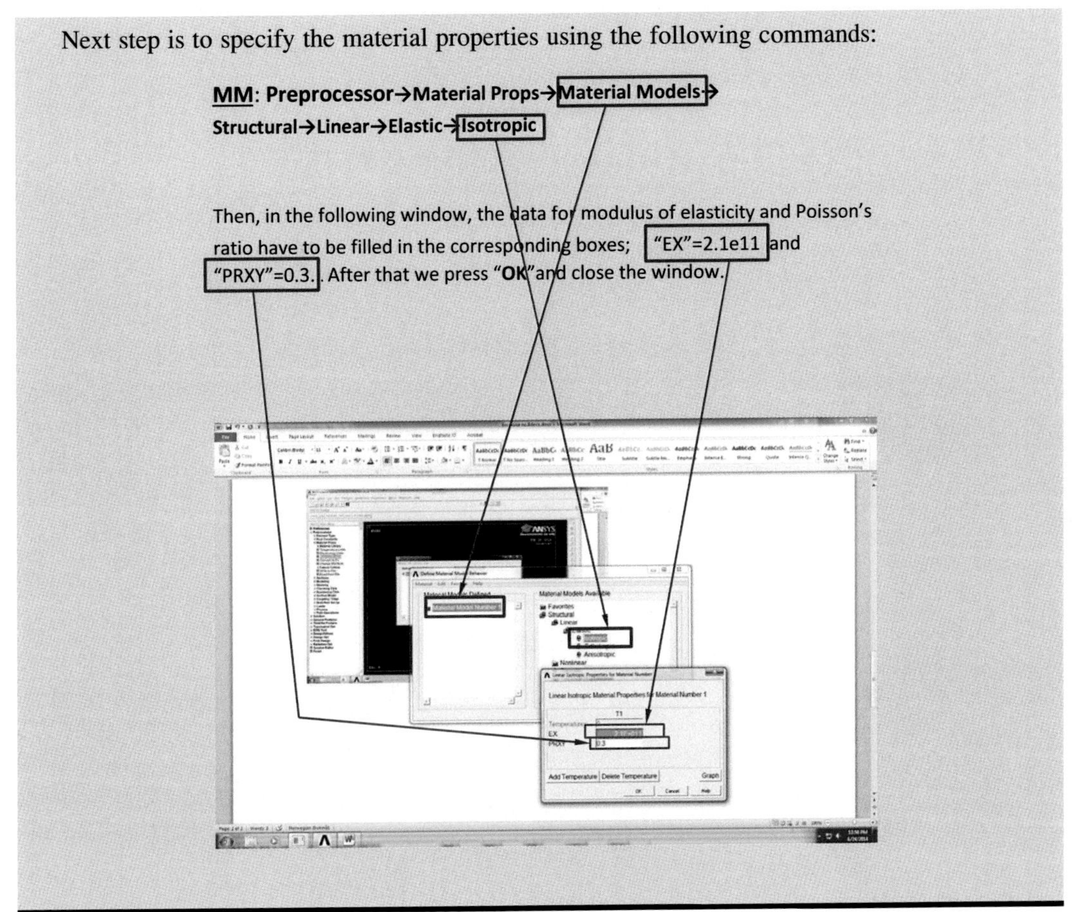
Next step is to specify the material properties using the following commands:
MM: Preprocessor→Material Props→Material Models→
Structural→Linear→Elastic→Isotropic
Then, in the following window, the data for modulus of elasticity and Poisson's ratio have to be filled in the corresponding boxes; "EX"=2.1e11 and "PRXY"=0.3. After that we press "OK" and close the window.
Material Models Available
Favorites
Structural
Linear
Anisotropic
Nonlinear
Linear Isotropic Material Properties for Material Number 1
T1
EX
PRXY
0.3
Add Temperature
Delete Temperature
Graph
OK
Cancel
Help

Continued

EXAMPLE 5.5: ANSYS IMPLEMENTATION—CONT'D

Now is the time to insert the coordinates of the Keypoints:

MM: Preprocessor → Modeling → Create → Keypoints → In Active CS

In the following window we have to fill the **Keypoint number** and the **x, y** coordinates of the first Keypoint, and then to press the button **Apply**. We must continue filling the successive windows for the rest Keypoints. After the completion of the coordinates of the last Keypoint, instead of **Apply** we must press the button **OK**.

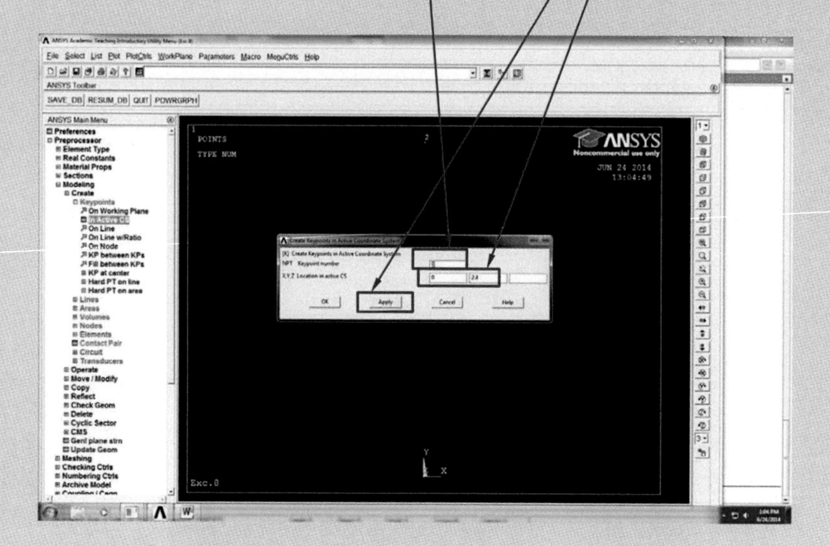

The above procedure yields a demonstration of locations of the Keypoints on the screen. Next step is to connect the Keypoints using the following commands in order to create the lines:

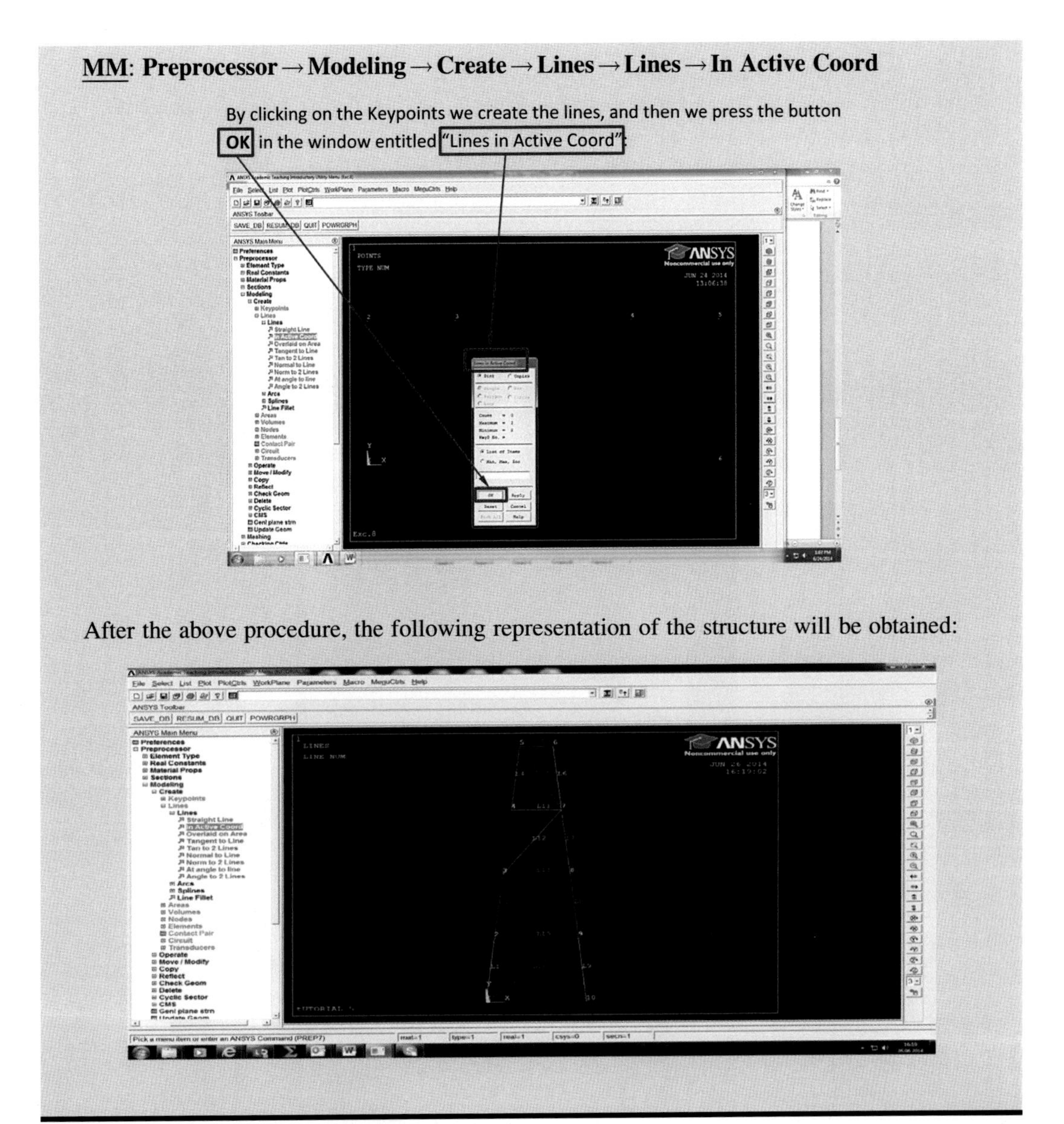
MM: Preprocessor → Modeling → Create → Lines → Lines → In Active Coord
By clicking on the Keypoints we create the lines, and then we press the button
OK in the window entitled "Lines in Active Coord":
After the above procedure, the following representation of the structure will be obtained:

Continued

EXAMPLE 5.5: ANSYS IMPLEMENTATION—CONT'D

Now is time to create the mesh of the solid. We can do it by the following commands:

MM: Preprocessor → Meshing → MeshTool

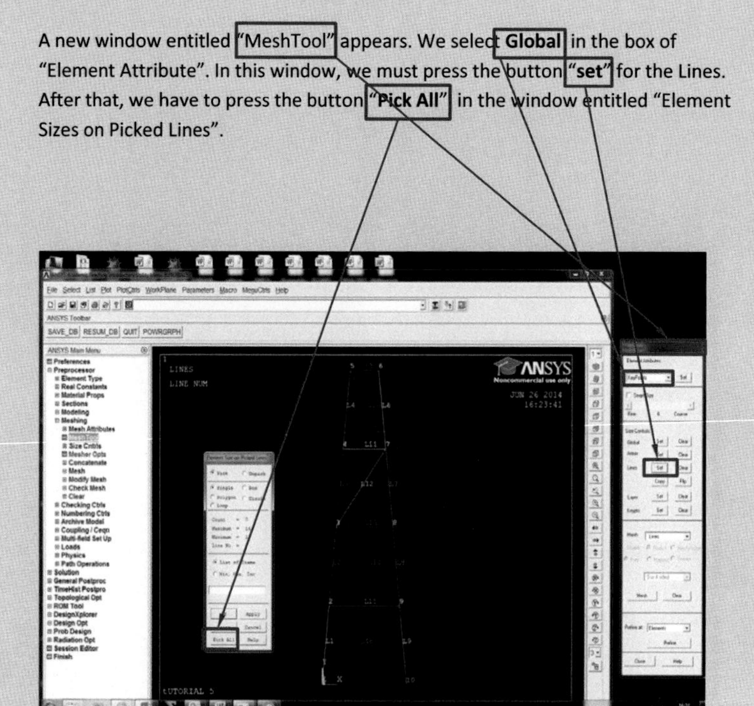

A new window entitled "MeshTool" appears. We select Global in the box of "Element Attribute". In this window, we must press the button "set" for the Lines. After that, we have to press the button "Pick All" in the window entitled "Element Sizes on Picked Lines".
ANSYS
LINES
LINE NUM
TUTORIAL 5

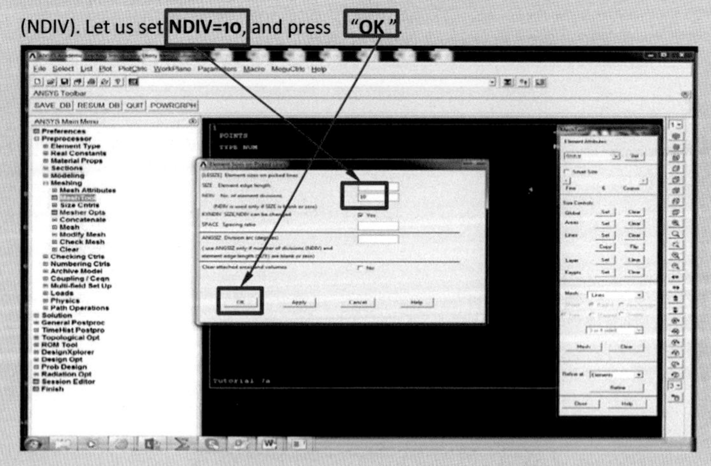

On the new window, we must fill the number of element divisions (NDIV). Let us set NDIV=10, and press "OK".

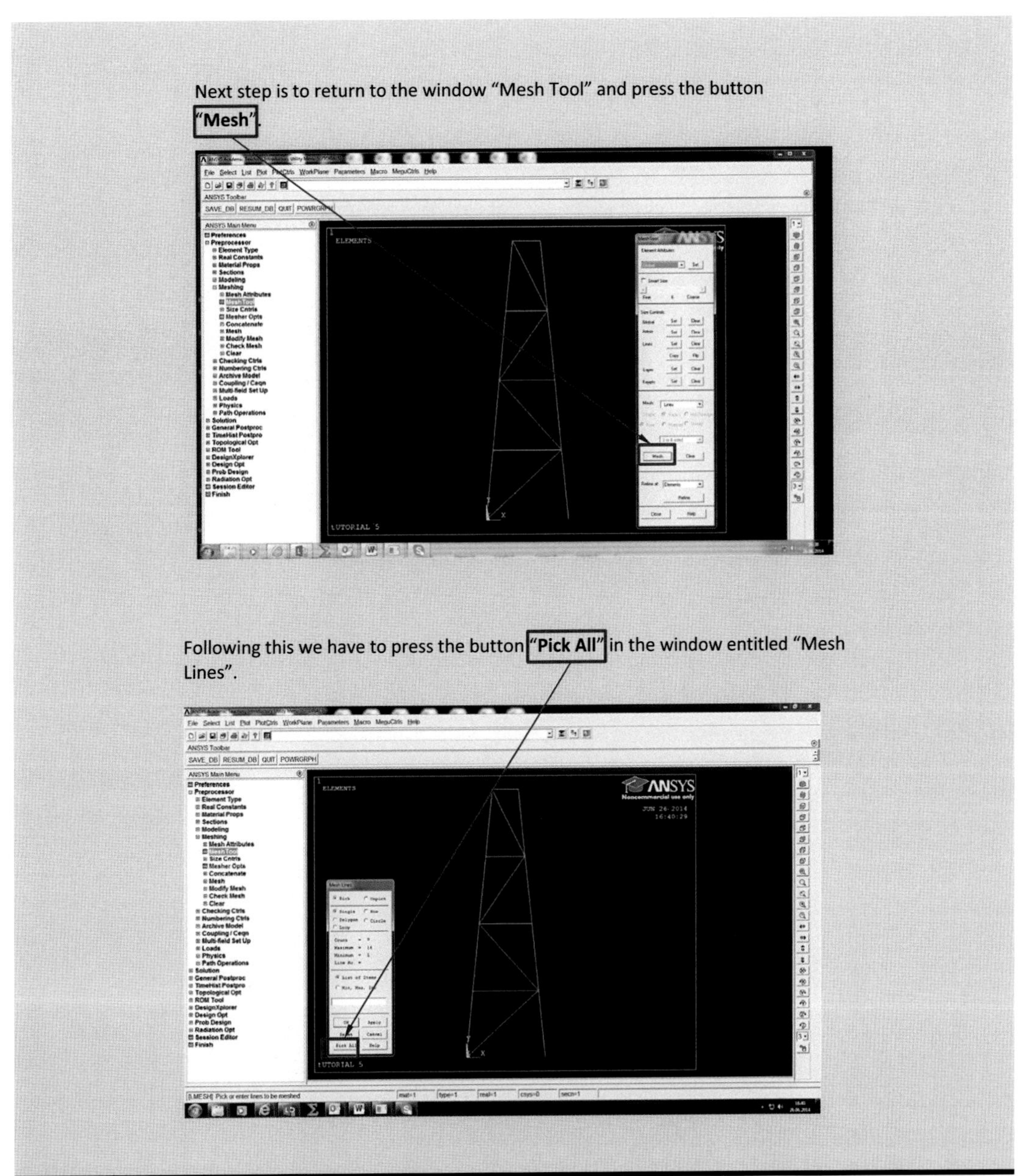
Next step is to return to the window "Mesh Tool" and press the button "Mesh".
Following this we have to press the button "Pick All" in the window entitled "Mesh Lines".
ELEMENTS
TUTORIAL 5
ANSYS

Continued

EXAMPLE 5.5: ANSYS IMPLEMENTATION—CONT'D

The following command should be typed in order to declare that the type of solution is "Static"

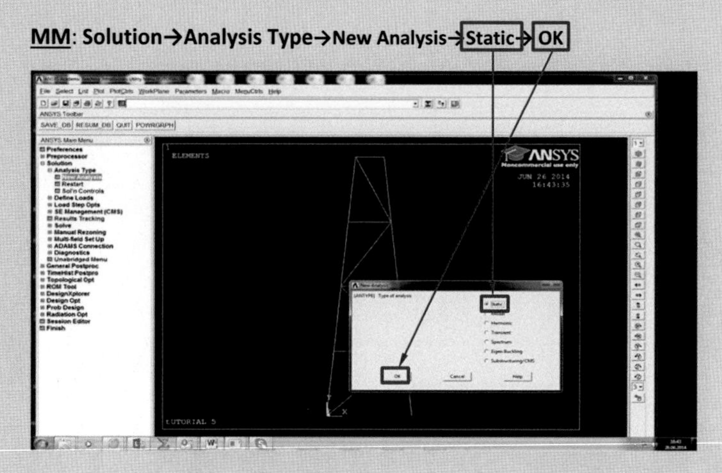

Now, using the following commands, is time to specify the boundary conditions on the supports of the structure.

MM: Solution → Define Loads → Apply → Structural → Displacements → On Nodes

We must pick the Keypoints **1** and **10** located on the pinned supports one by one, then click the button "**Apply**" and select the degrees of freedom **UX, UY** for setting the **"Displacement value"=0.** Then, click **"OK".**

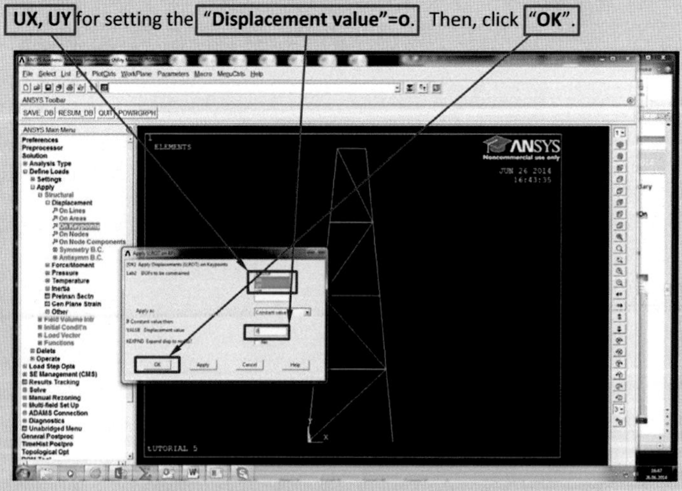

After the above procedure, we must specify the loads.

Therefore, we must type the commands:

MM: Solution → Define Loads → Apply → Structural → Force/Moment → On Keypoints

Then, using the cursor, we must pick the first Keypoint **2** subjected to concentrated load **H,** and then **"Apply"** in the new window entitled "Apply F/M on KPs"

Now the window entitled "Apply F/M on KPs", asks for the direction and the value of the concentrated load. We choose direction **FX** and value **10e3,** and then, click **"Apply"** so that the program displays the concentrated force.
We repeat the procedure for the remainderof the Keypoints, and after filling the value of the last one, instead of **"Apply"** we press **"OK"**.

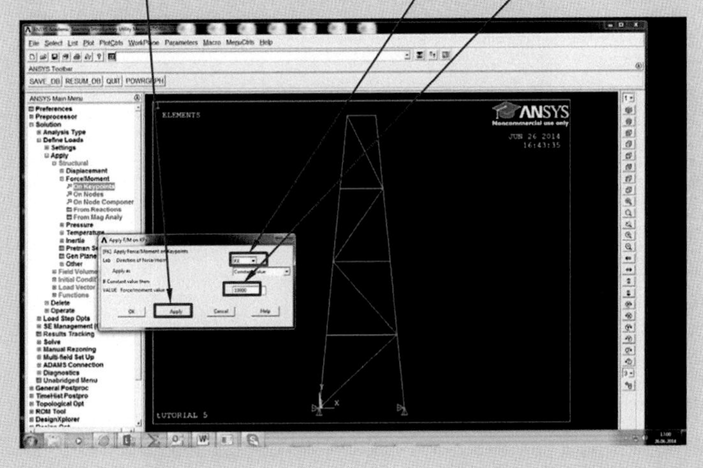

Now, using the following commands, is time to ask the program to solve the problem:
<u>MM</u>: Solution → Solve → Current LS → OK
The program indicates that "The solution is Done."
After the solution, we can now demonstrate the results. Using the following commands, the post processor has to read the results:
<u>MM</u>: General Postprocessor → Read results → First Set
Then, we have to continue with the following commands in order to print the deformed structure, and to compare it to the underformed one.

Continued

EXAMPLE 5.5: ANSYS IMPLEMENTATION—CONT'D

MM: General Postprocessor → Plot results → Deformed Shape

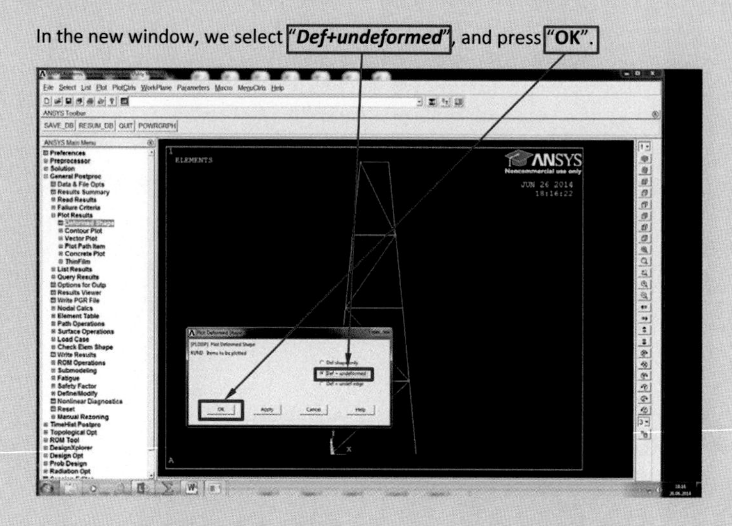

In the new window, we select "Def+undeformed", and press "OK".

Then the following diagram of the deformed structure will be obtained.

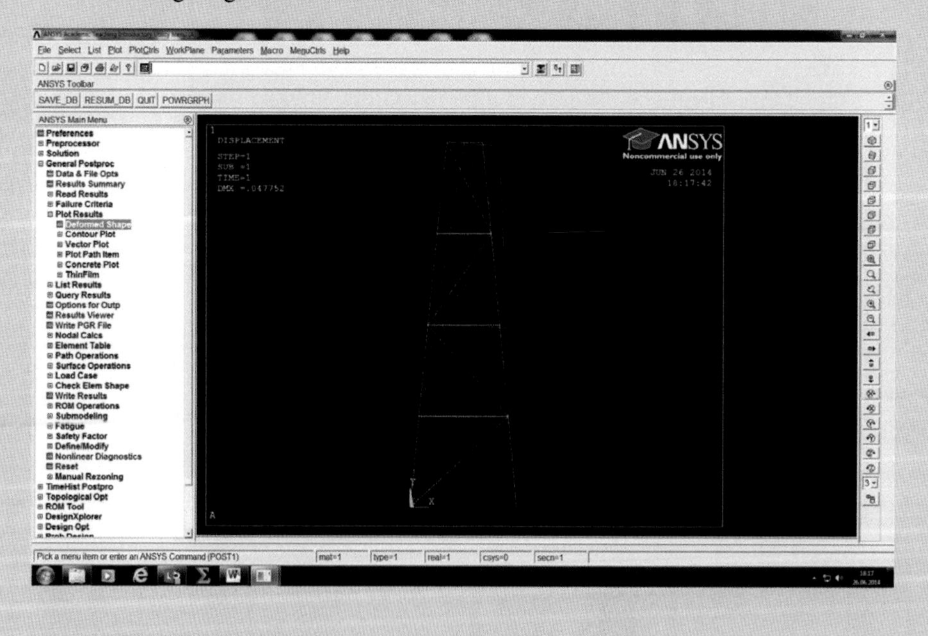

REFERENCES

[1] Bhatti MA. Fundamental finite element analysis and applications. Hoboken: John Wiley & Sons; 2005.

[2] Logan DL. A first course in the finite element method. Boston, MA: Gengage Learning; 2012.

[3] Oden JT, Becker EB, Carey GF. Finite elements: an introduction, volume I. New Jersey: Prentice Hall; 1981.

[4] Fish J, Belytschko T. A first course in finite elements. New York: Wiley; 2007.

[5] Zienkiewicz OC, Taylor RL, Fox DD. The finite element method for solid and structural mechanics. 7th ed. Oxford: Butterworth-Heinemann; 2013.

[6] Deif AS. Advanced matrix theory for scientists and engineers. 2nd edition. London: Abacus Press; 1991.

[7] Lawrence Kent. ANSYS Workbench Tutorial Release 14. Mission, KS: SDC Publications; 2012.

[8] Austrell P-E, Dahlblom O, Lindemann J, Olsson A, Olsson K-G, Persson K, et al. CALFEM a finite element toolbox, Version 3.4., Division of Structural Mechanics, Lund University; 2004.

[9] Hartmann F, Katz C. Structural analysis with finite elements. 2nd ed. Berlin: Springer; 2007.

[10] Cook RD, Malkus DS, Plesha ME, Witt RJ. Concepts and applications of finite element analysis. 4th ed. Hoboken: John Wiley & Sons; 2002.

CHAPTER

BEAMS

6

Even though beams and bars have similar geometric morphology, in addition to axial forces, beams carry bending moments and shear forces. As Figure 6.1 shows, beams are usually used in bridges, foundations, structures, etc.

Since the bending moment and shear forces cause rotation and deflection in a direction normal to the beam's axis, the nodal parameters in a beam element 1–2 are shown in Figure 6.2.

6.1 ELEMENT EQUATION OF A TWO-DIMENSIONAL BEAM SUBJECTED TO NODAL FORCES

6.1.1 THE DISPLACEMENT FUNCTION

Taking into account that a two-dimensional (2D) beam element has four degrees of freedom (u_1, θ_1, u_2, θ_2), a suitable polynomial $u(x)$ for displacements distribution along the beam's axis should contain four unknown constants. Therefore, the displacement function should have the following functional form:

$$u(x) = a_3x^3 + a_2x^2 + a_1x + a_0 \tag{6.1}$$

Using Equation (6.1), the following conditions

$$u(0) = u_1 \tag{6.2}$$

$$\left.\frac{\mathrm{d}u(x)}{\mathrm{d}x}\right|_{x=0} = \vartheta_1 \tag{6.3}$$

$$u(L) = u_2 \tag{6.4}$$

$$\left.\frac{\mathrm{d}u(x)}{\mathrm{d}x}\right|_{x=L} = \vartheta_2 \tag{6.5}$$

yield the following equations:

$$a_0 = u_1 \tag{6.6}$$

$$a_1 = \theta_1 \tag{6.7}$$

$$a_3L^3 + a_2L^2 + a_1L + a_0 = u_2 \tag{6.8}$$

Essentials of the Finite Element Method. http://dx.doi.org/10.1016/B978-0-12-802386-0.00006-2

FIGURE 6.1

Examples of structures consisting of beams.

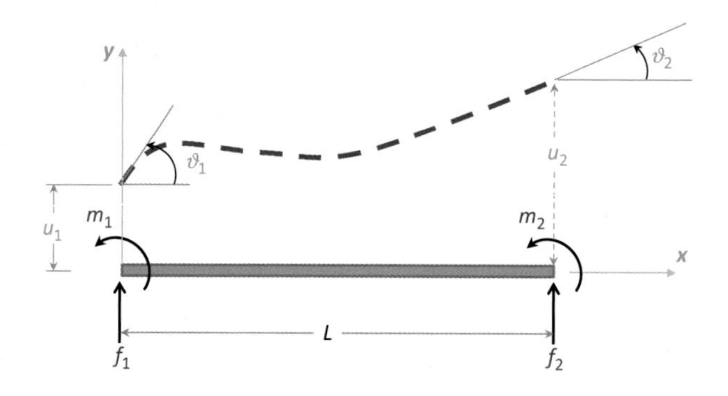

FIGURE 6.2

Nomenclature of the nodal parameters for the beam element 1-2.

$$3a_3L^2 + 2a_2L + a_1 = \theta_2 \tag{6.9}$$

The above equations can be written in the following matrix format with respect to the parameters a_0, a_1, a_2, a_3:

$$\begin{bmatrix} 1 & 0 & 0 & 0 \\ 0 & 1 & 0 & 0 \\ 1 & L & L^2 & L^3 \\ 0 & 1 & 2L & 3L^2 \end{bmatrix} \begin{Bmatrix} a_0 \\ a_1 \\ a_2 \\ a_3 \end{Bmatrix} = \begin{Bmatrix} u_1 \\ \theta_1 \\ u_2 \\ \theta_2 \end{Bmatrix} \tag{6.10}$$

The solution of the above matrix equation yields

$$\begin{Bmatrix} a_0 \\ a_1 \\ a_2 \\ a_3 \end{Bmatrix} = \begin{bmatrix} 1 & 0 & 0 & 0 \\ 0 & 1 & 0 & 0 \\ -\dfrac{3}{L^2} & -\dfrac{2}{L} & \dfrac{3}{L^2} & -\dfrac{1}{L} \\ \dfrac{2}{L^3} & \dfrac{1}{L^2} & -\dfrac{2}{L^3} & \dfrac{1}{L^2} \end{bmatrix} \begin{Bmatrix} u_1 \\ \theta_1 \\ u_2 \\ \theta_2 \end{Bmatrix} \tag{6.11}$$

Since the parameters a_0, a_1, a_2, a_3 are now known, Equation (6.1) can be written in the following form:

$$\begin{aligned} u(x) = {} & x^3\begin{bmatrix} \frac{2}{L^3} & \frac{1}{L^2} & -\frac{2}{L^3} & \frac{1}{L^2}\end{bmatrix}\begin{Bmatrix} u_1 \\ \vartheta_1 \\ u_2 \\ \vartheta_2 \end{Bmatrix} + x^2\begin{bmatrix} -\frac{3}{L^2} & -\frac{2}{L} & \frac{3}{L^2} & -\frac{1}{L}\end{bmatrix}\begin{Bmatrix} u_1 \\ \vartheta_1 \\ u_2 \\ \vartheta_2 \end{Bmatrix} \\ & + x\begin{bmatrix} 0 & 1 & 0 & 0 \end{bmatrix}\begin{Bmatrix} u_1 \\ \vartheta_1 \\ u_2 \\ \vartheta_2 \end{Bmatrix} + \begin{bmatrix} 1 & 0 & 0 & 0 \end{bmatrix}\begin{Bmatrix} u_1 \\ \vartheta_1 \\ u_2 \\ \vartheta_2 \end{Bmatrix} \end{aligned} \tag{6.12}$$

or

$$u(x) = \begin{bmatrix} N_1 & N_2 & N_3 & N_4 \end{bmatrix}\begin{Bmatrix} u_1 \\ \vartheta_1 \\ u_2 \\ \vartheta_2 \end{Bmatrix} \tag{6.13}$$

where N_1, N_2, N_3, and N_4 are shape functions given by

$$N_1 = \frac{1}{L^3}\left(2x^3 - 3x^2L + L^3\right) \tag{6.14}$$

$$N_2 = \frac{1}{L^3}\left(x^3L - 2x^2L^2 + xL^3\right) \tag{6.15}$$

$$N_3 = \frac{1}{L^3}\left(-2x^3 + 3x^2L\right) \tag{6.16}$$

$$N_4 = \frac{1}{L^3}\left(x^3L - x^2L^2\right) \tag{6.17}$$

6.1.2 THE ELEMENT STIFFNESS MATRIX

As it is known from the mechanics of solids, the internal forces, that is, bending moments $m(x)$ and shear forces $f(x)$ can be correlated to the displacement distribution $u(x)$:

$$f(x) = EI\frac{\mathrm{d}^3u(x)}{\mathrm{d}x^3} \tag{6.18}$$

$$m(x) = EI\frac{\mathrm{d}^2u(x)}{\mathrm{d}x^2} \tag{6.19}$$

Taking into account Equation (6.13), the above expressions yield:

$$f(x) = EI \cdot \frac{d^3}{dx^3}\begin{bmatrix} N_1 & N_2 & N_3 & N_4 \end{bmatrix} \cdot \begin{Bmatrix} u_1 \\ \vartheta_1 \\ u_2 \\ \vartheta_2 \end{Bmatrix} \tag{6.20}$$

$$m(x) = EI \cdot \frac{d^2}{dx^2}[N_1 \ N_2 \ N_3 \ N_4] \cdot \begin{Bmatrix} u_1 \\ \vartheta_1 \\ u_2 \\ \vartheta_2 \end{Bmatrix} \tag{6.21}$$

or

$$f(x) = \frac{EI}{L^3} \cdot [12 \ 6L \ -12 \ 6L] \cdot \begin{Bmatrix} u_1 \\ \vartheta_1 \\ u_2 \\ \vartheta_2 \end{Bmatrix} \tag{6.22}$$

$$m(x) = \frac{EI}{L^3} \cdot \left[12x - 6L \ \ 6Lx - 4L^2 \ \ -12x + 6L \ \ 6Lx - 2L^2\right] \cdot \begin{Bmatrix} u_1 \\ \vartheta_1 \\ u_2 \\ \vartheta_2 \end{Bmatrix} \tag{6.23}$$

According to the mechanics of solids, the sign convention for external forces in beams is demonstrated in Figure 6.3.

Therefore, taking into account the nomenclature of Figure 6.2, the following boundary conditions can be formulated:

$$f(0) = f_1 \tag{6.24}$$

$$m(0) = -m_1 \tag{6.25}$$

$$f(L) = -f_2 \tag{6.26}$$

$$m(L) = m_2 \tag{6.27}$$

Combining Equations (6.24)–(6.27) with Equations (6.22) and (6.23), the following formulae can be obtained:

$$f_1 = \frac{EI}{L^3} \cdot [12 \ 6L \ -12 \ 6L] \cdot \begin{Bmatrix} u_1 \\ \vartheta_1 \\ u_2 \\ \vartheta_2 \end{Bmatrix} \tag{6.28}$$

FIGURE 6.3

Sign convention for external forces at the ends of a beam.

$$-m_1 = \frac{EI}{L^3} \cdot \begin{bmatrix} -6L & -4L^2 & 6L & -2L^2 \end{bmatrix} \cdot \begin{Bmatrix} u_1 \\ \vartheta_1 \\ u_2 \\ \vartheta_2 \end{Bmatrix} \tag{6.29}$$

$$-f_2 = \frac{EI}{L^3} \cdot \begin{bmatrix} 12 & 6L & -12 & 6L \end{bmatrix} \cdot \begin{Bmatrix} u_1 \\ \vartheta_1 \\ u_2 \\ \vartheta_2 \end{Bmatrix} \tag{6.30}$$

$$m_2 = \frac{EI}{L^3} \cdot \begin{bmatrix} 6L & 2L^2 & -6L & 4L^2 \end{bmatrix} \cdot \begin{Bmatrix} u_1 \\ \vartheta_1 \\ u_2 \\ \vartheta_2 \end{Bmatrix} \tag{6.31}$$

Equations (6.28)–(6.31) can be written in matrix form providing the following element equation:

$$\begin{Bmatrix} f_1 \\ m_1 \\ f_2 \\ m_2 \end{Bmatrix} = \frac{EI}{L^3} \begin{bmatrix} 12 & 6L & -12 & 6L \\ 6L & 4L^2 & -6L & 2L^2 \\ -12 & -6L & 12 & -6L \\ 6L & 2L^2 & -6L & 4L^2 \end{bmatrix} \begin{Bmatrix} u_1 \\ \vartheta_1 \\ u_2 \\ \vartheta_2 \end{Bmatrix} \tag{6.32}$$

Once the nodal displacements $u_1, \vartheta_1, u_2, \vartheta_2$ are known, the distribution of displacements $u(x)$, shear forces $f(x)$, and bending moments $m(x)$ along the beam can be calculated using Equations (6.13), (6.22), and (6.23), respectively.

It should be noted that Equation (6.32) corresponds to a 2D beam element subjected to only nodal forces. In cases of beams loaded by varying loads between the nodes, equivalent nodal forces should be derived to simulate the effects of the varying loads. To this aim, the beam element should be assumed to rest on fixed supports on both ends, and then the equivalent nodal loadings simulating the varying load effects should be computed. These equivalent nodal loadings should be added to the ends of each element as external loads. Table 6.1 summarizes the equivalent nodal loads for common cases of varying loads. Taking into account the values F_1, M_1, F_2, M_2 of the equivalent nodal forces as well as the nomenclature of Figure 6.2, the element Equation (6.32) can now be written in the following form:

$$\begin{Bmatrix} f_1 \\ m_1 \\ f_2 \\ m_2 \end{Bmatrix} = \frac{EI}{L^3} \begin{bmatrix} 12 & 6L & -12 & 6L \\ 6L & 4L^2 & -6L & 2L^2 \\ -12 & -6L & 12 & -6L \\ 6L & 2L^2 & -6L & 4L^2 \end{bmatrix} \begin{Bmatrix} u_1 \\ \vartheta_1 \\ u_2 \\ \vartheta_2 \end{Bmatrix} - \begin{Bmatrix} -F_1 \\ -M_1 \\ -F_2 \\ M_2 \end{Bmatrix} \tag{6.33}$$

Table 6.1 Equivalent nodal loading for common cases of varying loads

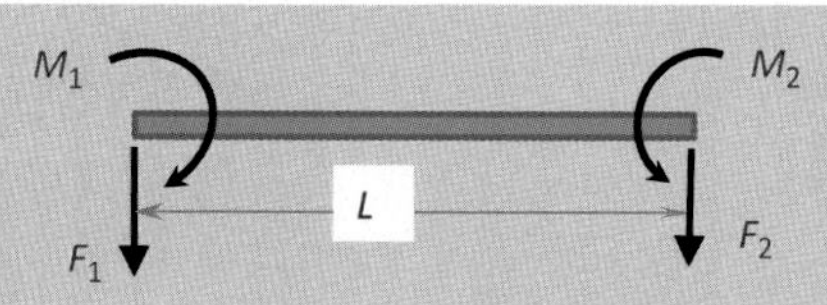

Loading case of a beam element	Equivalent nodal forces
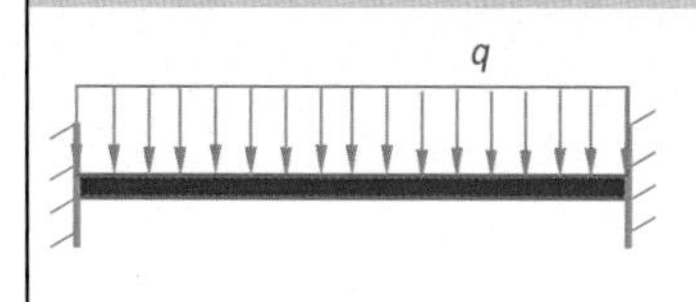	$F_1 = qL/2 \quad M_1 = qL^2/12$ $F_2 = qL/2 \quad M_2 = qL^2/12$
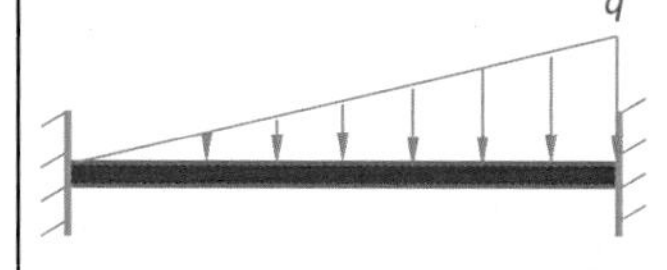	$F_1 = 3qL/20 \quad M_1 = qL^2/30$ $F_2 = 7qL/20 \quad M_2 = qL^2/20$
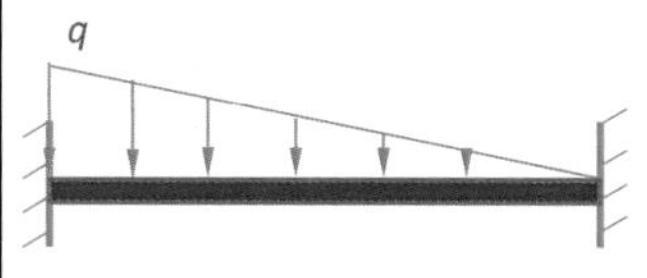	$F_1 = 7qL/20 \quad M_1 = qL^2/20$ $F_2 = 3qL/20 \quad M_2 = qL^2/30$
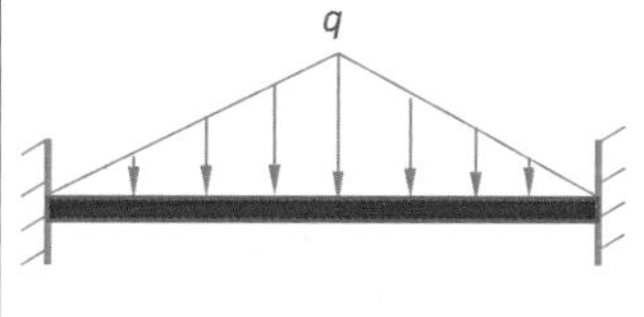	$F_1 = qL/4, \quad M_1 = 5qL^2/96$ $F_2 = qL/4, \quad M_2 = 5qL^2/96$
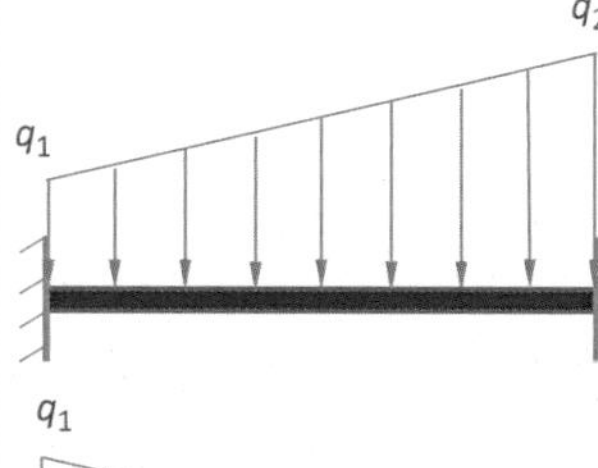 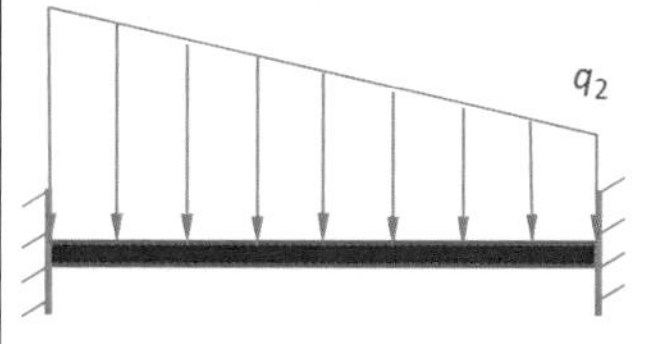	$F_1 = \frac{q_1 L}{2} + \frac{3(q_2 - q_1)L}{20}$ $M_1 = \frac{q_1 L^2}{12} + \frac{(q_2 - q_1)L^2}{30}$ $F_2 = \frac{(q_1 + q_2)L}{2} - F_1$ $M_2 = \frac{(q_2 - q_1)L^2}{6} + \frac{q_1 L^2}{2} - F_1 L - M_1$

EXAMPLE 6.1

Determine the displacements, bending moments, and shear forces for the following continuous beam.

Data

$$L = 8.0\text{m},\ \ q = 3\text{kN/m},\ \ E = 200\text{GPa},\ \ I = 125 \times 10^{-6}\text{m}^4$$

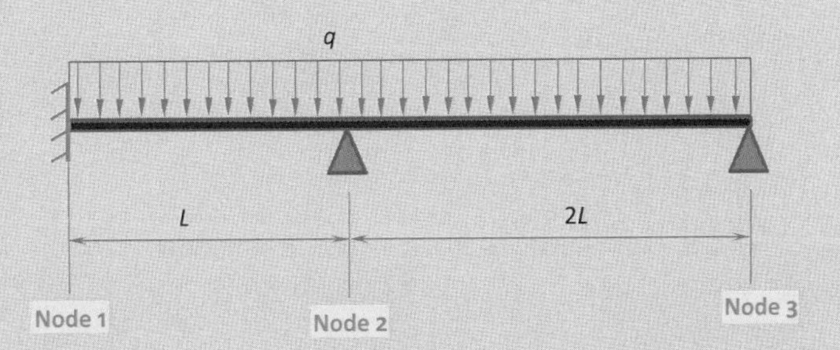

(a) Equivalent Nodal Forces

Since the element Equation (6.32) corresponds to a beam element subjected to only nodal forces, the first step for the solution of the above problem is the transformation of the uniform load q to equivalent nodal forces. Taking into account Table 6.1, the following simulation of the given structure should be used:

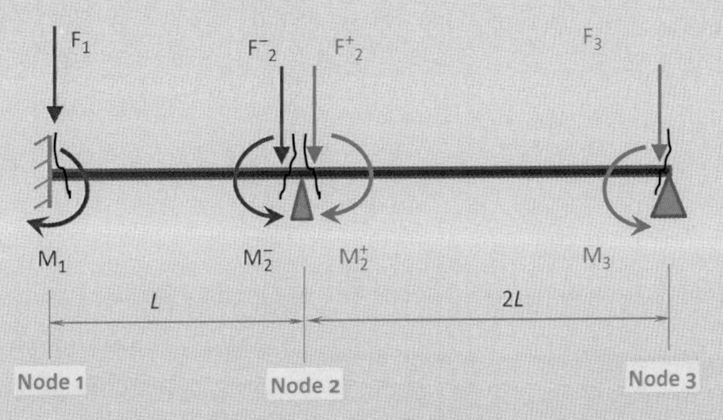

where the equivalent loads acting on the nodes of each element are given by the following formulae:

$$F_1 = qL/2 = 12{,}000\text{N},\ \ M_1 = qL^2/12 = 16{,}000\text{Nm}$$
$$F_2^- = qL/2 = 12{,}000\text{N},\ \ M_2^- = qL^2/12 = 16{,}000\text{Nm}$$

Continued

EXAMPLE 6.1—CONT'D

$$F_2^+ = q(2L)/2 = 24{,}000 \text{ N}, \quad M_2^+ = q(2L)^2/12 = 64{,}000 \text{ Nm}$$
$$F_3 = q(2L)/2 = 24{,}000 \text{ N}, \quad M_3 = q(2L)^2/12 = 64{,}000 \text{ Nm}$$

(b) Local Element Equations

The second step for the solution of the above problem is the derivation of the element equations. Taking into account Equation (6.33), as well as the nodal forces acting on each node, the following element equations can be obtained:

Element 1, nodes 1-2

$$\begin{Bmatrix} f_1 \\ m_1 \\ f_2 \\ m_2 \end{Bmatrix} = \frac{EI}{L^3}\begin{bmatrix} 12 & 6L & -12 & 6L \\ 6L & 4L^2 & -6L & 2L^2 \\ -12 & -6L & 12 & -6L \\ 6L & 2L^2 & -6L & 4L^2 \end{bmatrix}\begin{Bmatrix} u_1 \\ \vartheta_1 \\ u_2 \\ \vartheta_2 \end{Bmatrix} - \begin{Bmatrix} -qL/2 \\ -qL^2/12 \\ -qL/2 \\ +qL^2/12 \end{Bmatrix}$$

or

$$\begin{Bmatrix} f_1 \\ m_1 \\ f_2 \\ m_2 \end{Bmatrix} = \begin{bmatrix} 585{,}937.5 & 2{,}343{,}750 & -585{,}937.5 & 2{,}343{,}750 \\ 2{,}343{,}750 & 1.25\times10^7 & -2{,}343{,}750 & 6{,}250{,}000 \\ -585{,}937.5 & -2{,}343{,}750 & 585{,}937.5 & -2{,}343{,}750 \\ 2{,}343{,}750 & 6{,}250{,}000 & -2{,}343{,}750 & 1.25\times10^7 \end{bmatrix}\begin{Bmatrix} u_1 \\ \vartheta_1 \\ u_2 \\ \vartheta_2 \end{Bmatrix} - \begin{Bmatrix} -12{,}000 \\ -16{,}000 \\ -12{,}000 \\ 16{,}000 \end{Bmatrix}$$

Element 2, nodes 2-3

$$\begin{Bmatrix} f_2 \\ m_2 \\ f_3 \\ m_3 \end{Bmatrix} = \frac{EI}{(2L)^3}\begin{bmatrix} 12 & 6(2L) & -12 & 6(2L) \\ 6(2L) & 4(2L)^2 & -6(2L) & 2(2L)^2 \\ -12 & -6(2L) & 12 & -6(2L) \\ 6(2L) & 2(2L)^2 & -6(2L) & 4(2L)^2 \end{bmatrix}\begin{Bmatrix} u_2 \\ \vartheta_2 \\ u_3 \\ \vartheta_3 \end{Bmatrix} - \begin{Bmatrix} -q(2L)/2 \\ -q(2L)^2/12 \\ -q(2L)/2 \\ +q(2L)^2/12 \end{Bmatrix}$$

or

$$\begin{Bmatrix} f_2 \\ m_2 \\ f_3 \\ m_3 \end{Bmatrix} = \begin{bmatrix} 585{,}937.5 & 4{,}687{,}500 & -585{,}937.5 & 4{,}687{,}500 \\ 4{,}687{,}500 & 5\times10^7 & -4{,}687{,}500 & 2.5\times10^7 \\ -585{,}937.5 & -4{,}687{,}500 & 585{,}937.5 & -4{,}687{,}500 \\ 4{,}687{,}200 & 2.5\times10^7 & -4{,}687{,}500 & 5\times10^7 \end{bmatrix}\begin{Bmatrix} u_2 \\ \vartheta_2 \\ u_3 \\ \vartheta_3 \end{Bmatrix} - \begin{Bmatrix} -24{,}000 \\ -64{,}000 \\ -24{,}000 \\ 64{,}000 \end{Bmatrix}$$

(c) Expansion of Element Equations in Global Coordinates

The above local element equations should now be expanded to the degrees of freedom of the whole structure:

Element 1, nodes 1-2

$$\begin{bmatrix} f_1 \\ m_1 \\ f_2 \\ m_2 \\ f_3 \\ m_3 \end{bmatrix} = \begin{bmatrix} 585{,}937.5 & 2{,}343{,}750 & -585{,}937.5 & 2{,}343{,}750 & 0 & 0 \\ 2{,}343{,}750 & 1.25\times10^7 & -2{,}343{,}750 & 6{,}250{,}000 & 0 & 0 \\ -585{,}937.5 & -2{,}343{,}750 & 585{,}937.5 & -2{,}343{,}750 & 0 & 0 \\ 2{,}343{,}750 & 6{,}250{,}000 & -2{,}343{,}750 & 1.25\times10^7 & 0 & 0 \\ 0 & 0 & 0 & 0 & 0 & 0 \\ 0 & 0 & 0 & 0 & 0 & 0 \end{bmatrix}\begin{bmatrix} u_1 \\ \vartheta_1 \\ u_2 \\ \vartheta_2 \\ u_3 \\ \vartheta_3 \end{bmatrix} - \begin{bmatrix} -12{,}000 \\ -16{,}000 \\ -12{,}000 \\ 16{,}000 \\ 0 \\ 0 \end{bmatrix}$$

Element 2, nodes 2-3

$$\begin{Bmatrix} f_1 \\ m_1 \\ f_2 \\ m_2 \\ f_3 \\ m_3 \end{Bmatrix} = \begin{bmatrix} 0 & 0 & 0 & 0 & 0 & 0 \\ 0 & 0 & 0 & 0 & 0 & 0 \\ 0 & 0 & 585{,}937.5 & 4{,}687{,}500 & -585{,}937.5 & 4{,}687{,}500 \\ 0 & 0 & 4{,}687{,}500 & 5\times10^7 & -4{,}687{,}500 & 2.5\times10^7 \\ 0 & 0 & -585{,}937.5 & -4{,}687{,}500 & 585{,}937.5 & -4{,}687{,}500 \\ 0 & 0 & 4{,}687{,}200 & 2.5\times10^7 & -4{,}687{,}500 & 5\times10^7 \end{bmatrix} \begin{Bmatrix} u_1 \\ \vartheta_1 \\ u_2 \\ \vartheta_2 \\ u_3 \\ \vartheta_3 \end{Bmatrix} - \begin{Bmatrix} 0 \\ 0 \\ -24{,}000 \\ -64{,}000 \\ -24{,}000 \\ 64{,}000 \end{Bmatrix}$$

(d) Structural Equation in Global Coordinates

The structural equation in global coordinates can be obtained by superposition of the above expanded element equations. Therefore, adding the above element equations the following global equation can be obtained:

$$\begin{Bmatrix} F_1 \\ M_1 \\ F_2 \\ M_2 \\ F_3 \\ M_3 \end{Bmatrix} = \begin{bmatrix} 585{,}937.5 & 2{,}343{,}750 & -585{,}937.5 & 2{,}343{,}750 & 0 & 0 \\ 2{,}343{,}750 & 1.25\times10^7 & -2{,}343{,}750 & 6{,}250{,}000 & 0 & 0 \\ -585{,}937.5 & -2{,}343{,}750 & 1{,}171{,}875 & 2{,}343{,}750 & -585{,}937.5 & 4{,}687{,}500 \\ 2{,}343{,}750 & 6{,}250{,}000 & -7{,}031{,}250 & 6.25\times10^7 & -4{,}687{,}500 & 2.5\times10^7 \\ 0 & 0 & -585{,}937.5 & -4{,}687{,}500 & 585{,}937.5 & -4{,}687{,}500 \\ 0 & 0 & 4{,}687{,}200 & 2.5\times10^7 & -4{,}687{,}500 & 5\times10^7 \end{bmatrix} \begin{Bmatrix} u_1 \\ \vartheta_1 \\ u_2 \\ \vartheta_2 \\ u_3 \\ \vartheta_3 \end{Bmatrix} - \begin{Bmatrix} -12{,}000 \\ -160{,}000 \\ -36{,}000 \\ -48{,}000 \\ -24{,}000 \\ 64{,}000 \end{Bmatrix}$$

The above equation can be written in the following matrix form:

$$\{R\} = [K]\{d\} - \{f\} \tag{6.34}$$

or

$$[[K]\ [-I]]\begin{Bmatrix} \{d\} \\ \{R\} \end{Bmatrix} = \{f\} \tag{6.35}$$

(e) Boundary Conditions

For the derivation of the algebraic system providing the nodal displacement field, the boundary conditions of the problem should be incorporated to the above structural equation. Taking into account

Continued

EXAMPLE 6.1—CONT'D

the types of supports as well as the nodal forces, the six following boundary conditions can be specified:

(a) Boundary conditions for nodal displacements

$$\begin{aligned}&(1)\ u_1=0\\&(2)\ \vartheta_1=0\\&(3)\ u_2=0\\&(4)\ u_3=0\end{aligned}$$

(b) Boundary conditions for nodal forces

$$\begin{aligned}&(5)\ M_2=0\\&(6)\ M_3=0\end{aligned}$$

The above boundary conditions can now be expressed in a matrix format:

(a) Boundary conditions for nodal displacements

$$\begin{bmatrix}1&0&0&0&0&0\\0&1&0&0&0&0\\0&0&1&0&0&0\\0&0&0&0&1&0\\0&0&0&0&0&0\\0&0&0&0&0&0\end{bmatrix}\begin{Bmatrix}u_1\\\theta_1\\u_2\\\theta_2\\u_3\\\theta_3\end{Bmatrix}=\begin{Bmatrix}0\\0\\0\\0\\0\\0\end{Bmatrix} \tag{6.36}$$

or in an abbreviated format:

$$[BCd]\{d\}=\{DO\} \tag{6.37}$$

or

$$[[BCd]\ [O]]\begin{Bmatrix}\{d\}\\\{R\}\end{Bmatrix}=\{DO\} \tag{6.38}$$

(b) Boundary conditions for nodal forces

$$\begin{bmatrix}0&0&0&0&0&0\\0&0&0&0&0&0\\0&0&0&0&0&0\\0&0&0&0&0&0\\0&0&0&1&0&0\\0&0&0&0&0&1\end{bmatrix}\begin{Bmatrix}F_1\\M_1\\F_2\\M_2\\F_3\\M_3\end{Bmatrix}=\begin{Bmatrix}0\\0\\0\\0\\0\\0\end{Bmatrix}$$

or in an abbreviated format:

$$[BCR]\{R\} = \{RO\} \tag{6.39}$$

or

$$\begin{bmatrix}[O] & [BCR]\end{bmatrix}\begin{Bmatrix}\{d\}\\\{R\}\end{Bmatrix} = \{RO\} \tag{6.40}$$

Summation of Equations (6.38) and (6.40) yields

$$\begin{bmatrix}[BCd] & [BCR]\end{bmatrix}\begin{Bmatrix}\{d\}\\\{R\}\end{Bmatrix} = \{DO + RO\} \tag{6.41}$$

(f) Algebraic System for Nodal Displacements Derivation

Combining the derived structural Equation (6.35) with the above boundary conditions (Equation 6.41), the following algebraic system can be obtained:

$$\begin{bmatrix}[K] & [-I]\\ [BCd] & [BCR]\end{bmatrix}\begin{Bmatrix}\{d\}\\\{R\}\end{Bmatrix} = \begin{Bmatrix}\{f\}\\\{DO + RO\}\end{Bmatrix} \tag{6.42}$$

or in an abbreviated format

$$[A]\,\{X\} = \{B\} \tag{6.43}$$

where

$$\{X\} = \{u_1, \vartheta_1, u_2, \vartheta_2, u_3, \vartheta_3, F_1, M_1, F_2, M_2, F_3, M_3\}^{\mathrm{T}} \tag{6.44}$$

and

$$[A] = \begin{bmatrix}
585{,}937 & 2.34375\times10^6 & -585{,}937 & 2.34375\times10^6 & 0 & 0 & -1 & 0 & 0 & 0 & 0 & 0\\
2.34375\times10^6 & 1.25\times10^7 & -2.34375\times10^6 & 6.25\times10^6 & 0 & 0 & 0 & -1 & 0 & 0 & 0 & 0\\
-585{,}937 & -2.34375\times10^6 & 585{,}937 & 2.34375\times10^6 & -585{,}937 & 4.6875\times10^6 & 0 & 0 & -1 & 0 & 0 & 0\\
2.34375\times10^6 & 6.25\times10^6 & 2.34375\times10^6 & 6.25\times10^7 & -4.6875\times10^6 & 2.5\times10^7 & 0 & 0 & 0 & -1 & 0 & 0\\
0 & 0 & -585{,}937 & -4.6875\times10^6 & 585{,}937 & -4.6875\times10^6 & 0 & 0 & 0 & 0 & -1 & 0\\
0 & 0 & 4.6875\times10^6 & 2.5\times10^7 & -4.6875\times10^6 & 5\times10^7 & 0 & 0 & 0 & 0 & 0 & -1\\
1 & 0 & 0 & 0 & 0 & 0 & 0 & 0 & 0 & 0 & 0 & 0\\
0 & 1 & 0 & 0 & 0 & 0 & 0 & 0 & 0 & 0 & 0 & 0\\
0 & 0 & 1 & 0 & 0 & 0 & 0 & 0 & 0 & 0 & 0 & 0\\
0 & 0 & 0 & 0 & 1 & 0 & 0 & 0 & 0 & 0 & 0 & 0\\
0 & 0 & 0 & 0 & 0 & 0 & 0 & 0 & 0 & 1 & 0 & 0\\
0 & 0 & 0 & 0 & 0 & 0 & 0 & 0 & 0 & 0 & 0 & 1
\end{bmatrix} \tag{6.45}$$

Continued

EXAMPLE 6.1—CONT'D

$$\{B\}=\begin{Bmatrix} -12{,}000 \\ -16{,}000 \\ -36{,}000 \\ -48{,}000 \\ -24{,}000 \\ 64{,}000 \\ 0 \\ 0 \\ 0 \\ 0 \\ 0 \\ 0 \end{Bmatrix} \tag{6.46}$$

The solution of the above system yields the nodal displacements and the reactions on the beam's supports:

$$\{X\}=\begin{Bmatrix} u_1 \\ \vartheta_1 \\ u_2 \\ \vartheta_2 \\ u_3 \\ \vartheta_3 \\ F_1 \\ M_1 \\ F_2 \\ M_2 \\ F_3 \\ M_3 \end{Bmatrix}=\begin{Bmatrix} 0 \\ 0 \\ 0 \\ -0.0016 \\ 0 \\ 0.00208 \\ 8250 \\ 6000 \\ 42{,}000 \\ 0 \\ 21{,}750 \\ 0 \end{Bmatrix} \tag{6.47}$$

(g) Displacements, Shear Forces, and Bending Moments

Knowing the nodal displacements $\{u_1, \vartheta_1, u_2, \vartheta_2, u_3, \vartheta_3\}^T$, the displacements $u(x)$, shear forces $f(x)$, and bending moments $m(x)$ for any point x of the beam can be derived by Equations (6.13), (6.22), and (6.23), respectively:

Element 1, nodes 1-2

Derivation of displacements

$$N1=0.001953125\left(512-24x^2+2x^3\right)$$

$$N2=0.001953125\left(512x-128x^2+8x^3\right)$$

$$N3=0.001953125\left(24x^2-2x^3\right)$$

$$N4=0.001953125\left(-64x^2+8x^3\right)$$

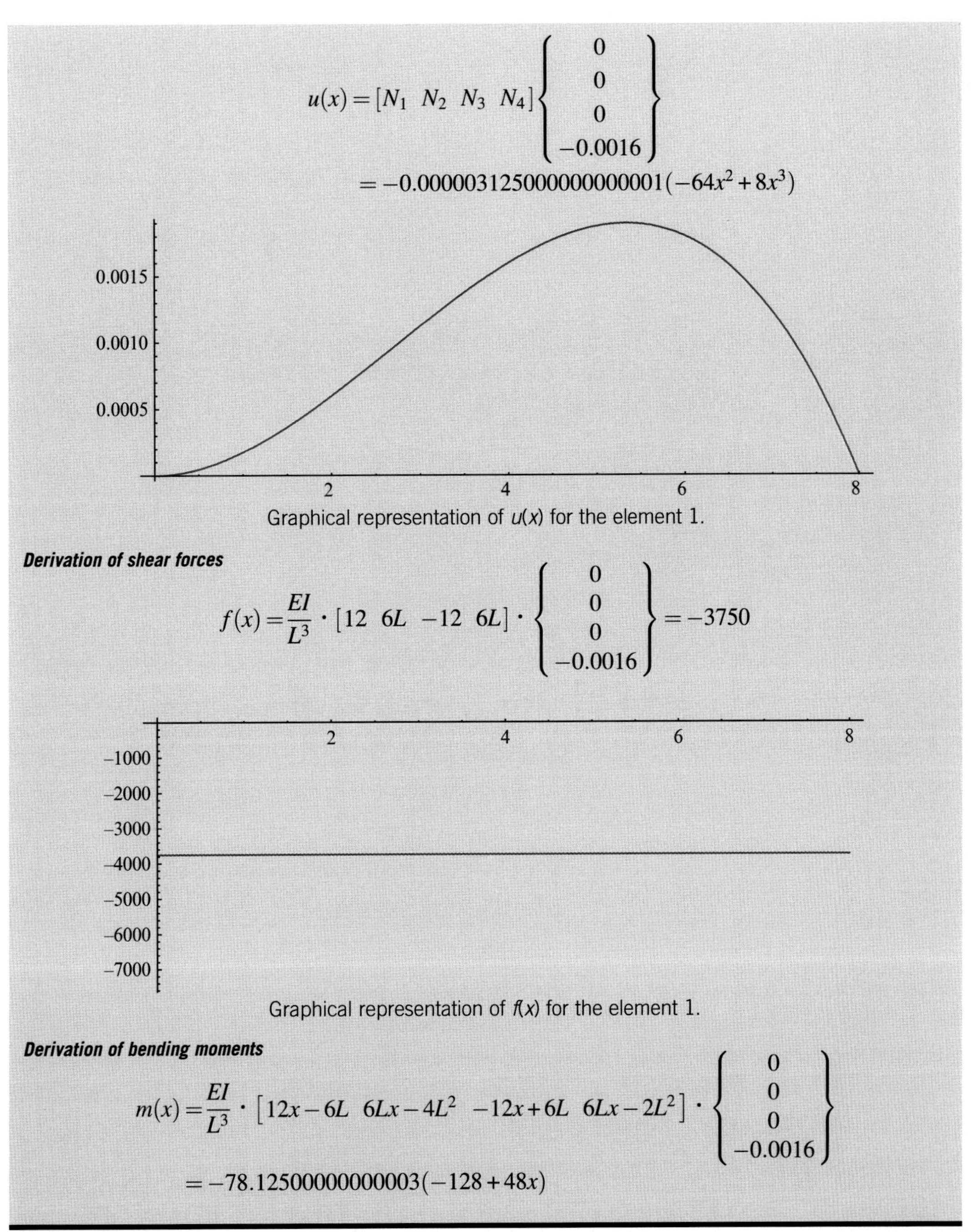

$$u(x) = [N_1 \;\; N_2 \;\; N_3 \;\; N_4]\begin{Bmatrix} 0 \\ 0 \\ 0 \\ -0.0016 \end{Bmatrix}$$

$$= -0.000003125000000000001(-64x^2 + 8x^3)$$

Graphical representation of $u(x)$ for the element 1.

Derivation of shear forces

$$f(x) = \frac{EI}{L^3} \cdot [12 \;\; 6L \;\; -12 \;\; 6L] \cdot \begin{Bmatrix} 0 \\ 0 \\ 0 \\ -0.0016 \end{Bmatrix} = -3750$$

Graphical representation of $f(x)$ for the element 1.

Derivation of bending moments

$$m(x) = \frac{EI}{L^3} \cdot \left[12x - 6L \;\; 6Lx - 4L^2 \;\; -12x + 6L \;\; 6Lx - 2L^2\right] \cdot \begin{Bmatrix} 0 \\ 0 \\ 0 \\ -0.0016 \end{Bmatrix}$$

$$= -78.12500000000003(-128 + 48x)$$

Continued

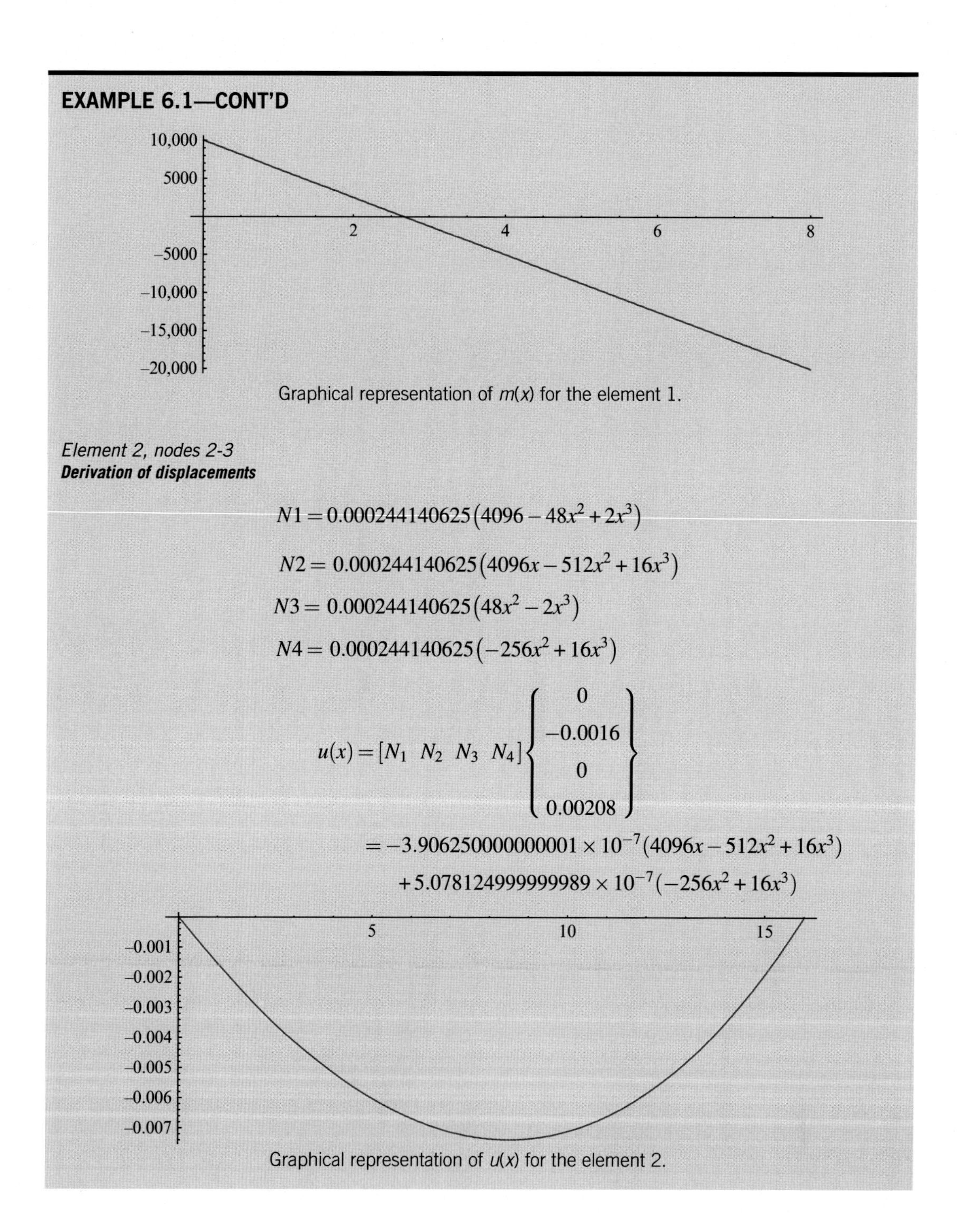

EXAMPLE 6.1—CONT'D

Graphical representation of $m(x)$ for the element 1.

Element 2, nodes 2-3

Derivation of displacements

$$N1 = 0.000244140625(4096 - 48x^2 + 2x^3)$$

$$N2 = 0.000244140625(4096x - 512x^2 + 16x^3)$$

$$N3 = 0.000244140625(48x^2 - 2x^3)$$

$$N4 = 0.000244140625(-256x^2 + 16x^3)$$

$$u(x) = [N_1 \;\; N_2 \;\; N_3 \;\; N_4]\begin{Bmatrix} 0 \\ -0.0016 \\ 0 \\ 0.00208 \end{Bmatrix}$$

$$= -3.906250000000001 \times 10^{-7}(4096x - 512x^2 + 16x^3)$$

$$+ 5.078124999999989 \times 10^{-7}(-256x^2 + 16x^3)$$

Graphical representation of $u(x)$ for the element 2.

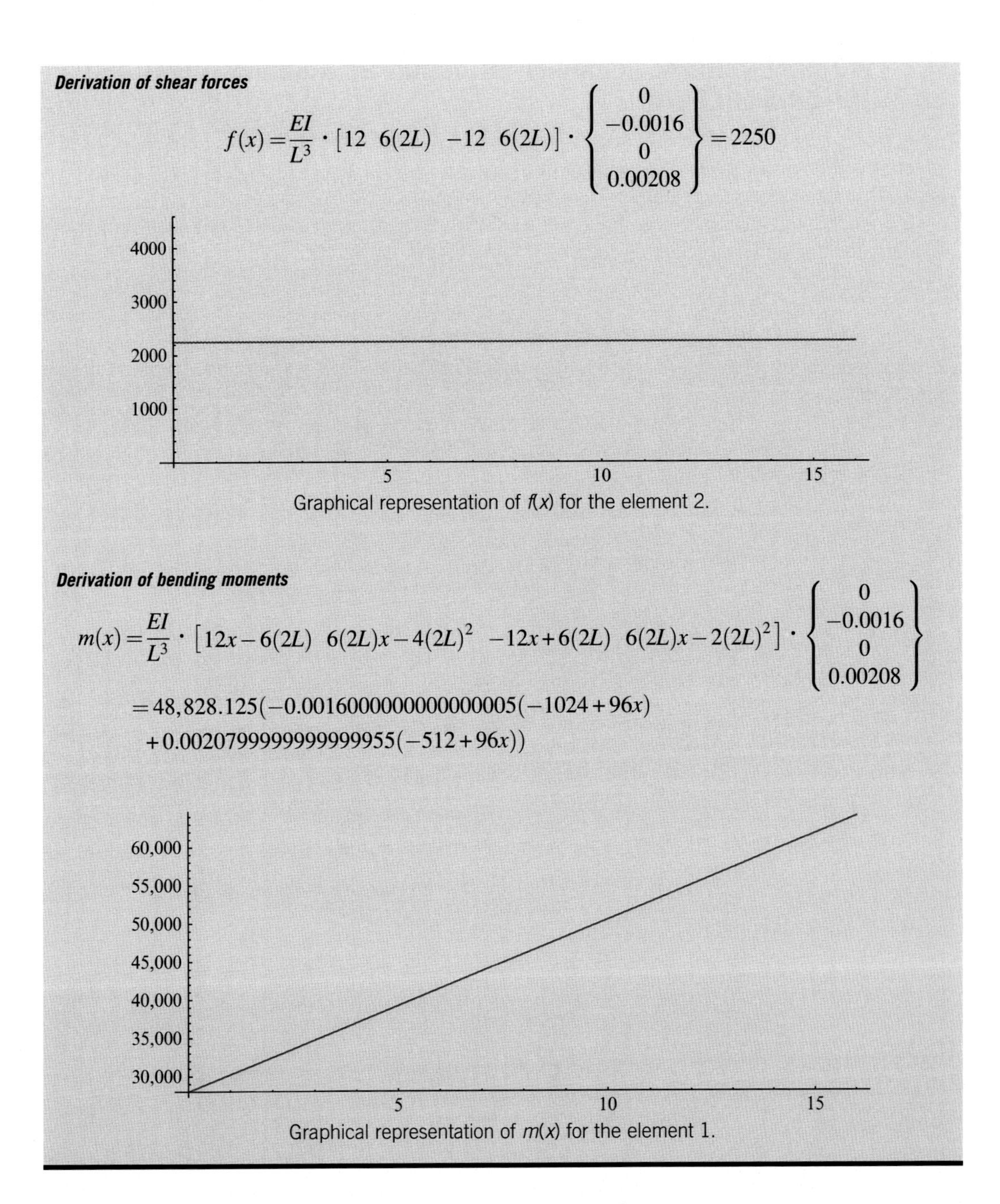

Derivation of shear forces

$$f(x)=\frac{EI}{L^3}\cdot[12\ \ 6(2L)\ \ -12\ \ 6(2L)]\cdot\begin{Bmatrix}0\\-0.0016\\0\\0.00208\end{Bmatrix}=2250$$

Graphical representation of $f(x)$ for the element 2.

Derivation of bending moments

$$m(x)=\frac{EI}{L^3}\cdot\left[12x-6(2L)\ \ 6(2L)x-4(2L)^2\ \ -12x+6(2L)\ \ 6(2L)x-2(2L)^2\right]\cdot\begin{Bmatrix}0\\-0.0016\\0\\0.00208\end{Bmatrix}$$

$$=48{,}828.125(-0.0016000000000000005(-1024+96x)$$
$$+0.0020799999999999955(-512+96x))$$

Graphical representation of $m(x)$ for the element 1.

6.2 TWO-DIMENSIONAL ELEMENT EQUATION OF A BEAM SUBJECTED TO A UNIFORM LOADING

Apart from using trial functions such as Equation (6.1), element equations can be derived by using the governing differential equation of beams. Following this concept, the element equation will be derived for a 2D beam subjected to a uniform loading.

It is already known from the mechanics of solids that the governing equation of a beam subjected to bending due to a uniformly varying loading (Figure 6.4) is

$$EI\frac{\mathrm{d}^4u(x)}{\mathrm{d}x^4} = -q \tag{6.48}$$

Furthermore, the shear forces $f(x)$, bending moments $m(x)$, and slope $\vartheta(x)$ are given by the following equations:

$$f(x) = EI\frac{\mathrm{d}^3u(x)}{\mathrm{d}x^3} \tag{6.49}$$

$$m(x) = EI\frac{\mathrm{d}^2u(x)}{\mathrm{d}x^2} \tag{6.50}$$

$$\vartheta(x) = \frac{\mathrm{d}u(x)}{\mathrm{d}x} \tag{6.51}$$

Integration of Equation (6.48) yields:

$$f(x) = EI\frac{\mathrm{d}^3u(x)}{\mathrm{d}x^3} = C_1 - qx \tag{6.52}$$

$$m(x) = EI\frac{\mathrm{d}^2u(x)}{\mathrm{d}x^2} = C_2 + C_1x - \frac{1}{2}qx^2 \tag{6.53}$$

$$\vartheta(x) = \frac{\mathrm{d}u(x)}{\mathrm{d}x} = \frac{C_3}{EI} + \frac{C_2}{EI}x + \frac{1}{2}\frac{C_1}{EI}x^2 - \frac{1}{6}\frac{q}{EI}x^3 \tag{6.54}$$

$$u(x) = \frac{C_4}{EI} + \frac{C_3}{EI}x + \frac{1}{2}\frac{C_2}{EI}x^2 + \frac{1}{6}\frac{C_1}{EI}x^3 - \frac{q}{24EI}x^4 \tag{6.55}$$

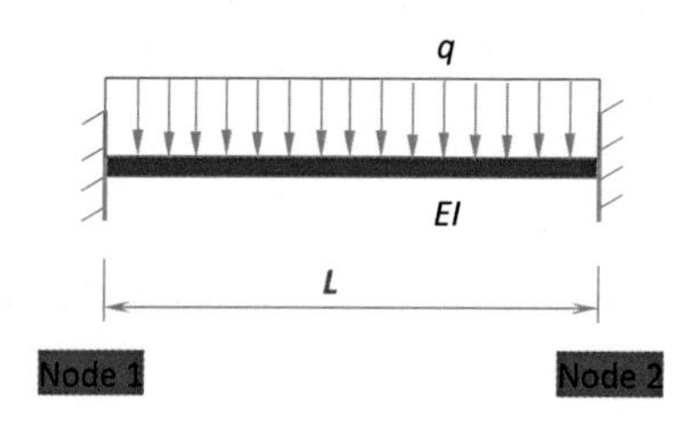

FIGURE 6.4

Beam subjected to uniformly varying loading.

It should be noted that the above equations are based on the sign convention of solid mechanics as it is demonstrated in Figure 6.3. Taking into account the sign convention of Figure 6.2, the above equation for $x=0$ yields:

$$f_1 = f(0) = C_1 \tag{6.56}$$

$$-m_1 = m(0) = C_2 \tag{6.57}$$

$$\vartheta_1 = \vartheta(0) = \frac{C_3}{EI} \tag{6.58}$$

$$u_1 = u(0) = \frac{C_4}{EI} \tag{6.59}$$

Using the above equations, the unknown constants C_1, C_2, C_3, C_4 can be expressed in terms of the physical quantities f_1, m_1, ϑ_1, u_1:

$$C_1 = f_1 \tag{6.60}$$

$$C_2 = -m_1 \tag{6.61}$$

$$C_3 = EI\vartheta_1 \tag{6.62}$$

$$C_4 = EIu_1 \tag{6.63}$$

Taking into account the above results, Equations (6.52)–(6.55) can now be written in the following form:

$$u(x) = u_1 + \vartheta_1 x - m_1\frac{x^2}{2EI} + f_1\frac{x^3}{6EI} - \frac{q}{24EI}x^4 \tag{6.64}$$

$$\vartheta(x) = 0.u_1 + \vartheta_1 - m_1\frac{x}{EI} + f_1\frac{x^2}{2EI} - \frac{q}{6EI}x^3 \tag{6.65}$$

$$m(x) = 0.u_1 + 0.\vartheta_1 - m_1 + f_1 x - \frac{qx^2}{2} \tag{6.66}$$

$$f(x) = 0.u_1 + 0.\vartheta_1 + 0.m_1 + f_1 - qx \tag{6.67}$$

Taking into account the sign convention of Figure 6.3, the physical quantities u_2, ϑ_2, m_2, f_2 for $x=L$ (node 2) can be obtained using the above equations. Therefore:

$$u_2 = u_1 + \vartheta_1 L - m_1\frac{L^2}{2EI} + f_1\frac{L^3}{6EI} - \frac{q}{24EI}L^4 \tag{6.68}$$

$$\vartheta_2 = 0.u_1 + \vartheta_1 - m_1\frac{L}{EI} + f_1\frac{L^2}{2EI} - \frac{q}{6EI}L^3 \tag{6.69}$$

$$m_2 = 0.u_1 + 0.\vartheta_1 - m_1 + f_1 L - \frac{qL^2}{2} \tag{6.70}$$

$$-f_2 = 0.u_1 + 0.\vartheta_1 + 0.m_1 + f_1 - qL \tag{6.71}$$

Rearrangement of the above equations yields:

$$\frac{L^3}{6EI}f_1 - \frac{L^2}{2EI}m_1 + 0.f_2 + 0.m_2 = -u_1 - L\vartheta_1 + u_2 + 0.\vartheta_2 + \frac{qL^4}{24EI} \tag{6.72}$$

$$\frac{L^2}{2EI}f_1 - \frac{L}{EI}m_1 + 0.f_2 + 0.m_2 = 0.u_1 - \vartheta_1 + 0.u_2 + \vartheta_2 + \frac{qL^3}{6EI} \tag{6.73}$$

$$Lf_1 - m_1 + 0.f_2 - m_2 = 0.u_1 + 0.\vartheta_1 + 0.u_2 + 0.\vartheta_2 + \frac{qL^2}{2} \tag{6.74}$$

$$f_1 + 0.m_1 + f_2 + 0.m_2 = 0.u_1 + 0.\vartheta_1 + 0.u_2 + 0.\vartheta_2 + qL \tag{6.75}$$

The above system of equations can now be written in the following matrix form:

$$\begin{bmatrix} \frac{L^3}{6EI} & -\frac{L^2}{2EI} & 0 & 0 \\ \frac{L^2}{2EI} & -\frac{L}{EI} & 0 & 0 \\ L & -1 & 0 & -1 \\ 1 & 0 & 1 & 0 \end{bmatrix} \begin{Bmatrix} f_1 \\ m_1 \\ f_2 \\ m_2 \end{Bmatrix} = \begin{bmatrix} -1 & -L & 1 & 0 \\ 0 & -1 & 0 & 1 \\ 0 & 0 & 0 & 0 \\ 0 & 0 & 0 & 0 \end{bmatrix} \begin{Bmatrix} u_1 \\ \vartheta_1 \\ u_2 \\ \vartheta_2 \end{Bmatrix} + \begin{Bmatrix} \frac{qL^4}{24EI} \\ \frac{qL^3}{6EI} \\ \frac{qL^2}{2} \\ qL \end{Bmatrix} \tag{6.76}$$

or in abbreviated form

$$[U_1]\begin{Bmatrix} f_1 \\ m_1 \\ f_2 \\ m_2 \end{Bmatrix} = [U_2]\begin{Bmatrix} u_1 \\ \vartheta_1 \\ u_2 \\ \vartheta_2 \end{Bmatrix} + \{U_3\} \tag{6.77}$$

where

$$[U_1] = \begin{bmatrix} \frac{L^3}{6EI} & -\frac{L^2}{2EI} & 0 & 0 \\ \frac{L^2}{2EI} & -\frac{L}{EI} & 0 & 0 \\ L & -1 & 0 & -1 \\ 1 & 0 & 1 & 0 \end{bmatrix} \tag{6.78}$$

$$[U_2] = \begin{bmatrix} -1 & -L & 1 & 0 \\ 0 & -1 & 0 & 1 \\ 0 & 0 & 0 & 0 \\ 0 & 0 & 0 & 0 \end{bmatrix} \tag{6.79}$$

$$\{U_3\} = \begin{Bmatrix} \frac{qL^4}{24EI} \\ \frac{qL^3}{6EI} \\ \frac{qL^2}{2} \\ qL \end{Bmatrix} \tag{6.80}$$

Therefore, Equation (6.77) yields

$$\begin{Bmatrix} f_1 \\ m_1 \\ f_2 \\ m_2 \end{Bmatrix} = [U_1]^{-1}[U_2]\begin{Bmatrix} u_1 \\ \vartheta_1 \\ u_2 \\ \vartheta_2 \end{Bmatrix} + [U_1]^{-1}\{U_3\} \tag{6.81}$$

Performing the required algebraic operations, the following beam element equation can now be obtained:

$$\begin{Bmatrix} f_1 \\ m_1 \\ f_2 \\ m_2 \end{Bmatrix} = \frac{EI}{L^3}\begin{bmatrix} 12 & 6L & -12 & 6L \\ 6L & 4L^2 & -6L & 2L^2 \\ -12 & -6L & 12 & -6L \\ 6L & 2L^2 & -6L & 4L^2 \end{bmatrix}\begin{Bmatrix} u_1 \\ \vartheta_1 \\ u_2 \\ \vartheta_2 \end{Bmatrix} + \begin{Bmatrix} \frac{qL}{2} \\ \frac{qL^2}{12} \\ \frac{qL}{2} \\ -\frac{qL^2}{12} \end{Bmatrix} \tag{6.82}$$

It should be mentioned that Equation (6.82) is same with the element equation used in Example 6.1, derived using the trial function (Equation 6.1) and the nodal forces obtained by Table 6.1.

6.3 TWO-DIMENSIONAL ELEMENT EQUATION OF A BEAM SUBJECTED TO AN ARBITRARY VARYING LOADING

Adopting a similar procedure, the element equation for a beam subjected to any arbitrary varying loading (Figure 6.5) will now be derived. In that case, Equation (6.48) can be written in a general form:

$$EI\frac{d^4u(x)}{dx^4} = -q(x) \tag{6.83}$$

where $q(x)$ can be a function of any type of varying loading.

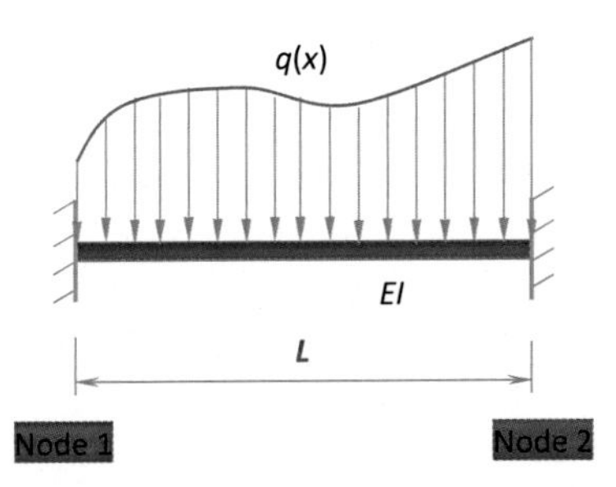

FIGURE 6.5

Beam subjected to an arbitrary varying loading.

The functions for shear force $f(x)$, bending moment $m(x)$, slope $\vartheta(x)$, and displacement $u(x)$ can be derived using Equations (6.49)–(6.51). Therefore, integrating Equation (6.83), the following formulae can be obtained:

$$f(x)=EI\frac{\mathrm{d}^3u(x)}{\mathrm{d}x^3}=C_1+I_1(x) \tag{6.84}$$

$$m(x)=EI\frac{\mathrm{d}^2u(x)}{\mathrm{d}x^2}=C_2+C_1x+I_2(x) \tag{6.85}$$

$$\vartheta(x)=\frac{\mathrm{d}u(x)}{\mathrm{d}x}=\frac{C_3}{EI}+\frac{C_2}{EI}x+\frac{1}{2}\frac{C_1}{EI}x^2+I_3(x) \tag{6.86}$$

$$u(x)=\frac{C_4}{EI}+\frac{C_3}{EI}x+\frac{1}{2}\frac{C_2}{EI}x^2+\frac{1}{6}\frac{C_1}{EI}x^3+I_4(x) \tag{6.87}$$

where

$$I_1(x)=-\int q(x)\mathrm{d}x \tag{6.88}$$

$$I_2(x)=\int I_1(x)\mathrm{d}x \tag{6.89}$$

$$I_3(x)=\int \frac{I_2(x)}{EI}\mathrm{d}x \tag{6.90}$$

$$I_4(x)=\int I_3(x)\mathrm{d}x \tag{6.91}$$

Translating the sign convention of Figure 6.3 to the sign convention of Figure 6.2, and applying Equations (6.84)–(6.87) on $x=0$, the following boundary conditions can be specified:

$$f_1=f(0)=C_1+I_1(0) \tag{6.92}$$

$$-m_1=m(0)=C_2+I_2(0) \tag{6.93}$$

$$\vartheta_1=\vartheta(0)=\frac{C_3}{EI}+I_3(0) \tag{6.94}$$

$$u_1=u(0)=\frac{C_4}{EI}+I_4(0) \tag{6.95}$$

Therefore, the arbitrary constants C_1, C_2, C_3, C_4 can be expressed in terms of the physical quantities f_1, m_1, ϑ_1, u_1:

$$C_1=f_1-I_1(0) \tag{6.96}$$

$$C_2=-m_1-I_2(0) \tag{6.97}$$

$$C_3=EI\vartheta_1-EI.I_3(0) \tag{6.98}$$

$$C_4=EIu_1-EI.I_4(0) \tag{6.99}$$

Using the above equations, Equations (6.84)–(6.87) can now be written:

$$u(x) = u_1 + \vartheta_1 x - m_1 \frac{x^2}{2EI} + f_1 \frac{x^3}{6EI} + g_1(x) \tag{6.100}$$

$$\vartheta(x) = 0.u_1 + \vartheta_1 - m_1 \frac{x}{EI} + f_1 \frac{x^2}{2EI} + g_2(x) \tag{6.101}$$

$$f(x) = 0.u_1 + 0.\vartheta_1 + 0.m_1 + f_1 + g_3(x) \tag{6.102}$$

$$m(x) = 0.u_1 + 0.\vartheta_1 - m_1 + f_1 x + g_4(x) \tag{6.103}$$

where

$$g_1(x) = -I_4(0) - xI_3(0) - x^2 \frac{I_2(0)}{2EI} - x^3 \frac{I_1(0)}{6EI} + I_4(x) \tag{6.104}$$

$$g_2(x) = -I_3(0) - \frac{x}{EI} I_2(0) - \frac{x^2}{2EI} I_1(0) + I_3(x) \tag{6.105}$$

$$g_3(x) = -I_1(0) + I_1(x) \tag{6.106}$$

$$g_4(x) = -I_2(0) - xI_1(0) + I_2(x) \tag{6.107}$$

Taking into account the sign convention of Figure 6.3, the physical quantities f_1, m_1, ϑ_1, u_1 for $x = L$ (node 2) can be obtained by the above equations:

$$u_2 = u_1 + \vartheta_1 L - m_1 \frac{L^2}{2EI} + f_1 \frac{L^3}{6EI} + g_1(L) \tag{6.108}$$

$$\vartheta_2 = 0.u_1 + \vartheta_1 - m_1 \frac{L}{EI} + f_1 \frac{L^2}{2EI} + g_2(L) \tag{6.109}$$

$$-f_2 = 0.u_1 + 0.\vartheta_1 + 0.m_1 + f_1 + g_3(L) \tag{6.110}$$

$$m_2 = 0.u_1 + 0.\vartheta_1 - m_1 + f_1 L + g_4(L) \tag{6.111}$$

The above four equations can be rearranged as follows:

$$\frac{L^3}{6EI} f_1 - \frac{L^2}{2EI} m_1 + 0.f_2 + 0.m_2 = -u_1 - L\vartheta_1 + u_2 + 0.\vartheta_2 - g_1(L) \tag{6.112}$$

$$\frac{L^2}{2EI} f_1 - \frac{L}{EI} m_1 + 0.f_2 + 0.m_2 = 0.u_1 - \vartheta_1 + 0.u_2 + \vartheta_2 - g_2(L) \tag{6.113}$$

$$L f_1 - m_1 + 0.f_2 - m_2 = 0.u_1 + 0.\vartheta_1 + 0.u_2 + 0.\vartheta_2 - g_4(L) \tag{6.114}$$

$$f_1 + 0.m_1 + f_2 + 0.m_2 = 0.u_1 + 0.\vartheta_1 + 0.u_2 + 0.\vartheta_2 - g_3(L) \tag{6.115}$$

The above system of equations can now be written in the following matrix form:

$$\begin{bmatrix} \frac{L^3}{6EI} & -\frac{L^2}{2EI} & 0 & 0 \\ \frac{L^2}{2EI} & -\frac{L}{EI} & 0 & 0 \\ L & -1 & 0 & -1 \\ 1 & 0 & 1 & 0 \end{bmatrix} \begin{Bmatrix} f_1 \\ m_1 \\ f_2 \\ m_2 \end{Bmatrix} = \begin{bmatrix} -1 & -L & 1 & 0 \\ 0 & -1 & 0 & 1 \\ 0 & 0 & 0 & 0 \\ 0 & 0 & 0 & 0 \end{bmatrix} \begin{Bmatrix} u_1 \\ \vartheta_1 \\ u_2 \\ \vartheta_2 \end{Bmatrix} - \begin{Bmatrix} g_1(L) \\ g_2(L) \\ g_4(L) \\ g_3(L) \end{Bmatrix} \tag{6.116}$$

or in an abbreviated form

$$[U_1]\begin{Bmatrix} f_1 \\ m_1 \\ f_2 \\ m_2 \end{Bmatrix} = [U_2]\begin{Bmatrix} u_1 \\ \vartheta_1 \\ u_2 \\ \vartheta_2 \end{Bmatrix} - \{G\} \tag{6.117}$$

Where $[U_1]$ and $[U_2]$ are given by Equations (6.78) and (6.79), respectively, and

$$\{G\} = \{g_1(L), g_2(L), g_4(L), g_3(L)\}^{\mathrm{T}} \tag{6.118}$$

Therefore, Equation (6.117) yields:

$$\begin{Bmatrix} f_1 \\ m_1 \\ f_2 \\ m_2 \end{Bmatrix} = \frac{EI}{L^3}\begin{bmatrix} 12 & 6L & -12 & 6L \\ 6L & 4L^2 & -6L & 2L^2 \\ -12 & -6L & 12 & -6L \\ 6L & 2L^2 & -6L & 4L^2 \end{bmatrix} \begin{Bmatrix} u_1 \\ \vartheta_1 \\ u_2 \\ \vartheta_2 \end{Bmatrix} + \{F\} \tag{6.118}$$

where

$$\{F\} = -[U_1]^{-1}.\{G\} \tag{6.119}$$

Performing the required algebraic operations, the following result can be obtained:

$$\{F\} = \begin{Bmatrix} \frac{12EI}{L^3}g_1 + \frac{6EI}{L^2}g_2 \\ -\frac{6EI}{L^2}g_1 + \frac{2EI}{L}g_2 \\ \frac{12EI}{L^3}g_1 - \frac{6EI}{L^2}g_2 + g_3 \\ -\frac{6EI}{L^2}g_1 + \frac{4EI}{L}g_2 - g_4 \end{Bmatrix} \tag{6.120}$$

The parameters $g_1 = g_1(L), g_2 = g_2(L), g_3 = g_3(L), g_4 = g_4(L)$ depends on the type of varying loading. For common cases of varying loadings, the parameters g_1, g_2, g_3, g_4 can be obtained using Table 6.2.

Table 6.2 Parameters g_i ($i=1,2,3,4$) and load matrix $\{F\}$ for common cases of varying loads

Loading case of a beam element	Parameters $\{G\}=\{g_1, g_2, g_3, g_4\}^T$ and load matrix $\{F\}=\{F_1, M_1, F_2, M_2\}^T$
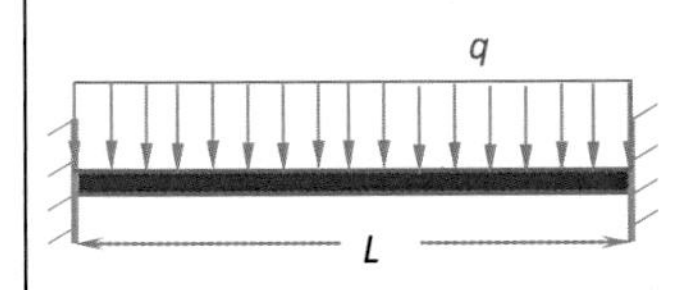	$g_1=-qL^4/24EI$, $g_2=-qL^3/6EI$ $g_3=-qL$, $g_4=-qL^2/2$ $F_1=qL/2$, $M_1=qL^2/12$ $F_2=qL/2$, $M_2=-qL^2/12$
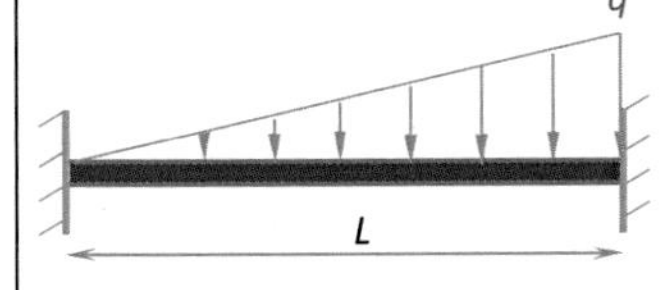	$g_1=-qL^4/120EI$, $g_2=-qL^3/24EI$ $g_3=-qL/2$, $g_4=-qL^2/6$ $F_1=3qL/20$, $M_1=qL^2/30$ $F_2=7qL/20$, $M_2=-qL^2/20$
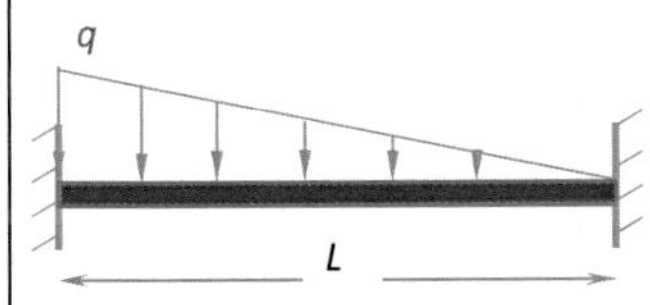	$g_1=-qL^4/30EI$, $g_2=-qL^3/8EI$ $g_3=-qL/2$, $g_4=-qL^2/3$ $F_1=7qL/20$, $M_1=qL^2/20$ $F_2=3qL/20$, $M_2=-qL^2/30$
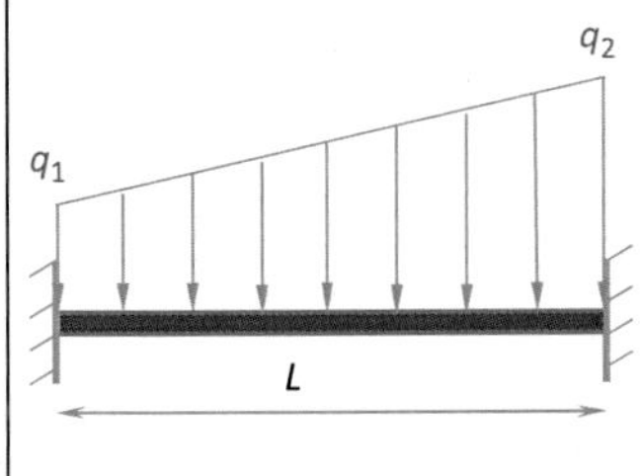	$g_1=-\dfrac{(4q_1+q_2)L^4}{120EI}$, $g_2=-\dfrac{(3q_1+q_2)L^3}{24EI}$ $g_3=-(q_1+q_2)L/2$, $g_4=(2q_1+q_2)L^2/3$ $F_1=\dfrac{q_1L}{2}+\dfrac{3(q_2-q_1)L}{20}$, $M_1=\dfrac{q_1L^2}{12}+\dfrac{(q_2-q_1)L^2}{30}$ $F_2=\dfrac{(q_1+q_2)L}{2}-F_1$, $M_2=-\dfrac{(q_2-q_1)L^2}{6}-\dfrac{q_1L^2}{2}+F_1L+M_1$
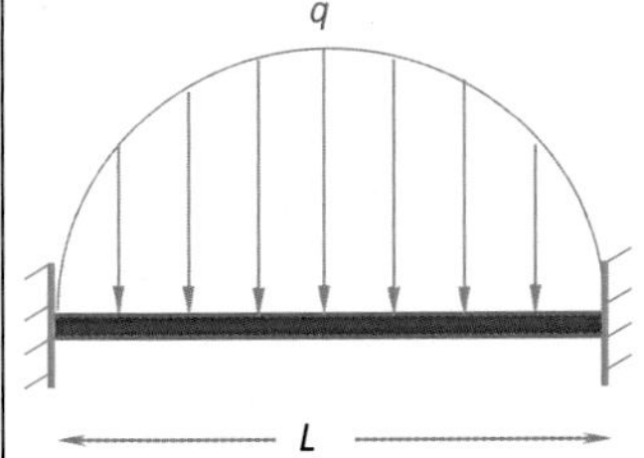	$g_1=41\pi qL^4/768EI$, $g_2=-11\pi qL^3/128EI$ $g_3=\pi qL/4$, $g_4=\pi qL/4$ $F_1=\pi qL/8$, $M_1=19\pi qL^2/128$ $F_2=\pi qL/8$, $M_2=19\pi qL^2/128$

EXAMPLE 6.2: ANALYSIS OF A BEAM USING MATLAB/CALFEM

Determine the displacements, reactions, bending moments and shear forces for the following beam by using MATLAB/CALFEM.

Data

$L=8.0$ m; $q=3$ KN/m; $E=200$ GPa; $I=125\times10^{-6}$ m^4 or $R=0.152$ (radius of circular cross-section); $A=40\times10^{-4}$ m^2; and $F=12$ KN

Continued

EXAMPLE 6.2: ANALYSIS OF A BEAM USING MATLAB/CALFEM—CONT'D

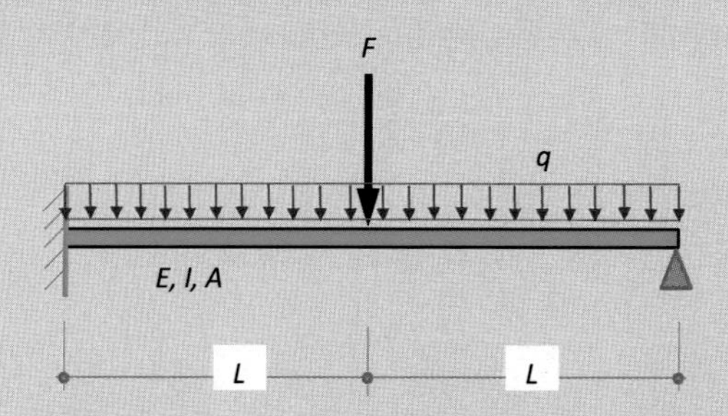

Step 1: Degrees of Freedom

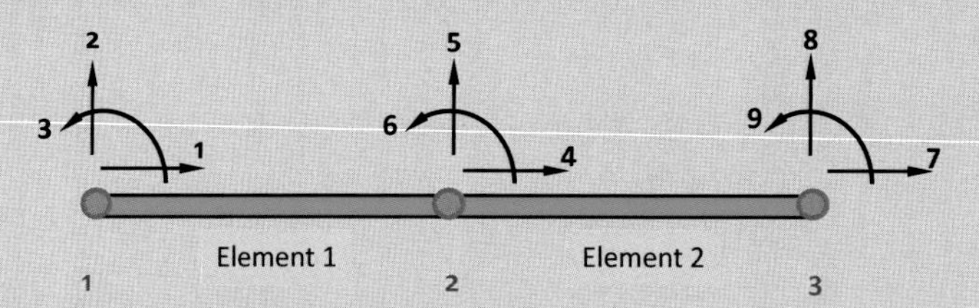

The structure has two elements, three nodes, and nine degrees of freedom.

In order to derive a solution for the above problem we need:

a) the global stiffness matrix $[K]$,
b) the load vector $\{f\}$,
c) the boundary conditions $[bc]$

Step 2: Load Vector {f}

f = zeros(9,1);f(5) = -12000;

The above vector **{f}** corresponds to concentrated loads acting on nodes. Apart from the above load vector, it should be taken into account that the distributed load **q** acts along elements 1 and 2. The corresponding vectors for distributed loads acting on the elements 1 and 2 are symbolized by **eq1** and **eq2**, respectively, and they will be derived in Step 4.

Step 3: Boundary Conditions {bc}

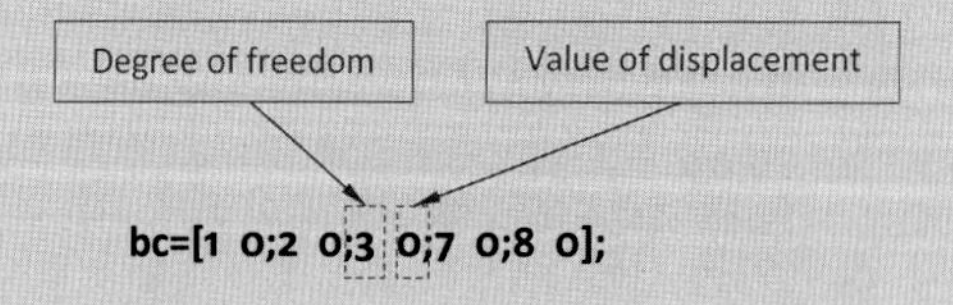

The above matrix means that the displacements in the degrees of freedom **1, 2, 7, 8** are **zero**, and the slope in the degree of freedom **3** is also **zero**.

Step 4: Global Stiffness Matrix {K}

Derivation of Element Matrices Kei

The command for deriving the element matrices in the global coordinate system for any 2D beam element is **beam2e(ex,ey,ep,eq)**. In this command, apart from the already known vectors ***ex, ey*** containing the coordinates of the beam element as well as the vector **ep** containing its material properties, we also need the distributed load vector **eq**. The syntax of **eq** is **eq** $= [\mathbf{q}_{\bar{x}}, \mathbf{q}_{\bar{y}}]$, where $q_{\bar{x}}, q_{\bar{y}}$ are the components of the distributed load in the local coordinate system of the element. The data ***exi, eyi, epi, eqi*** for any element *i* are:

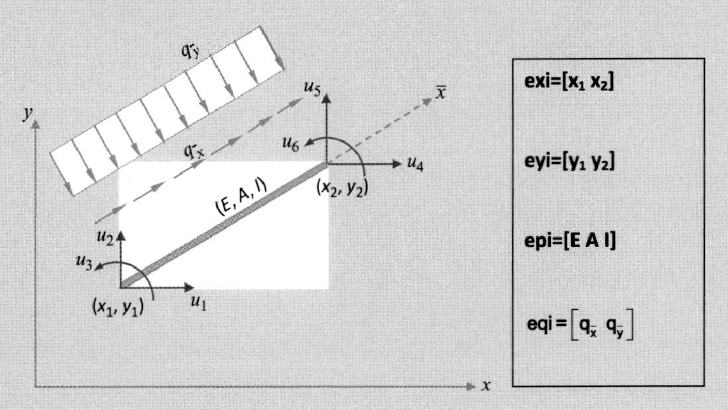

Then, the corresponding element matrix is:

```
[Kei,fei]=beam2e(exi,eyi,epi,eqi)
```

Therefore, the element matrices of the structure can be derived as follows:

```
L = 8;qx = 0;qy = 3000;E = 2e11;I = 125e-6;A = 40e-4;
ex1 = [0 L];ex2 = [L 2*L];
ey1 = [0 0];ey2 = [0 0];
ep1 = [E A I];ep2 = [E A I];
eq1 = [0 qy];eq2 = [0 qy];
[Ke1,fe1] = beam2e(ex1,ey1,ep1,eq1);
[Ke2,fe2] = beam2e(ex2,ey2,ep2,eq2);
```

In the above commands Ke1 and Ke2 are the element matrices of the elements 1 and 2, and fe1 and fe2 are their load vectors, respectively.

*Derivation of the Topology Matrix [**Edof**]*

The topology matrix **[Edof]** specifies the interconnections of the elements. The first component of each row indicates the number of the element, and the remaining six components of the row indicate the degrees of freedom of the ends of the corresponding element. Therefore, the matrix **[Edof]** of the structure can be derived as follows:

Continued

EXAMPLE 6.2: ANALYSIS OF A BEAM USING MATLAB/CALFEM—CONT'D

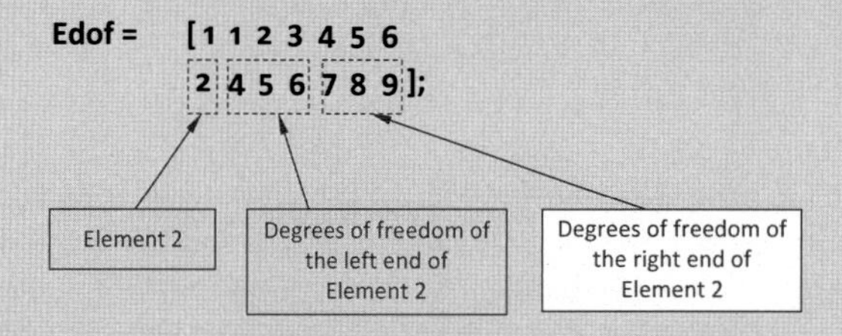

Assembly of the Element Matrices

Based on the topology information, the global stiffness matrix can be generated by assembling the element stiffness matrices using the command assem:

```
K=zeros(9);
[K,f]=assem(Edof(1,:),K,Ke1,f,fe1);
[K,f]=assem(Edof(2,:),K,Ke2,f,fe2);
```

Step 5: Computation of the Displacement Field [a] and the Reactions {r}

The system of equations for the derivation of [a] and $\{r\}$ are solved considering the global stiffness matrix $[K]$, the load vector $\{f\}$, and the boundary conditions $[bc]$:

```
[a,r]=solveq(K,f,bc)
```

Results

```
a =
        0
        0
        0
        0
  -0.0589
  -0.0035
        0
        0
   0.0141
r =
   1.0e + 05 *
        0
   0.3825
   1.3200
        0
  -0.0000
        0
        0
   0.2175
  -0.0000
```

Step 6: Computation of the Distributions of displacements Ed, Axial Forces es(:,1), shear Forces es(:,2), and Bending Moments es(:,3)

Distributions of displacements $\bar{u}$, $\bar{v}$ and section forces N, V, M along the local coordinate system $\bar{x}$ of each element are symbolized by the following nomenclature:

es = [N V M]

Ed = $[\bar{u}, \bar{v}]$

The computation of the es can be performed by the command **beam2s(ex,ey,ep,Ed,eq,n)**.

Required data for deriving the above matrices for each individual element are:

(a) the coordinate matrices **ex, ey** of the element ends;
(b) the matrix **ep** of the material properties of the element;
(c) the matrix **Ed** of the element deformation;
(d) the load matrix **eq** of the element;
(e) the number of evaluation points **n** (i.e., in how many points we need the results in order the derived diagrams to be smooth).

Therefore, for the two elements of the problem, the following results for **n = 10** points can be derived:

```
Ed=extract(Edof,a);
es1=beam2s(ex1,ey1,ep1,Ed(1,:),eq1,20)
es2=beam2s(ex2,ey2,ep2,Ed(2,:),eq2,20)
```

Step 7: Graphical Representation of Displacements Ed, Axial Forces es(:,1), Shear Forces es(:,2), and Bending Moments es(:,3)

Since the results **Ed, es1, es2** have been obtained, their graphical representation is now possible. These distributions should be put on the drawing of the undeformed elements.

Drawing of the Undeformed Elements

```
figure(1)                          % creates a window for graphics
   plotpar=[2 1 0];                % specifies the line type and color and the node mark
   eldraw2(ex1,ey1,plotpar);       % creates the drawing of the undeformed element 1
   eldraw2(ex2,ey2,plotpar);       % creates the drawing of the undeformed element 2
```

Drawing of the Displacement Distribution

First, we have to specify the scale factor in order the illustrated quantities be obvious within the area of the monitor. In order to specify a reasonable scale factor, we have to correlate the displayed quantities of each point with a reference, for example, the Ed(2,:). The scale factor can be specified using the following command:

```
sfac=scalfact2(ex2,ey2,Ed(2,:),0.1);   % the value 0.1 express the relation between the
                                       illustrated quantity and the element size.
```

Let us now derive the diagram of the displacement distributions of elements 1 and 2. The command to do this is eldisp2:

```
plotpar=[1 2 1];
eldisp2(ex1,ey1,Ed(1,:),plotpar,sfac);
eldisp2(ex2,ey2,Ed(2,:),plotpar,sfac);
```

Continued

EXAMPLE 6.2: ANALYSIS OF A BEAM USING MATLAB/CALFEM—CONT'D

Apart from the distributions, we have to specify the axes of coordinates:

```
axis([-1.0 18 -2.0 1.5]);   % this command specifies the x and y axes. The values of the ends
                            of the x-axis are -1.0 and 18.0. The values of the ends of the
                            y-axis are -0.5 and 6.0
```

The title of the drawing can be placed by the following command:

```
title('displacements')
```

Drawing of the Axial Force, Shear Force, and Bending Moment Distribution

The axial forces **N**, shear forces **V**, and bending moments **M** are displayed by using the function **eldia2**. For example, the following command

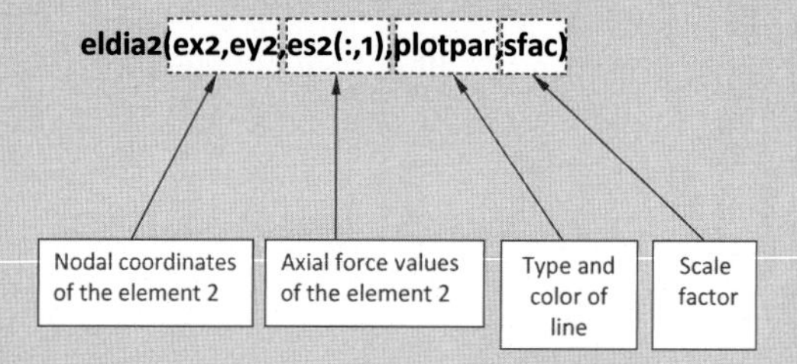

will display the diagram of axial forces **es2 (: , 1)** of element 2.

Similarly, the command

```
eldia2(ex2,ey2,es2(:,2),plotpar,sfac);
```

will display the diagram of shear forces **es2 (: , 2)** of the element 2, and the command

```
eldia2(ex2,ey2,es2(:,3),plotpar,sfac);
```

will display the diagram of bending moments **es2 (: , 3)** of the element 2.

Therefore, the list of commands for deriving the drawings of axial forces, shear forces, and bending moments is:

Drawing of the Axial Forces

```
figure(2)
  plotpar = [2 1];
  sfac = scalfact2(ex1,ey1,es1(:,1),0.2);
  eldia2(ex1,ey1,es1(:,1),plotpar,sfac);
  eldia2(ex2,ey2,es2(:,1),plotpar,sfac);
  axis([-1.5 18 -0.5 2.0]);
  title('axial force')
```

Drawing of the Shear Forces

```
figure(3)
  sfac = scalfact2(ex1,ey1,es1(:,2),0.2);
  eldia2(ex1,ey1,es1(:,2),plotpar,sfac);
  eldia2(ex2,ey2,es2(:,2),plotpar,sfac);
  axis([-1.5 18 -1.5 2.0]);
  title('shear force')
```

Drawing of the Bending Moments

```
figure(4)
   sfac = scalfact2(ex1,ey1,es1(:,3),0.2);
   eldia2(ex1,ey1,es1(:,3),plotpar,sfac);
   eldia2(ex2,ey2,es2(:,3),plotpar,sfac);
   axis([-1.5 18 -1.5 2.0]);
   title('bending moment')
```

Results

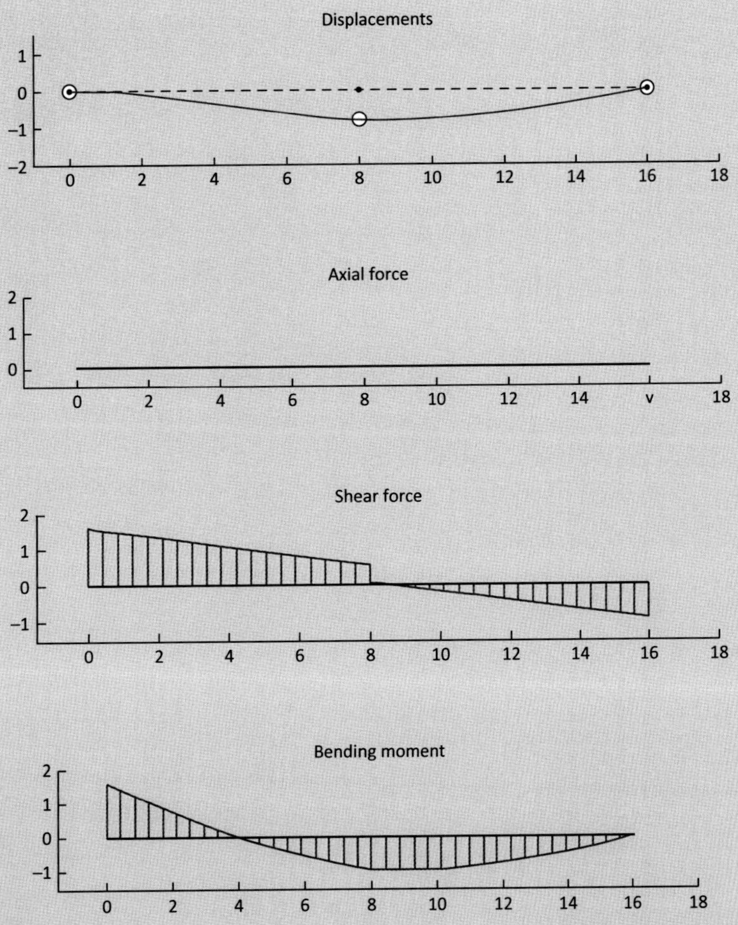

Displacements

Axial force

Shear force

Bending moment

Continued

EXAMPLE 6.2: ANALYSIS OF A BEAM USING MATLAB/CALFEM—CONT'D

The CALFEM/MATLAB computer code

```
f = zeros(9,1);f(5) = -12000;
bc = [1 0;2 0;3 0;7 0;8 0];
L = 8;qx = 0;qy = -3000;E = 2e11;I = 125e-6;
ex1 = [0 L];ex2 = [L 2*L];
ey1 = [0 0];ey2 = [0 0];
ep1 = [E 40e-4 I];ep2 = [E 40e-4 I];
eq1 = [0 qy];eq2 = [0 qy];
[Ke1,fe1] = beam2e(ex1,ey1,ep1,eq1);
[Ke2,fe2] = beam2e(ex2,ey2,ep2,eq2);
Edof = [1 1 2 3 4 5 6;2 4 5 6 7 8 9];
K = zeros(9);
[K,f] = assem(Edof(1,:),K,Ke1,f,fe1);
[K,f] = assem(Edof(2,:),K,Ke2,f,fe2);
[a,r] = solveq(K,f,bc)
Ed = extract(Edof,a);
es1 = beam2s(ex1,ey1,ep1,Ed(1,:),eq1,20)
es2 = beam2s(ex2,ey2,ep2,Ed(2,:),eq2,20)
figure(1)
plotpar = [2 1 0];
eldraw2(ex1,ey1,plotpar);
eldraw2(ex2,ey2,plotpar);
sfac = scalfact2(ex2,ey2,Ed(2,:),0.1);
plotpar = [1 2 1];
eldisp2(ex1,ey1,Ed(1,:),plotpar,sfac);
eldisp2(ex2,ey2,Ed(2,:),plotpar,sfac);
axis([-1.0 18 -2.0 1.5]);
title('displacements')
figure(2)
plotpar = [2 1];
sfac = scalfact2(ex1,ey1,es1(:,1),0.2);
eldia2(ex1,ey1,es1(:,1),plotpar,sfac);
eldia2(ex2,ey2,es2(:,1),plotpar,sfac);
axis([-1.5 18 -0.5 2.0]);
title('axial force')
figure(3)
sfac = scalfact2(ex1,ey1,es1(:,2),0.2);
eldia2(ex1,ey1,es1(:,2),plotpar,sfac);
eldia2(ex2,ey2,es2(:,2),plotpar,sfac);
axis([-1.5 18 -1.5 2.0]);
```

```
title('shear force')
figure(4)
sfac = scalfact2(ex1,ey1,es1(:,3),0.2);
eldia2(ex1,ey1,es1(:,3),plotpar,sfac);
eldia2(ex2,ey2,es2(:,3),plotpar,sfac);
axis([-1.5 18 -1.5 2.0]);
title('bending moment')
```

EXAMPLE 6.3: ANALYSIS OF A BEAM USING ANSYS

Determine the deformed shape of the following beam.

Data

$L = 8.0$ m
$q = 3$ KN/m
$E = 200$ GPa
$R = 0.152$ (radius of circular cross section)
$F = 12$ KN

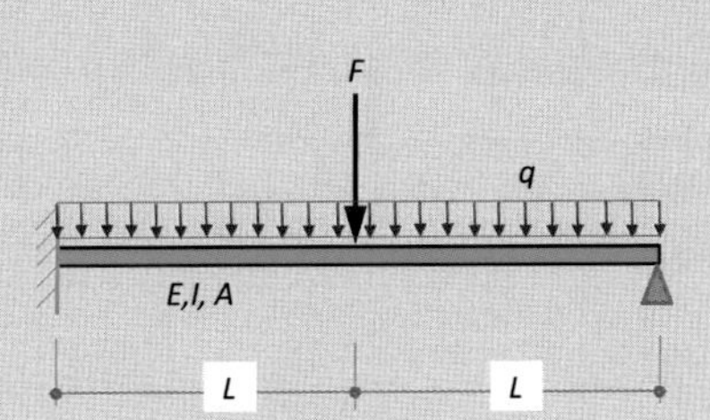

UM: **File → Change Directory**

Using this command, we select the directory to save all files generated for this exercise. Let us choose Directory **D**, and file **ANSYS TUTORIALS**.

INPUT: /UNITS, SI

This command defines the units to SI.

UM: File → Change Job Name

With this command, we specify the job name. Let us choose Example 6.3

UM: File → Change Title

We can choose again **BEAM**.

MM: Preferences

Continued

EXAMPLE 6.3: ANALYSIS OF A BEAM USING ANSYS—CONT'D

The following window will appear. We must choose *"Structural"*

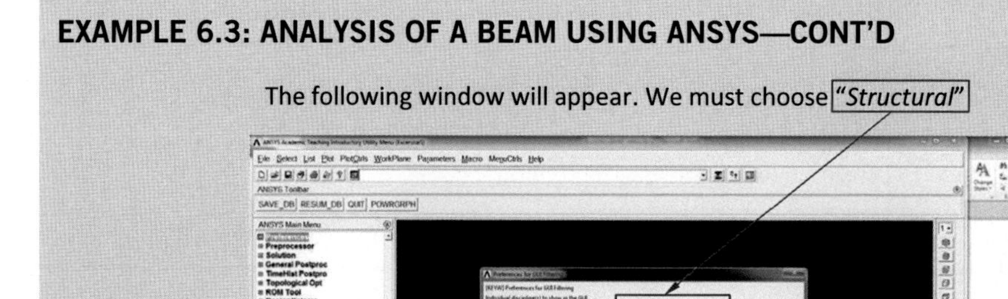

Now we have to start data entry through the preprocessor:

MM: **Preprocessor→Element Type→Add/Edit/Delete→Add→Beam→ 2 node 188→OK→Close**

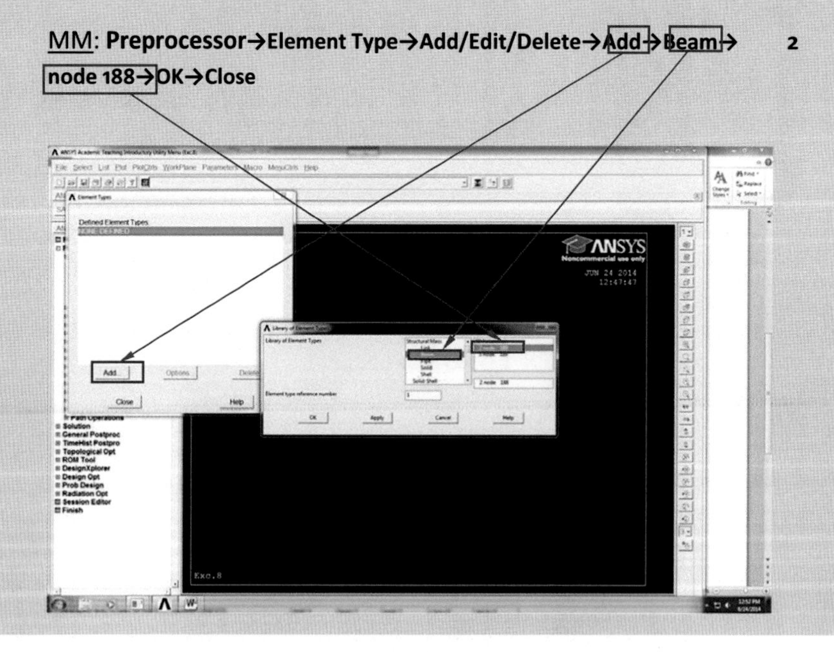

The next step is to specify the material properties using the following commands:

MM: Preprocessor → Material Props → Material Models → Structural → Linear → Elastic → Isotropic

Then, in the following window, the data for modulus of elasticity and Poisson's ratio have to be filled in the corresponding boxes; "EX" = 2.1e11 and "PRXY" = 0.3.

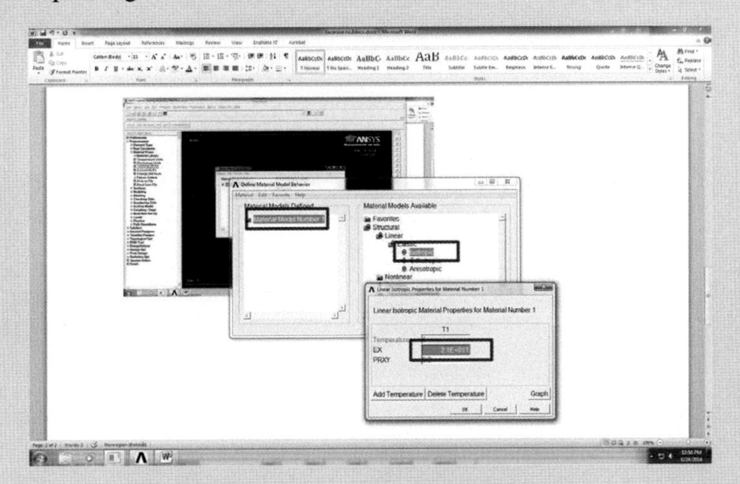

Next step is to specify the beam cross-section type and size:

MM: Preprocessor → Sections → Beam → Common Sections

A cross-sections library then appears, and a circular cross-section with radius **R = 0.152** is selected. Apart from the radius R, the number of divisions along the perimeter and radius should be specified. The program needs these divisions in order to calculate the stress distribution in the surface of the cross-sections. Let us choose **N = 12**, and **T = 8**. The completion of the following boxes

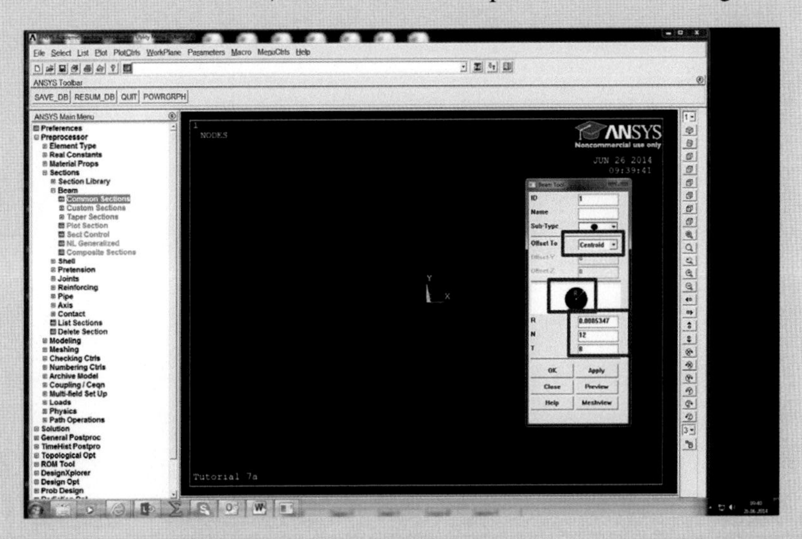

Continued

and then pressing **Meshview** yield the following cross-section mesh:

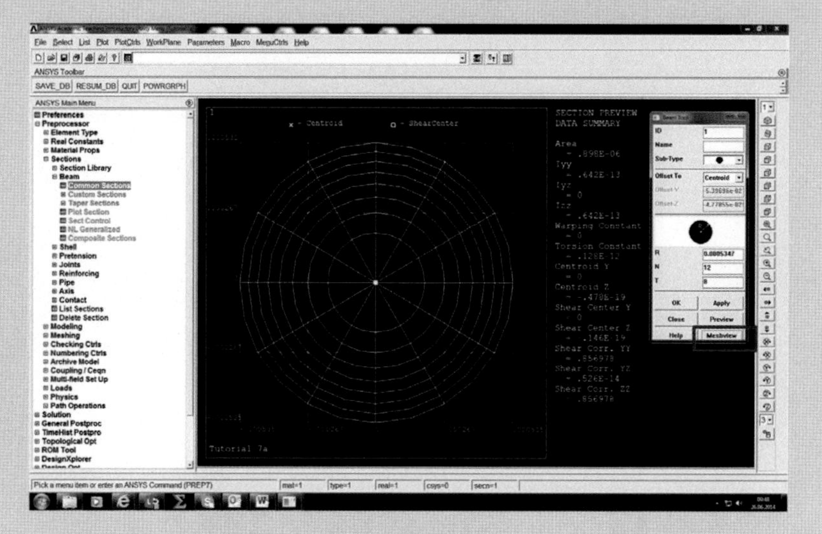

Now it is time to insert the coordinates of the Keypoints:

<u>MM</u>: Preprocessor → Modeling → Create → Keypoints → In Active CS

In the following window, we have to fill the **Keypoint number** and the **x, y** coordinates of the first Keypoint, and then to press the button **Apply**. We must continue filling the successive windows for the remainter of the Keypoints. After the completion of the coordinates of the last Keypoint, instead of **Apply** we must press the button **OK**.

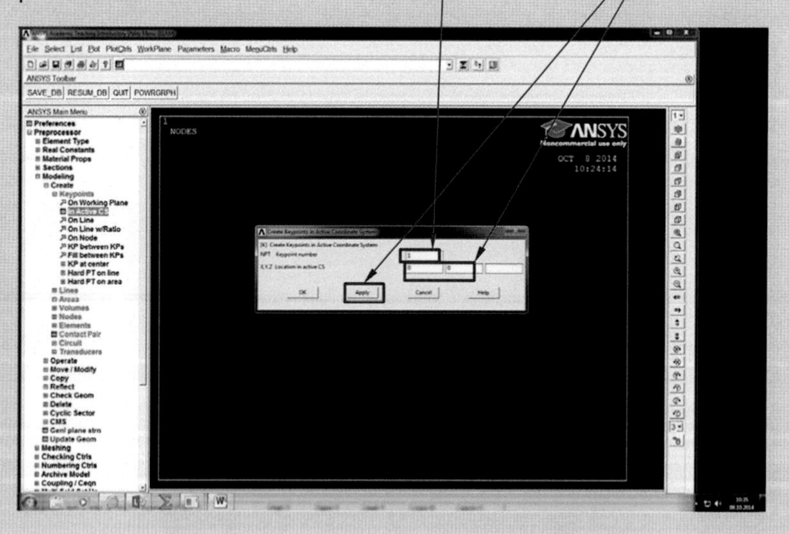

The above procedure yields a demonstration of locations of the keypoints on the screen. The next step is to connect the Keypoints using the following commands in order to create the lines:

MM: Preprocessor → Modeling → Create → Lines → Lines → In Active Coord

By clicking on the Keypoints, we create the lines, and then we press the button **OK** in the window entitled "Lines in Active Coord".

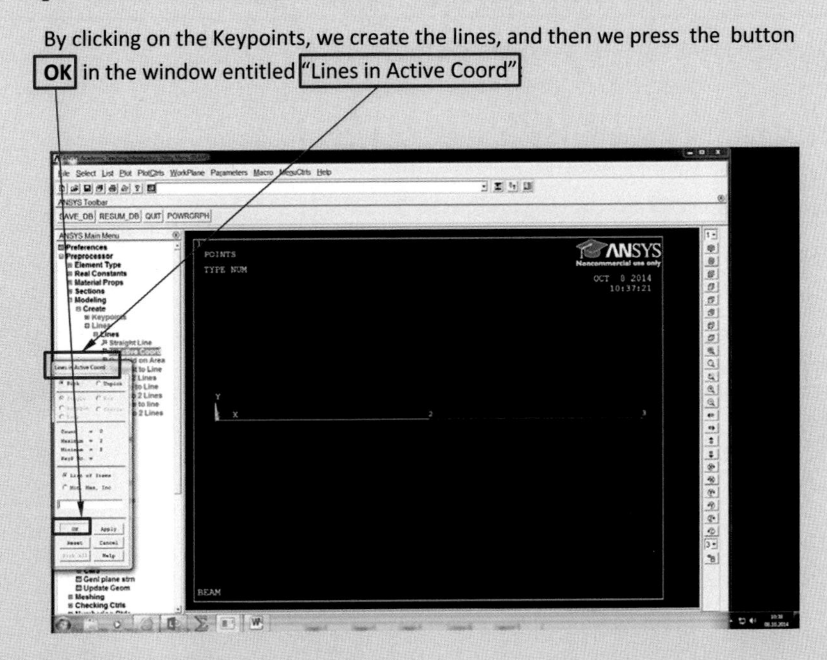

Now it is time to create the mesh of the solid. We can do this using the following commands:

MM: Preprocessor → Meshing → MeshTool

A window entitled "Mesh tool" will apear. In this window we must press the button **"set"** for the Lines. After that we have to press the button **"Pick All"** in the window entitled "Element Sizes on Picked Lines".

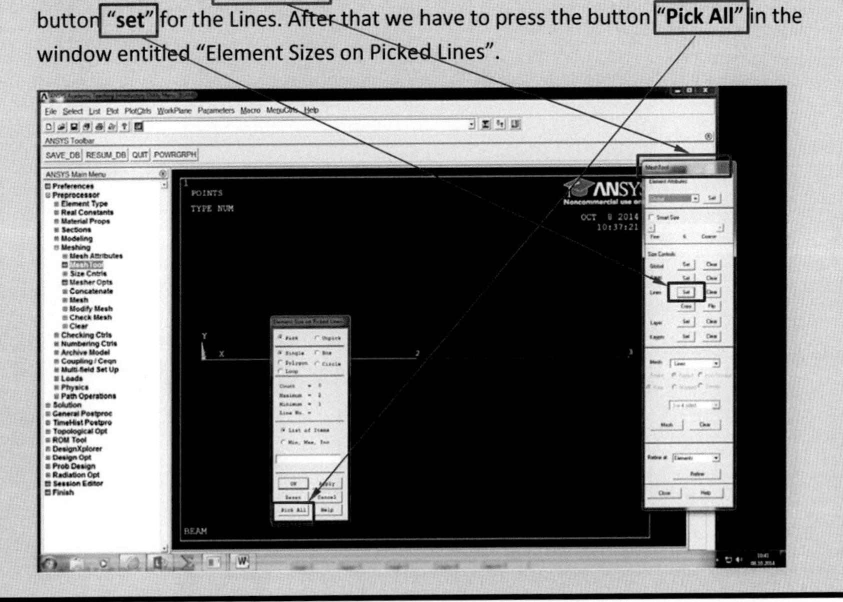

Continued

EXAMPLE 6.3: ANALYSIS OF A BEAM USING ANSYS—CONT'D

On the new window we must fill the number the number of element divisions (NDIV). Let's set **NDIV=10,** and press **"OK".**

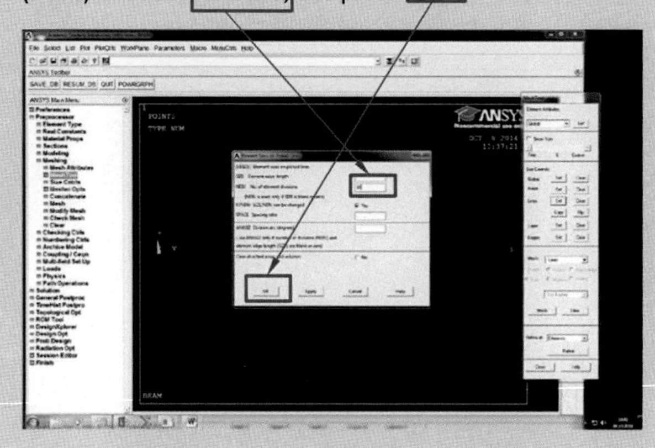

The next step is to return again in the window "Mesh Tool" and press the button **"Mesh"**.

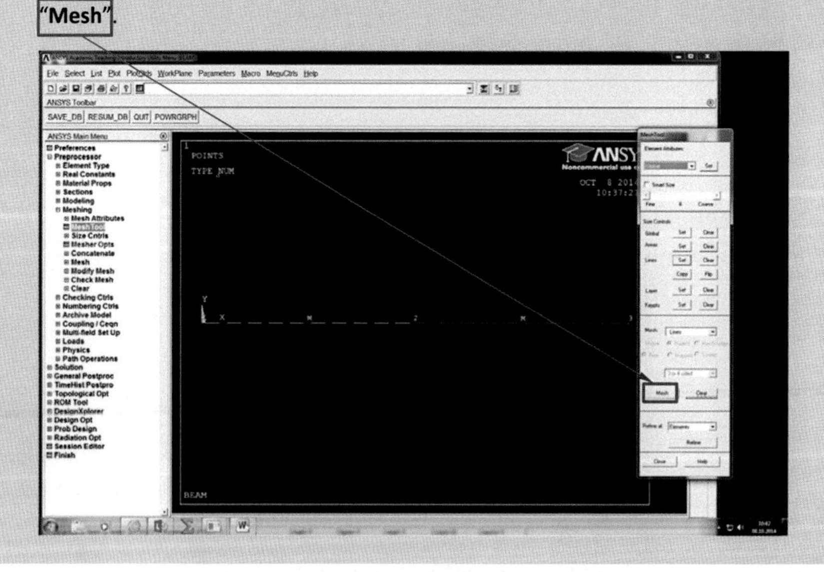

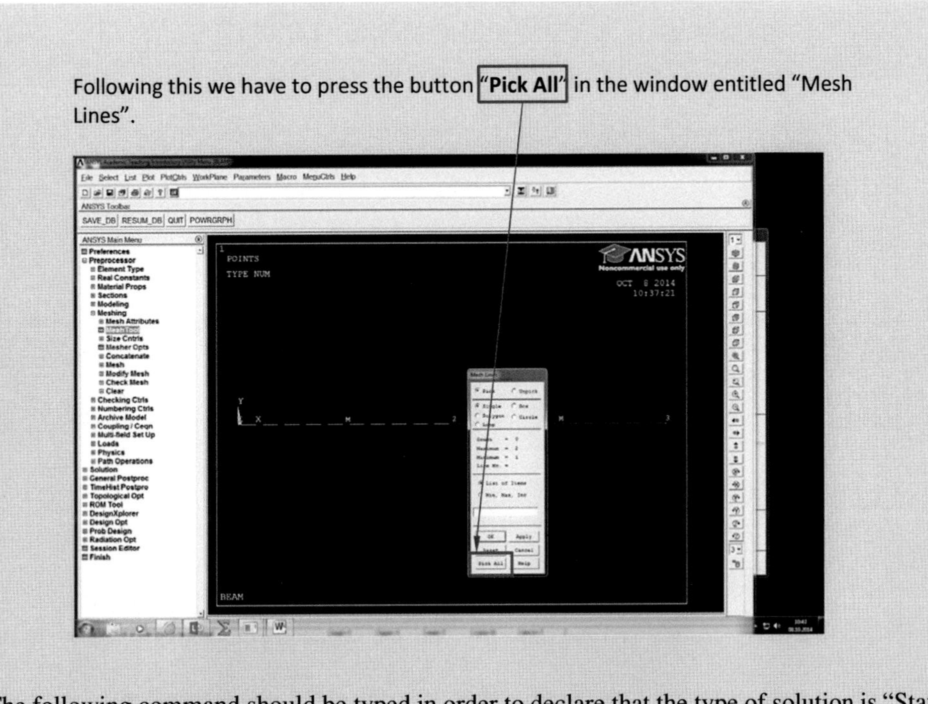

The following command should be typed in order to declare that the type of solution is "Static".

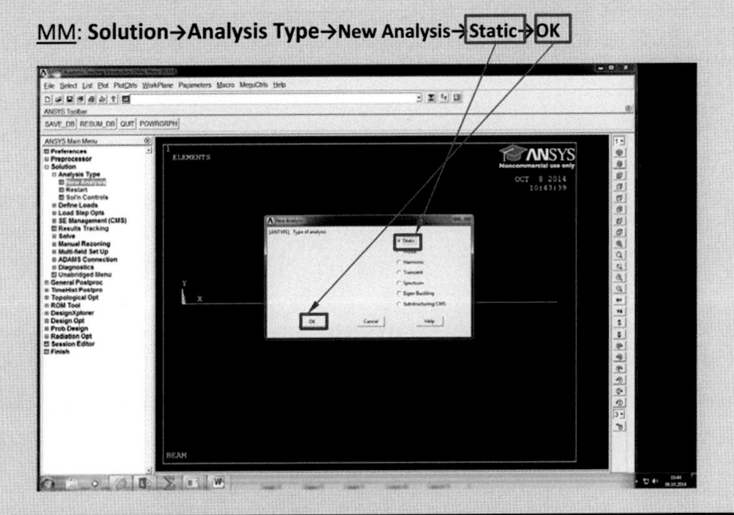

Continued

EXAMPLE 6.3: ANALYSIS OF A BEAM USING ANSYS—CONT'D

Now, using the following commands, it is time to specify the boundary conditions on the supports of the structure.

MM: Solution → Define Loads → Apply → Structural → Displacements → On Nodes

We must pick the Keypoint **1** located on the fixed support, then click the button "**Apply**" and select the degrees of freedom **All DOF** for setting the "**Displacement value"=0.** Then, click **"OK"**.

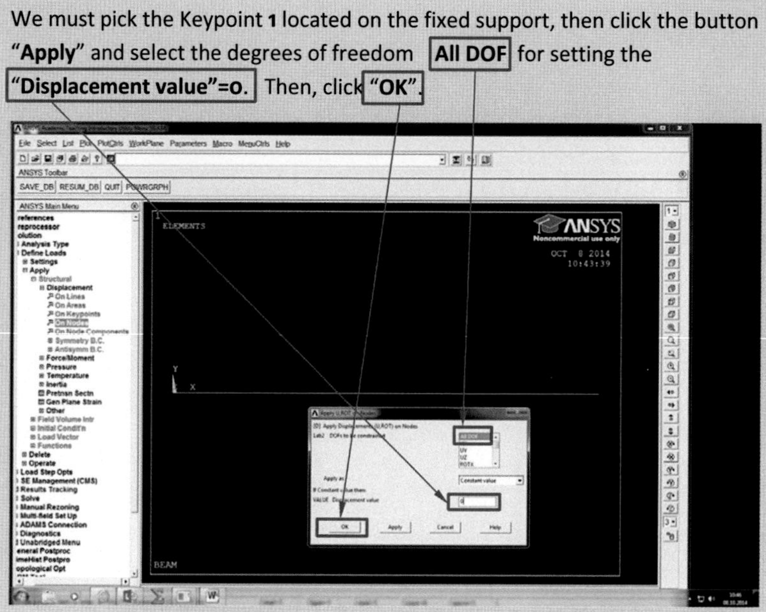

We have to continue with the commands

MM: Solution → Define Loads → Apply → Structural → Displacements → On Nodes

and pick Keypoint **3** located on the pinned support, then click the button "**Apply**" and select the degrees of freedom **UX, UY** for setting the "**Displacement value" = 0**. Then, click "**OK**".

After the above procedure, we must specify the loads.

Therefore, we must type the commands.

MM: Solution → Define Loads → Apply → Structural → Pressure → On Beams

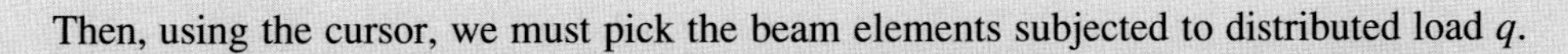
Then, using the cursor, we must pick the beam elements subjected to distributed load q.

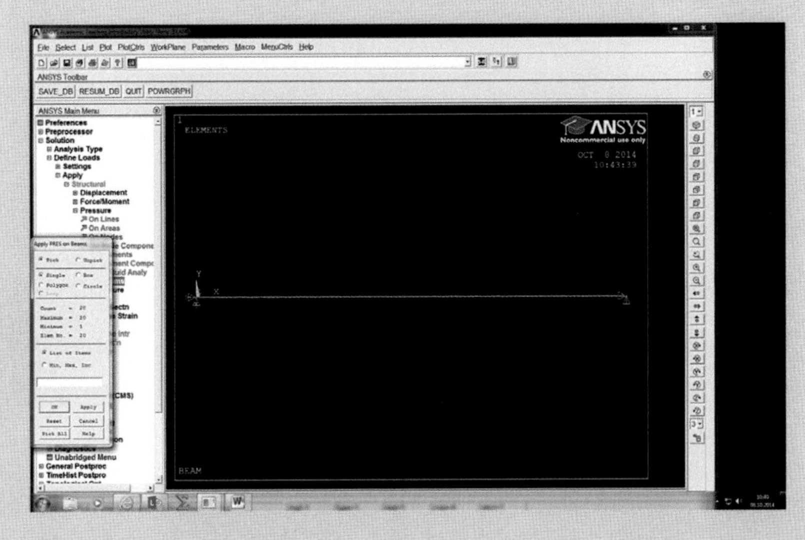

The next step is to press the button «**OK**», and then to insert the type of loading **"LKEY"=2,** and the value **3000** of the distributed load at the end points **i** and **j** of the beam:

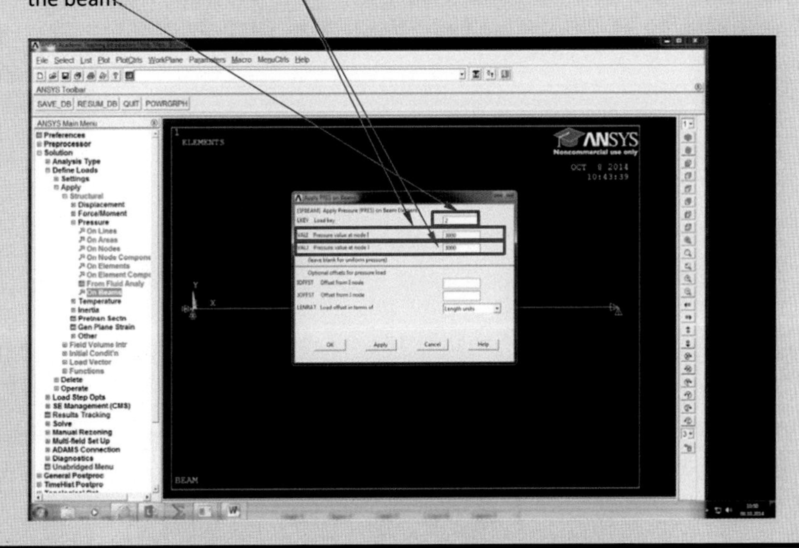

Continued

Then the following demonstration of the distributed load appears:

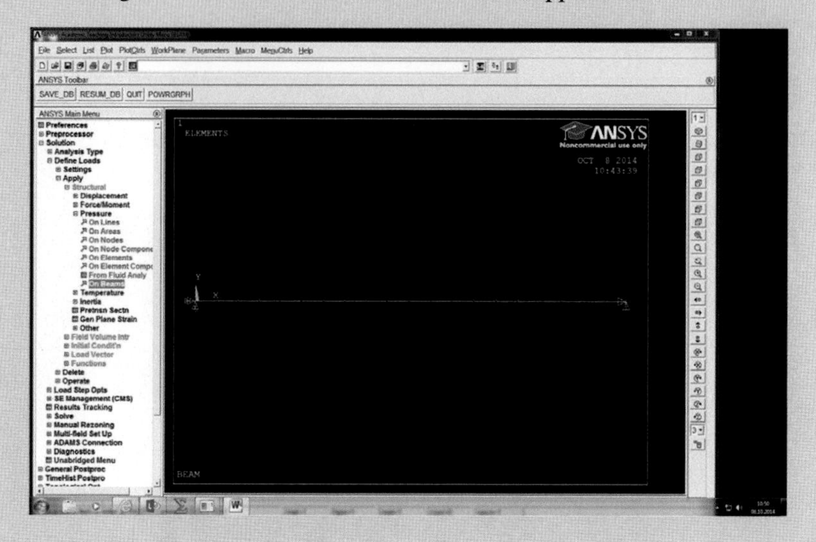

Apart from the distributed load, we also have to specify the concentrated load **F**.

MM: Solution → Define Loads → Apply → Structural → Force/Moment → On Keypoints

Then, using the cursor, we must pick Keypoint **3** subjected to concentrated load **F,** and press the button **"OK"** in the window entitled "Apply F/M on KPs".

The program opens now a new window entitled "Apply F/M on KPs", asking for the direction and the value of the concentrated load. We choose direction **FY** and value **-12000** and then, click **"OK"** in order the program to demonstrate the concentrated force as well.

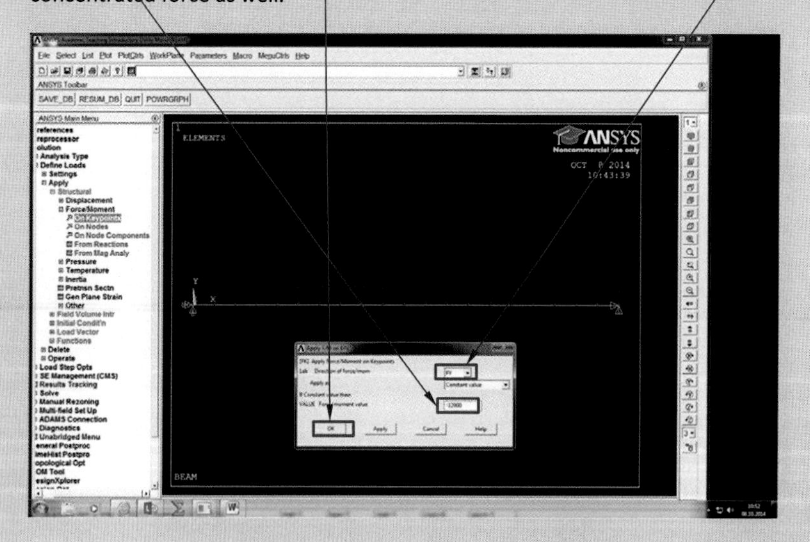

Now, using the following commands, it is time to ask the program to solve the problem.
MM: Solution → Solves → Current LS → OK

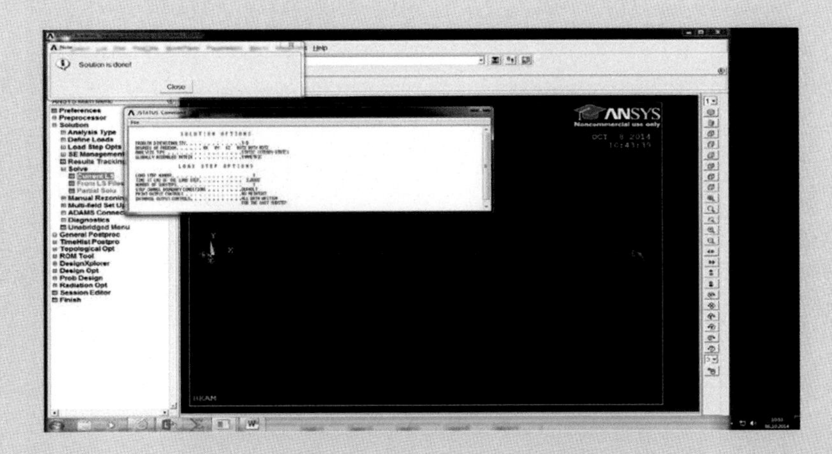

After the solution, we can demonstrate the results. Using the following commands, the post processor has to read the results:
MM: General Postprocessor → Read results → First Set
Then, we have to continue with the following commands in order to print the deformed structure, and compare it with the undeformed one.
MM: General Postprocessor → Plot results → Deformed Shape

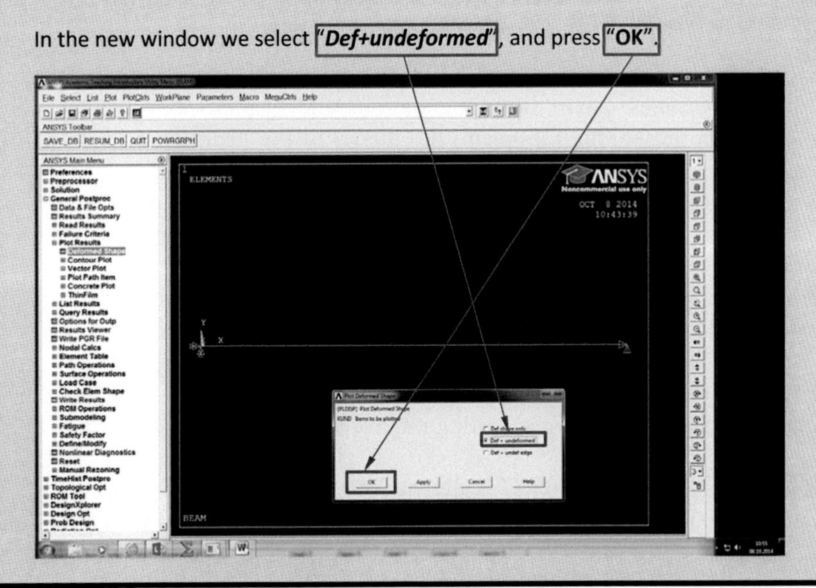

Continued

EXAMPLE 6.3: ANALYSIS OF A BEAM USING ANSYS—CONT'D

Then the following diagram of the deformed structure will be obtained:

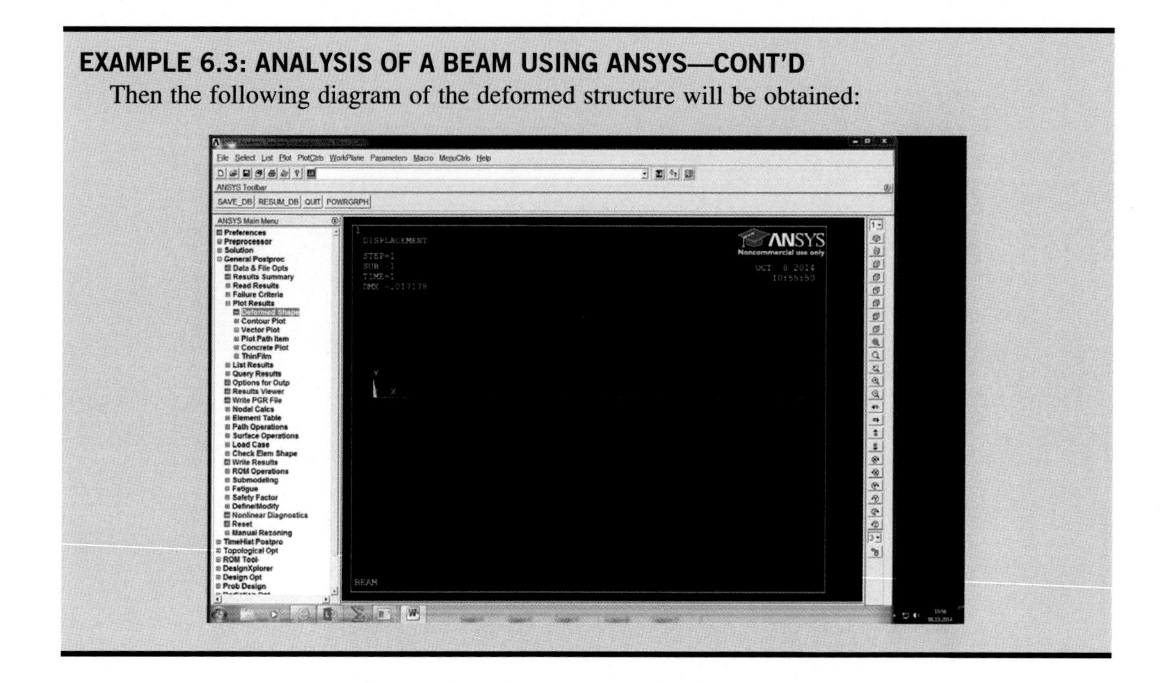

6.4 TWO-DIMENSIONAL ELEMENT EQUATION OF A BEAM ON ELASTIC FOUNDATION SUBJECTED TO UNIFORM LOADING

In many engineering applications, beams resting on elastic foundations are used to carry bending moments and shear forces. Apart from foundations and beams supported by identical springs that are spaced uniformly along the beam, the theory of the beam on an elastic foundation can be used to analyze cylindrical shells subjected to axisymmetric loading. In this section, focus will be given to the derivation of the stiffness matrix of a beam element on elastic foundation. Furthermore, applications of finite element methods on cylindrical tanks will be presented and discussed.

The main assumption of the theory of the beam on an elastic foundation is that the foundation resists the loads transmitted by the beam proportionally to the deflection varying load. This type of foundation is called "Winkler foundation" and is fairly accurate for small deflections.

It is known from the mechanics of solids that the mathematical model simulating the deflection of an infinite beam under uniform load q, resting on an elastic foundation, is:

$$EI\frac{\mathrm{d}^4u(x)}{\mathrm{d}x^4}+ku(x)=-q \tag{6.121}$$

where k is the modulus of the elastic foundation. The general solution of the above equation is:

$$u(x)=C_1G_1(x)+C_2G_2(x)+C_3G_3(x)+C_4G_4(x)+\frac{q}{k} \tag{6.122}$$

where

$$G_1(x)=\frac{e^{\beta x}+e^{-\beta x}}{2}\cos\beta x \tag{6.123}$$

$$G_2(x)=\frac{e^{\beta x}+e^{-\beta x}}{2}\sin\beta x \tag{6.124}$$

$$G_3(x)=\frac{e^{\beta x}-e^{-\beta x}}{2}\cos\beta x \tag{6.125}$$

$$G_4(x)=\frac{e^{\beta x}-e^{-\beta x}}{2}\sin\beta x \tag{6.126}$$

$$\beta=\sqrt[4]{\frac{k}{EI}} \tag{6.127}$$

Therefore, the variations of the shear force $f(x)$, bending moment $m(x)$, and slope $\vartheta(x)$ can be obtained using the following equations:

$$\vartheta(x)=\frac{du(x)}{dx}=C_1G_1'(x)+C_2G_2'(x)+C_3G_3'(x)+C_4G_4'(x) \tag{6.128}$$

$$m(x)=EI\frac{d^2u(x)}{dx^2}=EI\left[C_1G_1''(x)+C_2G_2''(x)+C_3G_3''(x)+C_4G_4''(x)\right] \tag{6.129}$$

$$f(x)=EI\frac{d^3u(x)}{dx^3}=EI\left[C_1G_1'''(x)+C_2G_2'''(x)+C_3G_3'''(x)+C_4G_4'''(x)\right] \tag{6.130}$$

Translating the sign convention of Figure 6.3 to the sign convention of Figure 6.2 and applying the above equations on $x=0$, the following boundary conditions can be specified:

$$f_1=f(0)=EI\left[C_1G_1'''(0)+C_2G_2'''(0)+C_3G_3'''(0)+C_4G_4'''(0)\right] \tag{6.131}$$

$$-m_1=m(0)=EI\left[C_1G_1''(0)+C_2G_2''(0)+C_3G_3''(0)+C_4G_4''(0)\right] \tag{6.132}$$

$$\vartheta_1=\vartheta(0)=C_1G_1'(0)+C_2G_2'(0)+C_3G_3'(0)+C_4G_4'(0) \tag{6.133}$$

$$u_1=u(0)=C_1G_1(0)+C_2G_2(0)+C_3G_3(0)+C_4G_4(0)+\frac{q}{k} \tag{6.134}$$

The above system of equations can be solved with respect to the four unknown constants C_1, C_2, C_3, and C_4. Therefore, these constants can be correlated to the physical quantities f_1, m_1, ϑ_1, and u_1. Formulating Equations (6.131)–(6.134) in the following matrix form,

$$\begin{bmatrix} EIG_1'''(0) & EIG_2'''(0) & EIG_3'''(0) & EIG_4'''(0) \\ EIG_1''(0) & EIG_2''(0) & EIG_3''(0) & EIG_4''(0) \\ G_1'(0) & G_2'(0) & G_3'(0) & G_4'(0) \\ G_1(0) & G_2(0) & G_3(0) & G_4(0) \end{bmatrix} \begin{Bmatrix} C_1 \\ C_2 \\ C_3 \\ C_4 \end{Bmatrix} = \begin{Bmatrix} f_1 \\ -m_1 \\ \vartheta_1 \\ u_1 - \frac{q}{k} \end{Bmatrix} \tag{6.135}$$

and performing required algebraic operations, the following solution can be obtained:

$$\begin{Bmatrix} C_1 \\ C_2 \\ C_3 \\ C_4 \end{Bmatrix} = \begin{Bmatrix} u_1 - \frac{q}{k} \\ \frac{1}{2\beta}\vartheta_1 + \frac{EI}{4\beta^3} f_1 \\ \frac{1}{2\beta}\vartheta_1 - \frac{EI}{4\beta^3} f_1 \\ -\frac{EI}{2\beta^2} m_1 \end{Bmatrix} \tag{6.136}$$

Taking into account the above solution, Equations (6.122), and (6.128)–(6.130) can now be written as follows:

$$u(x) = G_1(x)u_1 + \frac{G_2(x)+G_3(x)}{2\beta}\vartheta_1 - \frac{EIG_4(x)}{2\beta^2}m_1 + \frac{EI[G_2(x)-G_3(x)]}{4\beta^3} f_1 + \frac{q}{k}[1-G_1(x)] \tag{6.137}$$

$$\vartheta(x) = G_1'(x)u_1 + \frac{G_2'(x)+G_3'(x)}{2\beta}\vartheta_1 - \frac{EIG_4'(x)}{2\beta^2}m_1 + \frac{EI\left[G_2'(x)-G_3'(x)\right]}{4\beta^3} f_1 - \frac{q}{k}G_1'(x) \tag{6.138}$$

$$m(x) = EIG_1''(x)u_1 + EI\frac{G_2''(x)+G_3''(x)}{2\beta}\vartheta_1 - \frac{(EI)^2 G_4''(x)}{2\beta^2}m_1 + \frac{(EI)^2\left[G_2''(x)-G_3''(x)\right]}{4\beta^3} f_1 - \frac{qEI}{k}G_1''(x) \tag{6.139}$$

$$f(x) = EIG_1'''(x)u_1 + EI\frac{G_2'''(x)+G_3'''(x)}{2\beta}\vartheta_1 - \frac{(EI)^2 G_4'''(x)}{2\beta^2}m_1 + \frac{(EI)^2\left[G_2'''(x)-G_3'''(x)\right]}{4\beta^3} f_1 - \frac{qEI}{k}G_1'''(x) \tag{6.140}$$

Taking into account the sign convention of Figure 6.3, the physical quantities u_2, ϑ_2, f_2, and m_2 for $x=L$ can be obtained using the above equations. Performing the required algebraic equations, the following formulae can be obtained:

$$u_2 = u(L) = R_0 u_1 + R_1\vartheta_1 + \frac{R_2}{EI}f_1 + \frac{R_3}{EI}m_1 + \frac{q}{k}(1-R_0) \tag{6.141}$$

$$\vartheta_2 = \vartheta(L) = -4\beta^4 R_3 u_1 + R_0\vartheta_1 + \frac{R_1}{EI}f_1 + \frac{R_2}{EI}m_1 + \frac{q}{k}4\beta^4 R_3 \tag{6.142}$$

$$-f_2 = f(L) = -4\beta^4 EIR_2 u_1 - 4\beta^4 EIR_3\vartheta_1 + R_0 f_1 + R_1 m_1 + \frac{q}{k}4EI\beta^4 R_2 \tag{6.143}$$

$$m_2 = m(L) = -4\beta^4 EIR_1 u_1 - 4\beta^4 EIR_2\vartheta_1 - 4\beta^4 R_3 f_1 + R_0 m_1 + \frac{q}{k}4EI\beta^4 R_1 \tag{6.144}$$

where R_0, R_1, R_2, and R_3 are the Rayleigh functions given by:

$$R_0 = G_1(L) \tag{6.145}$$

$$R_1 = \frac{1}{2\beta}[G_2(L) + G_3(L)] \tag{6.146}$$

$$R_2 = \frac{1}{2\beta^2}G_4(L) \tag{6.147}$$

$$R_3 = \frac{1}{4\beta^3}[G_2(L) - G_3(L)] \tag{6.148}$$

Equations (6.141)–(6.144) can now be rearranged yielding the following system:

$$\frac{R_2}{EI}f_1 + \frac{R_3}{EI}m_1 + 0.f_2 + 0.m_2 = -R_0u_1 - R_1\vartheta_1 + u_2 + 0.\vartheta_2 - \frac{q}{k}(1 - R_0) \tag{6.149}$$

$$\frac{R_1}{EI}f_1 + \frac{R_2}{EI}m_1 + 0.f_2 + 0.m_2 = 4\beta^4R_3u_1 - R_0\vartheta_1 + 0.u_2 + \vartheta_2 - \frac{q}{k}4\beta^4R_3 \tag{6.150}$$

$$R_0f_1 + R_1m_1 + f_2 + 0.m_2 = 4\beta^4EIR_2u_1 + 4\beta^4EIR_3\vartheta_1 + 0.u_2 + 0.\vartheta_2 - \frac{q}{k}4EI\beta^4R_2 \tag{6.151}$$

$$-4\beta^4R_3f_1 + R_0m_1 + 0.f_2 - m_2 = 4\beta^4EIR_1u_1 + 4\beta^4EIR_2\vartheta_1 + 0.u_2 + 0.\vartheta_2 - \frac{q}{k}4EI\beta^4R_1 \tag{6.152}$$

The above system of equations can be written in the following matrix form:

$$\begin{bmatrix} \frac{R_2}{EI} & \frac{R_3}{EI} & 0 & 0 \\ \frac{R_1}{EI} & \frac{R_2}{EI} & 0 & 0 \\ R_0 & R_1 & 1 & 0 \\ -4\beta^4R_3 & R_0 & 0 & -1 \end{bmatrix} \begin{Bmatrix} f_1 \\ m_1 \\ f_2 \\ m_2 \end{Bmatrix} = \begin{bmatrix} -R_0 & -R_1 & 1 & 0 \\ 4\beta^4R_3 & -R_0 & 0 & 1 \\ 4\beta^4EIR_2 & 4\beta^4EIR_3 & 0 & 0 \\ 4\beta^4EIR_1 & 4\beta^4EIR_2 & 0 & 0 \end{bmatrix} \begin{Bmatrix} u_1 \\ \vartheta_1 \\ u_2 \\ \vartheta_2 \end{Bmatrix} - \begin{Bmatrix} \frac{q}{k}(1 - R_0) \\ \frac{q}{k}4\beta^4R^3 \\ \frac{q}{k}4EI\beta^4R_2 \\ \frac{q}{k}4EI\beta^4R_1 \end{Bmatrix} \tag{6.153}$$

or

$$[U_1]\begin{Bmatrix} f_1 \\ m_1 \\ f_2 \\ m_2 \end{Bmatrix} = [U_2]\begin{Bmatrix} u_1 \\ \vartheta_1 \\ u_2 \\ \vartheta_2 \end{Bmatrix} - \{G\} \tag{6.154}$$

where

$$[U_1] = \begin{bmatrix} \frac{R_2}{EI} & \frac{R_3}{EI} & 0 & 0 \\ \frac{R_1}{EI} & \frac{R_2}{EI} & 0 & 0 \\ R_0 & R_1 & 1 & 0 \\ -4\beta^4R_3 & R_0 & 0 & -1 \end{bmatrix} \tag{6.155}$$

$$[U_2] = \begin{bmatrix} -R_0 & -R_1 & 1 & 0 \\ 4\beta^4 R_3 & -R_0 & 0 & 1 \\ 4\beta^4 EIR_2 & 4\beta^4 EIR_3 & 0 & 0 \\ 4\beta^4 EIR_1 & 4\beta^4 EIR_2 & 0 & 0 \end{bmatrix} \begin{Bmatrix} u_1 \\ \vartheta_1 \\ u_2 \\ \vartheta_2 \end{Bmatrix} \tag{6.156}$$

$$\{G\} = \begin{Bmatrix} \frac{q}{k}(1 - R_0) \\ \frac{q}{k} 4\beta^4 R^3 \\ \frac{q}{k} 4EI\beta^4 R_2 \\ \frac{q}{k} 4EI\beta^4 R_1 \end{Bmatrix} \tag{6.157}$$

Therefore, Equation (6.154) yields:

$$\begin{Bmatrix} f_1 \\ m_1 \\ f_2 \\ m_2 \end{Bmatrix} = [U_1]^{-1}.[U_2] \begin{Bmatrix} u_1 \\ \vartheta_1 \\ u_2 \\ \vartheta_2 \end{Bmatrix} - [U_1]^{-1}.\{G\} \tag{6.158}$$

Performing the required algebraic operations, the following beam element equation can now be obtained:

$$\begin{Bmatrix} f_1 \\ m_1 \\ f_2 \\ m_2 \end{Bmatrix} = [k] \begin{Bmatrix} u_1 \\ \vartheta_1 \\ u_2 \\ \vartheta_2 \end{Bmatrix} + \{F\} \tag{6.159}$$

where:

$$[K] = EI \begin{bmatrix} \frac{R_0R_2 + 4\beta^4 R_3^2}{-R_2^2 + R_1R_3} & \frac{R_1R_2 - R_0R_3}{-R_2^2 + R_1R_3} & \frac{R_2}{R_2^2 - R_1R_3} & \frac{R_3}{-R_2^2 + R_1R_3} \\ \frac{R_0R_1 + 4\beta^4 R_2R_3}{R_2^2 - R_1R_3} & \frac{R_1^2 - R_0R_2}{R_2^2 - R_1R_3} & \frac{R_1}{-R_2^2 + R_1R_3} & \frac{R_2}{R_2^2 - R_1R_3} \\ \frac{R_0^2R_2 + 4\beta^4 R_2\left(R_2^2 - 2R_1R_3\right) - R_0\left(R_1^2 - 4\beta^4 R_3^2\right)}{R_2^2 - R_1R_3} & \frac{R_1^3 - 2R_0R_1R_2 + R_0^2R_3 - 4\beta^4 R_2^2R_3 + 4\beta^4 R_1R_3^2}{-R_2^2 + R_1R_3} & \frac{R_1^2 - R_0R_2}{R_2^2 - R_1R_3} & \frac{R_1R_2 - R_0R_3}{-R_2^2 + R_1R_3} \\ \frac{R_1\left(R_0^2 - 4\beta^4 R_2^2\right) + 4\beta^4 R_3\left(R_1^2 - 2R_0R_2\right) + 16\beta^8 R_3^3}{R_2^2 - R_1R_3} & \frac{-R_0^2R_2 - 4\beta^4 R_2\left(R_2^2 - 2R_1R_3\right) + R_0\left(R_1^2 - 4\beta^4 R_3^2\right)}{R_2^2 - R_1R_3} & \frac{R_0R_1 + 4\beta^4 R_2R_3}{-R_2^2 + R_1R_3} & \frac{R_0R_2 + 4\beta^4 R_3^2}{R_2^2 - R_1R_3} \end{bmatrix} \tag{6.160}$$

$$\{F\} = \frac{EIq}{k\left(R_2^2 - R_1R_3\right)} \begin{bmatrix} R_2\left(R_0 - 1\right) + 4\beta^4 R_3^2 \\ R_1 - R_0R_1 - 4\beta^4 R_2R_3 \\ R_1^2\left(R_0 - 1\right) - R_0^2R_2 - 4\beta^4 R_2^3 + 8\beta^4 R_1R_2R_3 + R_0\left(R_2 - 4\beta^4 R_3^2\right) \\ R_1\left(R_0 - R_0^2 + 4\beta^4 R_2^2\right) - 4\beta^4 R_3\left(R_1^2 + R_2\left(2R_0 - 1\right)\right) - 16\beta^8 R_3^3 \end{bmatrix} \tag{6.161}$$

It should be noted that once the nodal parameters u_1, ϑ_1, f_1, and m_1 are known the distribution along the beam element of displacements $u(x)$, slopes $\vartheta(x)$, bending moments $m(x)$, and shear forces $f(x)$ can be obtained by Equations (6.137)–(6.140), respectively.

6.5 ENGINEERING APPLICATIONS OF THE ELEMENT EQUATION OF THE BEAM ON ELASTIC FOUNDATION

6.5.1 BEAM SUPPORTED ON EQUISPACED ELASTIC SPRINGS

Sometimes beams are supported on discrete elastic springs with same constant k_s, equally spaced along the beam (Figure 6.6).

When the distance l is short, it means there are large numbers of springs per unit length, and many finite elements (FE) should be used to analyze this problem. However, there is an alternative procedure to reduce the number of FEs using the element equation of the beam on elastic foundation with modulus

$$k = \frac{k_s}{l} \tag{6.162}$$

where k_s is the elastic constant of the discrete spring and l is the distance between springs. Even though the above simulation yields approximate results, the accuracy is acceptable for engineering applications in case that

$$l = \frac{\pi}{4\beta} \tag{6.163}$$

where β is given by Equation (6.127). It should be noted that the error becomes small when the spacing l between the springs becomes short.

6.5.2 CYLINDRICAL SHELLS UNDER AXISYMMETRIC LOADING

It is known from the mechanics of solids that the theory of the beam on elastic foundations can be used to analyze cylindrical shells (Figure 6.7) subjected to axisymmetric loading (e.g., cylindrical tanks for oil storage, pipelines, etc.)

The following equilibrium equations of a cylindrical shell element (Figure 6.8) subjected to uniform pressure p

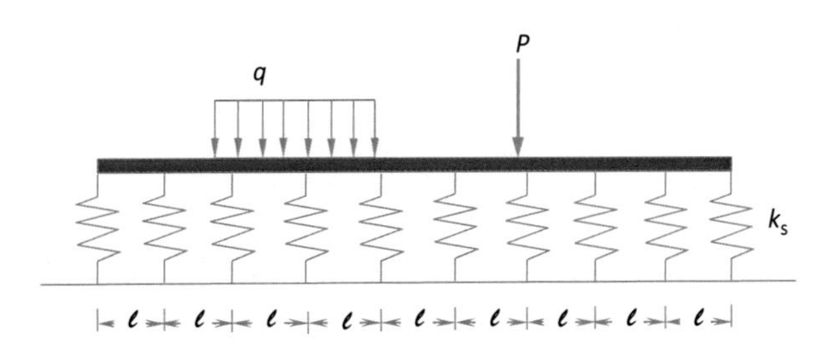

FIGURE 6.6

Beam resting on equispaced elastic springs.

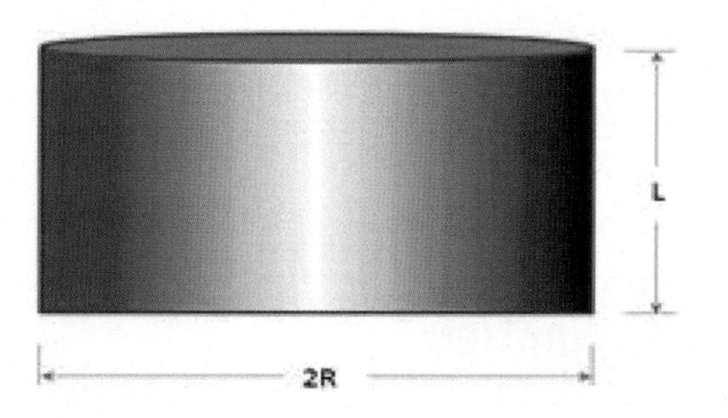

FIGURE 6.7

Cylindrical shell subjected to axisymmetric loads.

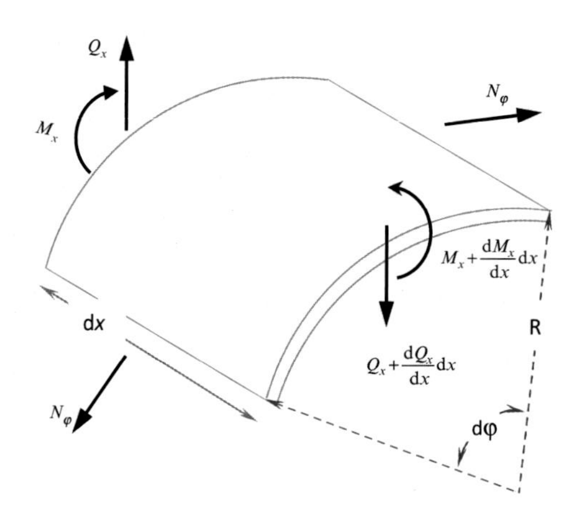

FIGURE 6.8

Equilibrium of a cylindrical shell element.

$$\frac{\mathrm{d}Q_x}{\mathrm{d}x}\cdot R\cdot \mathrm{d}\phi\cdot \mathrm{d}x+N_\phi\cdot \mathrm{d}\phi\cdot \mathrm{d}x-p\cdot R\cdot \mathrm{d}\phi\cdot \mathrm{d}x=0 \tag{6.164}$$

$$\frac{\mathrm{d}M_x}{\mathrm{d}x}\cdot R\cdot \mathrm{d}\phi\cdot \mathrm{d}x-Q_x\cdot R\cdot \mathrm{d}\phi\cdot \mathrm{d}x=0 \tag{6.165}$$

yields

$$\frac{\mathrm{d}Q_x}{\mathrm{d}x}+\frac{1}{R}N_\phi=p \tag{6.166}$$

$$\frac{\mathrm{d}M_x}{\mathrm{d}x}=Q_x \tag{6.167}$$

Combining the above two equations, the following formula can be obtained:

$$\frac{d^2M_x}{dx^2}+\frac{1}{R}N_\phi=p \tag{6.168}$$

Taking into account the equilibrium equation

$$N_\phi=\sigma_\phi\cdot t \tag{6.169}$$

the Hooke's law

$$\sigma_\phi=E\cdot\varepsilon_\phi \tag{6.170}$$

and the definition of the strain

$$\varepsilon_\phi=\frac{w}{R} \tag{6.171}$$

where σ_ϕ, ε_ϕ are the circumferential stress and strains, respectively, t is the shell's thickness, and w is its radial deflection, Equation (6.168) can be written:

$$\frac{d^2M_x}{dx^2}+\frac{Et}{R^2}w=p \tag{6.172}$$

According to the theory of shells (e.g., see Flügge, 1973), the bending moment M_x is correlated to the radial deflection w by the following equation:

$$M_x=D\frac{d^2w}{dx^2} \tag{6.173}$$

where D is the flexural rigidity given by

$$D=\frac{Et}{12(1-\nu^2)} \tag{6.174}$$

Combining Equations (6.172) and (6.173), the following governing equation can be obtained:

$$D\frac{d^4M_x}{dx^4}+\frac{Et}{R^2}w=p \tag{6.175}$$

The above equation is the equation of a beam resting on an elastic foundation with modulus

$$k=\frac{Et}{R^2} \tag{6.176}$$

Therefore, for analyzing cylindrical shells under axisymmetric loads, the element Equation (6.159) can be used. It should be noted that in stiffness matrix [k] and load matrix $\{F\}$ given by Equations (6.160) and (6.161), the rigidity EI should be replaced by D, and the foundation modulus k should be replaced by Et/R^2.

EXAMPLE 6.4: FE ANALYSIS OF A CYLINDRICAL TANK

Using beam on elastic foundation elements, determine the radial displacement, strain and the stress fields of the following oil storage tank.

Data

$t = 20$ mm (wall thickness)
$L = 10$ m
$2R = 30$ m
$E = 200$ GPa (modulus of elasticity)
$\nu = 0.29$ (Poisson's ratio)
$\rho = 900$ kg/m^3 (oil's density)

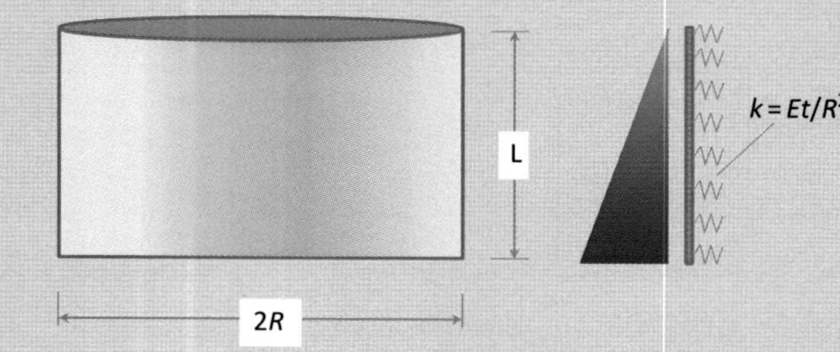

Solution

We will use five elements. The length of each element is $L = 2.0$ m. According to the theory, the following calculations should be performed:

Constants

```
k = E t/(R^2) = 5.73667x10^9
   EI = E t/(12 (1-n^2)) = 1.17439x10^11
   b = (k/EI)^(1/4) = 0.470123

   po = 900x10x10 = 90000

   q1 = 9 po/10 = 81000
   q2 = 7 po/10 = 63000
   q3 = 5 po/10 = 23684.2
   q4 = 3 po/10 = 27000
   q5 = po/10 = 9000
```

```
G1 = ((e^(b L) + e^(-b L))/2) Cos[b L] = 0.869981
G2 = (( e^(b L) + e^(-b L))/2) Sin[b L] = 1.19182
G3 = (( e^(b L)- e^(-b L))/2) Cos[b L] = 0.639728
G4 = (( e^(b L)- e^(-b L))/2) Sin[b L] = 0.876391

R0 = G1 = 0.869981
R1 = (1/(2 b)).(G2 + G3) = 1.94795
R2 = (1/(b^2)).G4 = 3.96529
R3 = (1/(4 b^3)).(G2-G3) = 1.32837
```

Element's Stiffness Matrix [k]

$$\begin{matrix} -3.392416118492208 \times 10^{10} & -5.8724730083321 \times 10^{10} & 3.545097932790122 \times 10^{10} & -1.187609968666608 \times 10^{10} \\ 2.435242166613687 \times 10^{10} & 3.082482628487138 \times 10^{9} & -1.74153201905627 \times 10^{10} & 3.545097932790122 \times 10^{10} \\ 7.30662256887665 \times 10^{10} & 7.556662973321651 \times 10^{10} & 3.082482628487138 \times 10^{9} & -5.8724730083321 \times 10^{10} \\ -1.470769395579627 \times 10^{10} & -7.30662256887665 \times 10^{10} & -2.435242166613687 \times 10^{10} & 3.392416118492208 \times 10^{10} \end{matrix}$$

Loading Matrices {F₁}, {F₂}, {F₃}, {F₄}, {F₅}

$$\{F_1\} = \begin{matrix} -21,558.197637621393 \\ -97,949.7167559567 \\ -1,075,196.094113588 \\ -79,618.88829053477 \end{matrix}$$

$$\{F_2\} = \begin{matrix} -16,767.487051483306 \\ -76,183.11303241077 \\ -836,263.6287550131 \\ -61,925.80200374927 \end{matrix}$$

$$\{F_3\} = \begin{matrix} -6303.566560708009 \\ -28,640.26805729728 \\ -314,384.8228402305 \\ -23,280.37669313882 \end{matrix}$$

Continued

$$\{F_4\} = \begin{matrix} -7186.065879207131 \\ -32,649.905585318902 \\ -358,398.6980378628 \\ -26,539.629430178262 \end{matrix}$$

$$\{F_5\} = \begin{matrix} -2395.3552930690435 \\ -10,883.301861772967 \\ -119,466.2326792876 \\ -8846.543143392752 \end{matrix}$$

Expanded Stiffness Matrices [k₁], [k₂], [k₃], [k₄], [k₅]

Element 1

$$\begin{matrix}
-3.392416118492208\times10^{10} & -5.8724730083321\times10^{10} & 3.545097932790122\times10^{10} & -1.187609968666608\times10^{10} & 0 & 0 & 0 & 0 & 0 & 0 & 0 \\
2.435242166613687\times10^{10} & 3.082482628487138\times10^{9} & -1.74153201905627\times10^{10} & 3.545097932790122\times10^{10} & 0 & 0 & 0 & 0 & 0 & 0 & 0 \\
7.30662256887665\times10^{10} & 7.556662973321651\times10^{10} & 3.082482628487138\times10^{9} & -5.8724730083321\times10^{10} & 0 & 0 & 0 & 0 & 0 & 0 & 0 \\
-1.470769395579627\times10^{10} & -7.30662256887665\times10^{10} & -2.435242166613687\times10^{10} & 3.392416118492208\times10^{10} & 0 & 0 & 0 & 0 & 0 & 0 & 0 \\
0 & 0 & 0 & 0 & 0 & 0 & 0 & 0 & 0 & 0 & 0 \\
0 & 0 & 0 & 0 & 0 & 0 & 0 & 0 & 0 & 0 & 0 \\
0 & 0 & 0 & 0 & 0 & 0 & 0 & 0 & 0 & 0 & 0 \\
0 & 0 & 0 & 0 & 0 & 0 & 0 & 0 & 0 & 0 & 0 \\
0 & 0 & 0 & 0 & 0 & 0 & 0 & 0 & 0 & 0 & 0 \\
0 & 0 & 0 & 0 & 0 & 0 & 0 & 0 & 0 & 0 & 0 \\
0 & 0 & 0 & 0 & 0 & 0 & 0 & 0 & 0 & 0 & 0
\end{matrix}$$

Element 2

$$\begin{matrix}
0 & 0 & 0 & 0 & 0 & 0 & 0 & 0 & 0 & 0 & 0 & 0 \\
0 & 0 & 0 & 0 & 0 & 0 & 0 & 0 & 0 & 0 & 0 & 0 \\
0 & 0 & -3.392416118492208\times10^{10} & -5.8724730083321\times10^{10} & 3.545097932790122\times10^{10} & -1.187609968666608\times10^{10} & 0 & 0 & 0 & 0 & 0 & 0 \\
0 & 0 & 2.435242166613687\times10^{10} & 3.082482628487138\times10^{9} & -1.74153201905627\times10^{10} & 3.545097932790122\times10^{10} & 0 & 0 & 0 & 0 & 0 & 0 \\
0 & 0 & 7.30662256887665\times10^{10} & 7.556662973321651\times10^{10} & 3.082482628487138\times10^{9} & -5.8724730083321\times10^{10} & 0 & 0 & 0 & 0 & 0 & 0 \\
0 & 0 & -1.470769395579627\times10^{10} & -7.30662256887665\times10^{10} & -2.435242166613687\times10^{10} & 3.392416118492208\times10^{10} & 0 & 0 & 0 & 0 & 0 & 0 \\
0 & 0 & 0 & 0 & 0 & 0 & 0 & 0 & 0 & 0 & 0 & 0 \\
0 & 0 & 0 & 0 & 0 & 0 & 0 & 0 & 0 & 0 & 0 & 0 \\
0 & 0 & 0 & 0 & 0 & 0 & 0 & 0 & 0 & 0 & 0 & 0 \\
0 & 0 & 0 & 0 & 0 & 0 & 0 & 0 & 0 & 0 & 0 & 0 \\
0 & 0 & 0 & 0 & 0 & 0 & 0 & 0 & 0 & 0 & 0 & 0 \\
0 & 0 & 0 & 0 & 0 & 0 & 0 & 0 & 0 & 0 & 0 & 0
\end{matrix}$$

Element 3

0	0	0	0	0	0	0	0	0	0	0	0
0	0	0	0	0	0	0	0	0	0	0	0
0	0	0	0	0	0	0	0	0	0	0	0
0	0	0	0	0	0	0	0	0	0	0	0
0	0	0	0	$-3.392416118492208 \times 10^{10}$	$-5.8724730083321 \times 10^{10}$	$3.545097932790122 \times 10^{10}$	$-1.187609968666608 \times 10^{10}$	0	0	0	0
0	0	0	0	$2.435242166613687 \times 10^{10}$	$3.082482628487138 \times 10^{9}$	$-1.74153201905627 \times 10^{10}$	$3.545097932790122 \times 10^{10}$	0	0	0	0
0	0	0	0	$7.30662256887665 \times 10^{10}$	$7.556662973321651 \times 10^{10}$	$3.082482628487138 \times 10^{9}$	$-5.8724730083321 \times 10^{10}$	0	0	0	0
0	0	0	0	$-1.470769395579627 \times 10^{10}$	$-7.30662256887665 \times 10^{10}$	$-2.435242166613687 \times 10^{10}$	$3.392416118492208 \times 10^{10}$	0	0	0	0
0	0	0	0	0	0	0	0	0	0	0	0
0	0	0	0	0	0	0	0	0	0	0	0
0	0	0	0	0	0	0	0	0	0	0	0
0	0	0	0	0	0	0	0	0	0	0	0

Element 4

0	0	0	0	0	0	0	0	0	0	0	0
0	0	0	0	0	0	0	0	0	0	0	0
0	0	0	0	0	0	0	0	0	0	0	0
0	0	0	0	0	0	0	0	0	0	0	0
0	0	0	0	0	0	0	0	0	0	0	0
0	0	0	0	0	0	0	0	0	0	0	0
0	0	0	0	0	0	$-3.392416118492208 \times 10^{10}$	$-5.8724730083321 \times 10^{10}$	$3.545097932790122 \times 10^{10}$	$-1.187609968666608 \times 10^{10}$	0	0
0	0	0	0	0	0	$2.435242166613687 \times 10^{10}$	$3.082482628487138 \times 10^{9}$	$-1.74153201905627 \times 10^{10}$	$3.545097932790122 \times 10^{10}$	0	0
0	0	0	0	0	0	$7.30662256887665 \times 10^{10}$	$7.556662973321651 \times 10^{10}$	$3.082482628487138 \times 10^{9}$	$-5.8724730083321 \times 10^{10}$	0	0
0	0	0	0	0	0	$-1.470769395579627 \times 10^{10}$	$-7.30662256887665 \times 10^{10}$	$-2.435242166613687 \times 10^{10}$	$3.392416118492208 \times 10^{10}$	0	0
0	0	0	0	0	0	0	0	0	0	0	0
0	0	0	0	0	0	0	0	0	0	0	0

Continued

EXAMPLE 6.4: FE ANALYSIS OF A CYLINDRICAL TANK—CONT'D

Element 5

$$\begin{bmatrix} 0&0&0&0&0&0&0&0&0&0&0&0\\ 0&0&0&0&0&0&0&0&0&0&0&0\\ 0&0&0&0&0&0&0&0&0&0&0&0\\ 0&0&0&0&0&0&0&0&0&0&0&0\\ 0&0&0&0&0&0&0&0&0&0&0&0\\ 0&0&0&0&0&0&0&0&0&0&0&0\\ 0&0&0&0&0&0&0&0&0&0&0&0\\ 0&0&0&0&0&0&0&0&0&0&0&0\\ 0&0&0&0&0&0&0&0&-3.392416118492208\times10^{10}&-5.8724730083321\times10^{10}&3.545097932790122\times10^{10}&-1.187609968666608\times10^{10}\\ 0&0&0&0&0&0&0&0&2.435242166613687\times10^{10}&3.082482628487138\times10^{9}&-1.74153201905627\times10^{10}&3.545097932790122\times10^{10}\\ 0&0&0&0&0&0&0&0&7.30662256887665\times10^{10}&7.556662973321651\times10^{10}&3.082482628487138\times10^{9}&-5.8724730083321\times10^{10}\\ 0&0&0&0&0&0&0&0&-1.470769395579627\times10^{10}&-7.30662256887665\times10^{10}&-2.435242166613687\times10^{10}&3.392416118492208\times10^{10} \end{bmatrix}$$

Expanded Loading Matrices $\{F_1\}, \{F_2\}, \{F_3\}, \{F_4\}, \{F_5\}$

Element 1

$$\begin{Bmatrix} -21{,}558.197637621393\\ -97{,}949.7167559567\\ -1{,}075{,}196.094113588\\ -79{,}618.88829053477\\ 0\\ 0\\ 0\\ 0\\ 0\\ 0\\ 0\\ 0 \end{Bmatrix}$$

Element 2	
	0
	0
	−16,767.487051483306
	−76,183.11303241077
	−836,263.6287550131
	−61,925.80200374927
	0
	0
	0
	0
	0
	0
Element 3	
	0
	0
	0
	0
	−6303.566560708009
	−28,640.26805729728
	−314,384.8228402305
	−23,280.37669313882
	0
	0
	0
	0

Continued

EXAMPLE 6.4: FE ANALYSIS OF A CYLINDRICAL TANK—CONT'D

Element 4

$$\begin{matrix} 0 \\ 0 \\ 0 \\ 0 \\ 0 \\ 0 \\ -7186.065879207131 \\ -32,649.905585318902 \\ -358,398.6980378628 \\ -26,539.629430178262 \\ 0 \\ 0 \end{matrix}$$

Element 5

$$\begin{matrix} 0 \\ 0 \\ 0 \\ 0 \\ 0 \\ 0 \\ 0 \\ 0 \\ -2395.3552930690435 \\ -10,883.301861772967 \\ -119,466.2326792876 \\ -8846.543143392752 \end{matrix}$$

Nodal Displacements

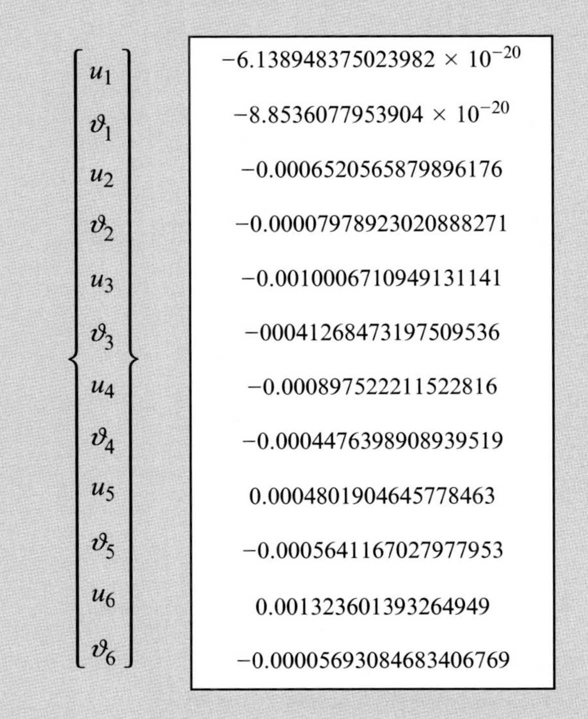

$$\begin{Bmatrix} u_1 \\ \vartheta_1 \\ u_2 \\ \vartheta_2 \\ u_3 \\ \vartheta_3 \\ u_4 \\ \vartheta_4 \\ u_5 \\ \vartheta_5 \\ u_6 \\ \vartheta_6 \end{Bmatrix} = \boxed{\begin{matrix} -6.138948375023982 \times 10^{-20} \\ -8.8536077953904 \times 10^{-20} \\ -0.000652056587989 6176 \\ -0.0000797892302088 8271 \\ -0.00100067109491311 41 \\ -00041268473197509536 \\ -0.000897522211522816 \\ -0.0004476398908939519 \\ 0.0004801904645778463 \\ -0.0005641167027977953 \\ 0.001323601393264949 \\ -0.00005693084683406769 \end{matrix}}$$

Reactions

$$F_1 = -2.214690157192107 \times 10^7\,\mathrm{N}$$
$$M_1 = 8625117.628236709\,\mathrm{Nm}$$

6.6 ELEMENT EQUATION FOR A BEAM SUBJECTED TO TORSION

Apart from bending moments, shear forces and axial forces, the beams can carry torsional moments. The aim of this step is the derivation of the element equation $\{m_x\} = [K]\{\theta_x\}$ correlating the nodal torsional moments m_{1x}, m_{2x} with the corresponding nodal angles of twist θ_{1x}, θ_{2x} (Figure 6.9).

The simplest procedure for the derivation of the element's equation for this loading case is based on the principle of the direct equilibrium.

6.6.1 THE MECHANICAL BEHAVIOR OF THE MATERIAL

As it is well known from the mechanics of solids, the mechanical behavior of a beam under torsion is governed by the following law:

$$\tau = G\gamma \tag{6.177}$$

where G is the shear modulus and τ, γ are the shear stress and shear strain, respectively (Figure 6.10).

For a cylindrical beam, the shear stress in the location $r = R$ is given by the following equation

$$\tau = \frac{M}{J}R \tag{6.178}$$

where M is the torsional moment and J is the polar moment of inertia.

Let $\Delta\theta$ symbolize the relative angle of twist of the node 2 with respect to the angle of twist of the node 1, that is,

$$\Delta\theta = \theta_{2x} - \theta_{1x} \tag{6.179}$$

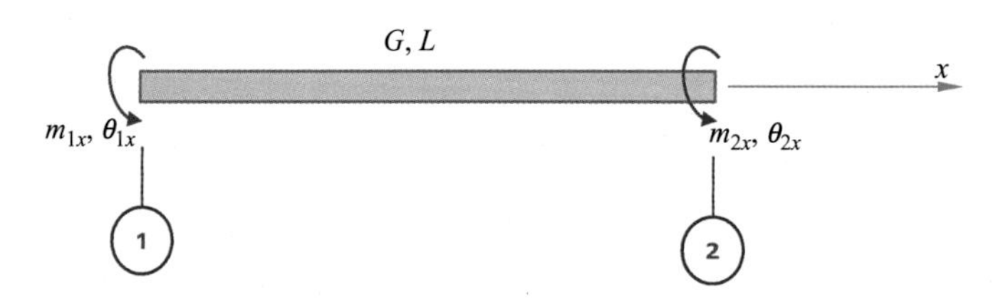

FIGURE 6.9

Beam element under nodal torsional moments.

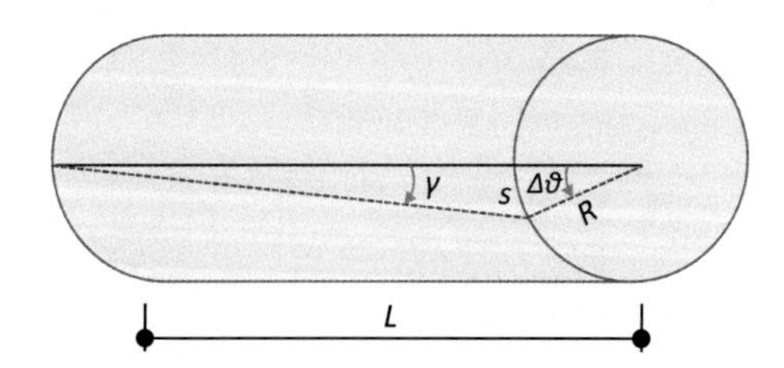

FIGURE 6.10

Demonstration of the shear strain for a cylindrical beam under torsion.

From the geometry of the cylinder shown in Figure 6.10, the following relations can be obtained:

$$s = \Delta\theta \cdot R \tag{6.180}$$

$$s = \gamma \cdot L \tag{6.181}$$

Combination of the above equations yields

$$\gamma = \frac{R}{L}\Delta\theta \tag{6.182}$$

Using Equations (6.178) and (6.182), Equation (6.177) yields:

$$\frac{M}{J}R = G\frac{R}{L}\Delta\theta \tag{6.183}$$

Using Equation (6.179), the above equation yields

$$M = \frac{GJ}{L}(\theta_{2x} - \theta_{1x}) \tag{6.184}$$

6.6.2 THE PRINCIPLE OF DIRECT EQUILIBRIUM

In order to correlate the torsional moment M, which causes the twist of the beam, the equilibrium equation will be applied in the two pieces of the beam element 1–2 (see Figure 6.11).

Therefore:

$$m_{1x} + M = 0 \tag{6.185}$$

$$-M + m_{2x} = 0 \tag{6.186}$$

The above equations can be written in the following matrix form:

$$\begin{Bmatrix} m_{1x} \\ m_{2x} \end{Bmatrix} = \begin{Bmatrix} -M \\ M \end{Bmatrix} \tag{6.187}$$

Using Equation (6.184), Equation (6.187) yields:

$$\begin{Bmatrix} m_{1x} \\ m_{2x} \end{Bmatrix} = \begin{bmatrix} (GJ/L) & -(GJ/L) \\ -(GJ/L) & (GJ/L) \end{bmatrix} \begin{Bmatrix} \theta_{1x} \\ \theta_{2x} \end{Bmatrix} \tag{6.188}$$

In common beams with noncircular cross-sections, the polar moment of inertia can be obtained from Table 6.3 (we assume that the direction of the loads is passing through the shear center (SC) in order to prevent twisting of the web flanges of the open sections).

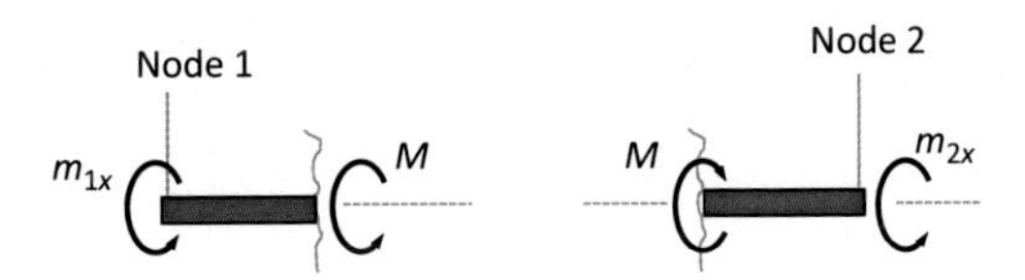

FIGURE 6.11

Equilibrium of the two pieces of the beam.

Table 6.3 Polar moment of inertia for common beam profiles with thickness t

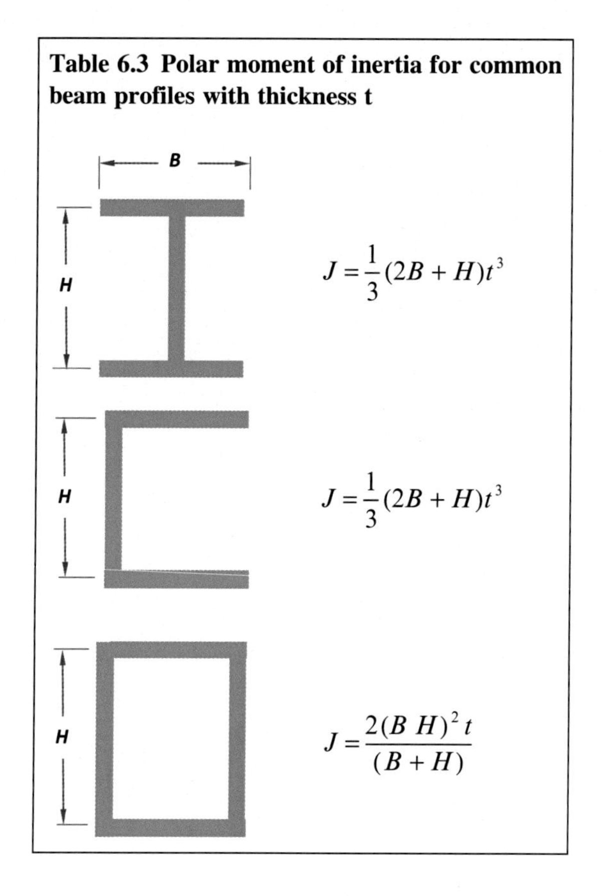

6.7 TWO-DIMENSIONAL ELEMENT EQUATION FOR A BEAM SUBJECTED TO NODAL AXIAL FORCES, SHEAR FORCES, BENDING MOMENTS, AND TORSIONAL MOMENTS

For the beam shown in Figure 6.12, we recall the element equations regarding nodal axial forces (from Chapter 4) as well as shear forces, bending moments, and torsional moments (from this chapter). Taking into account the nomenclature of Figure 6.12, the following matrix equations can be written:

$$\begin{Bmatrix} f_{1x} \\ f_{2x} \end{Bmatrix} = \begin{bmatrix} (EA/L) & -(EA/L) \\ -(EA/L) & (EA/L) \end{bmatrix} \begin{Bmatrix} u_{1x} \\ u_{2x} \end{Bmatrix} \tag{6.189}$$

$$\begin{Bmatrix} f_{1y} \\ m_{1z} \\ f_{2y} \\ m_{2z} \end{Bmatrix} = \frac{EI_z}{L^3} \begin{bmatrix} 12 & 6L & -12 & 6L \\ 6L & 4L^2 & -6L & 2L^2 \\ -12 & -6L & 12 & -6L \\ 6L & 2L^2 & -6L & 4L^2 \end{bmatrix} \begin{Bmatrix} u_{1y} \\ \theta_{1z} \\ u_{2y} \\ \theta_{2z} \end{Bmatrix} \tag{6.190}$$

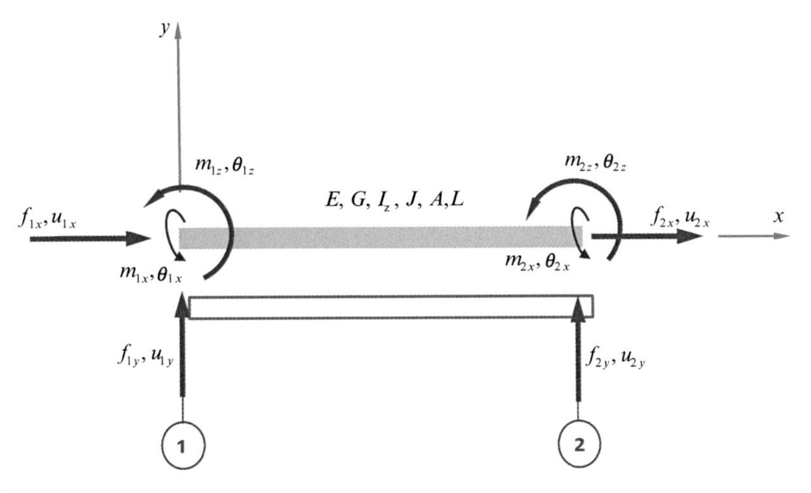

FIGURE 6.12

Beam element subjected to two-dimensional nodal forces.

$$\begin{Bmatrix} m_{1x} \\ m_{2x} \end{Bmatrix} = \begin{bmatrix} (GJ/L) & -(GJ/L) \\ -(GJ/L) & (GJ/L) \end{bmatrix} \begin{Bmatrix} \theta_{1x} \\ \theta_{2x} \end{Bmatrix} \tag{6.191}$$

Combination of Equations (6.189)–(6.191) yields the following element equation with respect to the degrees of freedom $u_{ix}, u_{iy}, \theta_{ix}, \theta_{iz} (i = 1,2)$:

$$\begin{Bmatrix} f_{1x} \\ f_{1y} \\ m_{1x} \\ m_{1z} \\ f_{2x} \\ f_{2y} \\ m_{2x} \\ m_{2z} \end{Bmatrix} = \begin{bmatrix} \frac{EA}{L} & 0 & 0 & 0 & \frac{-EA}{L} & 0 & 0 & 0 \\ 0 & \frac{12EI_z}{L^3} & 0 & \frac{6EI_z}{L^2} & 0 & \frac{-12EI_z}{L^3} & 0 & \frac{6EI_z}{L^2} \\ 0 & 0 & \frac{GJ}{L} & 0 & 0 & 0 & \frac{-GJ}{L} & 0 \\ 0 & \frac{6EI_z}{L^2} & 0 & \frac{4EI_z}{L} & 0 & \frac{-6EI_z}{L^2} & 0 & \frac{2EI_z}{L} \\ \frac{-EA}{L} & 0 & 0 & 0 & \frac{EA}{L} & 0 & 0 & 0 \\ 0 & \frac{-12EI_z}{L^3} & 0 & \frac{-6EI_z}{L^2} & 0 & \frac{12EI_z}{L^3} & 0 & \frac{-6EI_z}{L^2} \\ 0 & 0 & \frac{-GJ}{L} & 0 & 0 & 0 & \frac{GJ}{L} & 0 \\ 0 & \frac{6EI_z}{L^2} & 0 & \frac{2EI_z}{L} & 0 & \frac{-6EI_z}{L^2} & 0 & \frac{4EI_z}{L} \end{bmatrix} \begin{Bmatrix} u_{1x} \\ u_{1y} \\ \theta_{1x} \\ \theta_{1z} \\ u_{2x} \\ u_{2y} \\ \theta_{2x} \\ \theta_{2z} \end{Bmatrix} \tag{6.192}$$

The element Equation (6.192) corresponds to a beam's deformation in the x–y plane.

6.8 THREE-DIMENSIONAL ELEMENT EQUATION FOR A BEAM SUBJECTED TO NODAL AXIAL FORCES, SHEAR FORCES, BENDING MOMENTS, AND TORSIONAL MOMENTS

The element Equation (6.192) can now be expanded to the 3D loading system $f_{ix}, f_{iy}, f_{iz}, m_{ix}, m_{iy}, m_{iz}$ $(i=1,2)$ (see Figure 6.13). To achieve this target, Equation (6.192) should be combined with the beam element equation for the x–z plane.

$$\begin{Bmatrix} f_{1z} \\ m_{1y} \\ f_{2z} \\ m_{2y} \end{Bmatrix} = \begin{bmatrix} \frac{12EI_y}{L^3} & \frac{6EI_y}{L^2} & \frac{-12EI_y}{L^3} & \frac{6EI_y}{L^2} \\ \frac{6EI_y}{L^2} & \frac{4EI_y}{L} & \frac{-6EI_y}{L^2} & \frac{2EI_y}{L} \\ \frac{-12EI_y}{L^3} & \frac{-6EI_y}{L^2} & \frac{12EI_y}{L^3} & \frac{-6EI_y}{L^2} \\ \frac{6EI_y}{L^2} & \frac{2EI_y}{L} & \frac{-6EI_y}{L^2} & \frac{4EI_y}{L} \end{bmatrix} \begin{Bmatrix} u_{1z} \\ \theta_{1y} \\ u_{2z} \\ \theta_{2y} \end{Bmatrix} \tag{6.193}$$

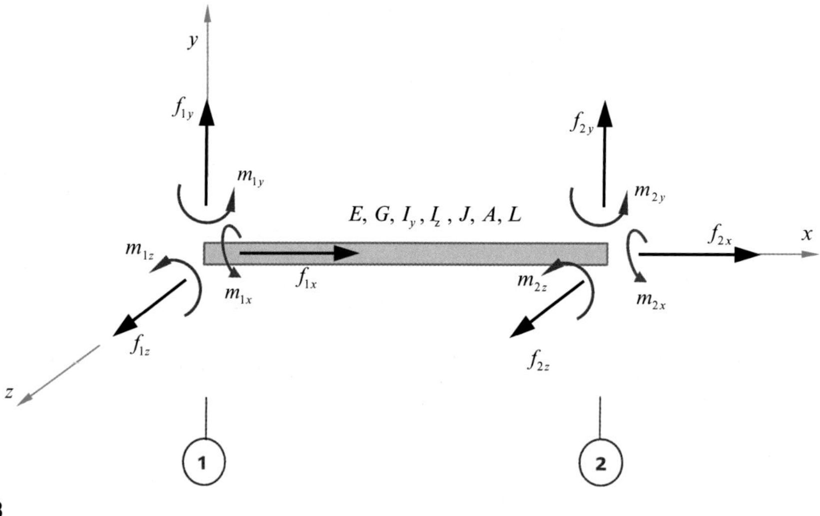

FIGURE 6.13

Beam element subjected to three-dimensional nodal forces.

yielding

$$\begin{Bmatrix} f_{1x} \\ f_{1y} \\ f_{1z} \\ m_{1x} \\ m_{1y} \\ m_{1z} \\ f_{2x} \\ f_{2y} \\ f_{2z} \\ m_{2x} \\ m_{2y} \\ m_{2z} \end{Bmatrix} = \begin{bmatrix} \left[K_{fu}^{11}\right] & \left[K_{f\theta}^{11}\right] & \left[K_{fu}^{12}\right] & \left[K_{f\theta}^{12}\right] \\ \left[K_{mu}^{11}\right] & \left[K_{m\theta}^{11}\right] & \left[K_{mu}^{12}\right] & \left[K_{m\theta}^{12}\right] \\ \left[K_{fu}^{21}\right] & \left[K_{f\theta}^{21}\right] & \left[K_{fu}^{22}\right] & \left[K_{f\theta}^{22}\right] \\ \left[K_{mu}^{21}\right] & \left[K_{m\theta}^{21}\right] & \left[K_{mu}^{22}\right] & \left[K_{m\theta}^{22}\right] \end{bmatrix} \begin{Bmatrix} u_{1x} \\ u_{1y} \\ u_{1z} \\ \vartheta_{1x} \\ \theta_{1y} \\ \theta_{1z} \\ u_{2x} \\ u_{2y} \\ u_{2z} \\ \theta_{2x} \\ \theta_{2y} \\ \theta_{2z} \end{Bmatrix} \tag{6.194}$$

where

$$\left[K_{fu}^{11}\right] = \begin{bmatrix} \frac{EA}{L} & 0 & 0 \\ 0 & \frac{12EI_z}{L^3} & 0 \\ 0 & 0 & \frac{12EI_y}{L^3} \end{bmatrix} \tag{6.195}$$

$$\left[K_{mu}^{11}\right] = \begin{bmatrix} 0 & 0 & 0 \\ 0 & 0 & \frac{6EI_y}{L^2} \\ 0 & \frac{6EI_z}{L^2} & 0 \end{bmatrix} \tag{6.196}$$

$$\left[K_{fu}^{21}\right] = \begin{bmatrix} \frac{-EA}{L} & 0 & 0 \\ 0 & \frac{-12EI_z}{L^3} & 0 \\ 0 & 0 & \frac{-12EI_y}{L^3} \end{bmatrix} \tag{6.197}$$

$$\left[K_{mu}^{21}\right] = \begin{bmatrix} 0 & 0 & 0 \\ 0 & 0 & \dfrac{6EI_y}{L^2} \\ 0 & \dfrac{6EI_z}{L^2} & 0 \end{bmatrix} \tag{6.198}$$

$$\left[K_{f\theta}^{11}\right] = \begin{bmatrix} 0 & 0 & 0 \\ 0 & 0 & \dfrac{6EI_z}{L^2} \\ 0 & \dfrac{6EI_y}{L^2} & 0 \end{bmatrix} \tag{6.199}$$

$$\left[K_{m\theta}^{11}\right] = \begin{bmatrix} \dfrac{GJ}{L} & 0 & 0 \\ 0 & \dfrac{4EI_y}{L} & 0 \\ 0 & 0 & \dfrac{4EI_z}{L} \end{bmatrix} \tag{6.200}$$

$$\left[K_{f\theta}^{21}\right] = \begin{bmatrix} 0 & 0 & 0 \\ 0 & 0 & \dfrac{-6EI_z}{L^2} \\ 0 & \dfrac{-6EI_y}{L^2} & 0 \end{bmatrix} \tag{6.201}$$

$$\left[K_{m\theta}^{21}\right] = \begin{bmatrix} \dfrac{-GJ}{L} & 0 & 0 \\ 0 & \dfrac{2EI_y}{L} & 0 \\ 0 & 0 & \dfrac{2EI_z}{L} \end{bmatrix} \tag{6.202}$$

$$\left[K_{fu}^{12}\right] = \begin{bmatrix} \dfrac{-EA}{L} & 0 & 0 \\ 0 & \dfrac{-12EI_z}{L^3} & 0 \\ 0 & 0 & \dfrac{-12EI_y}{L^3} \end{bmatrix} \tag{6.203}$$

$$\left[K_{mu}^{12}\right] = \begin{bmatrix} 0 & 0 & 0 \\ 0 & 0 & \dfrac{-6EI_y}{L^2} \\ 0 & \dfrac{-6EI_z}{L^2} & 0 \end{bmatrix} \tag{6.204}$$

$$\left[K_{fu}^{22}\right] = \begin{bmatrix} \dfrac{EA}{L} & 0 & 0 \\ 0 & \dfrac{12EI_z}{L^3} & 0 \\ 0 & 0 & \dfrac{12EI_y}{L^3} \end{bmatrix} \tag{6.205}$$

$$\left[K_{mu}^{22}\right] = \begin{bmatrix} 0 & 0 & 0 \\ 0 & 0 & \dfrac{-6EI_y}{L^2} \\ 0 & \dfrac{-6EI_z}{L^2} & 0 \end{bmatrix} \tag{6.206}$$

$$\left[K_{f\theta}^{12}\right] = \begin{bmatrix} 0 & 0 & 0 \\ 0 & 0 & \dfrac{6EI_z}{L^2} \\ 0 & \dfrac{6EI_y}{L^2} & 0 \end{bmatrix} \tag{6.207}$$

$$\left[K_{m\theta}^{12}\right] = \begin{bmatrix} \dfrac{-GJ}{L} & 0 & 0 \\ 0 & \dfrac{2EI_y}{L} & 0 \\ 0 & 0 & \dfrac{2EI_z}{L} \end{bmatrix} \tag{6.208}$$

$$\left[K_{f\theta}^{22}\right] = \begin{bmatrix} 0 & 0 & 0 \\ 0 & 0 & \dfrac{-6EI_z}{L^2} \\ 0 & \dfrac{-6EI_y}{L^2} & 0 \end{bmatrix} \tag{6.209}$$

$$\left[K_{m\theta}^{22}\right] = \begin{bmatrix} \dfrac{GJ}{L} & 0 & 0 \\ 0 & \dfrac{4EI_y}{L} & 0 \\ 0 & 0 & \dfrac{4EI_z}{L} \end{bmatrix} \tag{6.210}$$

EXAMPLE 6.5: FE ANALYSIS OF A 3D BEAM PROBLEM

Calculate the nodal displacement, bending moment, shear force, and torsional moment distributions for the following structure.

Data

$L = 1.0$ m

$d_a = 0.02$ m

$d_b = 0.03$ m

$E = 2 \times 10^{11}$ Pa

$G = 80 \times 10^9$ Pa

$F_z = 8000$ N

$F_y = 10{,}000$ N

$R = 0.04$ m

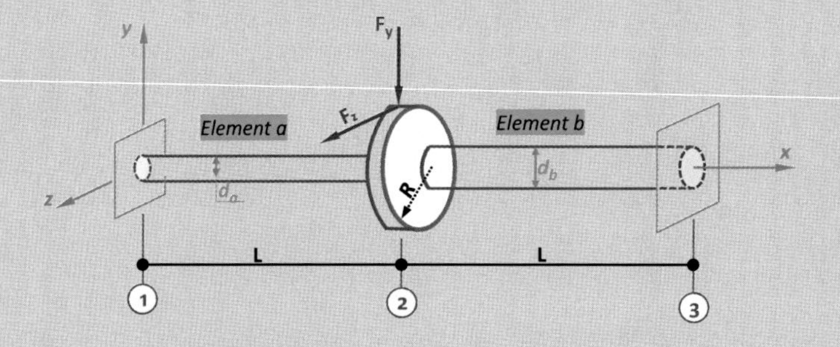

In this example, the structure is subjected to 3D loading conditions. Before starting the solution we should calculate the moments of inertia I with respect to axes y and z for the elements a and b, respectively, as well as the polar moment of inertia J. Since the cross-section of the beams is circular, the moments of inertia with respect to axes y and z are equal, that is,

$$I_y^a = I_z^a = \frac{\pi d_a^4}{64} = 7.85 \times 10^{-9}$$

$$I_y^b = I_z^b = \frac{\pi d_b^4}{64} = 39.74 \times 10^{-9}$$

$$J^a = \frac{\pi d_a^4}{32} = 15.7 \times 10^{-9}$$

$$J^b = \frac{\pi d_b^4}{32} = 79.48 \times 10^{-9}$$

Step 1: Local Element Equations

Element 1-2

Since there is an absence of axial forces, the general stiffness matrix K_a for the 3D beam element α correlates the following nodal parameters

$$\{d\}=\{u_{1y}\ u_{1z}\ \theta_{1x}\ \theta_{1y}\ \theta_{1z}\ u_{2y}\ u_{2z}\ \theta_{2x}\ \theta_{2y}\ \theta_{2z}\}^{\mathrm{T}}$$

$$\{R\}=\{f_{1y}\ f_{1z}\ m_{1x}\ m_{1y}\ m_{1z}\ f_{2y}\ f_{2z}\ m_{2x}\ m_{2y}\ m_{2z}\}^{\mathrm{T}}$$

Therefore, the stiffness matrix of Equation (6.194) can be simplified as follows:

$$\begin{bmatrix}
12EIaz/(L^\wedge 3) & 0 & 0 & 0 & 6EIaz/(L^\wedge 2) & -12EIaz/(L^\wedge 3) & 0 & 0 & 0 & 6EIaz/(L^\wedge 2)\\
0 & 12EIay/(L^\wedge 3) & 0 & 6EIay/(L^\wedge 2) & 0 & 0 & -12EIay/(L^\wedge 3) & 0 & 6EIay/(L^\wedge 2) & 0\\
0 & 0 & GJa/L & 0 & 0 & 0 & 0 & -GJa/L & 0 & 0\\
0 & 6EIay/(L^\wedge 2) & 0 & 4EIay/L & 0 & 0 & -6EIay/(L^\wedge 2) & 0 & 2EIay/L & 0\\
6EIaz/(L^\wedge 2) & 0 & 0 & 0 & 4EIaz/L & -6EIaz/(L^\wedge 2) & 0 & 0 & 0 & 2EIaz/L\\
-12EIaz/(L^\wedge 3) & 0 & 0 & 0 & -6EIaz/(L^\wedge 2) & 12EIaz/(L^\wedge 3) & 0 & 0 & 0 & -6EIaz/(L^\wedge 2)\\
0 & -12EIay/(L^\wedge 3) & 0 & -6EIay/(L^\wedge 2) & 0 & 0 & 12EIay/(L^\wedge 3) & 0 & -6EIay/(L^\wedge 2) & 0\\
0 & 0 & -GJa/L & 0 & 0 & 0 & 0 & GJa/L & 0 & 0\\
0 & 6EIay/(L^\wedge 2) & 0 & 2EIay/L & 0 & 0 & -6EIay/(L^\wedge 2) & 0 & 4EIay/L & 0\\
6EIaz/(L^\wedge 2) & 0 & 0 & 0 & 2EIaz/L & -6EIaz/(L^\wedge 2) & 0 & 0 & 0 & 4EIaz/L
\end{bmatrix}$$

yielding

$$\begin{Bmatrix} f_{1y}\\ f_{1z}\\ m_{1x}\\ m_{1y}\\ m_{1z}\\ f_{2y}\\ f_{2z}\\ m_{2x}\\ m_{2y}\\ m_{2z} \end{Bmatrix} =
\begin{bmatrix}
18840 & 0 & 0 & 0 & 9420 & -18840 & 0 & 0 & 0 & 9420\\
0 & 18840 & 0 & 9420 & 0 & 0 & -18840 & 0 & 9420 & 0\\
0 & 0 & 1256 & 0 & 0 & 0 & 0 & -1256 & 0 & 0\\
0 & 9420 & 0 & 6280 & 0 & 0 & -9420 & 0 & 3140 & 0\\
9420 & 0 & 0 & 0 & 6280 & -9420 & 0 & 0 & 0 & 3140\\
-18840 & 0 & 0 & 0 & -9420 & 18840 & 0 & 0 & 0 & -9420\\
0 & -18840 & 0 & -9420 & 0 & 0 & 18840 & 0 & -9420 & 0\\
0 & 0 & -1256 & 0 & 0 & 0 & 0 & 1256 & 0 & 0\\
0 & 9420 & 0 & 3140 & 0 & 0 & -9420 & 0 & 6280 & 0\\
9420 & 0 & 0 & 0 & 3140 & -9420 & 0 & 0 & 0 & 6280
\end{bmatrix}
\begin{Bmatrix} u_{1y}\\ u_{1z}\\ \vartheta_{1x}\\ \vartheta_{1y}\\ \vartheta_{1z}\\ u_{2y}\\ u_{2z}\\ \vartheta_{2x}\\ \vartheta_{2y}\\ \vartheta_{2z} \end{Bmatrix}$$

Element 2-3

Following same procedure, the corresponding local stiffness matrix for the beam b is

$$\begin{bmatrix}
12EIbz/(L^\wedge 3) & 0 & 0 & 0 & 6EIbz/(L^\wedge 2) & -12EIbz/(L^\wedge 3) & 0 & 0 & 0 & 6EIbz/(L^\wedge 2)\\
0 & 12EIby/(L^\wedge 3) & 0 & 6EIby/(L^\wedge 2) & 0 & 0 & -12EIby/(L^\wedge 3) & 0 & 6EIby/(L^\wedge 2) & 0\\
0 & 0 & GJb/L & 0 & 0 & 0 & 0 & -GJb/L & 0 & 0\\
0 & 6EIby/(L^\wedge 2) & 0 & 4EIby/L & 0 & 0 & -6EIby/(L^\wedge 2) & 0 & 2EIby/L & 0\\
6EIbz/(L^\wedge 2) & 0 & 0 & 0 & 4EIbz/L & -6EIbz/(L^\wedge 2) & 0 & 0 & 0 & 2EIbz/L\\
-12EIbz/(L^\wedge 3) & 0 & 0 & 0 & -6EIbz/(L^\wedge 2) & 12EIbz/(L^\wedge 3) & 0 & 0 & 0 & -6EIbz/(L^\wedge 2)\\
0 & -12EIby/(L^\wedge 3) & 0 & -6EIby/(L^\wedge 2) & 0 & 0 & 12EIby/(L^\wedge 3) & 0 & -6EIby/(L^\wedge 2) & 0\\
0 & 0 & -GJb/L & 0 & 0 & 0 & 0 & GJb/L & 0 & 0\\
0 & 6EIby/(L^\wedge 2) & 0 & 2EIby/L & 0 & 0 & -6EIby/(L^\wedge 2) & 0 & 4EIby/L & 0\\
6EIbz/(L^\wedge 2) & 0 & 0 & 0 & 2EIbz/L & -6EIbz/(L^\wedge 2) & 0 & 0 & 0 & 4EIbz/L
\end{bmatrix}$$

Continued

EXAMPLE 6.5: FE ANALYSIS OF A 3D BEAM PROBLEM—CONT'D

yielding

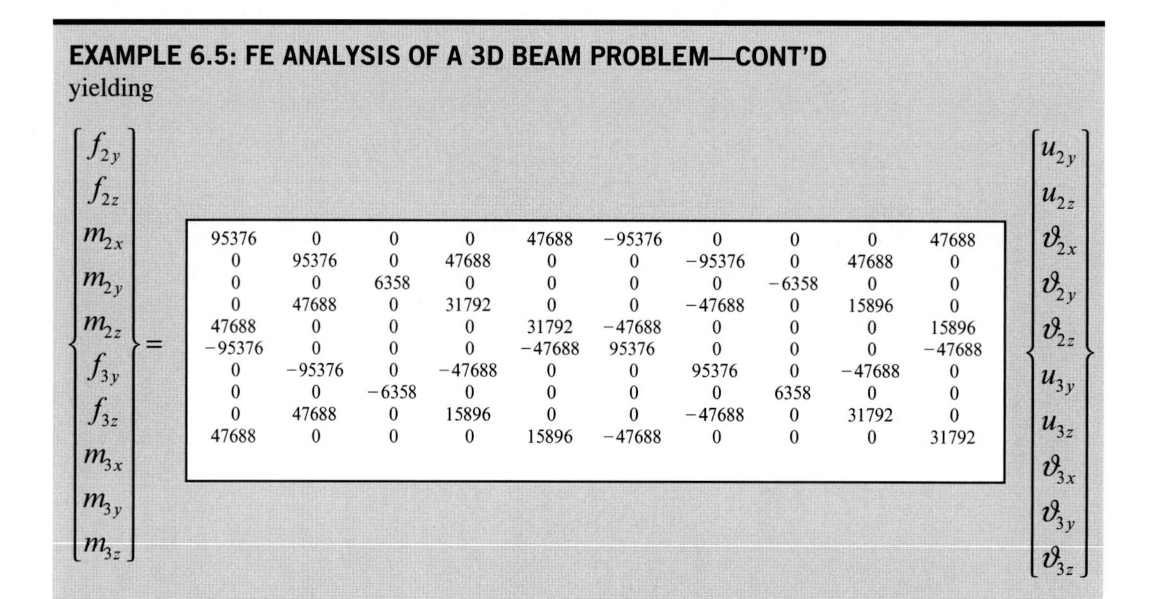

$$\begin{Bmatrix} f_{2y} \\ f_{2z} \\ m_{2x} \\ m_{2y} \\ m_{2z} \\ f_{3y} \\ f_{3z} \\ m_{3x} \\ m_{3y} \\ m_{3z} \end{Bmatrix} = \begin{bmatrix} 95376 & 0 & 0 & 0 & 47688 & -95376 & 0 & 0 & 0 & 47688 \\ 0 & 95376 & 0 & 47688 & 0 & 0 & -95376 & 0 & 47688 & 0 \\ 0 & 0 & 6358 & 0 & 0 & 0 & 0 & -6358 & 0 & 0 \\ 0 & 47688 & 0 & 31792 & 0 & 0 & -47688 & 0 & 15896 & 0 \\ 47688 & 0 & 0 & 0 & 31792 & -47688 & 0 & 0 & 0 & 15896 \\ -95376 & 0 & 0 & 0 & -47688 & 95376 & 0 & 0 & 0 & -47688 \\ 0 & -95376 & 0 & -47688 & 0 & 0 & 95376 & 0 & -47688 & 0 \\ 0 & 0 & -6358 & 0 & 0 & 0 & 0 & 6358 & 0 & 0 \\ 0 & 47688 & 0 & 15896 & 0 & 0 & -47688 & 0 & 31792 & 0 \\ 47688 & 0 & 0 & 0 & 15896 & -47688 & 0 & 0 & 0 & 31792 \end{bmatrix} \begin{Bmatrix} u_{2y} \\ u_{2z} \\ \vartheta_{2x} \\ \vartheta_{2y} \\ \vartheta_{2z} \\ u_{3y} \\ u_{3z} \\ \vartheta_{3x} \\ \vartheta_{3y} \\ \vartheta_{3z} \end{Bmatrix}$$

Step 2: Expanded Local Element Equations

Element 1-2

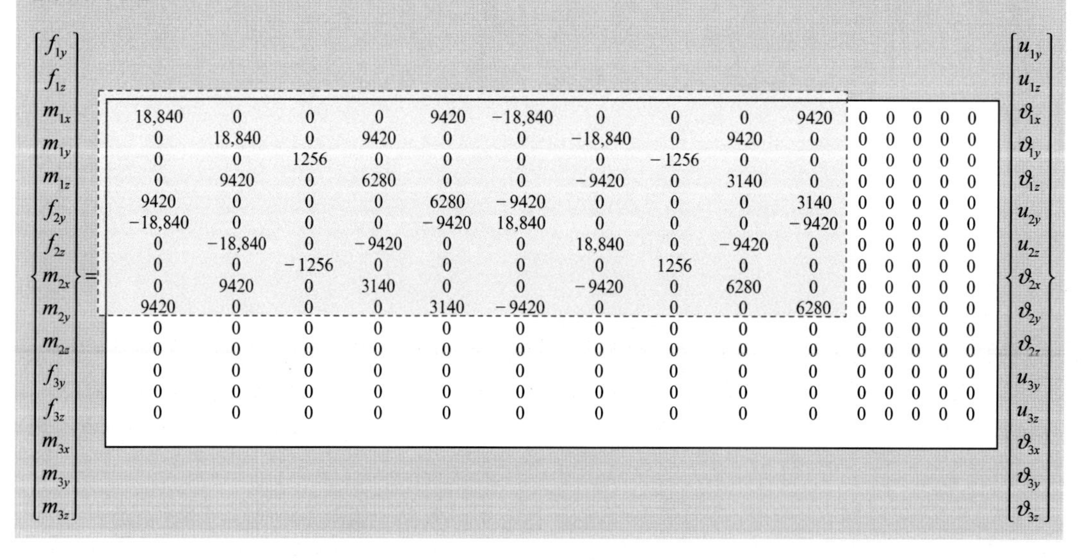

$$\begin{Bmatrix} f_{1y} \\ f_{1z} \\ m_{1x} \\ m_{1y} \\ m_{1z} \\ f_{2y} \\ f_{2z} \\ m_{2x} \\ m_{2y} \\ m_{2z} \\ f_{3y} \\ f_{3z} \\ m_{3x} \\ m_{3y} \\ m_{3z} \end{Bmatrix} = \begin{bmatrix} 18{,}840 & 0 & 0 & 0 & 9420 & -18{,}840 & 0 & 0 & 0 & 9420 & 0 & 0 & 0 & 0 & 0 \\ 0 & 18{,}840 & 0 & 9420 & 0 & 0 & -18{,}840 & 0 & 9420 & 0 & 0 & 0 & 0 & 0 & 0 \\ 0 & 0 & 1256 & 0 & 0 & 0 & 0 & -1256 & 0 & 0 & 0 & 0 & 0 & 0 & 0 \\ 0 & 9420 & 0 & 6280 & 0 & 0 & -9420 & 0 & 3140 & 0 & 0 & 0 & 0 & 0 & 0 \\ 9420 & 0 & 0 & 0 & 6280 & -9420 & 0 & 0 & 0 & 3140 & 0 & 0 & 0 & 0 & 0 \\ -18{,}840 & 0 & 0 & 0 & -9420 & 18{,}840 & 0 & 0 & 0 & -9420 & 0 & 0 & 0 & 0 & 0 \\ 0 & -18{,}840 & 0 & -9420 & 0 & 0 & 18{,}840 & 0 & -9420 & 0 & 0 & 0 & 0 & 0 & 0 \\ 0 & 0 & -1256 & 0 & 0 & 0 & 0 & 1256 & 0 & 0 & 0 & 0 & 0 & 0 & 0 \\ 0 & 9420 & 0 & 3140 & 0 & 0 & -9420 & 0 & 6280 & 0 & 0 & 0 & 0 & 0 & 0 \\ 9420 & 0 & 0 & 0 & 3140 & -9420 & 0 & 0 & 0 & 6280 & 0 & 0 & 0 & 0 & 0 \\ 0 & 0 & 0 & 0 & 0 & 0 & 0 & 0 & 0 & 0 & 0 & 0 & 0 & 0 & 0 \\ 0 & 0 & 0 & 0 & 0 & 0 & 0 & 0 & 0 & 0 & 0 & 0 & 0 & 0 & 0 \\ 0 & 0 & 0 & 0 & 0 & 0 & 0 & 0 & 0 & 0 & 0 & 0 & 0 & 0 & 0 \\ 0 & 0 & 0 & 0 & 0 & 0 & 0 & 0 & 0 & 0 & 0 & 0 & 0 & 0 & 0 \\ 0 & 0 & 0 & 0 & 0 & 0 & 0 & 0 & 0 & 0 & 0 & 0 & 0 & 0 & 0 \end{bmatrix} \begin{Bmatrix} u_{1y} \\ u_{1z} \\ \vartheta_{1x} \\ \vartheta_{1y} \\ \vartheta_{1z} \\ u_{2y} \\ u_{2z} \\ \vartheta_{2x} \\ \vartheta_{2y} \\ \vartheta_{2z} \\ u_{3y} \\ u_{3z} \\ \vartheta_{3x} \\ \vartheta_{3y} \\ \vartheta_{3z} \end{Bmatrix}$$

Element 2-3

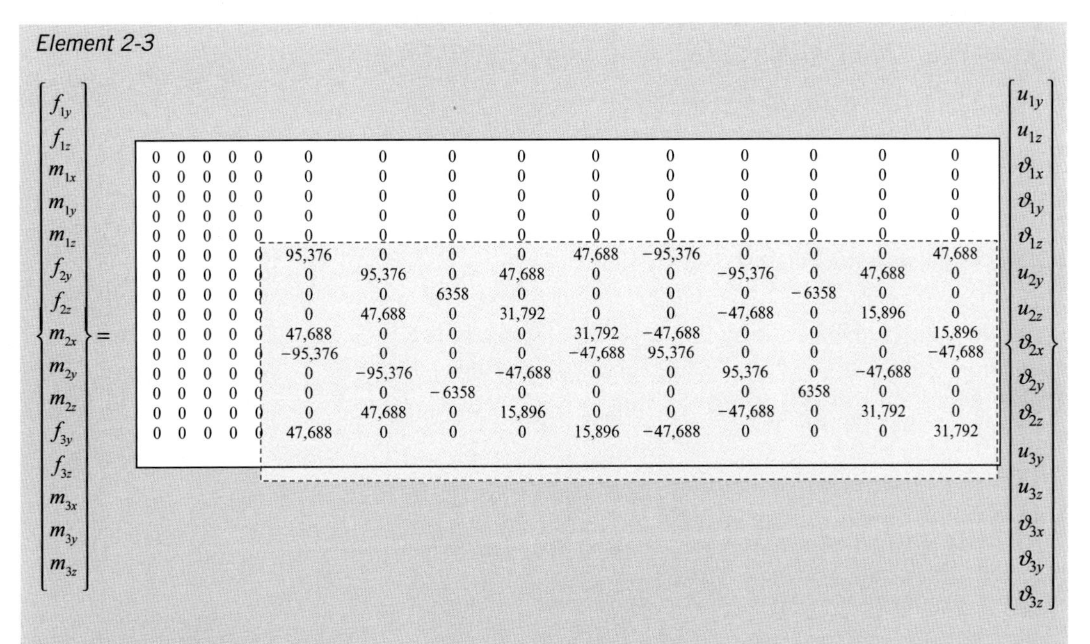

$$
\begin{Bmatrix} f_{1y} \\ f_{1z} \\ m_{1x} \\ m_{1y} \\ m_{1z} \\ f_{2y} \\ f_{2z} \\ m_{2x} \\ m_{2y} \\ m_{2z} \\ f_{3y} \\ f_{3z} \\ m_{3x} \\ m_{3y} \\ m_{3z} \end{Bmatrix} =
\begin{bmatrix}
0 & 0 & 0 & 0 & 0 & 0 & 0 & 0 & 0 & 0 & 0 & 0 & 0 & 0 & 0 \\
0 & 0 & 0 & 0 & 0 & 0 & 0 & 0 & 0 & 0 & 0 & 0 & 0 & 0 & 0 \\
0 & 0 & 0 & 0 & 0 & 0 & 0 & 0 & 0 & 0 & 0 & 0 & 0 & 0 & 0 \\
0 & 0 & 0 & 0 & 0 & 0 & 0 & 0 & 0 & 0 & 0 & 0 & 0 & 0 & 0 \\
0 & 0 & 0 & 0 & 0 & 0 & 0 & 0 & 0 & 0 & 0 & 0 & 0 & 0 & 0 \\
0 & 0 & 0 & 0 & 0 & 95{,}376 & 0 & 0 & 0 & 47{,}688 & -95{,}376 & 0 & 0 & 0 & 47{,}688 \\
0 & 0 & 0 & 0 & 0 & 0 & 95{,}376 & 0 & 47{,}688 & 0 & 0 & -95{,}376 & 0 & 47{,}688 & 0 \\
0 & 0 & 0 & 0 & 0 & 0 & 0 & 6358 & 0 & 0 & 0 & 0 & -6358 & 0 & 0 \\
0 & 0 & 0 & 0 & 0 & 0 & 47{,}688 & 0 & 31{,}792 & 0 & 0 & -47{,}688 & 0 & 15{,}896 & 0 \\
0 & 0 & 0 & 0 & 0 & 47{,}688 & 0 & 0 & 0 & 31{,}792 & -47{,}688 & 0 & 0 & 0 & 15{,}896 \\
0 & 0 & 0 & 0 & 0 & -95{,}376 & 0 & 0 & 0 & -47{,}688 & 95{,}376 & 0 & 0 & 0 & -47{,}688 \\
0 & 0 & 0 & 0 & 0 & 0 & -95{,}376 & 0 & -47{,}688 & 0 & 0 & 95{,}376 & 0 & -47{,}688 & 0 \\
0 & 0 & 0 & 0 & 0 & 0 & 0 & -6358 & 0 & 0 & 0 & 0 & 6358 & 0 & 0 \\
0 & 0 & 0 & 0 & 0 & 0 & 47{,}688 & 0 & 15{,}896 & 0 & 0 & -47{,}688 & 0 & 31{,}792 & 0 \\
0 & 0 & 0 & 0 & 0 & 47{,}688 & 0 & 0 & 0 & 15{,}896 & -47{,}688 & 0 & 0 & 0 & 31{,}792
\end{bmatrix}
\begin{Bmatrix} u_{1y} \\ u_{1z} \\ \vartheta_{1x} \\ \vartheta_{1y} \\ \vartheta_{1z} \\ u_{2y} \\ u_{2z} \\ \vartheta_{2x} \\ \vartheta_{2y} \\ \vartheta_{2z} \\ u_{3y} \\ u_{3z} \\ \vartheta_{3x} \\ \vartheta_{3y} \\ \vartheta_{3z} \end{Bmatrix}
$$

Step 3: Global Stiffness Matrix

Addition of the above expanded element matrices yields:

$$
\begin{Bmatrix} F_{1y} \\ F_{1z} \\ M_{1x} \\ M_{1y} \\ M_{1z} \\ F_{2y} \\ F_{2z} \\ M_{2x} \\ M_{2y} \\ M_{2z} \\ F_{3y} \\ F_{3z} \\ M_{3x} \\ M_{3y} \\ M_{3z} \end{Bmatrix} =
\begin{bmatrix}
1884 & 0 & 0 & 0 & 9420 & -18840 & 0 & 0 & 0 & 9420 & 0 & 0 & 0 & 0 & 0 \\
0 & 18840 & 0 & 9420 & 0 & 0 & -18840 & 0 & 9420 & 0 & 0 & 0 & 0 & 0 & 0 \\
0 & 0 & 1256 & 0 & 0 & 0 & 0 & -1256 & 0 & 0 & 0 & 0 & 0 & 0 & 0 \\
0 & 9420 & 0 & 6280 & 0 & 0 & -9420 & 0 & 3140 & 0 & 0 & 0 & 0 & 0 & 0 \\
9420 & 0 & 0 & 0 & 6280 & -9420 & 0 & 0 & 0 & 3140 & 0 & 0 & 0 & 0 & 0 \\
-18840 & 0 & 0 & 0 & -9420 & 114216 & 0 & 0 & 0 & 38268 & -95376 & 0 & 0 & 0 & 47688 \\
0 & -18840 & 0 & -9420 & 0 & 0 & 114216 & 0 & 38268 & 0 & 0 & -95376 & 0 & 47688 & 0 \\
0 & 0 & -1256 & 0 & 0 & 0 & 0 & 7614 & 0 & 0 & 0 & 0 & -6358 & 0 & 0 \\
0 & 9420 & 0 & 3140 & 0 & 0 & 38268 & 0 & 38072 & 0 & 0 & -47688 & 0 & 15896 & 0 \\
9420 & 0 & 0 & 0 & 3140 & 38268 & 0 & 0 & 0 & 38072 & -47688 & 0 & 0 & 0 & 15896 \\
0 & 0 & 0 & 0 & 0 & -95376 & 0 & 0 & 0 & -47688 & 95376 & 0 & 0 & 0 & -47688 \\
0 & 0 & 0 & 0 & 0 & 0 & -95376 & 0 & -47688 & 0 & 0 & 95376 & 0 & -47688 & 0 \\
0 & 0 & 0 & 0 & 0 & 0 & 0 & -6358 & 0 & 0 & 0 & 0 & 6358 & 0 & 0 \\
0 & 0 & 0 & 0 & 0 & 0 & 47688 & 0 & 15896 & 0 & 0 & -47688 & 0 & 31792 & 0 \\
0 & 0 & 0 & 0 & 0 & 47688 & 0 & 0 & 0 & 15896 & -47688 & 0 & 0 & 0 & 31792
\end{bmatrix}
\begin{Bmatrix} u_{1y} \\ u_{1z} \\ \vartheta_{1x} \\ \vartheta_{1y} \\ \vartheta_{1z} \\ u_{2y} \\ u_{2z} \\ \vartheta_{2x} \\ \vartheta_{2y} \\ \vartheta_{2z} \\ u_{3y} \\ u_{3z} \\ \vartheta_{3x} \\ \vartheta_{3y} \\ \vartheta_{3z} \end{Bmatrix}
$$

The above structure equation can be written in the following abbreviated form

$$\{R\} = [K] \cdot \{d\}$$

or

Continued

EXAMPLE 6.5: FE ANALYSIS OF A 3D BEAM PROBLEM—CONT'D

$$[[K] \quad -[I]] \cdot \begin{Bmatrix} \{d\} \\ \{R\} \end{Bmatrix} = \{O_{15x1}\} \tag{e1}$$

where $[K]$ is the global stiffness matrix of the structure.

Step 4: Boundary Conditions

(a) BC for Nodal Displacements

The boundary conditions regarding nodal displacements are:

$$(1) u_{1y} = 0 \qquad (6) u_{3x} = 0$$

$$(2) u_{1z} = 0 \qquad (7) u_{3z} = 0$$

$$(3) \theta_{1x} = 0 \qquad (8) \theta_{3x} = 0$$

$$(4) \theta_{1y} = 0 \qquad (9) \theta_{3y} = 0$$

$$(5) \theta_{1z} = 0 \qquad (10) \theta_{3z} = 0$$

The above boundary conditions can be written in the following matrix form:

$$\begin{bmatrix}
1 & 0 & 0 & 0 & 0 & 0 & 0 & 0 & 0 & 0 & 0 & 0 & 0 & 0 & 0 \\
0 & 1 & 0 & 0 & 0 & 0 & 0 & 0 & 0 & 0 & 0 & 0 & 0 & 0 & 0 \\
0 & 0 & 1 & 0 & 0 & 0 & 0 & 0 & 0 & 0 & 0 & 0 & 0 & 0 & 0 \\
0 & 0 & 0 & 1 & 0 & 0 & 0 & 0 & 0 & 0 & 0 & 0 & 0 & 0 & 0 \\
0 & 0 & 0 & 0 & 1 & 0 & 0 & 0 & 0 & 0 & 0 & 0 & 0 & 0 & 0 \\
0 & 0 & 0 & 0 & 0 & 0 & 0 & 0 & 0 & 0 & 1 & 0 & 0 & 0 & 0 \\
0 & 0 & 0 & 0 & 0 & 0 & 0 & 0 & 0 & 0 & 0 & 1 & 0 & 0 & 0 \\
0 & 0 & 0 & 0 & 0 & 0 & 0 & 0 & 0 & 0 & 0 & 0 & 1 & 0 & 0 \\
0 & 0 & 0 & 0 & 0 & 0 & 0 & 0 & 0 & 0 & 0 & 0 & 0 & 1 & 0 \\
0 & 0 & 0 & 0 & 0 & 0 & 0 & 0 & 0 & 0 & 0 & 0 & 0 & 0 & 1 \\
0 & 0 & 0 & 0 & 0 & 0 & 0 & 0 & 0 & 0 & 0 & 0 & 0 & 0 & 0 \\
0 & 0 & 0 & 0 & 0 & 0 & 0 & 0 & 0 & 0 & 0 & 0 & 0 & 0 & 0 \\
0 & 0 & 0 & 0 & 0 & 0 & 0 & 0 & 0 & 0 & 0 & 0 & 0 & 0 & 0 \\
0 & 0 & 0 & 0 & 0 & 0 & 0 & 0 & 0 & 0 & 0 & 0 & 0 & 0 & 0 \\
0 & 0 & 0 & 0 & 0 & 0 & 0 & 0 & 0 & 0 & 0 & 0 & 0 & 0 & 0
\end{bmatrix}
\begin{Bmatrix} u_{1y} \\ u_{1z} \\ \vartheta_{1x} \\ \vartheta_{1y} \\ \vartheta_{1z} \\ u_{2y} \\ u_{2z} \\ \vartheta_{2x} \\ \vartheta_{2y} \\ \vartheta_{2z} \\ u_{3y} \\ u_{3z} \\ \vartheta_{3x} \\ \vartheta_{3y} \\ \vartheta_{3z} \end{Bmatrix}
=
\begin{Bmatrix} 0 \\ 0 \\ 0 \\ 0 \\ 0 \\ 0 \\ 0 \\ 0 \\ 0 \\ 0 \\ 0 \\ 0 \\ 0 \\ 0 \\ 0 \end{Bmatrix}$$

or in an abbreviated form

$$[BCd]\{d\} = \{DO\}$$

or

$$[[BCd] \quad [O]] \begin{Bmatrix} \{d\} \\ \{R\} \end{Bmatrix} = \{DO\} \tag{e2}$$

(b) BC for Nodal Forces

Taking into account the physical model, the boundary conditions regarding the forces at the nodes are as follows:

$$
\begin{aligned}
&11)\quad F_{2y} = -F_y \\
&12)\quad F_{2z} = F_z \\
&13)\quad M_{2x} = F_z \cdot R \\
&14)\quad \mathrm{M}_{2y} = 0 \\
&15)\quad M_{2z} = 0
\end{aligned}
$$

The above boundary conditions can be written in a matrix format. It should be noted that the first 10 lines of the following matrix are occupied by the 10 boundary conditions regarding the nodal displacements. Therefore, the five boundary conditions regarding the nodal forces should be placed in the rest lines (11th to 15th):

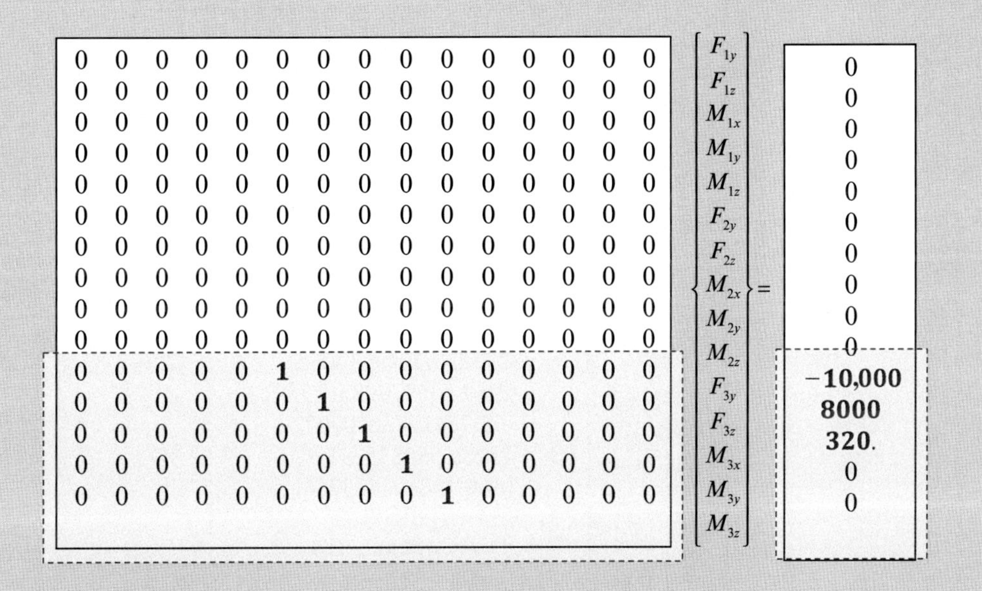

$$
\begin{bmatrix}
0&0&0&0&0&0&0&0&0&0&0&0&0&0&0\\
0&0&0&0&0&0&0&0&0&0&0&0&0&0&0\\
0&0&0&0&0&0&0&0&0&0&0&0&0&0&0\\
0&0&0&0&0&0&0&0&0&0&0&0&0&0&0\\
0&0&0&0&0&0&0&0&0&0&0&0&0&0&0\\
0&0&0&0&0&0&0&0&0&0&0&0&0&0&0\\
0&0&0&0&0&0&0&0&0&0&0&0&0&0&0\\
0&0&0&0&0&0&0&0&0&0&0&0&0&0&0\\
0&0&0&0&0&0&0&0&0&0&0&0&0&0&0\\
0&0&0&0&0&0&0&0&0&0&0&0&0&0&0\\
0&0&0&0&0&\mathbf{1}&0&0&0&0&0&0&0&0&0\\
0&0&0&0&0&0&\mathbf{1}&0&0&0&0&0&0&0&0\\
0&0&0&0&0&0&0&\mathbf{1}&0&0&0&0&0&0&0\\
0&0&0&0&0&0&0&0&\mathbf{1}&0&0&0&0&0&0\\
0&0&0&0&0&0&0&0&0&\mathbf{1}&0&0&0&0&0
\end{bmatrix}
\begin{Bmatrix}
F_{1y}\\ F_{1z}\\ M_{1x}\\ M_{1y}\\ M_{1z}\\ F_{2y}\\ F_{2z}\\ M_{2x}\\ M_{2y}\\ M_{2z}\\ F_{3y}\\ F_{3z}\\ M_{3x}\\ M_{3y}\\ M_{3z}
\end{Bmatrix}
=
\begin{bmatrix}
0\\0\\0\\0\\0\\0\\0\\0\\0\\0\\ \mathbf{-10{,}000}\\ \mathbf{8000}\\ \mathbf{320.}\\ 0\\ 0
\end{bmatrix}
$$

The above equation can be written in the following abbreviated form:

$$[BCR]\{R\} = \{RO\}$$

or

$$[[O]\ [BCR]]\begin{Bmatrix}\{d\}\\ \{R\}\end{Bmatrix} = \{RO\} \tag{e3}$$

Continued

EXAMPLE 6.5: FE ANALYSIS OF A 3D BEAM PROBLEM—CONT'D

Step 5: System of Equations

Combining Equations (e1)–(e3), the following 30×30 matrix equation can be composed:

$$\begin{bmatrix} [K] & [-I] \\ [BCd] & [BCR] \end{bmatrix} \begin{Bmatrix} \{d\} \\ \{R\} \end{Bmatrix} = \begin{Bmatrix} \{O\}_{15x1} \\ \{DO+RO\}_{15x1} \end{Bmatrix}$$

The solution of the above equation provides the following values of the nodal displacements and forces $\{d\}, \{R\}$, respectively:

$$\begin{Bmatrix} u_{1y} \\ u_{1z} \\ \vartheta_{1x} \\ \vartheta_{1y} \\ \vartheta_{1z} \\ u_{2y} \\ u_{2z} \\ \vartheta_{2x} \\ \vartheta_{2y} \\ \vartheta_{2z} \\ u_{3y} \\ u_{3z} \\ \vartheta_{3x} \\ \vartheta_{3y} \\ \vartheta_{3z} \\ F_{1y} \\ F_{1z} \\ M_{1x} \\ M_{1y} \\ M_{1z} \\ F_{2y} \\ F_{2z} \\ M_{2x} \\ M_{2y} \\ M_{2z} \\ F_{3y} \\ F_{3z} \\ M_{3x} \\ M_{3y} \\ M_{3z} \end{Bmatrix} = \boxed{\begin{matrix} 0 \\ 0 \\ 0 \\ 0 \\ 0 \\ -0.1320114743408167 \\ 0.10560917947265347 \\ 0.04202563563773901 \\ -0.10615287035247699 \\ 0.1326910879405961 \\ 0 \\ 0 \\ 0 \\ 0 \\ 0 \\ 3737.0462249814063 \\ -2989.6369799851236 \\ -52.78419836100022 \\ -1328.1584835391707 \\ 1660.198104423966 \\ -10000 \\ 8000 \\ 320 \\ 0 \\ 0 \\ 6262.953775018595 \\ -5010.363020014876 \\ -267.2158016389998 \\ 3348.8845235689255 \\ -4186.105654461156 \end{matrix}}$$

If the above results satisfy the following equilibrium equations, it can be concluded that the above solution is correct:

$$\sum F_y = 0$$
$$\sum F_z = 0$$
$$\sum M_x = 0$$
$$\sum M_y = 0$$
$$\sum M_z = 0$$

Indeed:

$$F_{1y} + F_{2y} + F_{3y} = 3737.05 - 10{,}000 + 6262.95 = 0$$
$$F_{1z} + F_{2z} + F_{3z} = -2989.64 + 8000 - 5010.36 = 0$$
$$M_{1x} + M_{2x} + M_{3x} = -52.78 + 320 - 267.22 = 0$$
$$M_{1y} + F_z \cdot L + M_{3y} + (F_{3z} \cdot 2L) = -1328.16 + (8000x1) + 3348.88 + (-5010.36x2) = 0$$
$$M_{1z} - F_y \cdot L + M_{3z} + (F_{3y} \cdot 2L) = 1660.2 - (10{,}000x1) - 4186.11 + (6262.95x2) = 0$$

Step 6: Bending Moment Distributions

Element α

(a) Plane x–y. Using the values of the nodal displacements in the x–y plane, namely,

$$\{d^\alpha\}_{x-y} = \begin{Bmatrix} u_{1y} \\ \theta_{1z} \\ u_{2y} \\ \theta_{2z} \end{Bmatrix}$$

the corresponding bending moment distribution $\boldsymbol{m_z}$ for the beam element α can be derived by the following known equation:

$$m_z^\alpha(x) = \frac{EI_z^\alpha}{L^3}\left[12x - 6L \quad 6Lx - 4L^2 \quad -12x + 6L \quad 6Lx - 4L^2\right]\begin{Bmatrix} u_{1y} \\ \theta_{1z} \\ u_{2y} \\ \theta_{2z} \end{Bmatrix}$$

or

$$m_z^\alpha(x) = \frac{EI_z^\alpha}{L^3}\left[12x - 6L \quad 6Lx - 4L^2 \quad -12x + 6L \quad 6Lx - 4L^2\right]\begin{Bmatrix} 0 \\ 0 \\ -0.132011 \\ 0.132691 \end{Bmatrix}$$

Continued

EXAMPLE 6.5: FE ANALYSIS OF A 3D BEAM PROBLEM—CONT'D

(b) Plane x–z. Using the values of the nodal displacements in the x–z plane, namely

$$\{d^{\alpha}\}_{x-z} = \begin{Bmatrix} u_{1z} \\ \theta_{1y} \\ u_{2z} \\ \theta_{2y} \end{Bmatrix},$$

the corresponding bending moment distribution $\boldsymbol{m_y}$ for the beam element α can be derived by the following known equation:

$$m_y^{\alpha}(x) = \frac{EI_y^{\alpha}}{L^3}\left[12x-6L \quad 6Lx-4L^2 \quad -12x+6L \quad 6Lx-4L^2\right]\begin{Bmatrix} u_{1z} \\ \theta_{1y} \\ u_{2z} \\ \theta_{2y} \end{Bmatrix}$$

or

$$m_y^{\alpha}(x) = \frac{EI_y^{\alpha}}{L^3}\left[12x-6L \quad 6Lx-4L^2 \quad -12x+6L \quad 6Lx-4L^2\right]\begin{Bmatrix} 0 \\ 0 \\ 0.105609 \\ -0.106152 \end{Bmatrix}$$

Element b

(a) Plane x–y. Using the values of the nodal displacements in the x–y plane, namely

$$\{d^{b}\}_{x-y} = \begin{Bmatrix} u_{2y} \\ \theta_{2z} \\ u_{3y} \\ \theta_{3z} \end{Bmatrix}$$

,the corresponding bending moment distribution $\boldsymbol{m_z}$ for the beam element b can be derived by the following known equation:

$$m_z^{b}(x) = \frac{EI_z^{b}}{L^3}\left[12x-6L \quad 6Lx-4L^2 \quad -12x+6L \quad 6Lx-4L^2\right]\begin{Bmatrix} u_{2y} \\ \theta_{2z} \\ u_{3y} \\ \theta_{3z} \end{Bmatrix}$$

or

$$m_z^{b}(x) = \frac{EI_z^{b}}{L^3}\left[12x-6L \quad 6Lx-4L^2 \quad -12x+6L \quad 6Lx-4L^2\right]\begin{Bmatrix} -0.132011 \\ 0.132691 \\ 0 \\ 0 \end{Bmatrix}$$

(b) Plane x–z. Using the values of the nodal displacements in the x–z plane, namely

$$\{d^{\mathrm{b}}\}_{x-z} = \begin{Bmatrix} u_{2z} \\ \theta_{2y} \\ u_{3z} \\ \theta_{3y} \end{Bmatrix}$$

the corresponding bending moment distribution $\boldsymbol{m_y}$ for the beam element b can be derived by the following known equation:

$$m_y^{\mathrm{b}}(x) = \frac{EI_y^{\mathrm{b}}}{L^3}\left[12x-6L \quad 6Lx-4L^2 \quad -12x+6L \quad 6Lx-4L^2\right]\begin{Bmatrix} u_{2z} \\ \theta_{2y} \\ u_{3z} \\ \theta_{3y} \end{Bmatrix}$$

or

$$m_y^{\mathrm{b}}(x) = \frac{EI_y^{\mathrm{b}}}{L^3}\left[12x-6L \quad 6Lx-4L^2 \quad -12x+6L \quad 6Lx-4L^2\right]\begin{Bmatrix} 0.105609 \\ -0.106152 \\ 0 \\ 0 \end{Bmatrix}$$

Step 7: Shear Force Distributions

Element α

(c) Plane x–y. Using the values of the nodal displacements in the x–y plane, namely

$$\{d^{\alpha}\}_{x-y} = \begin{Bmatrix} u_{1y} \\ \theta_{1z} \\ u_{2y} \\ \theta_{2z} \end{Bmatrix}$$

the corresponding shear force distribution $\boldsymbol{f_y}$ for the beam element α can be derived by the following known equation:

$$f_y^{\alpha}(x) = \frac{EI_z^{\alpha}}{L^3}\left[12 \quad 6L \quad -12 \quad 6L\right]\begin{Bmatrix} u_{1y} \\ \theta_{1z} \\ u_{2y} \\ \theta_{2z} \end{Bmatrix}$$

or

$$f_y^{\alpha}(x) = \frac{EI_z^{\alpha}}{L^3}\left[12 \quad 6L \quad -12 \quad 6L\right]\begin{Bmatrix} 0 \\ 0 \\ -0.132011 \\ 0.132691 \end{Bmatrix}$$

Continued

EXAMPLE 6.5: FE ANALYSIS OF A 3D BEAM PROBLEM—CONT'D

***(d) Plane* x–z.** Using the values of the nodal displacements in the x–z plane, namely

$$\{d^{\alpha}\}_{x-z}=\begin{Bmatrix} u_{1z} \\ \theta_{1y} \\ u_{2z} \\ \theta_{2y} \end{Bmatrix}$$

the corresponding shear force distribution f_z for the beam element α can be derived by the following known equation:

$$f_z^{\alpha}(x)=\frac{EI_y^{\alpha}}{L^3}[12\ \ 6L\ \ -12\ \ 6L]\begin{Bmatrix} u_{1z} \\ \theta_{1y} \\ u_{2z} \\ \theta_{2y} \end{Bmatrix}$$

or

$$f_z^{\alpha}(x)=\frac{EI_y^{\alpha}}{L^3}[12\ \ 6L\ \ -12\ \ 6L]\begin{Bmatrix} 0 \\ 0 \\ 0.105609 \\ -0.106152 \end{Bmatrix}$$

Element b

***(c) Plane* x–y.** Using the values of the nodal displacements in the x–y plane, namely

$$\{d^{b}\}_{x-y}=\begin{Bmatrix} u_{2y} \\ \theta_{2z} \\ u_{3y} \\ \theta_{3z} \end{Bmatrix}$$

the corresponding shear force distribution f_y for the beam element b can be derived by the following known equation:

$$f_y^{b}(x)=\frac{EI_z^{b}}{L^3}[12\ \ 6L\ \ -12\ \ 6L]\begin{Bmatrix} u_{2y} \\ \theta_{2z} \\ u_{3y} \\ \theta_{3z} \end{Bmatrix}$$

or

$$f_y^{b}(x)=\frac{EI_z^{b}}{L^3}[12\ \ 6L\ \ -12\ \ 6L]\begin{Bmatrix} -0.132011 \\ 0.132691 \\ 0 \\ 0 \end{Bmatrix}$$

(d) Plane x–z. Using the values of the nodal displacements in the x–z plane, namely

$$\{d^{\mathrm{b}}\}_{x-z}=\begin{Bmatrix} u_{2z} \\ \theta_{2y} \\ u_{3z} \\ \theta_{3y} \end{Bmatrix}$$

the corresponding shear force distribution f_z for the beam element b can be derived by the following known equation:

$$f_z^{\mathrm{b}}(x)=\frac{EI_y^{\mathrm{b}}}{L^3}\begin{bmatrix} 12 & 6L & -12 & 6L \end{bmatrix}\begin{Bmatrix} u_{2z} \\ \theta_{2y} \\ u_{3z} \\ \theta_{3y} \end{Bmatrix}$$

or

$$f_z^{\mathrm{b}}(x)=\frac{EI_y^{\mathrm{b}}}{L^3}\begin{bmatrix} 12 & 6L & -12 & 6L \end{bmatrix}\begin{Bmatrix} 0.105609 \\ -0.106152 \\ 0 \\ 0 \end{Bmatrix}$$

Step 8: Torsional Moment Distributions

Element α

Using the values of the nodal twists $\{\theta_{1x}\ \ \theta_{2x}\}^{\mathrm{T}}$, the corresponding torsional moment distribution $\boldsymbol{m_x}$ for the beam element α can be derived by the following known equation:

$$m_x^{\alpha}(x)=\begin{bmatrix} GJ^{\alpha}/L & -GJ^{\alpha}/L \\ -GJ^{\alpha}/L & GJ^{\alpha}/L \end{bmatrix}\begin{Bmatrix} \theta_{1x} \\ \theta_{2x} \end{Bmatrix}$$

or

$$m_x^{\alpha}(x)=\begin{bmatrix} GJ^{\alpha}/L & -GJ^{\alpha}/L \\ -GJ^{\alpha}/L & GJ^{\alpha}/L \end{bmatrix}\begin{Bmatrix} 0 \\ 0.042025 \end{Bmatrix}$$

Element b

Using the values of the nodal twists $\{\theta_{2x}\ \ \theta_{3x}\}^{\mathrm{T}}$, the corresponding torsional moment distribution $\boldsymbol{m_x}$ for the beam element b can be derived by the following known equation:

$$m_x^{\mathrm{b}}(x)=\begin{bmatrix} GJ^{\mathrm{b}}/L & -GJ^{\mathrm{b}}/L \\ -GJ^{\mathrm{b}}/L & GJ^{\mathrm{b}}/L \end{bmatrix}\begin{Bmatrix} \theta_{2x} \\ \theta_{3x} \end{Bmatrix}$$

or

$$m_x^{\mathrm{b}}(x)=\begin{bmatrix} GJ^{\mathrm{b}}/L & -GJ^{\mathrm{b}}/L \\ -GJ^{\mathrm{b}}/L & GJ^{\mathrm{b}}/L \end{bmatrix}\begin{Bmatrix} 0.042025 \\ 0 \end{Bmatrix}$$

REFERENCES

[1] Hartmann F, Katz C. Structural analysis with finite elements. Berlin: Springer; 2007.
[2] Cook RD, Malkus DS, Plesha ME, Witt RJ. Concepts and applications of finite element analysis. Hoboken: John Wiley & Sons; 2002.
[3] Logan DL. A first course in the finite element method. Boston, MA: Cengage Learning; 2012.
[4] Bhatti MA. Fundamental finite element analysis and applications. Hoboken: John Wiley& Sons; 2005.
[5] Wunderlich W, Pilkey W. Mechanics of structures—variational and computational methods. Boca Raton: CRC Press; 2003.
[6] Fish J, Belytschko T. A first course in finite elements. New York: John Wiley & Sons; 2007.
[7] Austrell P-E, Dahlblom O, Lindemann J, Olsson A, Olsson K-G, Persson K, Peterson H, Ristinmaa M, Sandberg G, Wernberg P-A. CALFEM—a finite element toolbox, version 3.4. Division of Structural Mechanics, Lund University; 2004.
[8] ANSYS, User e-manual, Version 13.
[9] Melosh RJ. Structural engineering analysis by finite elements. Upper Saddle River, NJ: Prentice-Hall; 1990.
[10] Shames IH, Dym CL. Energy and finite element methods in structural mechanics. New York: Hemisphere Publishing Corporation; 1985.
[11] Hetenyi M. Beams on elastic foundation. Michigan: The University of Michigan Press; 1961.
[12] Vlasov VZ, Leont'ev NN. Beams-plates-shells on elastic foundation. Gosudarstvennoe Izdatel'st vo Fiziko-Matematicheskoi Literatury Moskva; 1960 Israel Program for Scientific Translation 1966.
[13] Flugge W. Stresses in shells. New York: Springer-Verlag; 1973.

CHAPTER

FRAMES

7

7.1 FRAMED STRUCTURES

Frames are structures consisting of beams (Figure 7.1). The beams composing a frame generally have arbitrary directions. Therefore, all nodal parameters (displacements, rotations, forces, moments) should be expressed with respect to a common system of coordinates, the global coordinate system.

Since vertical deflections of node 2 of a horizontal beam 2-3 can cause axial deflections to the vertical beam 1-2 connected to the same node (Figure 7.2), the axial component of deflections should be taken into account to the beam element equation of a frame member.

7.2 TWO-DIMENSIONAL FRAME ELEMENT EQUATION SUBJECTED TO NODAL FORCES

Let us assume that an arbitrary node 1 moves to the location $1'$. Therefore, there is a vertical deflection d_{1y} and a horizontal deflection d_{1x} with respect to the global coordinate system $x-y$. According to Figure 7.3, the coordinates $d_{1\bar{y}}, d_{1\bar{x}}$ with respect to the local coordinate system $\bar{x}-\bar{y}$ can be correlated with the global coordinates d_{1y}, d_{1x}.

It is known from the geometry that:

$$d_{1\bar{x}} = d_{1x} \cos\vartheta + d_{1y} \sin\vartheta \tag{7.1}$$

$$d_{1\bar{y}} = -d_{1x} \sin\vartheta + d_{1y} \cos\vartheta \tag{7.2}$$

In contrast, any slope ϕ_1 on node 1 of the beam element with respect to the global system $x-y$ is equal to the slope $\overline{\phi}_1$ with respect to the local system $\bar{x}-\bar{y}$:

$$\overline{\phi}_1 = \phi_1 \tag{7.3}$$

Equations (7.1)–(7.3) can be written in the following matrix form:

$$\begin{Bmatrix} d_{1\bar{x}} \\ d_{1\bar{y}} \\ \overline{\phi}_1 \end{Bmatrix} = \begin{bmatrix} C & S & 0 \\ -S & C & 0 \\ 0 & 0 & 1 \end{bmatrix} \begin{Bmatrix} d_{1x} \\ d_{1y} \\ \phi_1 \end{Bmatrix} \tag{7.4}$$

where $C = \cos\vartheta$ and $S = \sin\vartheta$.

Essentials of the Finite Element Method. http://dx.doi.org/10.1016/B978-0-12-802386-0.00007-4

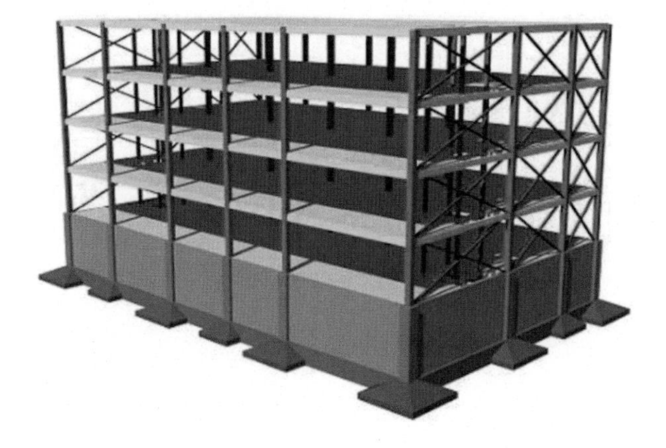

FIGURE 7.1

Framed structure.

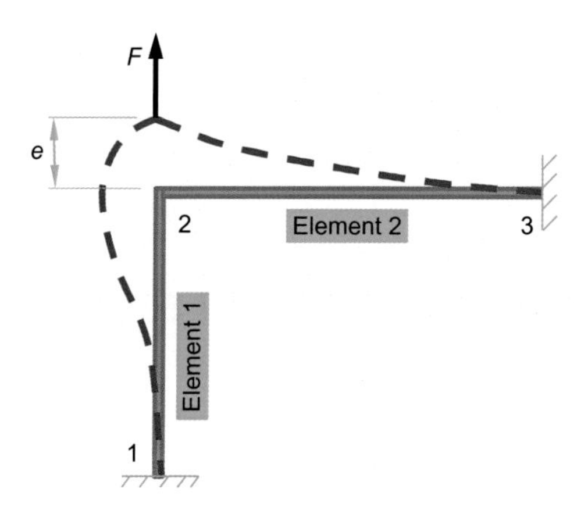

FIGURE 7.2

The displacement e of the node 2 causes axial deflection on the element 1 and vertical deflection on the element 2.

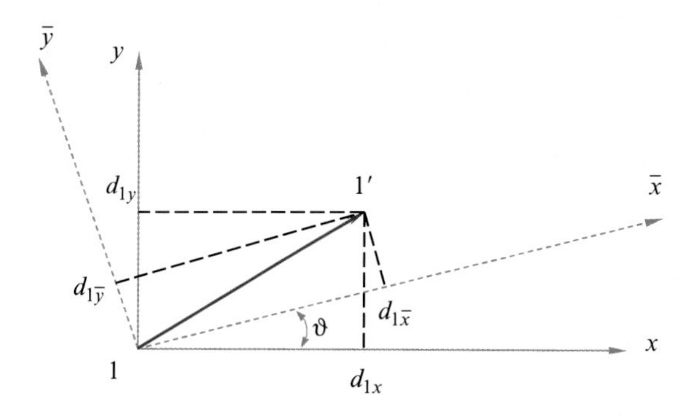

FIGURE 7.3

Transformation of displacements from the global coordinate system $x-y$ to the local system $\bar{x}-\bar{y}$.

Similarly, any set of nodal forces $f_{1\bar{x}}, f_{1\bar{y}}, \bar{m}_1$ with respect to the local system $\bar{x}-\bar{y}$ can be correlated to the nodal forces f_{1x}, f_{1y}, m_1 with respect to the global system $x-y$:

$$\begin{Bmatrix} f_{1\bar{x}} \\ f_{1\bar{y}} \\ \bar{m}_1 \end{Bmatrix} = \begin{bmatrix} C & S & 0 \\ -S & C & 0 \\ 0 & 0 & 1 \end{bmatrix} \begin{Bmatrix} f_{1x} \\ f_{1y} \\ m_1 \end{Bmatrix} \tag{7.5}$$

Let us now consider a frame member 1-2. For the displacements of node 2 we can formulate similar equations, that is,

$$\begin{Bmatrix} d_{2\bar{x}} \\ d_{2\bar{y}} \\ \bar{\phi}_2 \end{Bmatrix} = \begin{bmatrix} C & S & 0 \\ -S & C & 0 \\ 0 & 0 & 1 \end{bmatrix} \begin{Bmatrix} d_{2x} \\ d_{2y} \\ \phi_2 \end{Bmatrix} \tag{7.6}$$

$$\begin{Bmatrix} f_{2\bar{x}} \\ f_{2\bar{y}} \\ \bar{m}_2 \end{Bmatrix} = \begin{bmatrix} C & S & 0 \\ -S & C & 0 \\ 0 & 0 & 1 \end{bmatrix} \begin{Bmatrix} f_{2x} \\ f_{2y} \\ m_2 \end{Bmatrix} \tag{7.7}$$

Combination of Equations (7.4) and (7.6) yields

$$\begin{Bmatrix} d_{1\bar{x}} \\ d_{1\bar{y}} \\ \bar{\phi}_1 \\ d_{2\bar{x}} \\ d_{2\bar{y}} \\ \bar{\phi}_2 \end{Bmatrix} = \begin{bmatrix} C & S & 0 & 0 & 0 & 0 \\ -S & C & 0 & 0 & 0 & 0 \\ 0 & 0 & 1 & 0 & 0 & 0 \\ 0 & 0 & 0 & C & S & 0 \\ 0 & 0 & 0 & -S & C & 0 \\ 0 & 0 & 0 & 0 & 0 & 1 \end{bmatrix} \begin{Bmatrix} d_{1x} \\ d_{1y} \\ \phi_1 \\ d_{2x} \\ d_{2y} \\ \phi_2 \end{Bmatrix} \tag{7.8}$$

In the same way, combining Equations (7.5) and (7.7), the following matrix equation can be obtained:

$$\begin{Bmatrix} f_{1\bar{x}} \\ f_{1\bar{y}} \\ \bar{m}_1 \\ f_{2\bar{x}} \\ f_{2\bar{y}} \\ \bar{m}_2 \end{Bmatrix} = \begin{bmatrix} C & S & 0 & 0 & 0 & 0 \\ -S & C & 0 & 0 & 0 & 0 \\ 0 & 0 & 1 & 0 & 0 & 0 \\ 0 & 0 & 0 & C & S & 0 \\ 0 & 0 & 0 & -S & C & 0 \\ 0 & 0 & 0 & 0 & 0 & 1 \end{bmatrix} \begin{Bmatrix} f_{1x} \\ f_{1y} \\ m_1 \\ f_{2x} \\ f_{2y} \\ m_2 \end{Bmatrix} \tag{7.9}$$

We recall now the beam element equation (Equation 6.32) for a beam 1-2 aligned to the local axis $\bar{x}$. Therefore, its nodal forces $f_{1\bar{y}}, f_{2\bar{y}}$ and moments $\bar{m}_1, \bar{m}_2$ are correlated to the corresponding vertical deflections $d_{1\bar{y}}, d_{2\bar{y}}$ and rotations $\bar{\phi}_1, \bar{\phi}_2$ by the following already known equation:

$$\begin{Bmatrix} f_{1\bar{y}} \\ \bar{m}_1 \\ f_{2\bar{y}} \\ \bar{m}_2 \end{Bmatrix} = \begin{bmatrix} 12EI/L^3 & 6EI/L^2 & -12EI/L^3 & 6EI/L^2 \\ 6EI/L^2 & 4EI/L & -6EI/L^2 & 2EI/L \\ -12EI/L^3 & -6EI/L^2 & 12EI/L^3 & -6EI/L^2 \\ 6EI/L^2 & 2EI/L & -6EI/L^2 & 4EI/L \end{bmatrix} \begin{Bmatrix} d_{1\bar{y}} \\ \bar{\phi}_1 \\ d_{2\bar{y}} \\ \bar{\phi}_2 \end{Bmatrix} \tag{7.10}$$

However, as has been mentioned, in frame members it should be taken into account the axial deflections $d_{1\bar{x}}, d_{2\bar{x}}$ and of course the corresponding nodal forces $f_{1\bar{x}}, f_{2\bar{x}}$. To this scope, we recall the bar element equation (Equation 4.24):

$$\begin{Bmatrix} f_{1\bar{x}} \\ f_{2\bar{x}} \end{Bmatrix} = \begin{bmatrix} EA/L & -EA/L \\ -EA/L & EA/L \end{bmatrix} \begin{Bmatrix} d_{1\bar{x}} \\ d_{2\bar{x}} \end{Bmatrix} \tag{7.11}$$

Combining Equations (7.10) and (7.11), the following formula can be obtained:

$$\begin{Bmatrix} f_{1\bar{x}} \\ f_{1\bar{y}} \\ \bar{m}_1 \\ f_{2\bar{x}} \\ f_{2\bar{y}} \\ \bar{m}_2 \end{Bmatrix} = \begin{bmatrix} EA/L & 0 & 0 & -EA/L & 0 & 0 \\ 0 & 12EI/L^3 & 6EI/L^2 & 0 & -12EI/L^3 & 6EI/L^2 \\ 0 & 6EI/L^2 & 4EI/L & 0 & -6EI/L^2 & 2EI/L \\ -EA/L & 0 & 0 & EA/L & 0 & 0 \\ 0 & -12EI/L^3 & -6EI/L^2 & 0 & 12EI/L^3 & -6EI/L^2 \\ 0 & 6EI/L^2 & 2EI/L & 0 & -6EI/L^2 & 4EI/L \end{bmatrix} \begin{Bmatrix} d_{1\bar{x}} \\ d_{1\bar{y}} \\ \bar{\phi}_1 \\ d_{2\bar{x}} \\ d_{2\bar{y}} \\ \bar{\phi}_2 \end{Bmatrix} \tag{7.12}$$

Let us now write Equations (7.8), (7.9), and (7.12) in an abbreviated form, respectively:

$$\{\bar{d}\} = [T]\{d\} \tag{7.13}$$

$$\{\bar{f}\} = [T]\{f\} \tag{7.14}$$

$$\{\bar{f}\} = [\bar{k}]\{\bar{d}\} \tag{7.15}$$

where

$$[T] = \begin{bmatrix} C & S & 0 & 0 & 0 & 0 \\ -S & C & 0 & 0 & 0 & 0 \\ 0 & 0 & 1 & 0 & 0 & 0 \\ 0 & 0 & 0 & C & S & 0 \\ 0 & 0 & 0 & -S & C & 0 \\ 0 & 0 & 0 & 0 & 0 & 1 \end{bmatrix} \tag{7.16}$$

$$[\bar{k}] = \begin{bmatrix} EA/L & 0 & 0 & -EA/L & 0 & 0 \\ 0 & 12EI/L^3 & 6EI/L^2 & 0 & -12EI/L^3 & 6EI/L^2 \\ 0 & 6EI/L^2 & 4EI/L & 0 & -6EI/L^2 & 2EI/L \\ -EA/L & 0 & 0 & EA/L & 0 & 0 \\ 0 & -12EI/L^3 & -6EI/L^2 & 0 & 12EI/L^3 & -6EI/L^2 \\ 0 & 6EI/L^2 & 2EI/L & 0 & -6EI/L^2 & 4EI/L \end{bmatrix} \tag{7.17}$$

$$\{\bar{d}\} = \{d_{1\bar{x}}, d_{1\bar{y}}, \bar{\phi}_1, d_{2\bar{x}}, d_{2\bar{y}}, \bar{\phi}_2\}^{\mathrm{T}} \tag{7.18}$$

$$\{\bar{f}\} = \{f_{1\bar{x}}, f_{1\bar{y}}, \bar{m}_1, f_{2\bar{x}}, f_{2\bar{y}}, \bar{m}_2\}^{\mathrm{T}} \tag{7.19}$$

$$\{d\} = \{d_{1x}, d_{1y}, \phi_1, d_{2x}, d_{2y}, \phi_2\}^{\mathrm{T}} \tag{7.20}$$

$$\{f\} = \{f_{1x}, f_{1y}, m_1, f_{2x}, f_{2y}, m_2\}^{\mathrm{T}} \tag{7.21}$$

Then, using Equations (7.13) and (7.14), Equation (7.15) can now be written:

$$[T]\{f\} = [\bar{k}][T]\{d\} \tag{7.22}$$

Therefore,

$$\{f\} = [T]^{-1}[\bar{k}][T]\{d\}$$

or

$$\{f\} = [k]\{d\} \tag{7.23}$$

where

$$[k] = [T]^{-1}[\bar{k}]\,[T] \tag{7.24}$$

yielding:

$$[k] = \frac{E}{L}\begin{bmatrix} AC^2 + \frac{12I}{L^2}S^2 & \left(A - \frac{12I}{L^2}\right)CS & \frac{-6I}{L}S & -\left(AC^2 + \frac{12I}{L^2}S^2\right) & -\left(A - \frac{12I}{L^2}\right)CS & -\frac{6I}{L}C \\ & AS^2 + \frac{12I}{L^2}C^2 & \frac{6I}{L}C & -\left(A - \frac{12I}{L^2}\right)CS & -\left(AS^2 + \frac{12I}{L^2}C^2\right) & \frac{6I}{L}C \\ & & 4I & \frac{6I}{L}S & -\frac{6I}{L}C & 2I \\ & & & AC^2 + \frac{12I}{L^2}S^2 & \left(A - \frac{12I}{L^2}\right)CS & \frac{6I}{L}S \\ & \text{symmetric} & & & AS^2 + \frac{12I}{L^2}C^2 & -\frac{6I}{L}C \\ & & & & & 4I \end{bmatrix} \tag{7.25}$$

With Equation (7.23), the nodal parameters of any inclined frame element aligned to the local coordinate system $\bar{x} - \bar{y}$ can be transformed to the global coordinate system $x - y$.

7.3 TWO-DIMENSIONAL FRAME ELEMENT EQUATION SUBJECTED TO ARBITRARY VARYING LOADING

Frame members are often subjected to varying loadings (e.g., Figure 7.4). In these cases, the varying loading should be simulated with the corresponding equivalent nodal forces. Table 7.1 provides the equivalent nodal forces for several common varying loading cases. However, these nodal forces corresponds to the local coordinate system, that is, the coordinate system that is aligned with the beam element. When a frame element is horizontal, the local and the global coordinate systems are identical. However, for inclined frame elements, a transformation of the equivalent nodal forces from the local to the global coordinate system should be performed.

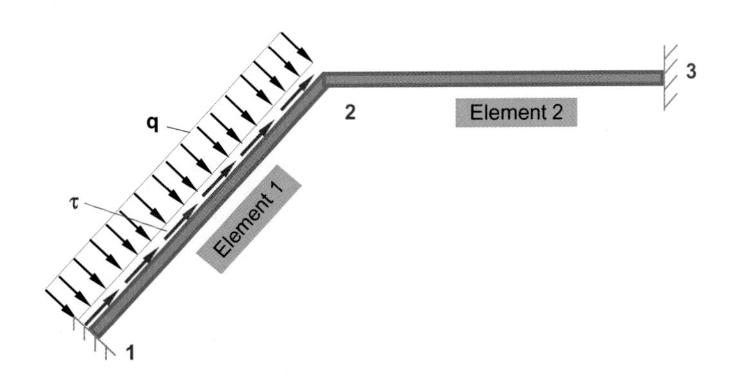

FIGURE 7.4

Frame containing inclined element subjected to varying loading.

Table 7.1 Force matrix {F} for inclined frame elements of length L subjected to common cases of varying loads

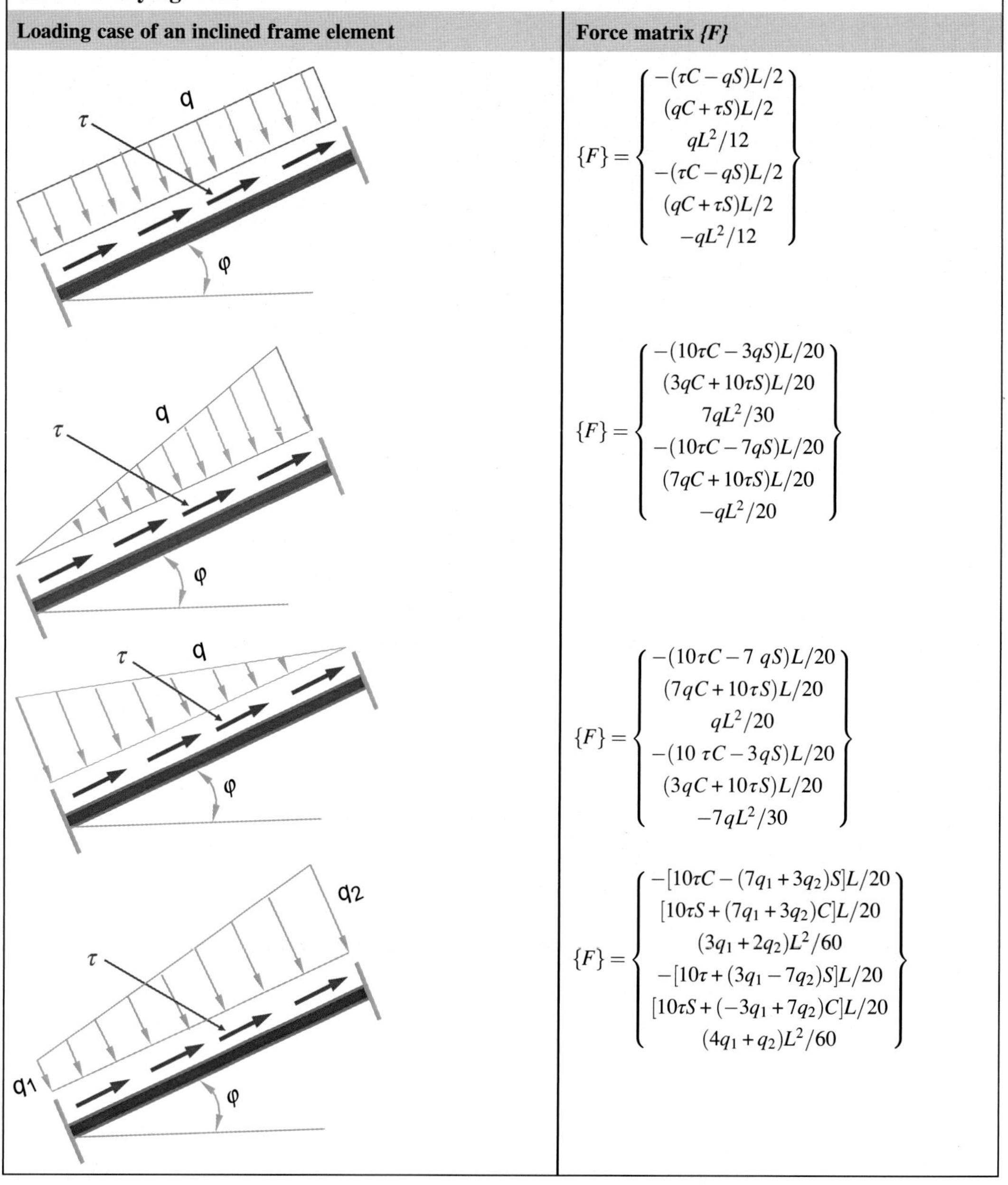

Loading case of an inclined frame element	Force matrix *{F}*
	$\{F\}=\begin{Bmatrix} -(\tau C-qS)L/2 \\ (qC+\tau S)L/2 \\ qL^2/12 \\ -(\tau C-qS)L/2 \\ (qC+\tau S)L/2 \\ -qL^2/12 \end{Bmatrix}$
	$\{F\}=\begin{Bmatrix} -(10\tau C-3qS)L/20 \\ (3qC+10\tau S)L/20 \\ 7qL^2/30 \\ -(10\tau C-7qS)L/20 \\ (7qC+10\tau S)L/20 \\ -qL^2/20 \end{Bmatrix}$
	$\{F\}=\begin{Bmatrix} -(10\tau C-7\,qS)L/20 \\ (7qC+10\tau S)L/20 \\ qL^2/20 \\ -(10\,\tau C-3qS)L/20 \\ (3qC+10\tau S)L/20 \\ -7qL^2/30 \end{Bmatrix}$
	$\{F\}=\begin{Bmatrix} -[10\tau C-(7q_1+3q_2)S]L/20 \\ [10\tau S+(7q_1+3q_2)C]L/20 \\ (3q_1+2q_2)L^2/60 \\ -[10\tau+(3q_1-7q_2)S]L/20 \\ [10\tau S+(-3q_1+7q_2)C]L/20 \\ (4q_1+q_2)L^2/60 \end{Bmatrix}$

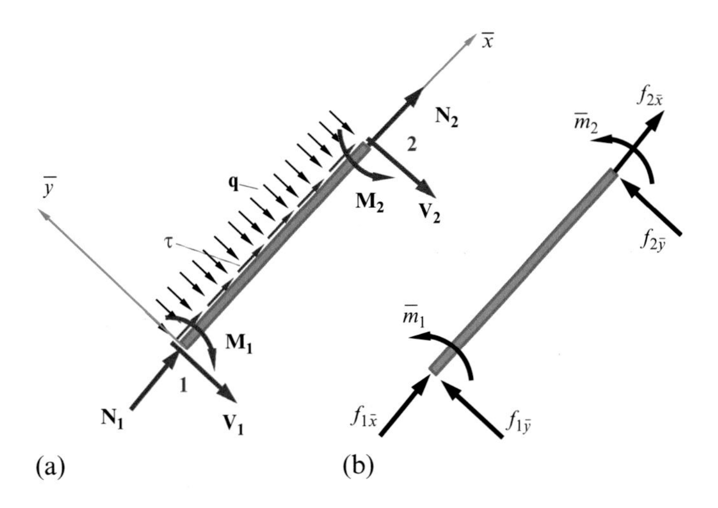

FIGURE 7.5

(a) Equivalent nodal forces for an inclined frame element, (b) sign convention for the finite element method.

We recall that the element equations of beams subjected to varying loading contains both stiffness and load matrix. Let us assume an incline frame element (e.g., element 1 of Figure 7.4) subjected to varying loading. It is already known that the element equation with respect to its local system $\bar{x}-\bar{y}$ (Figure 7.5a) has the following form:

$$\begin{Bmatrix} f_{1\bar{x}} \\ f_{1\bar{y}} \\ \bar{m}_1 \\ f_{2\bar{x}} \\ f_{2\bar{y}} \\ \bar{m}_2 \end{Bmatrix} = \begin{bmatrix} EA/L & 0 & 0 & -EA/L & 0 & 0 \\ 0 & 12EI/L^3 & 6EI/L^2 & 0 & -12EI/L^3 & 6EI/L^2 \\ 0 & 6EI/L^2 & 4EI/L & 0 & -6EI/L^2 & 2EI/L \\ -EA/L & 0 & 0 & EA/L & 0 & 0 \\ 0 & -12EI/L^3 & -6EI/L^2 & 0 & 12EI/L^3 & -6EI/L^2 \\ 0 & 6EI/L^2 & 2EI/L & 0 & -6EI/L^2 & 4EI/L \end{bmatrix} \begin{Bmatrix} d_{1\bar{x}} \\ d_{1\bar{y}} \\ \bar{\phi}_1 \\ d_{2\bar{x}} \\ d_{2\bar{y}} \\ \bar{\phi}_2 \end{Bmatrix} + \begin{Bmatrix} -N_1 \\ V_1 \\ M_1 \\ -N_2 \\ V_2 \\ -M_2 \end{Bmatrix} \tag{7.26}$$

Actually Equation (7.26) is a modification of Equation (6.118) in order to take into account the axial components $f_{1\bar{x}}, d_{1\bar{x}}, N_1$ and $f_{2\bar{x}}, d_{2\bar{x}}, N_2$ according to the sign convention of Figure 7.5b. Using abbreviated formulation, Equation (7.26) can be written as:

$$\{\bar{f}\} = [\bar{k}]\{\bar{d}\} + \{\bar{F}\} \tag{7.27}$$

where $[\bar{k}]$ is the matrix given by Equation (7.17) and

$$\{\bar{F}\} = \{-N_1, V_1, M_1, -N_2, V_2, -M_2\}^{\mathrm{T}} \tag{7.28}$$

Taking into account Equations (7.13) and (7.14), Equation (7.27) yields

$$[T]\{f\} = [\bar{k}][T]\{d\} + \{\bar{F}\} \tag{7.29}$$

The matrix $[T]$ is given by Equation (7.16).

Therefore:

$$\{f\} = [T]^{-1}[\bar{k}][T]\{d\} + [T]^{-1}\{\bar{F}\} \tag{7.30}$$

or

$$\{f\} = [k]\{d\} + \{F\} \tag{7.31}$$

The matrix $[k]$ is given by Equation (7.25), and the matrix $\{F\}$ is given by the following equation:

$$\{F\} = [T]^{-1}\{\bar{F}\} \tag{7.32}$$

For several common types of varying loading acting on an inclined frame member, the matrix $[F]$ can be obtained using Table 7.1.

EXAMPLE 7.1

Determine the nodal displacements and slopes as well as the reactions on the supports of the following plane frame subjected to two-dimensional loading conditions.

Data

$$q = 13\text{kN/m}$$
$$L = 1.5\text{m}$$
$$A = 3.8 \times 10^{-2}\text{m}^2$$
$$I = 4.2 \times 10^{-4}\text{m}^4$$
$$E = 210\text{GPa}$$

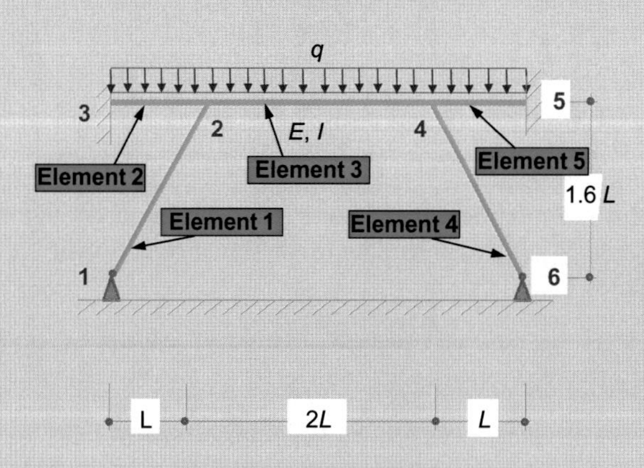

1. Calculation of the Slopes and Lengths for the Frame Elements

$$\theta_1 = \arctan(1.6L/L) = 58°$$
$$L_1 = \sqrt{L^2 + (1.6L)^2} = 3.56L$$

$$\theta_2 = 0$$
$$L_2 = L$$

$$\theta_3 = 0$$
$$L_3 = 2L$$

$$\theta_4 = 180 - \theta_1 = 122°$$
$$L_4 = \sqrt{L^2 + (1.6L)^2} = 3.56L$$

$$\theta_5 = 0$$
$$L_5 = L$$

Taking into account the above values of the angles $\theta_1, \ldots, \theta_5$, the following values of C and S for the five elements can be obtained:

$$C_1 = \cos\theta_1 = 0.5 \quad C_2 = \cos\theta_2 = 1 \quad C_3 = \cos\theta_3 = 1 \quad C_4 = \cos\theta_4 = -0.53 \quad C_5 = \cos\theta_5 = 1$$
$$S_1 = \sin\theta_1 = 0.85 \quad S_2 = \sin\theta_2 = 0 \quad S_3 = \sin\theta_3 = 0 \quad S_4 = \sin\theta_4 = 0.85 \quad S_5 = \sin\theta_5 = 0$$

2. *Local Stiffness Matrices* $[k_e]$ *of the Frame Elements* e (e = *1–5*)

Using Equation (7.25), the following local stiffness matrices for the five beam elements can be derived:

ELEMENT 1

8.255939×10^8	1.246250×10^9	-5.602505×10^7	-6.03116×10^8	-1.246250×10^9	-3.50157×10^7
1.246250×10^9	2.040687×10^9	3.501566×10^7	-1.246250×10^9	-2.040687×10^9	3.501566×10^7
-5.602505×10^7	3.501566×10^7	1.246557×10^8	5.602505×10^7	-3.501566×10^7	6.232787×10^7
-6.03116×10^8	-1.246250×10^9	5.602505×10^7	8.255939×10^8	1.246250×10^9	5.602505×10^7
-1.246250×10^9	-2.040687×10^9	-3.501566×10^7	1.246250×10^9	2.040687×10^9	-3.501566×10^7
-3.50157×10^7	3.501566×10^7	6.232787×10^7	5.602505×10^7	-3.501566×10^7	1.246557×10^8

ELEMENT 2

5.32×10^9	0.	0.	0	0.	-2.352×10^8
0.	3.136×10^8	2.352×10^8	0.	-3.136×10^8	2.352×10^8
0.	2.352×10^8	2.352×10^8	0.	-2.352×10^8	1.176×10^8
0	0.	0.	5.32×10^9	0.	0.
0.	-3.136×10^8	-2.352×10^8	0.	3.136×10^8	-2.352×10^8
-2.352×10^8	2.352×10^8	1.176×10^8	0.	-2.352×10^8	2.352×10^8

Continued

EXAMPLE 7.1—CONT'D

ELEMENT 3

$$
\begin{array}{cccccc}
2.66\times10^{9} & 0. & 0. & 0 & 0. & -5.88\times10^{7}\\
0. & 3.92\times10^{7} & 5.88\times10^{7} & 0. & -3.92\times10^{7} & 5.88\times10^{7}\\
0. & 5.88\times10^{7} & 1.176\times10^{8} & 0. & -5.88\times10^{7} & 5.88\times10^{7}\\
0 & 0. & 0. & 2.66\times10^{9} & 0. & 0.\\
0. & -3.92\times10^{7} & -5.88\times10^{7} & 0. & 3.92\times10^{7} & -5.88\times10^{7}\\
-5.88\times10^{7} & 5.88\times10^{7} & 5.88\times10^{7} & 0. & -5.88\times10^{7} & 1.176\times10^{8}
\end{array}
$$

ELEMENT 4

$$
\begin{array}{cccccc}
8.255939\times10^{8} & -1.246250\times10^{9} & -5.602505\times10^{7} & -6.03116\times10^{8} & 1.246250\times10^{9} & 3.501566\times10^{7}\\
-1.246250\times10^{9} & 2.040687\times10^{9} & -3.501566\times10^{7} & 1.246250\times10^{9} & -2.040687\times10^{9} & -3.501566\times10^{7}\\
-5.602505\times10^{7} & -3.501566\times10^{7} & 1.246557\times10^{8} & 5.602505\times10^{7} & 3.501566\times10^{7} & 6.232787\times10^{7}\\
-6.03116\times10^{8} & 1.246250\times10^{9} & 5.602505\times10^{7} & 8.255939\times10^{8} & -1.246250\times10^{9} & 5.602505\times10^{7}\\
1.246250\times10^{9} & -2.040687\times10^{9} & 3.501566\times10^{7} & -1.246250\times10^{9} & 2.040687\times10^{9} & 3.501566\times10^{7}\\
3.501566\times10^{7} & -3.501566\times10^{7} & 6.232787\times10^{7} & 5.602505\times10^{7} & 3.501566\times10^{7} & 1.246557\times10^{8}
\end{array}
$$

ELEMENT 5

$$
\begin{array}{cccccc}
5.32\times10^{9} & 0. & 0. & 0 & 0. & -2.352\times10^{8}\\
0. & 3.136\times10^{8} & 2.352\times10^{8} & 0. & -3.136\times10^{8} & 2.352\times10^{8}\\
0. & 2.352\times10^{8} & 2.352\times10^{8} & 0. & -2.352\times10^{8} & 1.176\times10^{8}\\
0 & 0. & 0. & 5.32\times10^{9} & 0. & 0.\\
0. & -3.136\times10^{8} & -2.352\times10^{8} & 0. & 3.136\times10^{8} & -2.352\times10^{8}\\
-2.352\times10^{8} & 2.352\times10^{8} & 1.176\times10^{8} & 0. & -2.352\times10^{8} & 2.352\times10^{8}
\end{array}
$$

3. Local Load Vectors $[r_e]$ of the Frame Elements e (e = 1–5)

Since only elements 2, 3, and 5 are subjected to varying loading q, the load vectors (Table 7.1) for elements 1 and 4 are zero, that is,

$$
[r_1]=[r_4]=\begin{Bmatrix}0\\0\\0\\0\\0\\0\end{Bmatrix}
$$

The nonzero load vectors (Table 7.1) for elements 2, 3, and 5 can be derived as:

$$[r_2] = \begin{Bmatrix} 0. \\ 9750. \\ 2437.5 \\ 0. \\ 9750. \\ -2437.5 \end{Bmatrix}$$

$$[r_3] = \begin{Bmatrix} 0. \\ 19,500. \\ 9750. \\ 0. \\ 19,500. \\ -9750. \end{Bmatrix}$$

$$[r_5] = \begin{Bmatrix} 0. \\ 9750. \\ 2437.5 \\ 0. \\ 9750. \\ -2437.5 \end{Bmatrix}$$

4. Expanded Stiffness Matrix [K_j] and Force Matrix [F_j] of the Frame Elements j (j = 1–5)

The expanded stiffness matrix $[K_j]$ and load vector $\{F_j\}$ can be derived by placing the local ones $[k_e]$ and $\{r_e\}$ within the following matrices with respect to the degrees of freedom of the whole structure. Since the number of degrees of freedom is large, we can write the expanded element equation of any element j to the following abbreviated form

$$\begin{Bmatrix} \{f_1\} \\ \{f_2\} \\ \{f_3\} \\ \{f_4\} \\ \{f_5\} \\ \{f_6\} \end{Bmatrix} = \underbrace{\begin{bmatrix} \; & \; & \; & \; & \; \\ \; & \; & \; & \; & \; \\ \; & \; & \; & \; & \; \\ \; & \; & \; & \; & \; \\ \; & \; & \; & \; & \; \\ \; & \; & \; & \; & \; \end{bmatrix}}_{[K_j]} \begin{Bmatrix} \{d_1\} \\ \{d_2\} \\ \{d_3\} \\ \{d_4\} \\ \{d_5\} \\ \{d_6\} \end{Bmatrix} + \{F_j\}$$

or

$$\{f\} = [K_j]\{d\} + \{F_j\}$$

Continued

EXAMPLE 7.1—CONT'D

In the above equation the matrices

$$\{f\}=[\{f_1\}\ \{f_2\}\ \{f_3\}\ \{f_4\}\ \{f_5\}\ \{f_6\}]$$

and

$$\{d\}=[\{d_1\}\ \{d_2\}\ \{d_3\}\ \{d_4\}\ \{d_5\}\ \{d_6\}]$$

contain the following submatrices

$$\{f_1\}=\{f_{1x}\ f_{1y}\ m_1\}^{\mathrm{T}},\ \{d_1\}=\{d_{1x}\ d_{1y}\ \varphi_1\}^{\mathrm{T}}$$
$$\{f_2\}=\{f_{2x}\ f_{2y}\ m_2\}^{\mathrm{T}},\ \{d_2\}=\{d_{2x}\ d_{2y}\ \varphi_2\}^{\mathrm{T}}$$
$$\{f_3\}=\{f_{3x}\ f_{3y}\ m_3\}^{\mathrm{T}},\ \{d_3\}=\{d_{3x}\ d_{3y}\ \varphi_3\}^{\mathrm{T}}$$
$$\{f_4\}=\{f_{4x}\ f_{4y}\ m_4\}^{\mathrm{T}},\ \{d_4\}=\{d_{4x}\ d_{4y}\ \varphi_4\}^{\mathrm{T}}$$
$$\{f_5\}=\{f_{5x}\ f_{5y}\ m_5\}^{\mathrm{T}},\ \{d_5\}=\{d_{5x}\ d_{5y}\ \varphi_5\}^{\mathrm{T}}$$
$$\{f_6\}=\{f_{6x}\ f_{6y}\ m_6\}^{\mathrm{T}},\ \{d_6\}=\{d_{6x}\ d_{6y}\ \varphi_6\}^{\mathrm{T}}$$

Using the following nomenclature for the local stiffness matrix of any element e:

$$[k_e]=\begin{bmatrix}[k_e^{\mathrm{I}}] & [k_e^{\mathrm{II}}]\\ [k_e^{\mathrm{III}}] & [k_e^{\mathrm{IV}}]\end{bmatrix}$$

and the local load vector of any element e

$$\{r_e\}=\begin{Bmatrix}F_e^{\mathrm{I}}\\ F_e^{\mathrm{II}}\end{Bmatrix}$$

the following expanded stiffness matrices $[K_j]$ and load vectors $\{F_j\}$ of the elements $j=1$–5 can be obtained:

- $j=1$: nodes 1 and 2

$$[K_1]=\begin{bmatrix}[k_1^{\mathrm{I}}] & [k_1^{\mathrm{II}}] & [O_{3\times3}] & [O_{3\times3}] & [O_{3\times3}] & [O_{3\times3}]\\ [k_1^{\mathrm{III}}] & [k_1^{\mathrm{IV}}] & [O_{3\times3}] & [O_{3\times3}] & [O_{3\times3}] & [O_{3\times3}]\\ [O_{3\times3}] & [O_{3\times3}] & [O_{3\times3}] & [O_{3\times3}] & [O_{3\times3}] & [O_{3\times3}]\\ [O_{3\times3}] & [O_{3\times3}] & [O_{3\times3}] & [O_{3\times3}] & [O_{3\times3}] & [O_{3\times3}]\\ [O_{3\times3}] & [O_{3\times3}] & [O_{3\times3}] & [O_{3\times3}] & [O_{3\times3}] & [O_{3\times3}]\\ [O_{3\times3}] & [O_{3\times3}] & [O_{3\times3}] & [O_{3\times3}] & [O_{3\times3}] & [O_{3\times3}]\end{bmatrix},\quad \{F_1\}=\begin{bmatrix}[F_1^{\mathrm{I}}]\\ [F_1^{\mathrm{II}}]\\ [O_{3\times1}]\\ [O_{3\times1}]\\ [O_{3\times1}]\\ [O_{3\times1}]\end{bmatrix}$$

- $j=2$: nodes 2 and 3

$$[K_2]=\begin{bmatrix}[O_{3\times3}] & [O_{3\times3}] & [O_{3\times3}] & [O_{3\times3}] & [O_{3\times3}] & [O_{3\times3}]\\ [O_{3\times3}] & [k_2^{\mathrm{I}}] & [k_2^{\mathrm{II}}] & [O_{3\times3}] & [O_{3\times3}] & [O_{3\times3}]\\ [O_{3\times3}] & [k_2^{\mathrm{III}}] & [k_2^{\mathrm{IV}}] & [O_{3\times3}] & [O_{3\times3}] & [O_{3\times3}]\\ [O_{3\times3}] & [O_{3\times3}] & [O_{3\times3}] & [O_{3\times3}] & [O_{3\times3}] & [O_{3\times3}]\\ [O_{3\times3}] & [O_{3\times3}] & [O_{3\times3}] & [O_{3\times3}] & [O_{3\times3}] & [O_{3\times3}]\\ [O_{3\times3}] & [O_{3\times3}] & [O_{3\times3}] & [O_{3\times3}] & [O_{3\times3}] & [O_{3\times3}]\end{bmatrix}, \quad \{F_2\}\begin{Bmatrix}[O_{3\times1}]\\ [F_2^{\mathrm{I}}]\\ [F_2^{\mathrm{II}}]\\ [O_{3\times1}]\\ [O_{3\times1}]\\ [O_{3\times1}]\end{Bmatrix}$$

- $j=3$: nodes 2 and 4

$$[K_3]=\begin{bmatrix}[O_{3\times3}] & [O_{3\times3}] & [O_{3\times3}] & [O_{3\times3}] & [O_{3\times3}] & [O_{3\times3}]\\ [O_{3\times3}] & [k_3^{\mathrm{I}}] & [O_{3\times3}] & [k_3^{\mathrm{II}}] & [O_{3\times3}] & [O_{3\times3}]\\ [O_{3\times3}] & O_{3x3} & [O_{3\times3}] & [O_{3\times3}] & [O_{3\times3}] & [O_{3\times3}]\\ [O_{3\times3}] & [k_3^{\mathrm{III}}] & [O_{3\times3}] & [k_3^{\mathrm{IV}}] & [O_{3\times3}] & [O_{3\times3}]\\ [O_{3\times3}] & [O_{3\times3}] & [O_{3\times3}] & [O_{3\times3}] & [O_{3\times3}] & [O_{3\times3}]\\ [O_{3\times3}] & [O_{3\times3}] & [O_{3\times3}] & [O_{3\times3}] & [O_{3\times3}] & [O_{3\times3}]\end{bmatrix}, \quad \{F_3\}=\begin{Bmatrix}[O_{3\times1}]\\ [F_3^{\mathrm{I}}]\\ [O_{3\times1}]\\ [F_3^{\mathrm{II}}]\\ [O_{3\times1}]\\ [O_{3\times1}]\end{Bmatrix}$$

- $j=4$: nodes 4 and 6

$$[K_4]=\begin{bmatrix}[O_{3\times3}] & [O_{3\times3}] & [O_{3\times3}] & [O_{3\times3}] & [O_{3\times3}] & [O_{3\times3}]\\ [O_{3\times3}] & [O_{3\times3}] & [O_{3\times3}] & [O_{3\times3}] & [O_{3\times3}] & [O_{3\times3}]\\ [O_{3\times3}] & [O_{3\times3}] & [O_{3\times3}] & [O_{3\times3}] & [O_{3\times3}] & [O_{3\times3}]\\ [O_{3\times3}] & [O_{3\times3}] & [O_{3\times3}] & [k_4^{\mathrm{I}}] & [O_{3\times3}] & [k_4^{\mathrm{II}}]\\ [O_{3\times3}] & [O_{3\times3}] & [O_{3\times3}] & [O_{3\times3}] & [O_{3\times3}] & [O_{3\times3}]\\ [O_{3\times3}] & [O_{3\times3}] & [O_{3\times3}] & [k_4^{\mathrm{III}}] & [O_{3\times3}] & [k_4^{\mathrm{IV}}]\end{bmatrix}, \quad \{F_4\}=\begin{Bmatrix}[O_{3\times1}]\\ [O_{3\times1}]\\ [O_{3\times1}]\\ [F_4^{\mathrm{I}}]\\ [O_{3\times1}]\\ [F_4^{\mathrm{II}}]\end{Bmatrix}$$

- $j=5$: nodes 4 and 5

$$[K_5]=\begin{bmatrix}[O_{3\times3}] & [O_{3\times3}] & [O_{3\times3}] & [O_{3\times3}] & [O_{3\times3}] & [O_{3\times3}]\\ [O_{3\times3}] & [O_{3\times3}] & [O_{3\times3}] & [O_{3\times3}] & [O_{3\times3}] & [O_{3\times3}]\\ [O_{3\times3}] & [O_{3\times3}] & [O_{3\times3}] & [O_{3\times3}] & [O_{3\times3}] & [O_{3\times3}]\\ [O_{3\times3}] & [O_{3\times3}] & [O_{3\times3}] & [k_5^{\mathrm{I}}] & [k_5^{\mathrm{II}}] & [O_{3\times3}]\\ [O_{3\times3}] & [O_{3\times3}] & [O_{3\times3}] & [k_5^{\mathrm{III}}] & [k_5^{\mathrm{IV}}] & [O_{3\times3}]\\ [O_{3\times3}] & [O_{3\times3}] & [O_{3\times3}] & [O_{3\times3}] & [O_{3\times3}] & [O_{3\times3}]\end{bmatrix}, \quad \{F_5\}=\begin{Bmatrix}[O_{3\times1}]\\ [O_{3\times1}]\\ [O_{3\times1}]\\ [F_5^{\mathrm{I}}]\\ [F_5^{\mathrm{II}}]\\ [O_{3\times1}]\end{Bmatrix}$$

Continued

EXAMPLE 7.1—CONT'D

5. Global Stiffness Matrix [K] and Load Vector {F} of the Structure

The global stiffness matrix $[K]$ and the load vector $\{F\}$ of the whole structure can be derived by adding the above extended matrices $[K_i]$ and vectors $\{F_i\}$ of the individual beams, respectively:

$$[K]=[K_1]+[K_2]+[K_3]+[K_4]+[K_5]$$

$$\{F\}=\{F_1\}+\{F_2\}+\{F_3\}+\{F_4\}+\{F_5\}$$

Taking into account the above results, the structure equation can now be written as follows:

$$\{R\}=[K]\{d\}+\{F\}$$

where $\{R\}$ is the vector containing the nodal forces of the six nodes:

$$\{R\}=[\{R_1\}\ \{R_2\}\ \{R_3\}\ \{R_4\}\ \{R_5\}\ \{R_6\}]^{\mathrm{T}} \text{ where } \{R_i\}=\{R_{ix}\ R_{iy}\ M_i\}^{\mathrm{T}}\ (i=1,\ldots,6)$$

It is more convenient to express the above structure equation in the following form:

$$[[K]\ \ -[I]]\begin{Bmatrix}\{d\}\\ \{R\}\end{Bmatrix}=-\{F\} \tag{e-1}$$

6. Boundary Conditions

For the derivation of the algebraic system providing the nodal displacement and force field $[\{d\}\{R\}]^{\mathrm{T}}$, the boundary conditions of the problem should be incorporated into the above 18×36 structural equation (e-1). Taking into account the type of supports, as well as the external nodal forces, the following 18 boundary conditions can be specified:

(a) Boundary conditions for nodal displacements

$$(1)\,d_{1x}=0,\quad (6)\,d_{5x}=0$$
$$(2)\,d_{1y}=0,\quad (7)\,d_{5y}=0$$
$$(3)\,d_{3x}=0,\quad (8)\,\phi_5=0$$
$$(4)\,d_{3y}=0,\quad (9)\,d_{6x}=0$$
$$(5)\,\phi_3=0,\quad (10)\,d_{6y}=0$$

(b) Boundary conditions for nodal forces

$$(11)\,M_1=0,\quad (15)\,F_{4x}=0$$
$$(12)\,F_{2x}=0,\quad (16)\,F_{4y}=0$$
$$(13)\,F_{2y}=0,\quad (17)\,M_4=0$$
$$(14)\,M_2=0,\quad (18)\,M_6=0$$

The above two groups of boundary conditions can be expressed in the following matrix formats:

(c) Boundary conditions for nodal displacements

$$
\underbrace{\begin{bmatrix}
1&0&0&0&0&0&0&0&0&0&0&0&0&0&0&0&0&0\\
0&1&0&0&0&0&0&0&0&0&0&0&0&0&0&0&0&0\\
0&0&0&0&0&0&1&0&0&0&0&0&0&0&0&0&0&0\\
0&0&0&0&0&0&0&1&0&0&0&0&0&0&0&0&0&0\\
0&0&0&0&0&0&0&0&1&0&0&0&0&0&0&0&0&0\\
0&0&0&0&0&0&0&0&0&0&0&0&1&0&0&0&0&0\\
0&0&0&0&0&0&0&0&0&0&0&0&0&1&0&0&0&0\\
0&0&0&0&0&0&0&0&0&0&0&0&0&0&1&0&0&0\\
0&0&0&0&0&0&0&0&0&0&0&0&0&0&0&1&0&0\\
0&0&0&0&0&0&0&0&0&0&0&0&0&0&0&0&1&0\\
0&0&0&0&0&0&0&0&0&0&0&0&0&0&0&0&0&0\\
0&0&0&0&0&0&0&0&0&0&0&0&0&0&0&0&0&0\\
0&0&0&0&0&0&0&0&0&0&0&0&0&0&0&0&0&0\\
0&0&0&0&0&0&0&0&0&0&0&0&0&0&0&0&0&0\\
0&0&0&0&0&0&0&0&0&0&0&0&0&0&0&0&0&0\\
0&0&0&0&0&0&0&0&0&0&0&0&0&0&0&0&0&0\\
0&0&0&0&0&0&0&0&0&0&0&0&0&0&0&0&0&0\\
0&0&0&0&0&0&0&0&0&0&0&0&0&0&0&0&0&0
\end{bmatrix}}_{[BCd]}
\underbrace{\begin{Bmatrix}
d_{1x}\\ d_{1y}\\ \varphi_1\\ d_{2x}\\ d_{2y}\\ \varphi_2\\ d_{3x}\\ d_{3y}\\ \varphi_3\\ d_{4x}\\ d_{4y}\\ \varphi_4\\ d_{5x}\\ d_{5y}\\ \varphi_5\\ d_{6x}\\ d_{6y}\\ \varphi_6
\end{Bmatrix}}_{\{d\}}
=
\underbrace{\begin{bmatrix}
0\\0\\0\\0\\0\\0\\0\\0\\0\\0\\0\\0\\0\\0\\0\\0\\0\\0
\end{bmatrix}}_{\{DO\}}
$$

The above equation can be written in an abbreviated format

$$[BCd]\{d\} = \{DO\}$$

or

$$[[BCd]\ \ [O]]\begin{Bmatrix}\{d\}\\ \{R\}\end{Bmatrix} = \{DO\} \tag{e-2}$$

Continued

EXAMPLE 7.1—CONT'D

(d) Boundary conditions for nodal forces

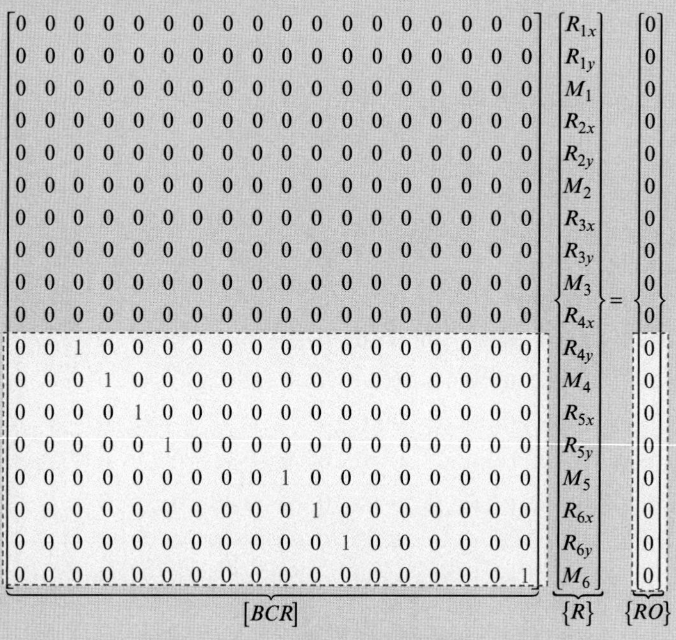

The above equation can be written in an abbreviated format

$$[BCR]\{R\} = \{RO\}$$

or

$$[[O]\ \ [BCR]]\begin{Bmatrix} \{d\} \\ \{R\} \end{Bmatrix} = \{RO\} \tag{e-3}$$

Addition of Equations (e-2) and (e-3) yields

$$[[BCd]\ \ [BCR]]\begin{Bmatrix} \{d\} \\ \{R\} \end{Bmatrix} = \{DO + RO\} \tag{e-4}$$

7. *Algebraic System*

Combining the structural equation (e-1) with the above equation (e-4), the following algebraic system can be obtained:

$$\underbrace{\begin{bmatrix} [K] & -[I] \\ [BCd] & [BCR] \end{bmatrix}}_{[A]} \underbrace{\begin{Bmatrix} \{d\} \\ \{R\} \end{Bmatrix}}_{\{X\}} = \underbrace{\begin{Bmatrix} -\{F\} \\ \{DO + RO\} \end{Bmatrix}}_{\{B\}}$$

or in an abbreviated form

$$[A]\{X\} = \{B\}$$

where the vector $\{X\}$ contain the 36 unknown nodal values

$$\{X\} = \{\{d\}\ \ \{R\}\}^{\mathrm{T}}$$

Solution of the above system yields the following results:

Results

$$
\left\{\begin{array}{c}
d_{1x} \\ d_{1y} \\ \varphi_1 \\ d_{2x} \\ d_{2y} \\ \varphi_2 \\ d_{3x} \\ d_{3y} \\ \varphi_3 \\ d_{4x} \\ d_{4y} \\ \varphi_4 \\ d_{5x} \\ d_{5y} \\ \varphi_5 \\ d_{6x} \\ d_{6y} \\ \varphi_6 \\ F_{1x} \\ F_{1y} \\ M_1 \\ F_{2x} \\ F_{2y} \\ M_2 \\ F_{3x} \\ F_{3y} \\ M_3 \\ F_{4x} \\ F_{4y} \\ M_4 \\ F_{5x} \\ F_{5y} \\ M_5 \\ F_{6x} \\ F_{6y} \\ M6
\end{array}\right\} =
\begin{array}{|c|}
\hline
-1.694721891768517 \times 10^{-19} \\
-6.55301959750068 \times 10^{-19} \\
0.000009109382490054363 \\
0.000001866087822363125 \\
-0.000011149993302324947 \\
-0.000026160188473426458 \\
0. \\
-2.182706162234896 \times 10^{-19} \\
-3.154992629804221 \times 10^{-19} \\
-0.000002015552160486833 \\
-0.000015937630005825552 \\
0.000026888729633116008 \\
0. \\
-2.654557689772808 \times 10^{-18} \\
4.988310221954764 \times 10^{-19} \\
7.913421091529113 \times 10^{-19} \\
-8.131516293641162 \times 10^{-20} \\
-0.000017355060840529865 \\
13175.87682748078 \\
19830.999350407954 \\
0. \\
0. \\
0. \\
0. \\
0. \\
19399.514228559015 \\
-8575.320445001606 \\
0. \\
0. \\
0. \\
0. \\
8423.811560117043 \\
-2549.858104368475 \\
-18112.5389847136 \\
30345.674860915977 \\
0. \\
\hline
\end{array}
$$

7.4 THREE-DIMENSIONAL BEAM ELEMENT EQUATION SUBJECTED TO NODAL FORCES

Let us now consider a beam element 1-2, arbitrarily oriented in space (Figure 7.6). We select the $\bar{y}$ axis to be perpendicular to the $\bar{x}-z$ plane.

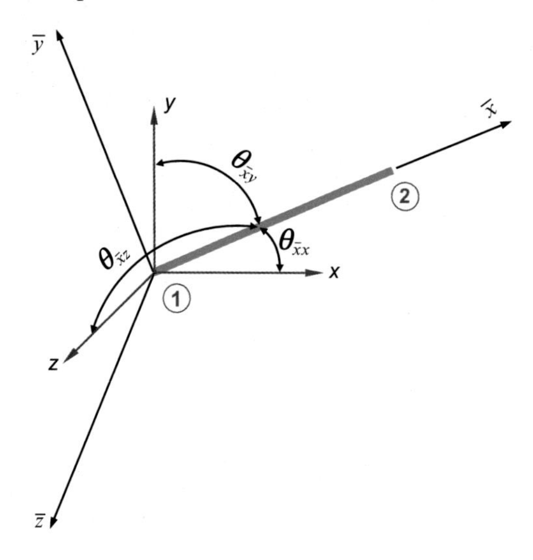

FIGURE 7.6

A beam element arbitrarily oriented in space.

Equation (6.194) expresses the element equation with respect to the local coordinate system. Since the local coordinates that are aligned to the element are now symbolized by $\bar{x}-\bar{y}-\bar{z}$, Equation (6.194) should be expressed as follows:

$$\begin{Bmatrix} f_{1\bar{x}} \\ f_{1\bar{y}} \\ f_{1\bar{z}} \\ m_{1\bar{x}} \\ m_{1\bar{y}} \\ m_{1\bar{z}} \\ f_{2\bar{x}} \\ f_{2\bar{y}} \\ f_{2\bar{z}} \\ m_{2\bar{x}} \\ m_{2\bar{y}} \\ m_{2\bar{z}} \end{Bmatrix} = \begin{bmatrix} [K_{fu}^{11}] & [K_{f\theta}^{11}] & [K_{fu}^{12}] & [K_{f\theta}^{12}] \\ [K_{mu}^{11}] & [K_{m\theta}^{11}] & [K_{mu}^{12}] & [K_{m\theta}^{12}] \\ [K_{fu}^{21}] & [K_{f\theta}^{21}] & [K_{fu}^{22}] & [K_{f\theta}^{22}] \\ [K_{mu}^{21}] & [K_{m\theta}^{21}] & [K_{mu}^{22}] & [K_{m\theta}^{22}] \end{bmatrix} \begin{Bmatrix} u_{1\bar{x}} \\ u_{1\bar{y}} \\ u_{1\bar{z}} \\ \vartheta_{1\bar{x}} \\ \theta_{1\bar{y}} \\ \theta_{1\bar{z}} \\ u_{2\bar{x}} \\ u_{2\bar{y}} \\ u_{2\bar{z}} \\ \theta_{2\bar{x}} \\ \theta_{2\bar{y}} \\ \theta_{2\bar{z}} \end{Bmatrix} \tag{7.33}$$

or in an abbreviated form

$$[\bar{K}]\{\bar{u}\} = \{\bar{f}\} \tag{7.34}$$

Our target is to perform the transformation of Equation (7.33) to the global coordinate system $x-y-z$. Since we have selected the $\bar{y}$ axis to be always perpendicular to the plane $\bar{x}-z$, the specification of the position of the local coordinate system $\bar{x}-\bar{y}-\bar{z}$ with respect to the global $x-y-z$ can be sufficiently performed by the direction cosines (with respect to the axes x,y,z) of the axis $\bar{x}$ only, namely $C_{\bar{x}x}, C_{\bar{x}y}, C_{\bar{x}z}$. As it is already known (see Chapter 5), these direction cosines can be calculated using the following formulae:

$$C_{\bar{x}x}=\frac{\vec{a}\cdot\vec{i}}{|\vec{a}|\cdot|\vec{i}|}=l \tag{7.35}$$

$$C_{\bar{x}y}=\frac{\vec{a}\cdot\vec{j}}{|\vec{a}|\cdot|\vec{j}|}=m \tag{7.36}$$

$$C_{\bar{x}z}=\frac{\vec{a}\cdot\vec{k}}{|\vec{a}|\cdot|\vec{k}|}=n \tag{7.37}$$

where $\vec{a}$ is the vectorial form of the element 1–2 given by the following equation:

$$\vec{a}=(x_2-x_1)\,\vec{i}+(y_2-y_1)\,\vec{j}+(z_2-z_1)\,\vec{k} \tag{7.38}$$

and $|\vec{a}|$ is its magnitude (i.e., the length)

$$|\vec{a}|=\sqrt{(x_2-x_1)^2+(y_2-y_1)^2+(z_2-z_1)^2}=L \tag{7.39}$$

Since

$$\vec{a}\cdot\vec{i}=(x_2-x_1) \tag{7.40}$$

$$\vec{a}\cdot\vec{j}=(y_2-y_1) \tag{7.41}$$

$$\vec{a}\cdot\vec{k}=(z_2-z_1) \tag{7.42}$$

Equations (7.35)–(7.37) yield

$$(x_2-x_1)=l\cdot|\vec{a}| \tag{7.43}$$

$$(y_2-y_1)=m\cdot|\vec{a}| \tag{7.44}$$

$$(z_2-z_1)=n\cdot|\vec{a}| \tag{7.45}$$

Therefore, using the above equations, Equation (7.38) can be written as:

$$\frac{\vec{a}}{|\vec{a}|}=l\cdot\vec{i}+m\cdot\vec{j}+n\cdot\vec{k} \tag{7.46}$$

Since the element is lying on the local axis $\bar{x}$ that means $\vec{a}=|\vec{a}|\cdot\vec{i}'$, the ratio $\vec{a}/|\vec{a}|$ represents the unit vector $\vec{i}'$ of the inclined axis $\vec{x}$, that is,

$$\vec{i'} = l \cdot \vec{i} + m \cdot \vec{j} + n \cdot \vec{k} \tag{7.47}$$

The unit vectors $\vec{j'}$ and $\vec{k'}$ corresponding to the local axes $\bar{y}$ and $\bar{z}$, respectively, can be obtained by the following formulae (see Chapter 2):

$$\vec{j'} = \vec{k} \times \vec{i'} = \frac{1}{D} \begin{vmatrix} \vec{i} & \vec{j} & \vec{k} \\ 0 & 0 & 1 \\ l & m & n \end{vmatrix} = \frac{1}{D}\left(-m \cdot \vec{i} + l \cdot \vec{j}\right) \tag{7.48}$$

$$\vec{k'} = \vec{i'} \times \vec{j'} = \frac{1}{D} \begin{vmatrix} \vec{i} & \vec{j} & \vec{k} \\ l & m & n \\ -m & l & 0 \end{vmatrix} = \frac{1}{D}\left(-l \cdot n \cdot \vec{i} - m \cdot n \cdot \vec{j} + D^2 \cdot \vec{k}\right) \tag{7.49}$$

where

$$D = \sqrt{l^2 + m^2} \tag{7.50}$$

Combining Equations (7.47)–(7.49), the following transformation formula can be derived:

$$\begin{Bmatrix} \vec{i'} \\ \vec{j'} \\ \vec{k'} \end{Bmatrix} = [\Phi] \begin{Bmatrix} \vec{i} \\ \vec{j} \\ \vec{k} \end{Bmatrix} \tag{7.51}$$

where

$$[\Phi] = \begin{bmatrix} l & m & n \\ -m/D & l/D & 0 \\ -n \cdot l/D & -n \cdot m/D & D \end{bmatrix} \tag{7.52}$$

Using Equation (7.51), the local stiffness $[\bar{K}]$ corresponding to the local coordinate system $\bar{x} - \bar{y} - \bar{z}$ can be transformed to the matrix $[K]$ corresponding to the global coordinate system $x - y - z$:

$$[K] = [T]^{-1} \cdot [\bar{K}] \cdot [T] \tag{7.53}$$

In the above equation the transformation matrix $[T]$ can be obtained using the following equation:

$$[T] = \begin{bmatrix} [\Phi] & [O_{3\times3}] & [O_{3\times3}] & [O_{3\times3}] \\ [O_{3\times3}] & [\Phi] & [O_{3\times3}] & [O_{3\times3}] \\ [O_{3\times3}] & [O_{3\times3}] & [\Phi] & [O_{3\times3}] \\ [O_{3\times3}] & [O_{3\times3}] & [O_{3\times3}] & [\Phi] \end{bmatrix} \tag{7.54}$$

where the submatrix $[\Phi]$ can be obtained from Equation (7.52) and the submatrix $[O_{3\times3}]$ has size 3×3 and contains zeros.

The stiffness matrix $[K] = [T]^{-1} \cdot [\bar{K}] \cdot [T]$ correlates with the nodal parameters of the global coordinate system x–y–z:

$$[K]\{u\} = \{f\} \tag{7.55}$$

where

$$\{u\} = [u_{1x}\ u_{1y}\ u_{1z}\ \theta_{1x}\ \theta_{1y}\ \theta_{1z}\ u_{2x}\ u_{2y}\ u_{2z}\ \theta_{2x}\ \theta_{2y}\ \theta_{2z}]^T \tag{7.56}$$

$$\{f\} = [f_{1x}\ f_{1y}\ f_{1z}\ m_{1x}\ m_{1y}\ m_{1z}\ f_{2x}\ f_{2y}\ f_{2z}\ m_{2x}\ m_{2y}\ m_{2z}]^T \tag{7.57}$$

For the specific case that $l=0, m=0$ (i.e., the element is parallel to the z-axis), the value of D is $D=0$. Therefore, the $[\Phi]$ matrix members $-m/D, l/D, -nl/D, -nm/D$ cannot be determined. The reason for this uncertainty is the fact that the local axis $\bar{y}$ can have any direction. In this case, we can select a direction for $\bar{y}$ which is coincided to the direction of y (see Figure 7.7).

Therefore, if the direction of $\bar{x}$ is same with the direction of z (Figure 7.7a), then

$$[\Phi] = \begin{bmatrix} 0 & 0 & 1 \\ 0 & 1 & 0 \\ -1 & 0 & 0 \end{bmatrix} \tag{7.58}$$

For the opposite direction of $\bar{x}$ (Figure 7.7b) the matrix $[\Phi]$ is:

$$[\Phi] = \begin{bmatrix} 0 & 0 & -1 \\ 0 & 1 & 0 \\ 1 & 0 & 0 \end{bmatrix} \tag{7.59}$$

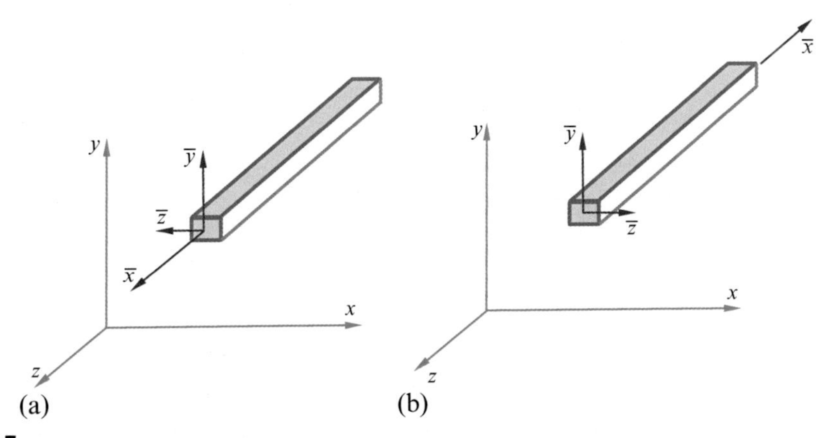

FIGURE 7.7

(a) The local axis $\bar{x}$ has same direction with the global axis z, (b) the local axis $\bar{x}$ has opposite direction with the global axis z.

7.5 DISTRIBUTION OF BENDING MOMENTS, SHEAR FORCES, AXIAL FORCES, AND TORSIONAL MOMENTS OF EACH ELEMENT

The matrix equation (7.55) along with the boundary conditions of the problem will allow us to calculate the nodal displacements $\{u\}$. From the vector $\{u\}$, we can calculate the corresponding displacements with respect to the local coordinate system $\bar{x}-\bar{y}-\bar{z}$ using the following equation:

$$\{\bar{u}\} = [T]\{u\} \tag{7.60}$$

Bending moment and shear force distributions

Then, the bending moment and shear force distributions with respect to the planes $\bar{x}-\bar{y}$ and $\bar{x}-\bar{z}$ can be derived from the following known equations (see Equations 6.22 and 6.23) by using the components $u_{i\bar{y}}, \theta_{i\bar{z}}, u_{i\bar{z}}, \theta_{i\bar{y}} (i=1,2)$ of the above matrix $\{u\}$

- for the plane $\bar{x}-\bar{y}$

$$\begin{Bmatrix} m_{\bar{z}}(\bar{x}) \\ m_{\bar{y}}(\bar{x}) \end{Bmatrix} = \frac{EI_{\bar{y}}}{L^3}\begin{bmatrix} 12x-6L & 6Lx-4L^2 & -12x+6L & 6Lx-2L^2 \\ 12 & 6L & -12 & 6L \end{bmatrix}\begin{Bmatrix} u_{1\bar{y}} \\ \theta_{1\bar{z}} \\ u_{2\bar{y}} \\ \theta_{2\bar{z}} \end{Bmatrix} \tag{7.61}$$

- for the plane $\bar{x}-\bar{z}$

$$\begin{Bmatrix} m_{\bar{y}}(\bar{x}) \\ m_{\bar{z}}(\bar{x}) \end{Bmatrix} = \frac{EI_{\bar{z}}}{L^3}\begin{bmatrix} 12x-6L & 6Lx-4L^2 & -12x+6L & 6Lx-2L^2 \\ 12 & 6L & -12 & 6L \end{bmatrix}\begin{Bmatrix} u_{1\bar{z}} \\ \theta_{1\bar{y}} \\ u_{2\bar{z}} \\ \theta_{2\bar{y}} \end{Bmatrix} \tag{7.62}$$

Axial force and torsional moment distributions

Finally, the axial force and torsional moment distributions along the axis $\bar{x}$ can be determined by the following equations:

$$f_{\bar{x}}(\bar{x}) = \begin{bmatrix} -\frac{EA}{L} & \frac{EA}{L} \end{bmatrix}\begin{Bmatrix} u_{1\bar{x}} \\ u_{2\bar{x}} \end{Bmatrix} \tag{7.63}$$

$$m_{\bar{x}}(\bar{x}) = \begin{bmatrix} -\frac{GJ}{L} & \frac{GJ}{L} \end{bmatrix}\begin{Bmatrix} \theta_{1\bar{x}} \\ \theta_{2\bar{x}} \end{Bmatrix} \tag{7.64}$$

EXAMPLE 7.2

Calculate the nodal displacements and the reactions on the supports of the following frame subjected to three-dimensional loading conditions. The cross-sectional area for both beams is $A=3.8\times10^{-2}$ m^2, the material properties are $E=210$ GPa, $G=79.3$ GPa, and the forces acting on the node nr. 3 are $R_x=10$kN, $R_y=15$kN, $R_z=5$kN.

Geometric Parameters of the Beam Cross-sections

$$I_a^z = 4.0 \times 10^{-4}\ \text{m}^4 \quad I_b^z = 5.0 \times 10^{-4}\ \text{m}^4$$
$$I_a^y = 4.5 \times 10^{-4}\ \text{m}^4 \quad I_b^y = 5.5 \times 10^{-4}\ \text{m}^4$$
$$J_a = 8.5 \times 10^{-4}\ \text{m}^4 \quad J_b = 10.5 \times 10^{-4}\ \text{m}^4$$

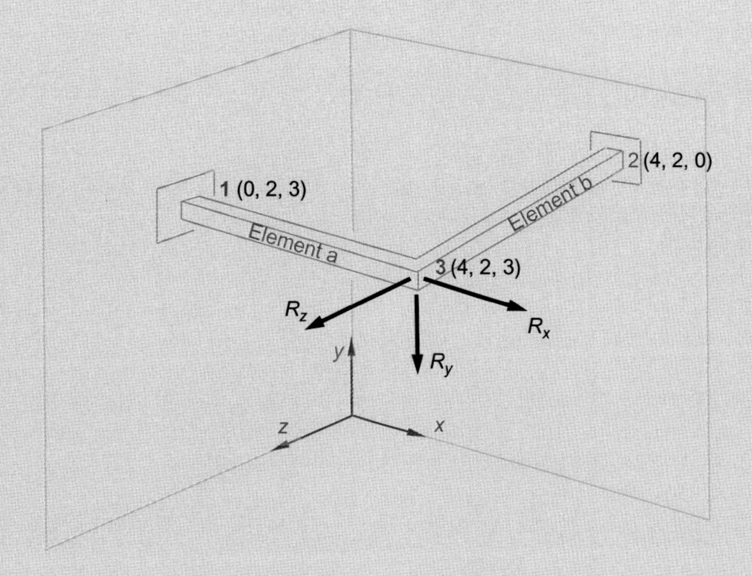

Solution

Step 1: Derivation of the Matrix [T] for the Elements a and b

First, we should identify the local coordinate systems of the beam elements with respect to the global coordinate system:

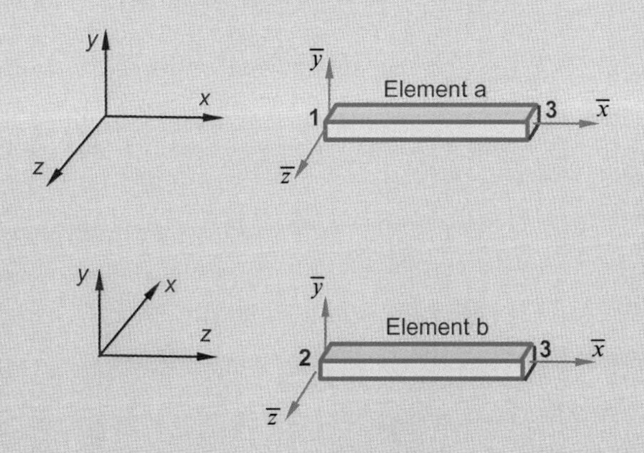

Continued

EXAMPLE 7.2—CONT'D

According to sub-chapter 7.4, the matrix [T] for the elements a and b, namely [T_a] and [T_b], can be derived as follows:

***Element* a**

$$l_a = (x_3 - x_1)/L_a = (4-0)/4 = 1$$
$$m_a = (y_3 - y_1)/L_a = (2-2)/4 = 0$$
$$n_a = (z_3 - z_1)/L_a = (3-3)/4 = 0$$
$$D_a = \sqrt{(l_a)^2 + (m_a)^2} = 1$$

$$[\Phi_a] = \begin{bmatrix} l_a & m_a & n_a \\ -m_a/D_a & l_a/D_a & 0 \\ -n_a l_a/D_a & -n_a m_a/D_a & D_a \end{bmatrix} = \begin{bmatrix} 1 & 0 & 0 \\ 0 & 1 & 0 \\ 0 & 0 & 1 \end{bmatrix}$$

Then,

$$[T_a] = \begin{bmatrix} [\Phi_a] & [O_{3\times3}] & [O_{3\times3}] & [O_{3\times3}] \\ [O_{3\times3}] & [\Phi_a] & [O_{3\times3}] & [O_{3\times3}] \\ [O_{3\times3}] & [O_{3\times3}] & [\Phi_a] & [O_{3\times3}] \\ [O_{3\times3}] & [O_{3\times3}] & [O_{3\times3}] & [\Phi_a] \end{bmatrix}$$

where

$$[O_{3\times3}] = \begin{bmatrix} 0 & 0 & 0 \\ 0 & 0 & 0 \\ 0 & 0 & 0 \end{bmatrix}$$

***Element* b**

$$l_b = (x_3 - x_2)/L_b = (4-4)/3 = 0$$
$$m_b = (y_3 - y_2)/L_b = (2-2)/3 = 0$$
$$n_b = (z_3 - z_2)/L_b = (3-0)/3 = 0$$
$$D_b = \sqrt{(l_b)^2 + (m_b)^2} = 0$$

Since $D_b = 0$ (i.e., the element b is parallel to the z-axis), according to Equation (7.58), the following expression for [Φ_b] and [T_b] can be obtained:

$$[\Phi_b] = \begin{bmatrix} 0 & 0 & 1 \\ 0 & 1 & 0 \\ -1 & 0 & 0 \end{bmatrix}$$

$$[T_b] = \begin{bmatrix} [\Phi_b] & [O_{3\times3}] & [O_{3\times3}] & [O_{3\times3}] \\ [O_{3\times3}] & [\Phi_b] & [O_{3\times3}] & [O_{3\times3}] \\ [O_{3\times3}] & [O_{3\times3}] & [\Phi_b] & [O_{3\times3}] \\ [O_{3\times3}] & [O_{3\times3}] & [O_{3\times3}] & [\Phi_b] \end{bmatrix}$$

***Step 2: Derivation of the Expanded Stiffness Matrix [Ke] for the Elements* a *and* b**

***Element* a**

***Stiffness matrix in the local coordinate system* $\bar{x}-\bar{y}-\bar{z}$.** Taking into account Equation (7.33), the following element equation should be derived for the element a (nodes 1–3):

$$\begin{Bmatrix} f_{1\bar{x}} \\ f_{1\bar{y}} \\ f_{1\bar{z}} \\ m_{1\bar{x}} \\ m_{1\bar{y}} \\ m_{1\bar{z}} \\ f_{3\bar{x}} \\ f_{3\bar{y}} \\ f_{3\bar{z}} \\ m_{3\bar{x}} \\ m_{3\bar{y}} \\ m_{3\bar{z}} \end{Bmatrix} = [\bar{k}_a] \begin{Bmatrix} u_{1\bar{x}} \\ u_{1\bar{y}} \\ u_{1\bar{z}} \\ \vartheta_{1\bar{x}} \\ \theta_{1\bar{y}} \\ \theta_{1\bar{z}} \\ u_{3\bar{x}} \\ u_{3\bar{y}} \\ u_{3\bar{z}} \\ \theta_{3\bar{x}} \\ \theta_{3\bar{y}} \\ \theta_{3\bar{z}} \end{Bmatrix}$$

where

$$[\bar{k}_a] = \begin{bmatrix} \left[Ka_{fu}^{11}\right] & \left[Ka_{f\theta}^{11}\right] & \left[Ka_{fu}^{12}\right] & \left[Ka_{f\theta}^{12}\right] \\ \left[Ka_{mu}^{11}\right] & \left[Ka_{m\theta}^{11}\right] & \left[Ka_{mu}^{12}\right] & \left[Ka_{m\theta}^{12}\right] \\ \left[Ka_{fu}^{21}\right] & \left[Ka_{f\theta}^{21}\right] & \left[Ka_{fu}^{22}\right] & \left[Ka_{f\theta}^{22}\right] \\ \left[Ka_{mu}^{21}\right] & \left[Ka_{m\theta}^{21}\right] & \left[Ka_{mu}^{22}\right] & \left[Ka_{m\theta}^{22}\right] \end{bmatrix}$$

Using Equations (6.195)–(6.210), the following results can be obtained for the submatrices composing the stiffness matrix of the element "a" in its local coordinate system $\bar{x}-\bar{y}-\bar{z}$

$$\left[Ka_{fu}^{11}\right] = \begin{pmatrix} 1.995\times10^{9} & 0 & 0 \\ 0 & 1.575\times10^{7} & 0 \\ 0 & 0 & 1.771875\times10^{7} \end{pmatrix}$$

$$\left[Ka_{mu}^{11}\right] = \begin{pmatrix} 0 & 0 & 0 \\ 0 & 0 & 3.54375\times10^{7} \\ 0 & 3.15\times10^{7} & 0 \end{pmatrix}$$

$$\left[Ka_{fu}^{21}\right] = \begin{pmatrix} -1.995\times10^{9} & 0 & 0 \\ 0 & -1.575\times10^{7} & 0 \\ 0 & 0 & -1.771875\times10^{7} \end{pmatrix}$$

$$\left[Ka_{mu}^{21}\right] = \begin{pmatrix} 0 & 0 & 0 \\ 0 & 0 & 3.54375\times10^{7} \\ 0 & 3.15\times10^{7} & 0 \end{pmatrix}$$

Continued

EXAMPLE 7.2—CONT'D

$$\left[Ka_{f\theta}^{11}\right]=\begin{pmatrix}0 & 0 & 0\\ 0 & 0 & 3.15\times10^{7}\\ 0 & 3.54375\times10^{7} & 0\end{pmatrix}$$

$$\left[Ka_{m\theta}^{11}\right]=\begin{pmatrix}1.685125\times10^{7} & 0 & 0\\ 0 & 9.450000000000003\times10^{7} & 0\\ 0 & 0 & 8.4\times10^{7}\end{pmatrix}$$

$$\left[Ka_{f\theta}^{21}\right]=\begin{pmatrix}0 & 0 & 0\\ 0 & 0 & -3.15\times10^{7}\\ 0 & -3.54375\times10^{7} & 0\end{pmatrix}$$

$$\left[Ka_{m\theta}^{21}\right]=\begin{pmatrix}-1.685125\times10^{7} & 0 & 0\\ 0 & 4.725000000000001\times10^{7} & 0\\ 0 & 0 & 4.2\times10^{7}\end{pmatrix}$$

$$\left[Ka_{fu}^{12}\right]=\begin{pmatrix}-1.995\times10^{9} & 0 & 0\\ 0 & -1.575\times10^{7} & 0\\ 0 & 0 & -1.771875\times10^{7}\end{pmatrix}$$

$$\left[Ka_{mu}^{12}\right]=\begin{pmatrix}0 & 0 & 0\\ 0 & 0 & -3.54375\times10^{7}\\ 0 & -3.15\times10^{7} & 0\end{pmatrix}$$

$$\left[Ka_{fu}^{22}\right]=\begin{pmatrix}1.995\times10^{9} & 0 & 0\\ 0 & 1.575\times10^{7} & 0\\ 0 & 0 & 1.771875\times10^{7}\end{pmatrix}$$

$$\left[Ka_{mu}^{22}\right]=\begin{pmatrix}0 & 0 & 0\\ 0 & 0 & -3.54375\times10^{7}\\ 0 & -3.15\times10^{7} & 0\end{pmatrix}$$

$$\left[Ka_{f\theta}^{12}\right]=\begin{pmatrix}0 & 0 & 0\\ 0 & 0 & 3.15\times10^{7}\\ 0 & 3.54375\times10^{7} & 0\end{pmatrix}$$

$$\left[Ka_{m\theta}^{12}\right]=\begin{pmatrix}-1.685125\times10^{7} & 0 & 0\\ 0 & 4.725000000000001\times10^{7} & 0\\ 0 & 0 & 4.2\times10^{7}\end{pmatrix}$$

$$\left[Ka_{f\theta}^{22}\right]=\begin{pmatrix}0 & 0 & 0\\ 0 & 0 & -3.15\times10^{7}\\ 0 & -3.54375\times10^{7} & 0\end{pmatrix}$$

$$\left[Ka_{m\theta}^{22}\right]=\begin{pmatrix}1.685125\times10^{7} & 0 & 0\\ 0 & 9.450000000000003\times10^{7} & 0\\ 0 & 0 & 8.4\times10^{7}\end{pmatrix}$$

***Stiffness matrix of the element "a" in the global coordinate system* x − y − z.** Taking into account Equation (7.53), the matrix $[\bar{k}_a]$ should be transformed to the global coordinate system $x-y-z$ as follows:

$$[K_a] = [T_a]^{-1} \cdot [\bar{k}_a] \cdot [T_a]$$

The result of this transformation can be written as:

$$[K_a] = \begin{bmatrix} K_a^{\mathrm{I}} & K_a^{\mathrm{II}} \\ K_a^{\mathrm{III}} & K_a^{\mathrm{IV}} \end{bmatrix}$$

where

$$[K_a^{\mathrm{I}}] = \begin{bmatrix} 1.995\times10^9 & 0. & 0. & 0. & 0. & 0. \\ 0. & 1.575\times10^7 & 0. & 0. & 0. & 3.15\times10^7 \\ 0. & 0. & 1.771875\times10^7 & 0. & 3.54375\times10^7 & 0. \\ 0. & 0. & 0. & 1.685125\times10^7 & 0. & 0. \\ 0. & 0. & 3.54375\times10^7 & 0. & 9.450000000000003\times10^7 & 0. \\ 0. & 3.15\times10^7 & 0. & 0. & 0. & 8.4\times10^7 \end{bmatrix}$$

$$[K_a^{\mathrm{II}}] = \begin{bmatrix} -1.995\times10^9 & 0. & 0. & 0. & 0. & 0. \\ 0. & -1.575\times10^7 & 0. & 0. & 0. & 3.15\times10^7 \\ 0. & 0. & -1.771875\times10^7 & 0. & 3.54375\times10^7 & 0. \\ 0. & 0. & 0. & -1.685125\times10^7 & 0. & 0. \\ 0. & 0. & -3.54375\times10^7 & 0. & 4.725000000000001\times10^7 & 0. \\ 0. & -3.15\times10^7 & 0. & 0. & 0. & 4.2\times10^7 \end{bmatrix}$$

$$[K_a^{\mathrm{III}}] = \begin{bmatrix} -1.995\times10^9 & 0. & 0. & 0. & 0. & 0. \\ 0. & -1.575\times10^7 & 0. & 0. & 0. & -3.15\times10^7 \\ 0. & 0. & -1.771875\times10^7 & 0. & -3.54375\times10^7 & 0. \\ 0. & 0. & 0. & -1.685125\times10^7 & 0. & 0. \\ 0. & 0. & 3.54375\times10^7 & 0. & 4.725000000000001\times10^7 & 0. \\ 0. & 3.15\times10^7 & 0. & 0. & 0. & 4.2\times10^7 \end{bmatrix}$$

$$[K_a^{\mathrm{IV}}] = \begin{bmatrix} 1.995\times10^9 & 0. & 0. & 0. & 0. & 0. \\ 0. & 1.575\times10^7 & 0. & 0. & 0. & -3.15\times10^7 \\ 0. & 0. & 1.771875\times10^7 & 0. & -3.54375\times10^7 & 0. \\ 0. & 0. & 0. & 1.685125\times10^7 & 0. & 0. \\ 0. & 0. & -3.54375\times10^7 & 0. & 9.450000000000003\times10^7 & 0. \\ 0. & -3.15\times10^7 & 0. & 0. & 0. & 8.4\times10^7 \end{bmatrix}$$

Continued

EXAMPLE 7.2—CONT'D

Expanded stiffness matrix of the element in the degrees of freedom of the whole structure. Since the number of degrees of freedom of the structure is large, we can use the following nomenclature to represent the nodal displacements and forces in an abbreviated form:

$$\{f_1\} = \begin{Bmatrix} f_{1x} \\ f_{1y} \\ f_{1z} \\ m_{1x} \\ m_{1y} \\ m_{1z} \end{Bmatrix}, \quad \{f_2\} = \begin{Bmatrix} f_{2x} \\ f_{2y} \\ f_{2z} \\ m_{2x} \\ m_{2y} \\ m_{2z} \end{Bmatrix}, \quad \{f_3\} = \begin{Bmatrix} f_{3x} \\ f_{3y} \\ f_{3z} \\ m_{3x} \\ m_{3y} \\ m_{3z} \end{Bmatrix}$$

$$\{u_1\} = \begin{Bmatrix} u_{1x} \\ u_{1y} \\ u_{1z} \\ \theta_{1x} \\ \theta_{1y} \\ \theta_{1z} \end{Bmatrix}, \quad \{u_2\} = \begin{Bmatrix} u_{2x} \\ u_{2y} \\ u_{2z} \\ \theta_{2x} \\ \theta_{2y} \\ \theta_{2z} \end{Bmatrix}, \quad \{f_3\} = \begin{Bmatrix} u_{3x} \\ u_{3y} \\ u_{3z} \\ \theta_{3x} \\ \theta_{3y} \\ \theta_{3z} \end{Bmatrix}$$

Therefore, as the ends of the element a are located on the nodes 1 and 3, the following element equation can be derived after the expansion of the matrix $[K_a]$ to the degrees of freedom of the whole structure:

$$\begin{Bmatrix} \{f_1\} \\ \{f_2\} \\ \{f_3\} \end{Bmatrix} = \begin{bmatrix} [k_a^{\mathrm{I}}] & [O_{6\times6}] & [k_a^{\mathrm{II}}] \\ [O_{6\times6}] & [O_{6\times6}] & [O_{6\times6}] \\ [k_a^{\mathrm{III}}] & [O_{6\times6}] & [k_a^{\mathrm{IV}}] \end{bmatrix} \begin{Bmatrix} \{u_1\} \\ \{u_2\} \\ \{u_3\} \end{Bmatrix}$$

Let us symbolize the above expanded stiffness by $[K_a^e]$, that is,

$$[K_a^e] = \begin{bmatrix} [K_a^{\mathrm{I}}] & [O_{6\times6}] & [K_a^{\mathrm{II}}] \\ [O_{6\times6}] & [O_{6\times6}] & [O_{6\times6}] \\ [K_a^{\mathrm{III}}] & [O_{6\times6}] & [K_a^{\mathrm{IV}}] \end{bmatrix}$$

Element b

Stiffness matrix in the local coordinate system $\bar{x}-\bar{y}-\bar{z}$. Taking into account Equation (7.33), the following element equation should be derived for the element b (nodes 2–3):

$$\begin{Bmatrix} f_{2\bar{x}} \\ f_{2\bar{y}} \\ f_{2\bar{z}} \\ m_{2\bar{x}} \\ m_{2\bar{y}} \\ m_{2\bar{z}} \\ f_{3\bar{x}} \\ f_{3\bar{y}} \\ f_{3\bar{z}} \\ m_{3\bar{x}} \\ m_{3\bar{y}} \\ m_{3\bar{z}} \end{Bmatrix} = [\bar{k}_b] \begin{Bmatrix} u_{2\bar{x}} \\ u_{2\bar{y}} \\ u_{2\bar{z}} \\ \vartheta_{2\bar{x}} \\ \theta_{2\bar{y}} \\ \theta_{2\bar{z}} \\ u_{3\bar{x}} \\ u_{3\bar{y}} \\ u_{3\bar{z}} \\ \theta_{3\bar{x}} \\ \theta_{3\bar{y}} \\ \theta_{3\bar{z}} \end{Bmatrix}$$

where

$$[\bar{k}_b] = \begin{bmatrix} [Kb_{fu}^{11}] & [Kb_{f\theta}^{11}] & [Kb_{fu}^{12}] & [Kb_{f\theta}^{12}] \\ [Kb_{mu}^{11}] & [Kb_{m\theta}^{11}] & [Kb_{mu}^{12}] & [Kb_{m\theta}^{12}] \\ [Kb_{fu}^{21}] & [Kb_{f\theta}^{21}] & [Kb_{fu}^{22}] & [Kb_{f\theta}^{22}] \\ [Kb_{mu}^{21}] & [Kb_{m\theta}^{21}] & [Kb_{mu}^{22}] & [Kb_{m\theta}^{22}] \end{bmatrix}$$

Using Equations (6.195)–(6.210), the following results can be obtained for the submatrices composing the stiffness matrix of the element a in its local coordinate system $\bar{x}-\bar{y}-\bar{z}$

$$[Kb_{fu}^{11}] = \begin{pmatrix} 2.66\times10^9 & 0 & 0 \\ 0 & 4.666666666666666\times10^7 & 0 \\ 0 & 0 & 5.133333333333333\times10^7 \end{pmatrix}$$

$$[Kb_{mu}^{11}] = \begin{pmatrix} -2.66\times10^9 & 0 & 0 \\ 0 & -4.666666666666666\times10^7 & 0 \\ 0 & 0 & -5.133333333333333\times10^7 \end{pmatrix}$$

$$[Kb_{fu}^{21}] = \begin{pmatrix} 0 & 0 & 0 \\ 0 & 0 & 7.7\times10^7 \\ 0 & 7.\times10^7 & 0 \end{pmatrix}$$

$$[Kb_{mu}^{21}] = \begin{pmatrix} 0 & 0 & 0 \\ 0 & 0 & 7.7\times10^7 \\ 0 & 7.\times10^7 & 0 \end{pmatrix}$$

Continued

EXAMPLE 7.2—CONT'D

$$\left[Kb_{f\theta}^{11}\right]=\begin{pmatrix}0 & 0 & 0\\ 0 & 0 & 7.\times 10^{7}\\ 0 & 7.7\times 10^{7} & 0\end{pmatrix}$$

$$\left[Kb_{m\theta}^{11}\right]=\begin{pmatrix}2.7755\times 10^{7} & 0 & 0\\ 0 & 1.54\times 10^{8} & 0\\ 0 & 0 & 1.4\times 10^{8}\end{pmatrix}$$

$$\left[Kb_{f\theta}^{21}\right]=\begin{pmatrix}0 & 0 & 0\\ 0 & 0 & -7.\times 10^{7}\\ 0 & -7.7\times 10^{7} & 0\end{pmatrix}$$

$$\left[Kb_{m\theta}^{21}\right]=\begin{pmatrix}-2.7755\times 10^{7} & 0 & 0\\ 0 & 7.7\times 10^{7} & 0\\ 0 & 0 & 7.\times 10^{7}\end{pmatrix}$$

$$\left[Kb_{fu}^{12}\right]=\begin{pmatrix}-2.66\times 10^{9} & 0 & 0\\ 0 & -4.666666666666666\times 10^{7} & 0\\ 0 & 0 & -5.133333333333333\times 10^{7}\end{pmatrix}$$

$$\left[Kb_{mu}^{12}\right]=\begin{pmatrix}0 & 0 & 0\\ 0 & 0 & -7.7\times 10^{7}\\ 0 & -7.\times 10^{7} & 0\end{pmatrix}$$

$$\left[Kb_{fu}^{22}\right]=\begin{pmatrix}2.66\times 10^{9} & 0 & 0\\ 0 & 4.666666666666666\times 10^{7} & 0\\ 0 & 0 & 5.133333333333333\times 10^{7}\end{pmatrix}$$

$$\left[Kb_{mu}^{22}\right]=\begin{pmatrix}0 & 0 & 0\\ 0 & 0 & -7.7\times 10^{7}\\ 0 & -7.\times 10^{7} & 0\end{pmatrix}$$

$$\left[Kb_{f\theta}^{12}\right]=\begin{pmatrix}0 & 0 & 0\\ 0 & 0 & 7.\times 10^{7}\\ 0 & 7.7\times 10^{7} & 0\end{pmatrix}$$

$$\left[Kb_{m\theta}^{12}\right]=\begin{pmatrix}-2.7755\times 10^{7} & 0 & 0\\ 0 & 7.7\times 10^{7} & 0\\ 0 & 0 & 7.\times 10^{7}\end{pmatrix}$$

$$\left[Kb_{f\theta}^{22}\right]=\begin{pmatrix}0 & 0 & 0\\ 0 & 0 & -7.\times 10^{7}\\ 0 & -7.7\times 10^{7} & 0\end{pmatrix}$$

$$\left[Kb_{m\theta}^{22}\right]=\begin{pmatrix}2.7755\times 10^{7} & 0 & 0\\ 0 & 1.54\times 10^{8} & 0\\ 0 & 0 & 1.4\times 10^{8}\end{pmatrix}$$

***Stiffness matrix of the element b in the global coordinate system* x − y − z.** Taking into account Equation (7.53), the matrix $[\bar{k}_b]$ should be transformed to the global coordinate system $x-y-z$ as follows:

$$[K_b] = [T_b]^{-1} \cdot [\bar{k}_b] \cdot [T_b]$$

The result of this transformation can be written as:

$$[K_b] = \begin{bmatrix} K_b^{\mathrm{I}} & K_b^{\mathrm{II}} \\ K_b^{\mathrm{III}} & K_b^{\mathrm{IV}} \end{bmatrix}$$

where

$$[K_b^{\mathrm{I}}] = \begin{bmatrix} 1.995\times10^9 & 0. & 0. & 0. & 0. & 0. \\ 0. & 1.575\times10^7 & 0. & 0. & 0. & 3.15\times10^7 \\ 0. & 0. & 1.771875\times10^7 & 0. & 3.54375\times10^7 & 0. \\ 0. & 0. & 0. & 1.685125\times10^7 & 0. & 0. \\ 0. & 0. & 3.54375\times10^7 & 0. & 9.450000000000003\times10^7 & 0. \\ 0. & 3.15\times10^7 & 0. & 0. & 0. & 8.4\times10^7 \end{bmatrix}$$

$$[K_b^{\mathrm{II}}] = \begin{bmatrix} -1.995\times10^9 & 0. & 0. & 0. & 0. & 0. \\ 0. & -1.575\times10^7 & 0. & 0. & 0. & 3.15\times10^7 \\ 0. & 0. & -1.771875\times10^7 & 0. & 3.54375\times10^7 & 0. \\ 0. & 0. & 0. & -1.685125\times10^7 & 0. & 0. \\ 0. & 0. & -3.54375\times10^7 & 0. & 4.725000000000001\times10^7 & 0. \\ 0. & -3.15\times10^7 & 0. & 0. & 0. & 4.2\times10^7 \end{bmatrix}$$

$$[K_b^{\mathrm{III}}] = \begin{bmatrix} -1.995\times10^9 & 0. & 0. & 0. & 0. & 0. \\ 0. & -1.575\times10^7 & 0. & 0. & 0. & -3.15\times10^7 \\ 0. & 0. & -1.771875\times10^7 & 0. & -3.54375\times10^7 & 0. \\ 0. & 0. & 0. & -1.685125\times10^7 & 0. & 0. \\ 0. & 0. & 3.54375\times10^7 & 0. & 4.725000000000001\times10^7 & 0. \\ 0. & 3.15\times10^7 & 0. & 0. & 0. & 4.2\times10^7 \end{bmatrix}$$

$$[K_b^{\mathrm{IV}}] = \begin{bmatrix} 1.995\times10^9 & 0. & 0. & 0. & 0. & 0. \\ 0. & 1.575\times10^7 & 0. & 0. & 0. & -3.15\times10^7 \\ 0. & 0. & 1.771875\times10^7 & 0. & -3.54375\times10^7 & 0. \\ 0. & 0. & 0. & 1.685125\times10^7 & 0. & 0. \\ 0. & 0. & -3.54375\times10^7 & 0. & 9.450000000000003\times10^7 & 0. \\ 0. & -3.15\times10^7 & 0. & 0. & 0. & 8.4\times10^7 \end{bmatrix}$$

Expanded stiffness matrix of the element in the degrees of freedom of the whole structure. Since the number of degrees of freedom of the structure is large, we can use the following nomenclature to represent the nodal displacements and forces in an abbreviated form:

$$\{f_1\} = \begin{Bmatrix} f_{1x} \\ f_{1y} \\ f_{1z} \\ m_{1x} \\ m_{1y} \\ m_{1z} \end{Bmatrix}, \quad \{f_2\} = \begin{Bmatrix} f_{2x} \\ f_{2y} \\ f_{2z} \\ m_{2x} \\ m_{2y} \\ m_{2z} \end{Bmatrix}, \quad \{f_3\} = \begin{Bmatrix} f_{3x} \\ f_{3y} \\ f_{3z} \\ m_{3x} \\ m_{3y} \\ m_{3z} \end{Bmatrix}$$

Continued

EXAMPLE 7.2—CONT'D

$$\{u_1\} = \begin{Bmatrix} u_{1x} \\ u_{1y} \\ u_{1z} \\ \theta_{1x} \\ \theta_{1y} \\ \theta_{1z} \end{Bmatrix}, \quad \{u_2\} = \begin{Bmatrix} u_{2x} \\ u_{2y} \\ u_{2z} \\ \theta_{2x} \\ \theta_{2y} \\ \theta_{2z} \end{Bmatrix}, \quad \{f_3\} = \begin{Bmatrix} u_{3x} \\ u_{3y} \\ u_{3z} \\ \theta_{3x} \\ \theta_{3y} \\ \theta_{3z} \end{Bmatrix}$$

Therefore, as the ends of the element b are located on the nodes 2 and 3, the following element equation can be derived after the expansion of the matrix $[K_b]$ to the degrees of freedom of the whole structure:

$$\begin{Bmatrix} \{f_1\} \\ \{f_2\} \\ \{f_3\} \end{Bmatrix} = \begin{bmatrix} [O_{6\times6}] & [O_{6\times6}] & [O_{6\times6}] \\ [O_{6\times6}] & [k_b^{\mathrm{I}}] & [k_b^{\mathrm{II}}] \\ [O_{6\times6}] & [k_b^{\mathrm{III}}] & [k_b^{\mathrm{IV}}] \end{bmatrix} \begin{Bmatrix} \{u_1\} \\ \{u_2\} \\ \{u_3\} \end{Bmatrix}$$

Let us symbolize the above expanded stiffness by $[K_b^e]$, that is,

$$[K_b^e] = \begin{bmatrix} [O_{6\times6}] & [O_{6\times6}] & [O_{6\times6}] \\ [O_{6\times6}] & [K_b^{\mathrm{I}}] & [K_b^{\mathrm{II}}] \\ [O_{6\times6}] & [K_b^{\mathrm{III}}] & [K_b^{\mathrm{IV}}] \end{bmatrix}$$

Step 3: Derivation of the Global Stiffness Matrix [K] for the Structure

The stiffness matrix of the whole structure can be derived by the addition of the expanded stiffness matrices of the elements a and b, respectively:

$$[K] = [K_a^e] + [K_b^e]$$

Then, the element equation can be written as follows:

$$[K]\{d\} = \{R\}$$

or

$$[[K] \ \ [-I]]\begin{Bmatrix} \{d\} \\ \{R\} \end{Bmatrix} = \{O_{18\times1}\} \quad \text{(e1)}$$

Step 4: Specification of the Boundary Conditions

Boundary conditions regarding nodal displacements. Taking into account the type of supports the following boundary conditions can be specified:

$$\begin{aligned} u_{1x} &= 0, \quad u_{2x} = 0 \\ u_{1y} &= 0, \quad u_{2y} = 0 \\ u_{1z} &= 0, \quad u_{2z} = 0 \\ \theta_{1x} &= 0, \quad \theta_{2x} = 0 \\ \vartheta_{1y} &= 0, \quad \vartheta_{2y} = 0 \\ \vartheta_{1z} &= 0, \quad \vartheta_{2z} = 0 \end{aligned}$$

The above 12 boundary conditions can be written in the following matrix format:

$$\underbrace{\begin{bmatrix} 1&0&0&0&0&0&0&0&0&0&0&0&0&0&0&0&0&0\\ 0&1&0&0&0&0&0&0&0&0&0&0&0&0&0&0&0&0\\ 0&0&1&0&0&0&0&0&0&0&0&0&0&0&0&0&0&0\\ 0&0&0&1&0&0&0&0&0&0&0&0&0&0&0&0&0&0\\ 0&0&0&0&1&0&0&0&0&0&0&0&0&0&0&0&0&0\\ 0&0&0&0&0&1&0&0&0&0&0&0&0&0&0&0&0&0\\ 0&0&0&0&0&0&1&0&0&0&0&0&0&0&0&0&0&0\\ 0&0&0&0&0&0&0&1&0&0&0&0&0&0&0&0&0&0\\ 0&0&0&0&0&0&0&0&1&0&0&0&0&0&0&0&0&0\\ 0&0&0&0&0&0&0&0&0&1&0&0&0&0&0&0&0&0\\ 0&0&0&0&0&0&0&0&0&0&1&0&0&0&0&0&0&0\\ 0&0&0&0&0&0&0&0&0&0&0&1&0&0&0&0&0&0\\ 0&0&0&0&0&0&0&0&0&0&0&0&0&0&0&0&0&0\\ 0&0&0&0&0&0&0&0&0&0&0&0&0&0&0&0&0&0\\ 0&0&0&0&0&0&0&0&0&0&0&0&0&0&0&0&0&0\\ 0&0&0&0&0&0&0&0&0&0&0&0&0&0&0&0&0&0\\ 0&0&0&0&0&0&0&0&0&0&0&0&0&0&0&0&0&0\\ 0&0&0&0&0&0&0&0&0&0&0&0&0&0&0&0&0&0 \end{bmatrix}}_{[BCd]} \underbrace{\begin{Bmatrix} u_{1x}\\ u_{1y}\\ u_{1z}\\ \theta_{1x}\\ \theta_{1y}\\ \theta_{1z}\\ u_{2x}\\ u_{2y}\\ u_{2z}\\ \theta_{2x}\\ \theta_{2y}\\ \theta_{2z}\\ u_{3x}\\ u_{3y}\\ u_{3z}\\ \theta_{3x}\\ \theta_{3y}\\ \theta_{3z} \end{Bmatrix}}_{\{d\}} = \underbrace{\begin{Bmatrix} 0\\0\\0\\0\\0\\0\\0\\0\\0\\0\\0\\0\\0\\0\\0\\0\\0\\0 \end{Bmatrix}}_{\{DO\}}$$

The above matrix equation can be written in the following abbreviated form:

$$[BCd]\{d\} = \{DO\}$$

or

$$[[BCd]\ \ [O_{18\times1}]]\begin{Bmatrix} \{d\} \\ \{R\} \end{Bmatrix} = \{DO\} \tag{e2}$$

Boundary conditions regarding nodal forces. Taking into account the external forces acting on the nodes, the following boundary conditions can be specified:

$$\begin{aligned} R_{3x} &= 10{,}000 \\ R_{3y} &= -15{,}000 \\ R_{3z} &= 5000 \\ M_{3x} &= 0 \\ M_{3y} &= 0 \\ M_{3z} &= 0 \end{aligned}$$

Continued

EXAMPLE 7.2—CONT'D

The above 6 boundary conditions can be written in the following matrix format:

$$
\underbrace{\begin{bmatrix}
0&0&0&0&0&0&0&0&0&0&0&0&0&0&0&0&0&0\\
0&0&0&0&0&0&0&0&0&0&0&0&0&0&0&0&0&0\\
0&0&0&0&0&0&0&0&0&0&0&0&0&0&0&0&0&0\\
0&0&0&0&0&0&0&0&0&0&0&0&0&0&0&0&0&0\\
0&0&0&0&0&0&0&0&0&0&0&0&0&0&0&0&0&0\\
0&0&0&0&0&0&0&0&0&0&0&0&0&0&0&0&0&0\\
0&0&0&0&0&0&0&0&0&0&0&0&0&0&0&0&0&0\\
0&0&0&0&0&0&0&0&0&0&0&0&0&0&0&0&0&0\\
0&0&0&0&0&0&0&0&0&0&0&0&0&0&0&0&0&0\\
0&0&0&0&0&0&0&0&0&0&0&0&0&0&0&0&0&0\\
0&0&0&0&0&0&0&0&0&0&0&0&0&0&0&0&0&0\\
0&0&0&0&0&0&0&0&0&0&0&0&0&0&0&0&0&0\\
0&0&0&0&0&0&0&0&0&0&0&0&1&0&0&0&0&0\\
0&0&0&0&0&0&0&0&0&0&0&0&0&1&0&0&0&0\\
0&0&0&0&0&0&0&0&0&0&0&0&0&0&1&0&0&0\\
0&0&0&0&0&0&0&0&0&0&0&0&0&0&0&1&0&0\\
0&0&0&0&0&0&0&0&0&0&0&0&0&0&0&0&1&0\\
0&0&0&0&0&0&0&0&0&0&0&0&0&0&0&0&0&1
\end{bmatrix}}_{[BCR]}
\underbrace{\begin{Bmatrix}
R_{1x}\\ R_{1y}\\ R_{1z}\\ M_{1x}\\ M_{1y}\\ M_{1z}\\ R_{2x}\\ R_{2y}\\ R_{2z}\\ M_{2x}\\ M_{2y}\\ M_{2z}\\ R_{3x}\\ R_{3y}\\ R_{3z}\\ M_{3x}\\ M_{3y}\\ M_{3z}
\end{Bmatrix}}_{\{R\}}
=
\underbrace{\begin{Bmatrix}
0\\0\\0\\0\\0\\0\\0\\0\\0\\0\\0\\0\\10{,}000\\-15{,}000\\5000\\0\\0\\0
\end{Bmatrix}}_{\{RO\}}
$$

The above matrix equation can be written in the following abbreviated form:

$$[BCR]\ \{R\} = \{RO\}$$

or

$$[[O_{18\times1}]\ \ [BCR]]\begin{Bmatrix}\{d\}\\ \{R\}\end{Bmatrix} = \{RO\} \tag{e3}$$

Addition of Equations (e2) and (e3) yields

$$\begin{bmatrix}[K] & [-I]\\ [BCd] & [BCR]\end{bmatrix}\begin{Bmatrix}\{d\}\\ \{R\}\end{Bmatrix} = \begin{Bmatrix}\{O_{18\times1}\}\\ \{DO\}+\{RO\}\end{Bmatrix} \tag{e4}$$

The solution of the above 36×36 algebraic system yields the following results:

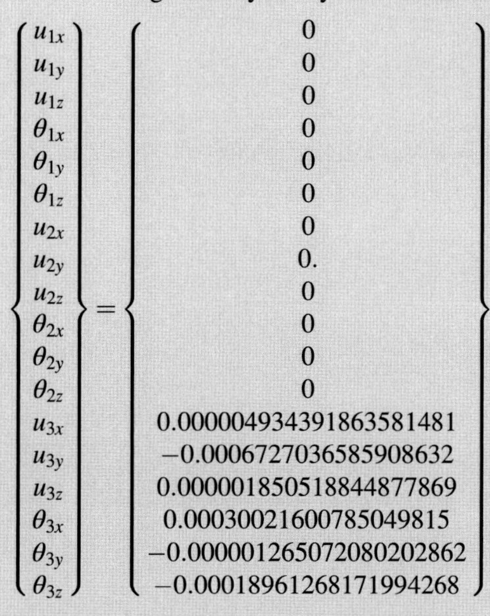

$$\begin{Bmatrix} u_{1x} \\ u_{1y} \\ u_{1z} \\ \theta_{1x} \\ \theta_{1y} \\ \theta_{1z} \\ u_{2x} \\ u_{2y} \\ u_{2z} \\ \theta_{2x} \\ \theta_{2y} \\ \theta_{2z} \\ u_{3x} \\ u_{3y} \\ u_{3z} \\ \theta_{3x} \\ \theta_{3y} \\ \theta_{3z} \end{Bmatrix} = \begin{Bmatrix} 0 \\ 0 \\ 0 \\ 0 \\ 0 \\ 0 \\ 0 \\ 0. \\ 0 \\ 0 \\ 0 \\ 0 \\ 0.000004934391863581481 \\ -0.0006727036585908632 \\ 0.000001850518844877869 \\ 0.00030021600785049815 \\ -0.000001265072080202862 \\ -0.00018961268171994268 \end{Bmatrix}$$

and

$$\begin{Bmatrix} R_{1x} \\ R_{1y} \\ R_{1z} \\ M_{1x} \\ M_{1y} \\ M_{1z} \\ R_{2x} \\ R_{2y} \\ R_{2z} \\ M_{2x} \\ M_{2y} \\ M_{2z} \\ R_{3x} \\ R_{3y} \\ R_{3z} \\ M_{3x} \\ M_{3y} \\ M_{3z} \end{Bmatrix} = \begin{Bmatrix} -9844.111767845105 \\ 4622.283148627903 \\ -77.61987262487169 \\ -5059.015002290706 \\ -125.35241735495285 \\ 13,226.432613374602 \\ -155.88823215489634 \\ 10,377.716851372097 \\ -4922.380127375129 \\ -26,074.13555182559 \\ 282.5376233201551 \\ 5262.6999811370115 \\ 10,000. \\ -15,000. \\ 5000. \\ 0. \\ 0. \\ 0. \end{Bmatrix}$$

EXAMPLE 7.3: ANALYSIS OF A PLANE FRAME USING CALFEM/MATLAB

Determine the distributions of displacements, bending moments, and shear forces in the following plane frame.

Data

$$q = 13 \text{kN/m}$$

$$F = 25 \text{kN}$$

Continued

EXAMPLE 7.3: ANALYSIS OF A PLANE FRAME USING CALFEM/MATLAB—CONT'D

$$L = 1.5\text{m}$$

$$A = 0.0725\text{m}^2$$

$$I = 4.2 \times 10^{-4}\text{m}^4$$

$$E = 210\text{GPa}$$

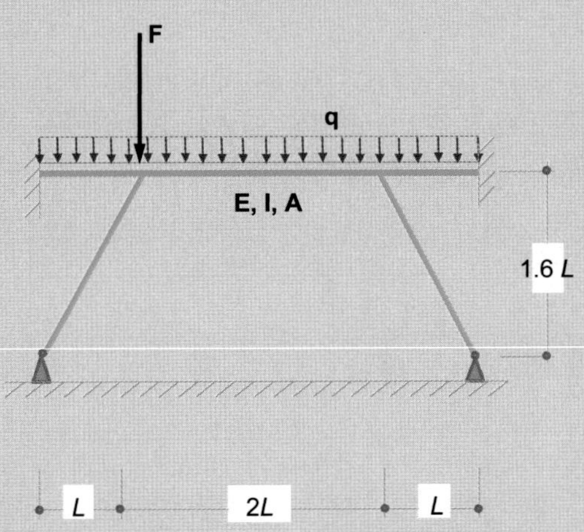

Step 1: Degrees of Freedom

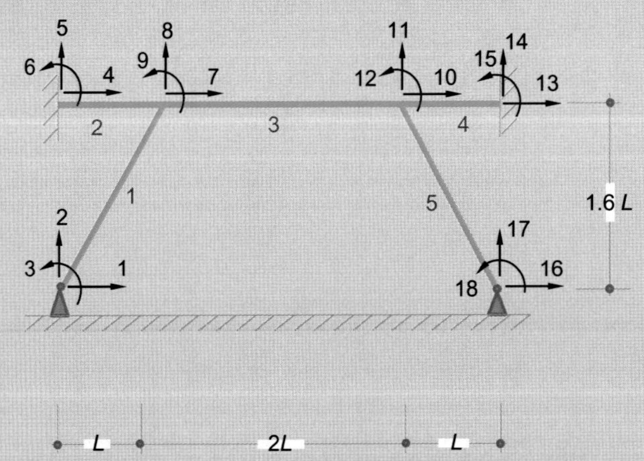

The structure has five elements, six nodes, and 18 degrees of freedom.
In order to derive a solution for the above problem we need:

- The global stiffness matrix $[K]$.
- The load vector $\{f\}$.
- The boundary conditions $[bc]$.

Step 2: Load Vector {f}

```
f = zeros(18,1);f(8) = -25000;
```

The above vector {**f**} corresponds to concentrated loads acting on nodes. Apart from the above load vector, the distributed load **q** acting along the elements 2, 3, and 4 should be taken into account. The corresponding vectors for distributed loads acting on the elements are symbolized by the symbol **eq** followed by the number of the element, and they will be derived in the Step 4.

Step 3: Boundary Conditions {bc}

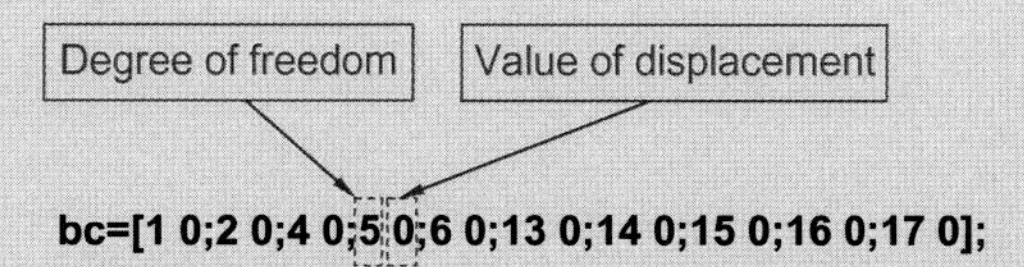

The above matrix means that the displacements in the degrees of freedom **1, 2, 4, 5, 6, 13, 14, 15, 16, 17** are **zero**.

Step 4: Global Stiffness Matrix {K}

Derivation of element matrices Kei

```
L = 1.5;qx = 0;qy = -13000;E = 2e11;I = 4.2e-4;A = 0.0725;
  ex1 = [0 L];ex2 = [0 L]; ex3 = [L 3*L];ex4 = [3*L 4*L]; ex5 = [3*L 4*L];
  ey1 = [0 1.6*L];ey2 = [1.6*L 1.6*L]; ey3 = [1.6*L 1.6*L];ey4 = [1.6*L 1.6*L];
ey5 = [1.6*L 0];
  ep = [E A I];
  eq1 = [0 0];eq2 = [0 qy]; eq3 = [0 qy];eq4 = [0 qy]; eq5 = [0 0];
  [Ke1,fe1] = beam2e(ex1,ey1,ep,eq1);
  [Ke2,fe2] = beam2e(ex2,ey2,ep,eq2);
  [Ke3,fe3] = beam2e(ex3,ey3,ep,eq3);
  [Ke4,fe4] = beam2e(ex4,ey4,ep,eq4);
  [Ke5,fe5] = beam2e(ex5,ey5,ep,eq5);
```

Derivation of the topology matrix [Edof]

```
Edof = [1 1 2 3 7 8 9;2 4 5 6 7 8 9;3 7 8 9 10 11 12;4 10 11 12 13 14 15;5 10 11 12 16 17 18];
```

Assembly of the element matrices

Based on the topology information, the global stiffness matrix can be generated by assembling the element stiffness matrices by the use of the command assem:

```
K = zeros(18);
[K,f] = assem(Edof(1,:),K,Ke1,f,fe1);
[K,f] = assem(Edof(2,:),K,Ke2,f,fe2);
[K,f] = assem(Edof(3,:),K,Ke3,f,fe3);
[K,f] = assem(Edof(4,:),K,Ke4,f,fe4);
[K,f] = assem(Edof(5,:),K,Ke5,f,fe5);
```

Step 5: Computation of the Displacement Field [a] and the Reactions {r}

The system of equations for the derivation of **[a]** and **{r}** are solved considering the global stiffness matrix **[K]**, the load vector **{f}**, and the boundary conditions **[bc]**:

```
[a,r] = solveq(K,f,bc)
```

Continued

EXAMPLE 7.3: ANALYSIS OF A PLANE FRAME USING CALFEM/MATLAB—CONT'D

Results

```
a =

  1.0e-04 *

     0
     0
  0.0906
     0
     0
     0
  0.0221
 -0.1657
 -0.2941
 -0.0075
 -0.0952
  0.2745
     0
     0
     0
     0
     0
 -0.1071
```

```
r =

  1.0e+04 *

  3.5650
  5.5517
  0.0000
 -2.1356
  0.8110
  0.2854
  0.0000
     0
  0.0000
  0.0000
  0.0000
  0.0000
  0.7230
  0.6446
 -0.1496
 -2.1523
  3.2927
     0
```

Step 6: Computation of the Distributions of Displacements Ed, Axial Forces es(:,1), Shear Forces es(:,2), and Bending Moments es(:,3)

Distributions of displacements $\bar{u}$, $\bar{v}$ and section forces N, V, and M along the local coordinate system $\bar{x}$ of each element are symbolized by the following nomenclature:

es = [N V M]

Ed = [$\bar{u}$, $\bar{v}$]

The computation of the es can be performed by the command **beam2s(ex,ey,ep,Ed,eq,n)**.

Required data for deriving the above matrices for each individual element are:

1. the coordinate matrices **ex, ey** of the element ends;
2. the matrix **ep** of the material properties of the element;
3. the matrix **Ed** of the element deformation;
4. the load matrix **eq** of the element;
5. the number of evaluation points **n** (i.e., in how many points we need the results in order the derived diagrams to be smooth).

Therefore, for the five elements of the problem, the following results for **n = 20** points can be derived:

```
Ed = extract(Edof,a);
es1 = beam2s(ex1,ey1,ep,Ed(1,:),eq1,20)
es2 = beam2s(ex2,ey2,ep,Ed(2,:),eq2,20)
es3 = beam2s(ex3,ey3,ep,Ed(3,:),eq3,20)
es4 = beam2s(ex4,ey4,ep,Ed(4,:),eq4,20)
es5 = beam2s(ex5,ey5,ep,Ed(5,:),eq5,20)
```

EXAMPLE 7.3: ANALYSIS OF A PLANE FRAME USING CALFEM/MATLAB—CONT'D

Results

es1 = 1.0e+04 *			es2 = 1.0e+04 *			es3 = 1.0e+04 *			es4 = 1.0e+04 *			es5 = 1.0e+04 *		
-6.5973	0.0807	-0.0000	2.1356	-0.8110	-0.2854	-1.4293	-1.9127	-0.7598	0.7230	-1.3054	-0.6453	-3.9330	-0.0800	-0.2265
-6.5973	0.0807	-0.0120	2.1356	-0.7083	-0.2255	-1.4293	-1.7074	-0.4740	0.7230	-1.2028	-0.5463	-3.9330	-0.0800	-0.2146
-6.5973	0.0807	-0.0240	2.1356	-0.6057	-0.1736	-1.4293	-1.5022	-0.2206	0.7230	-1.1002	-0.4554	-3.9330	-0.0800	-0.2027
-6.5973	0.0807	-0.0361	2.1356	-0.5031	-0.1298	-1.4293	-1.2969	0.0003	0.7230	-0.9975	-0.3725	-3.9330	-0.0800	-0.1907
-6.5973	0.0807	-0.0481	2.1356	-0.4004	-0.0942	-1.4293	-1.0916	0.1889	0.7230	-0.8949	-0.2978	-3.9330	-0.0800	-0.1788
-6.5973	0.0807	-0.0601	2.1356	-0.2978	-0.0666	-1.4293	-0.8864	0.3451	0.7230	-0.7923	-0.2312	-3.9330	-0.0800	-0.1669
-6.5973	0.0807	-0.0721	2.1356	-0.1952	-0.0471	-1.4293	-0.6811	0.4688	0.7230	-0.6896	-0.1728	-3.9330	-0.0800	-0.1550
-6.5973	0.0807	-0.0841	2.1356	-0.0925	-0.0358	-1.4293	-0.4758	0.5601	0.7230	-0.5870	-0.1224	-3.9330	-0.0800	-0.1431
-6.5973	0.0807	-0.0962	2.1356	0.0101	-0.0325	-1.4293	-0.2706	0.6191	0.7230	-0.4844	-0.0801	-3.9330	-0.0800	-0.1311
-6.5973	0.0807	-0.1082	2.1356	0.1127	-0.0374	-1.4293	-0.0653	0.6456	0.7230	-0.3818	-0.0459	-3.9330	-0.0800	-0.1192
-6.5973	0.0807	-0.1202	2.1356	0.2154	-0.0503	-1.4293	0.1399	0.6397	0.7230	-0.2791	-0.0198	-3.9330	-0.0800	-0.1073
-6.5973	0.0807	-0.1322	2.1356	0.3180	-0.0714	-1.4293	0.3452	0.6014	0.7230	-0.1765	-0.0018	-3.9330	-0.0800	-0.0954
-6.5973	0.0807	-0.1442	2.1356	0.4206	-0.1005	-1.4293	0.5505	0.5307	0.7230	-0.0739	0.0081	-3.9330	-0.0800	-0.0834
-6.5973	0.0807	-0.1562	2.1356	0.5232	-0.1378	-1.4293	0.7557	0.4276	0.7230	0.0288	0.0099	-3.9330	-0.0800	-0.0715
-6.5973	0.0807	-0.1683	2.1356	0.6259	-0.1831	-1.4293	0.9610	0.2920	0.7230	0.1314	0.0035	-3.9330	-0.0800	-0.0596
-6.5973	0.0807	-0.1803	2.1356	0.7285	-0.2366	-1.4293	1.1663	0.1241	0.7230	0.2340	-0.0109	-3.9330	-0.0800	-0.0477
-6.5973	0.0807	-0.1923	2.1356	0.8311	-0.2982	-1.4293	1.3715	-0.0763	0.7230	0.3367	-0.0334	-3.9330	-0.0800	-0.0358
-6.5973	0.0807	-0.2043	2.1356	0.9338	-0.3678	-1.4293	1.5768	-0.3090	0.7230	0.4393	-0.0640	-3.9330	-0.0800	-0.0238
-6.5973	0.0807	-0.2163	2.1356	1.0364	-0.4456	-1.4293	1.7820	-0.5742	0.7230	0.5419	-0.1028	-3.9330	-0.0800	-0.0119
-6.5973	0.0807	-0.2284	2.1356	1.1390	-0.5315	-1.4293	1.9873	-0.8718	0.7230	0.6446	-0.1496	-3.9330	-0.0800	-0.0000

Continued

EXAMPLE 7.3: ANALYSIS OF A PLANE FRAME USING CALFEM/MATLAB—CONT'D

Step 7: Graphical Representation of Displacements Ed, Axial Forces es(:,1), Shear Forces es(:,2), and Bending Moments es(:,3)

Since the results **Ed**, **es1**, and **es2** have been obtained, their graphical representation is now possible. These distributions should be put on the drawing of the undeformed elements.

Drawing of the undeformed elements

```
figure(1)  % creates a window for graphics
   plotpar = [2 1 0];  % specifies the line type and color and the node mark
   eldraw2(ex1,ey1,plotpar);  % creates the drawing of the undeformed element 1
   eldraw2(ex2,ey2,plotpar);  % creates the drawing of the undeformed element 2
   eldraw2(ex3,ey3,plotpar);  % creates the drawing of the undeformed element 3
   eldraw2(ex4,ey4,plotpar);  % creates the drawing of the undeformed element 4
   eldraw2(ex5,ey5,plotpar);  % creates the drawing of the undeformed element 5
```

Drawing of the displacement distribution

First, we have to specify the scale factor in order to ensure that the illustrated quantities will be obvious within the area of the monitor. In order to specify a reasonable scale factor, we have to correlate the displayed quantities of each point with a reference, for example, the Ed(2,:). The scale factor can be specified by the following command:

```
   sfac = scalfact2(ex2,ey2,Ed(2,:),0.1);  % the value 0.1 expresses the relation between
                                           the illustrated quantity and the element size.
```

Let us now derive the diagram of the displacement distributions of the five elements. The command to do this is eldisp2:

```
   plotpar = [1 2 1];
   eldisp2(ex1,ey1,Ed(1,:),plotpar,sfac);
   eldisp2(ex2,ey2,Ed(2,:),plotpar,sfac);
   eldisp2(ex3,ey3,Ed(3,:),plotpar,sfac);
   eldisp2(ex4,ey4,Ed(4,:),plotpar,sfac);
   eldisp2(ex5,ey5,Ed(5,:),plotpar,sfac);
```

Apart from the distributions, we have to specify the axes of coordinates:

```
   axis([-1 7 -1 4]); % this command specifies the x and y axes. The values of the ends of the x-
                        axis are -1 and 18. The values of the ends of the y-axis are -1 and 4
```

The title of the drawing can be placed by the following command:

```
   title('displacements')
```

Drawing of the axial force, shear force, and bending moment distribution

The axial forces **N**, shear forces **V**, and bending moments **M** are displayed by using the function **eldia2**. For example, the following command will display the diagram of axial forces **es2(:,1)** of the element 2.

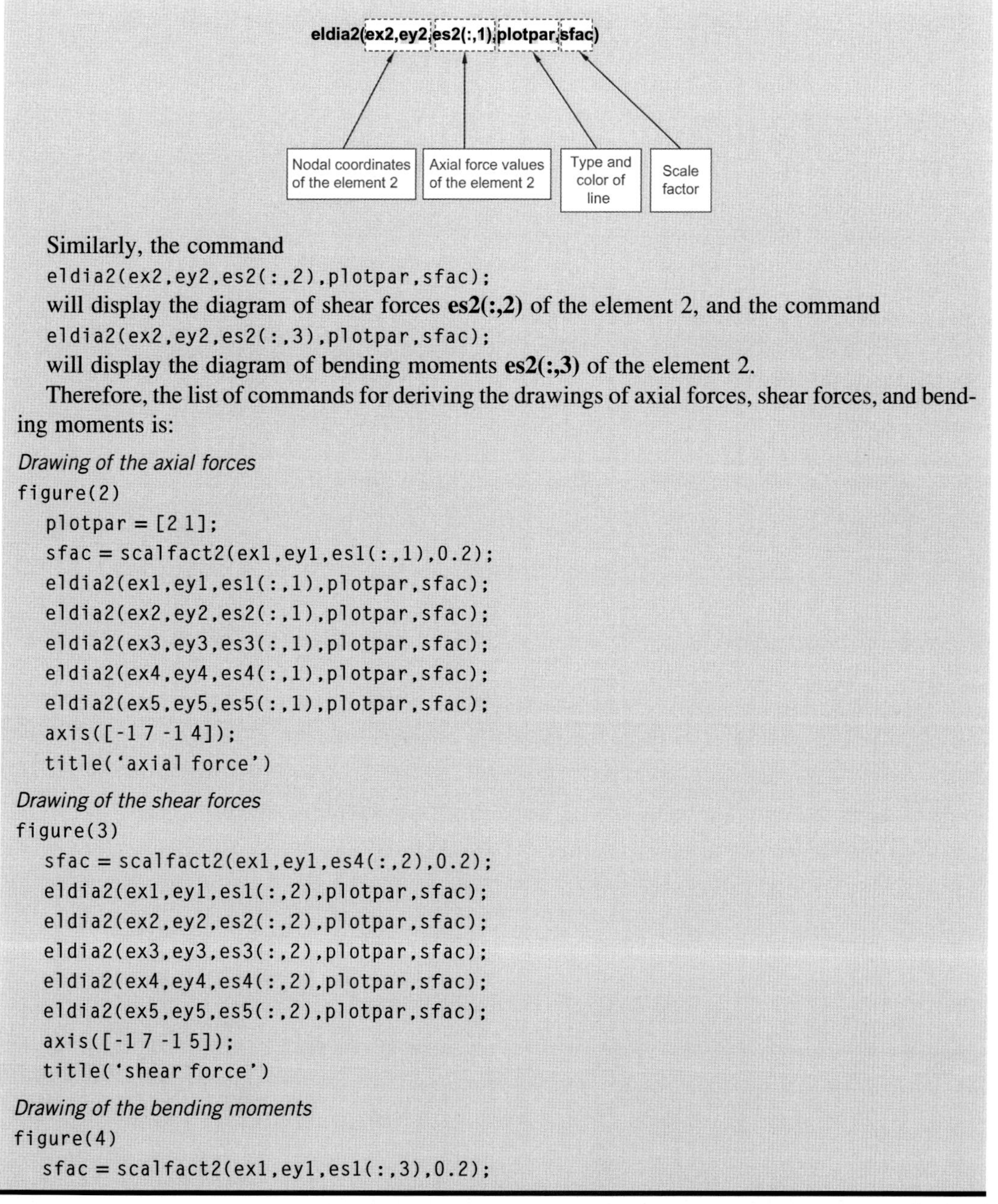

Similarly, the command

```
eldia2(ex2,ey2,es2(:,2),plotpar,sfac);
```

will display the diagram of shear forces **es2(:,2)** of the element 2, and the command

```
eldia2(ex2,ey2,es2(:,3),plotpar,sfac);
```

will display the diagram of bending moments **es2(:,3)** of the element 2.

Therefore, the list of commands for deriving the drawings of axial forces, shear forces, and bending moments is:

Drawing of the axial forces

```
figure(2)
   plotpar = [2 1];
   sfac = scalfact2(ex1,ey1,es1(:,1),0.2);
   eldia2(ex1,ey1,es1(:,1),plotpar,sfac);
   eldia2(ex2,ey2,es2(:,1),plotpar,sfac);
   eldia2(ex3,ey3,es3(:,1),plotpar,sfac);
   eldia2(ex4,ey4,es4(:,1),plotpar,sfac);
   eldia2(ex5,ey5,es5(:,1),plotpar,sfac);
   axis([-1 7 -1 4]);
   title('axial force')
```

Drawing of the shear forces

```
figure(3)
   sfac = scalfact2(ex1,ey1,es4(:,2),0.2);
   eldia2(ex1,ey1,es1(:,2),plotpar,sfac);
   eldia2(ex2,ey2,es2(:,2),plotpar,sfac);
   eldia2(ex3,ey3,es3(:,2),plotpar,sfac);
   eldia2(ex4,ey4,es4(:,2),plotpar,sfac);
   eldia2(ex5,ey5,es5(:,2),plotpar,sfac);
   axis([-1 7 -1 5]);
   title('shear force')
```

Drawing of the bending moments

```
figure(4)
   sfac = scalfact2(ex1,ey1,es1(:,3),0.2);
```

Continued

EXAMPLE 7.3: ANALYSIS OF A PLANE FRAME USING CALFEM/MATLAB—CONT'D

```
eldia2(ex1,ey1,es1(:,3),plotpar,sfac);
eldia2(ex2,ey2,es2(:,3),plotpar,sfac);
eldia2(ex3,ey3,es3(:,3),plotpar,sfac);
eldia2(ex4,ey4,es4(:,3),plotpar,sfac);
eldia2(ex5,ey5,es5(:,3),plotpar,sfac);
axis([-1 7 -1 5]);
title('bending moment')
```

Results

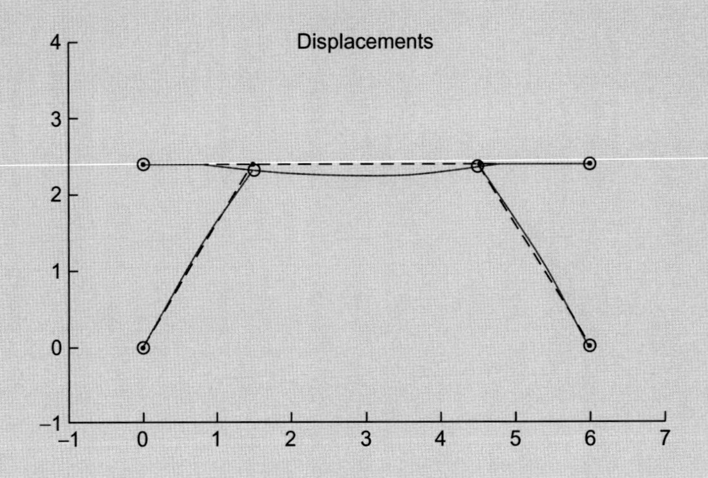

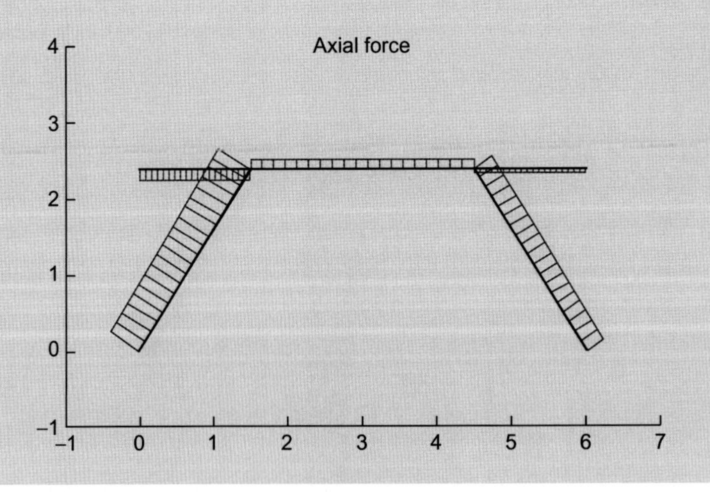

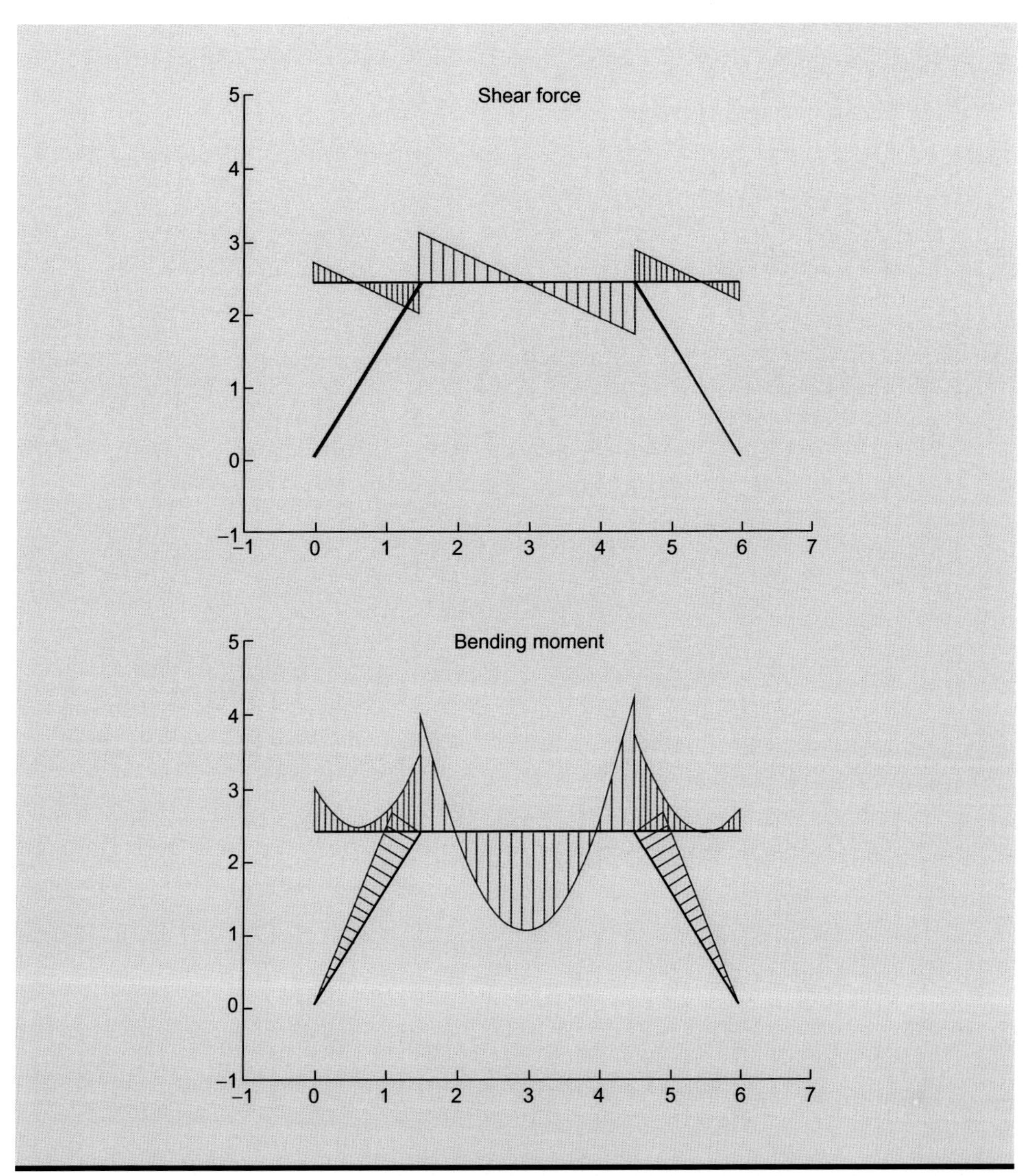

Continued

EXAMPLE 7.3: ANALYSIS OF A PLANE FRAME USING CALFEM/MATLAB—CONT'D

The CALFEM/MATLAB Computer Code

```
f = zeros(18,1);f(8) = -25000;
   bc = [1 0;2 0;4 0;5 0;6 0;13 0;14 0;15 0;16 0;17 0];
   L = 1.5;qx = 0;qy = -13000;E = 2e11;I = 4.2e-4;A = 0.0725;
   ex1 = [0 L];ex2 = [0 L]; ex3 = [L 3*L];ex4 = [3*L 4*L]; ex5 = [3*L 4*L];
   ey1 = [0 1.6*L];ey2 = [1.6*L 1.6*L]; ey3 = [1.6*L 1.6*L];ey4 = [1.6*L 1.6*L];
ey5 = [1.6*L 0];
   ep = [E A I];
   eq1 = [0 0];eq2 = [0 qy]; eq3 = [0 qy];eq4 = [0 qy]; eq5 = [0 0];
   [Ke1,fe1] = beam2e(ex1,ey1,ep,eq1);
   [Ke2,fe2] = beam2e(ex2,ey2,ep,eq2);
   [Ke3,fe3] = beam2e(ex3,ey3,ep,eq3);
   [Ke4,fe4] = beam2e(ex4,ey4,ep,eq4);
   [Ke5,fe5] = beam2e(ex5,ey5,ep,eq5);
   Edof = [1 1 2 3 7 8 9;2 4 5 6 7 8 9;3 7 8 9 10 11 12;4 10 11 12 13 14 15;5 10 11 12 16 17 18];
   K = zeros(18);
   [K,f] = assem(Edof(1,:),K,Ke1,f,fe1);
   [K,f] = assem(Edof(2,:),K,Ke2,f,fe2);
   [K,f] = assem(Edof(3,:),K,Ke3,f,fe3);
   [K,f] = assem(Edof(4,:),K,Ke4,f,fe4);
   [K,f] = assem(Edof(5,:),K,Ke5,f,fe5);
   [a,r] = solveq(K,f,bc)
   Ed = extract(Edof,a);
   es1 = beam2s(ex1,ey1,ep,Ed(1,:),eq1,20)
   es2 = beam2s(ex2,ey2,ep,Ed(2,:),eq2,20)
   es3 = beam2s(ex3,ey3,ep,Ed(3,:),eq3,20)
   es4 = beam2s(ex4,ey4,ep,Ed(4,:),eq4,20)
   es5 = beam2s(ex5,ey5,ep,Ed(5,:),eq5,20)
   figure(1)
   plotpar = [2 1 0];
   eldraw2(ex1,ey1,plotpar);
   eldraw2(ex2,ey2,plotpar);
   eldraw2(ex3,ey3,plotpar);
   eldraw2(ex4,ey4,plotpar);
   eldraw2(ex5,ey5,plotpar);
   sfac = scalfact2(ex2,ey2,Ed(2,:),0.1);
   plotpar = [1 2 1];
   eldisp2(ex1,ey1,Ed(1,:),plotpar,sfac);
   eldisp2(ex2,ey2,Ed(2,:),plotpar,sfac);
   eldisp2(ex3,ey3,Ed(3,:),plotpar,sfac);
   eldisp2(ex4,ey4,Ed(4,:),plotpar,sfac);
   eldisp2(ex5,ey5,Ed(5,:),plotpar,sfac);
   axis([-1 7 -1 4]);
```

```
title('displacements')
figure(2)
plotpar = [2 1];
sfac = scalfact2(ex1,ey1,es1(:,1),0.2);
eldia2(ex1,ey1,es1(:,1),plotpar,sfac);
eldia2(ex2,ey2,es2(:,1),plotpar,sfac);
eldia2(ex3,ey3,es3(:,1),plotpar,sfac);
eldia2(ex4,ey4,es4(:,1),plotpar,sfac);
eldia2(ex5,ey5,es5(:,1),plotpar,sfac);
axis([-1 7 -1 4]);
title('axial force')
figure(3)
sfac = scalfact2(ex1,ey1,es4(:,2),0.2);
eldia2(ex1,ey1,es1(:,2),plotpar,sfac);
eldia2(ex2,ey2,es2(:,2),plotpar,sfac);
eldia2(ex3,ey3,es3(:,2),plotpar,sfac);
eldia2(ex4,ey4,es4(:,2),plotpar,sfac);
eldia2(ex5,ey5,es5(:,2),plotpar,sfac);
axis([-1 7 -1 5]);
title('shear force')
figure(4)
sfac = scalfact2(ex1,ey1,es1(:,3),0.2);
eldia2(ex1,ey1,es1(:,3),plotpar,sfac);
eldia2(ex2,ey2,es2(:,3),plotpar,sfac);
eldia2(ex3,ey3,es3(:,3),plotpar,sfac);
eldia2(ex4,ey4,es4(:,3),plotpar,sfac);
eldia2(ex5,ey5,es5(:,3),plotpar,sfac);
axis([-1 7 -1 5]);
title('bending moment')
```

EXAMPLE 7.4: ANALYSIS OF A PLANE FRAME USING ANSYS

Determine the displacement distribution in the following plane frame.

Data

$$q = 13\,\text{kN/m}$$

$$F = 25\,\text{kN}$$

$$L = 1.5\,\text{m}$$

$$A = 0.0072\,\text{m}^2$$

$$I = 4.2 \times 10^{-4}\,\text{m}^4$$

$$E = 210\,\text{GPa}$$

Continued

EXAMPLE 7.4: ANALYSIS OF A PLANE FRAME USING ANSYS—CONT'D

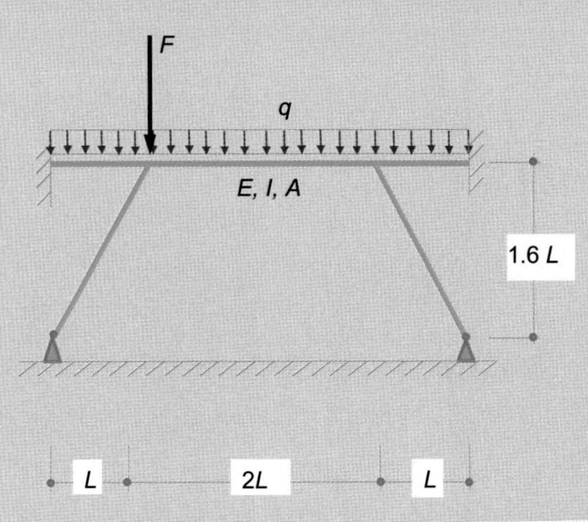

UM: File → Change Directory

Using this command, we select the directory to save all files generated for this exercise. Let us choose Directory **F**, and file **ANSYS TUTORIALS**.

INPUT: /UNITS, SI

This command defines the units to SI.

UM: File → Change Job Name

With this command we specify the job name. Let's choose **Tutorial 6**.

UM: File → Change Title

We can choose again **Tutorial 6**.

MM: Preferences

The following window will appear. We must choose *"Structural"*

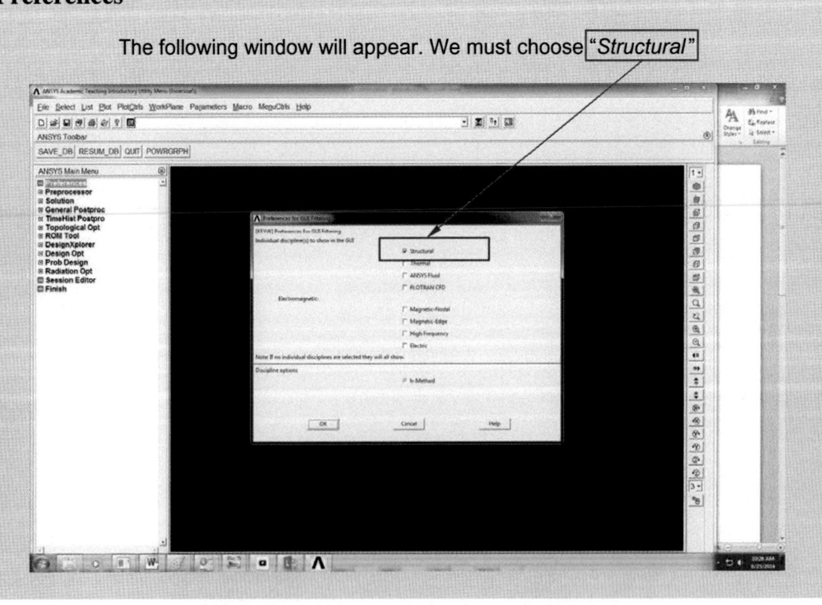

Now we have to start data entry through the preprocessor:

MM: Preprocessor→Element Type→Add/Edit/Delete→Add→Beam→node 188→OK→Close

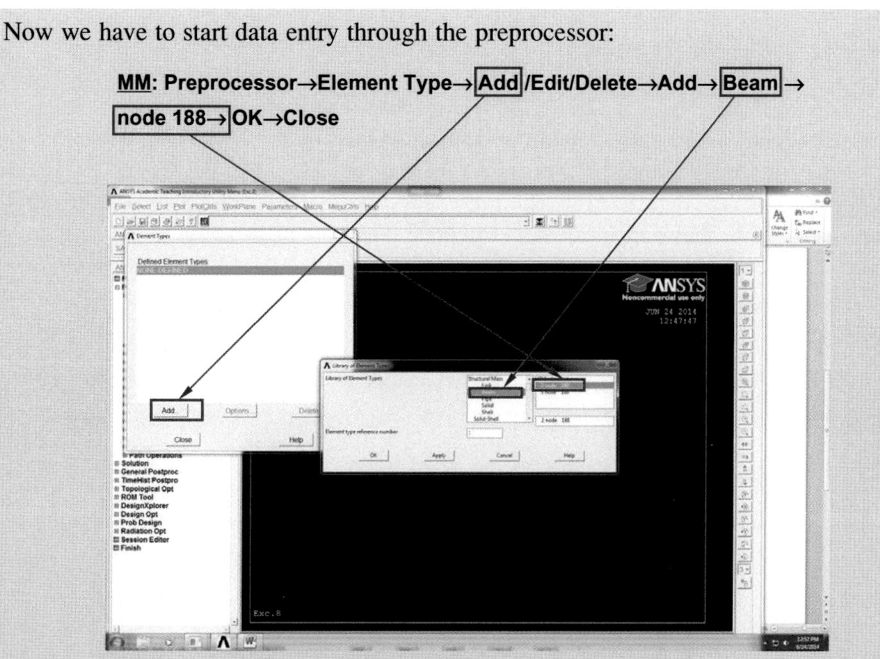

The next step is to specify the material properties using the following commands:

MM: Preprocessor→Material Props→Material Models→Structural→Linear→Elastic→Isotropic

Then, in the following window, the data for modulus of elasticity and Poisson's ratio have to be filled in the corresponding boxes; "EX" = 2.1e11 and "PRXY" = 0.3.

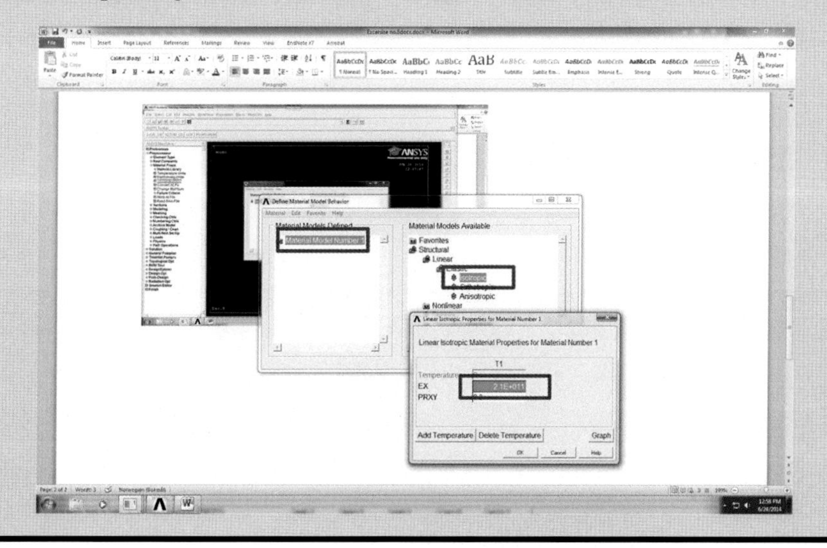

Continued

EXAMPLE 7.4: ANALYSIS OF A PLANE FRAME USING ANSYS—CONT'D

The next step is to specify the beam cross-section type and size:

MM: Preprocessor → Sections → Beam → Common Sections

A cross-sections library then appears, and a circular cross-section with radius **R = 0.152** is selected. Apart from the radius R, the number of divisions along the perimeter and radius should be specified. The program needs these divisions in order to calculate the stress distribution in the surface of the cross-sections. Let us choose **N = 12**, and **T = 8**. The completion of the following boxes

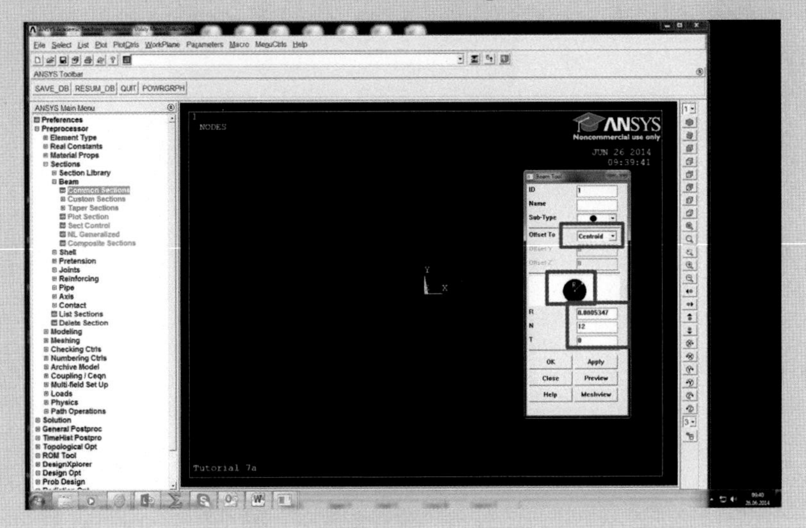

and then pressing **Meshview** yield the following cross-section mesh:

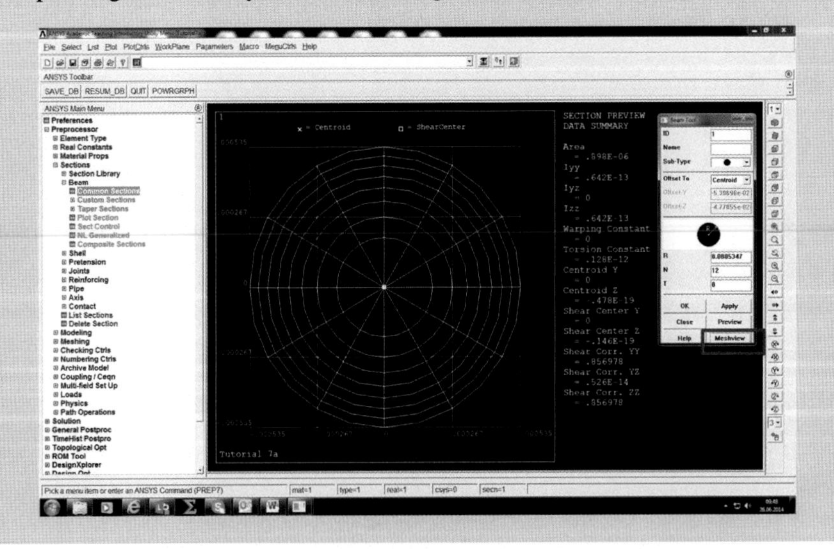

Now it is time to insert the coordinates of the Keypoints:

MM: Preprocessor → Modeling → Create → Keypoints → In Active CS

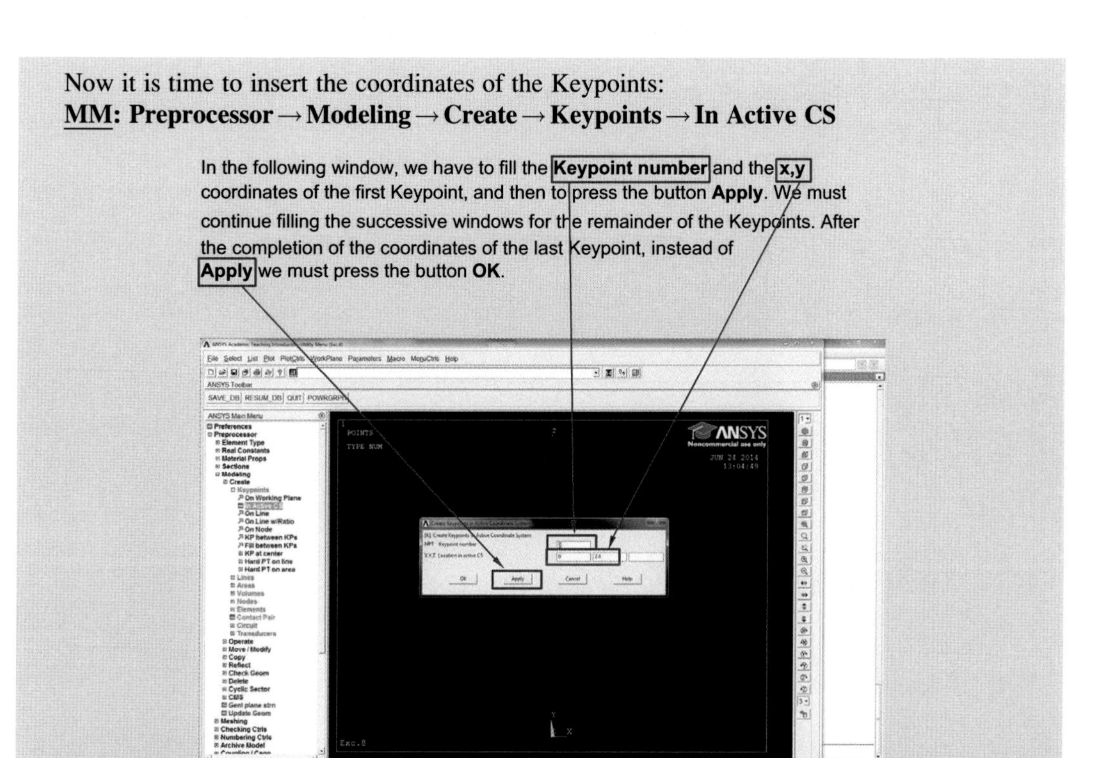

The above procedure yields a demonstration of locations of the Keypoints on the screen. The next step is to connect the Keyponts using the following commands in order to create the lines:

MM: Preprocessor → Modeling → Create → Lines → Lines → In Active Coord

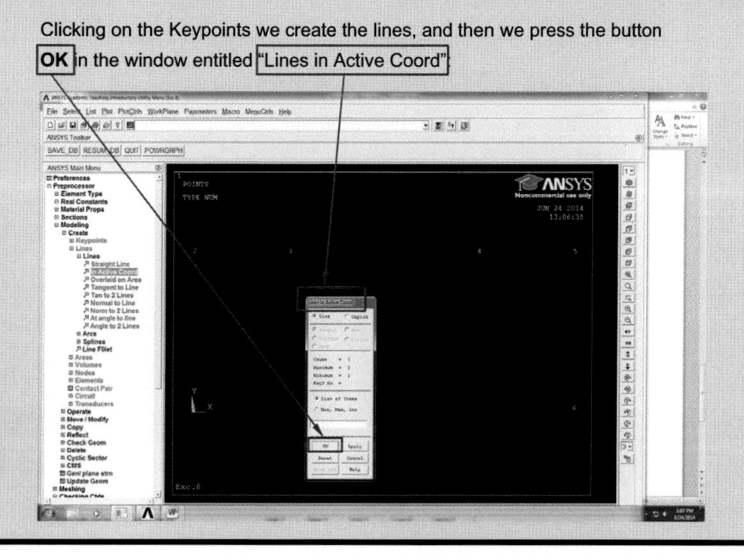

Clicking on the Keypoints we create the lines, and then we press the button OK in the window entitled "Lines in Active Coord".
ANSYS
Noncommercial use only

Continued

EXAMPLE 7.4: ANALYSIS OF A PLANE FRAME USING ANSYS—CONT'D

After the above procedure, the following representation of the structure will be obtained:

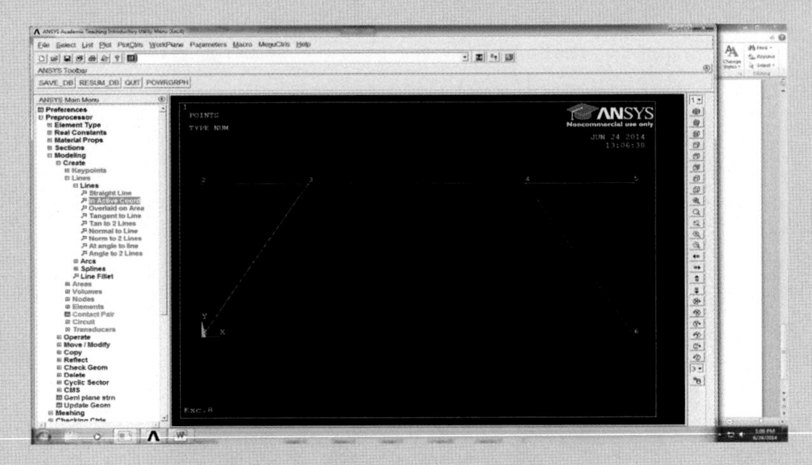

Now it is time to create the mesh of the solid. We can do it using the following commands:

MM: Preprocessor → Meshing → MeshTool

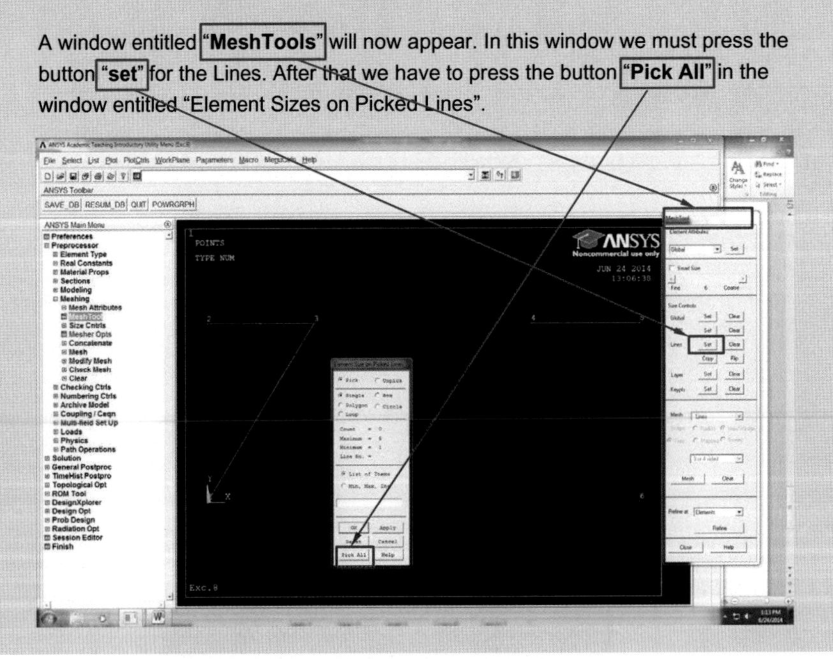

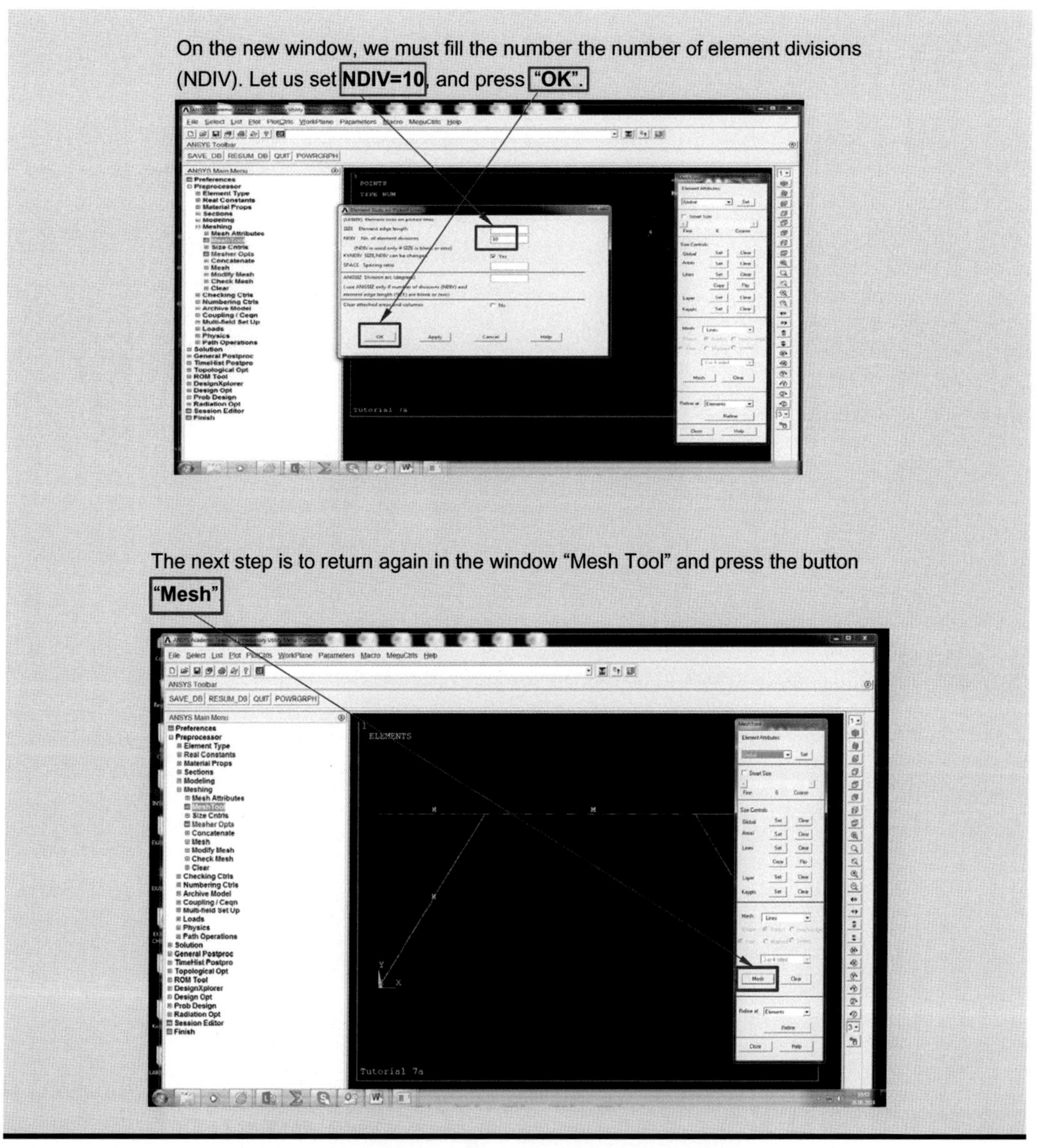
On the new window, we must fill the number the number of element divisions (NDIV). Let us set NDIV=10, and press "OK".
The next step is to return again in the window "Mesh Tool" and press the button "Mesh".
ANSYS Main Menu
Preferences
Preprocessor
Element Type
Real Constants
Material Props
Sections
Modeling
Meshing
Mesh Attributes
MeshTool
Size Cntrls
Mesher Opts
Concatenate
Mesh
Modify Mesh
Check Mesh
Clear
Checking Ctrls
Numbering Ctrls
Archive Model
Coupling / Ceqn
Multi-field Set Up
Loads
Physics
Path Operations
Solution
General Postproc
TimeHist Postpro
Topological Opt
ROM Tool
DesignXplorer
Design Opt
Prob Design
Radiation Opt
Session Editor
Finish
ELEMENTS
Tutorial 7a

Continued

EXAMPLE 7.4: ANALYSIS OF A PLANE FRAME USING ANSYS—CONT'D

Following this, we have to press the button **"Pick All"** in the window entitled "Mesh Lines".

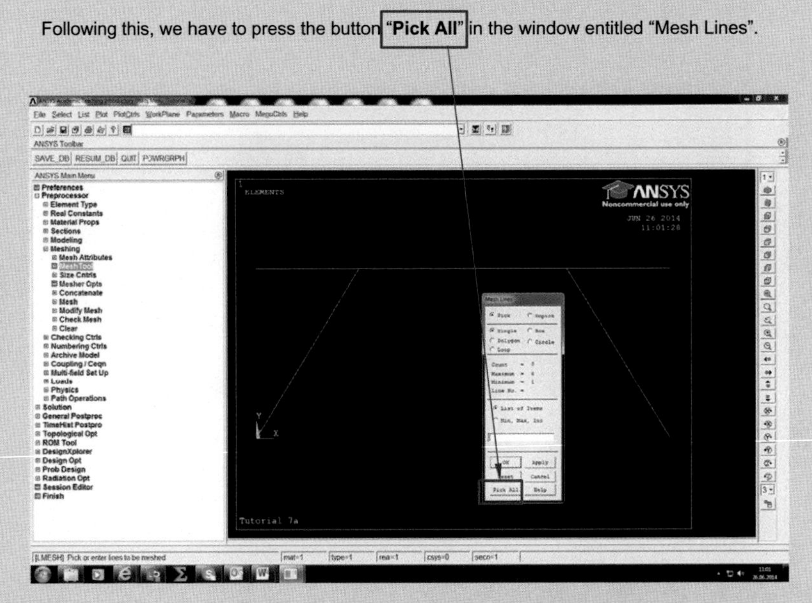

The following command should be typed in order to declare that the type of solution is "Static."

<u>MM</u>: Solution→Analysis Type→New Analysis→Static→OK

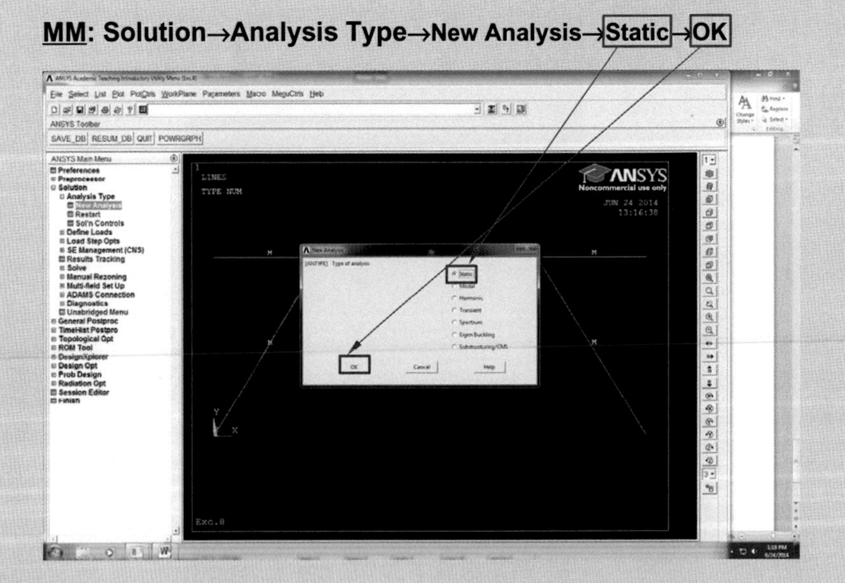

Now, using the following commands, it is time to specify the boundary conditions on the supports of the structure:

MM: Solution → Define Loads → Apply → Structural → Displacements → On Nodes

We must pick the Keypoints **1** and **6** located on the pinned supports one by one then click the button "**Apply**" and select the degrees of freedom **UX, UY** for setting the **"Displacement value"=0**. Then, click **"OK"**

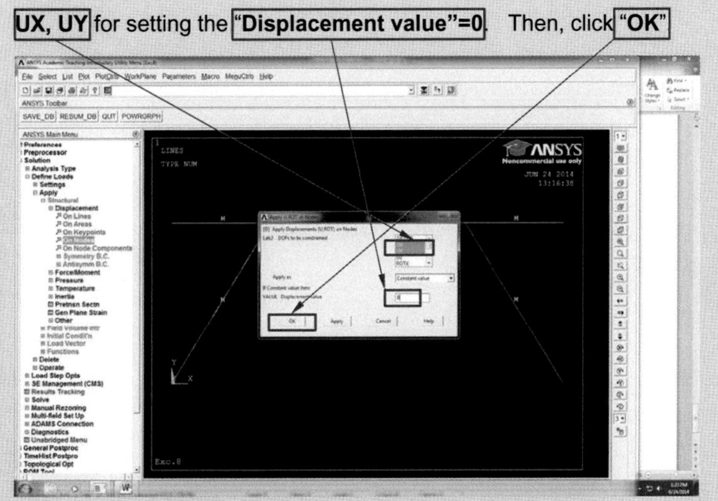

We have to continue with the commands

MM: Solution → Define Loads → Apply → Structural → Displacements → On Nodes

and pick one by one the Keypoints **2** and **5** located on the fixed supports, then click the button "**Apply**" and select the degrees of freedom **All DOF** for setting the "**Displacement value" = 0.** Then, click "**OK.**"

After the above procedure, we must specify the loads.

Therefore, we must type the commands

MM: Solution → Define Loads → Apply → Structural → Pressure → On Beams

Then, using the cursor, we must pick the beam elements subjected to distributed load q.

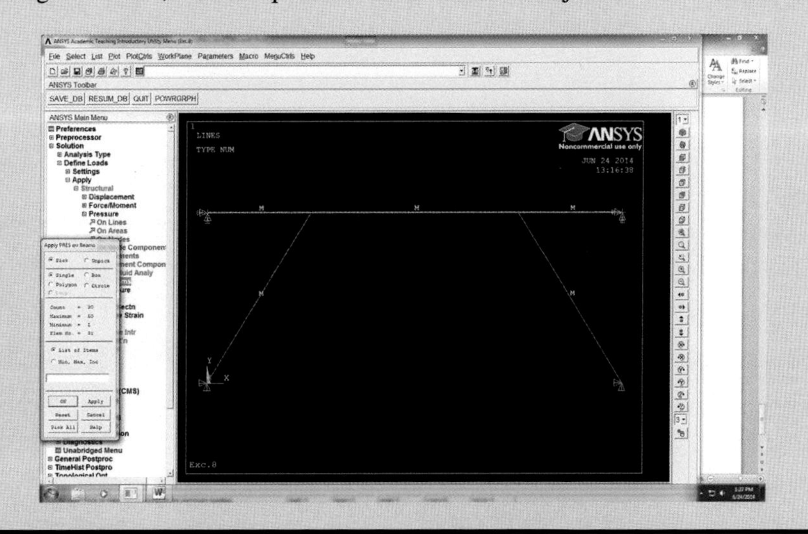

Continued

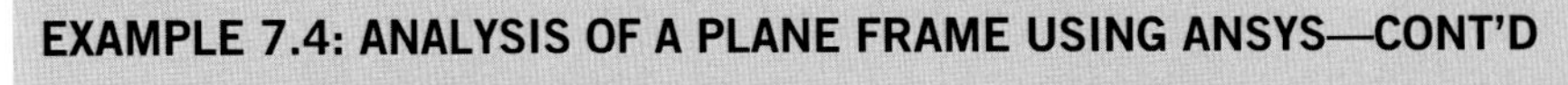

EXAMPLE 7.4: ANALYSIS OF A PLANE FRAME USING ANSYS—CONT'D

The next step is to press the button «**OK**», and then to insert the type of loading **"LKEY"=2**, and the value **13e3** of the distributed load at the end points **i** and **j** of the beam:

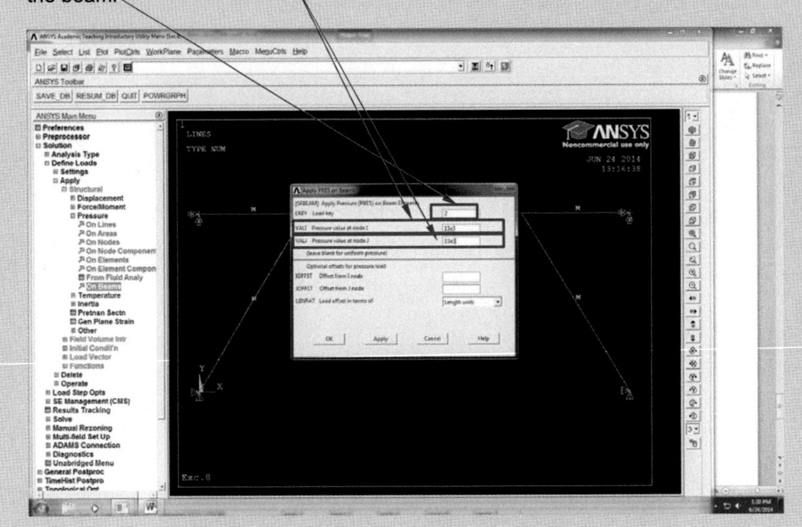

Then the following demonstration of the distributed load appears:

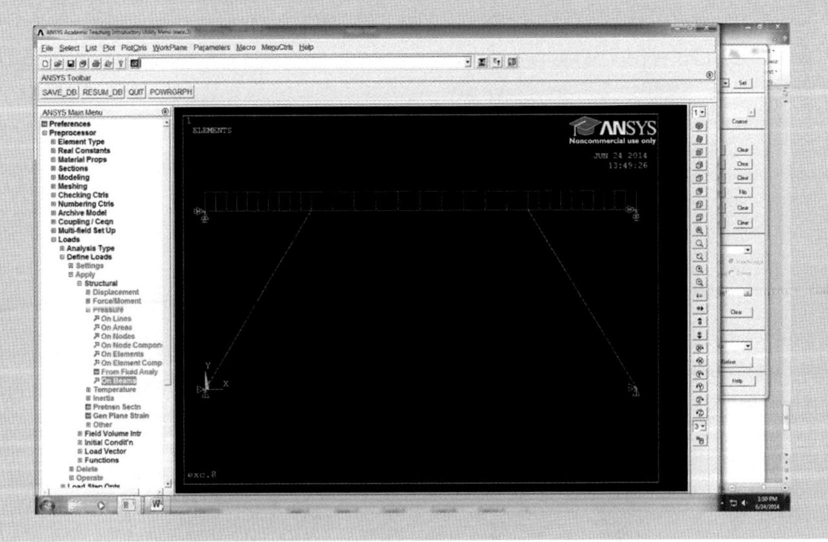

In addition to the distributed load, we have to specify the concentrated load **F**.

MM: Solution → Define Loads → Apply → Structural → Force/Moment → On Keypoints

Then, using the cursor, we must pick the Keypoint **3** subjected to concentrated load **F,** and press the button **"OK"** in the window entitled "Apply F/M on KPs."

The program now opens a new window entitled "Apply F/M on KPs", asking for the direction and the value of the concentrated load. We choose direction **FY** and value **-25e3,** and then, click **"OK"** in order the program to demonstrate the concentrated force as well.

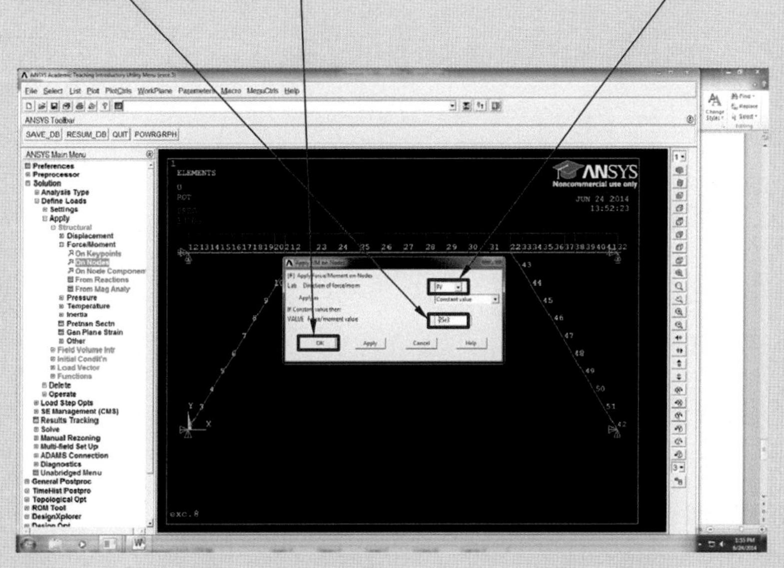

Now, using the following commands, it is time to ask the program to solve the problem:

MM: Solution → Solves → Current LS → OK

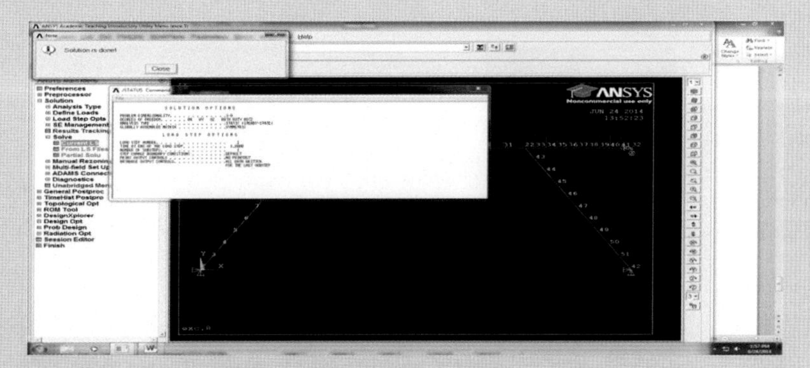

After the solution, we can now demonstrate the results. Using the following commands, the post processor has to read the results:

MM: General Postprocessor → Read results → First Set

Continued

EXAMPLE 7.4: ANALYSIS OF A PLANE FRAME USING ANSYS—CONT'D

Then, we have to continue with the following commands in order to print the deformed structure, and compare it with the undeformed one.

MM: General Postprocessor → Plot results → Deformed Shape

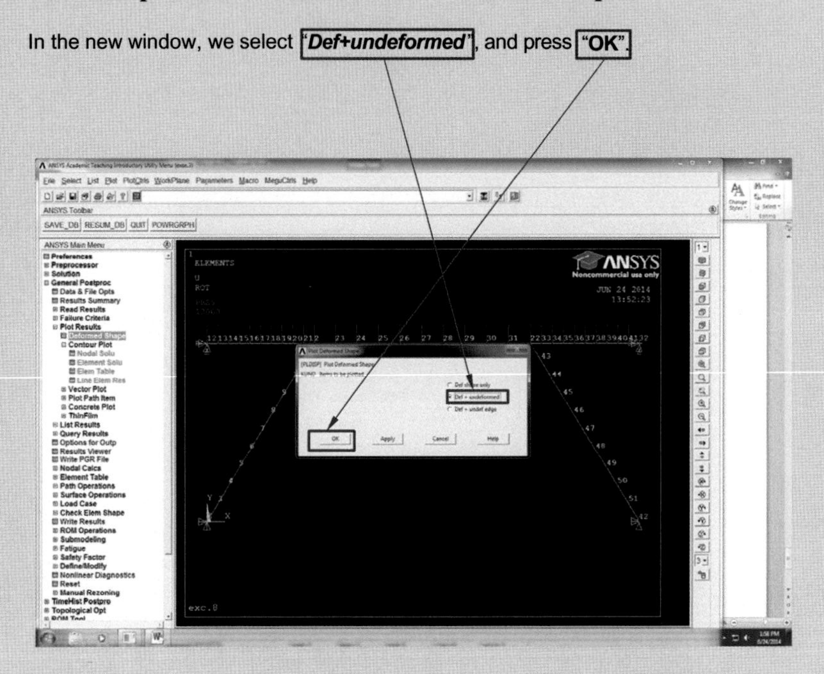

Then the following diagram of the deformed structure will be obtained:

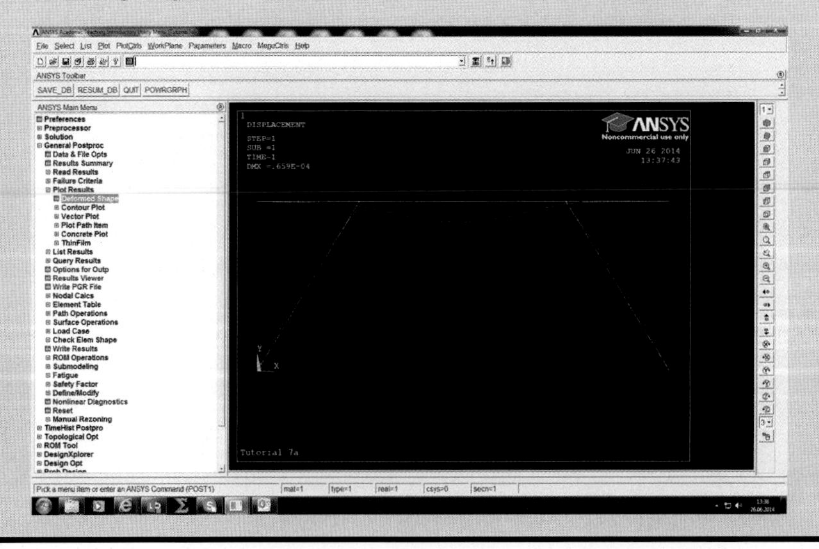

EXAMPLE 7.5: ANALYSIS OF A CABLE BRIDGE USING MATLAB/CALFEM

Determine the reactions on the supports, displacements, axial and shear forces, and bending moments of the following bridge. Data: $\boldsymbol{L} = \mathbf{15\ m}$, $\boldsymbol{q} = \mathbf{5200\ N/m}$.

Material and Cross-section Data

Beams: cross-section area $A_1 = 0.36\ \text{m}^2$, moment of inertia $I_1 = 0.0432\ \text{m}^4$, modulus of elasticity $E_1 = 2 \times 10^{11}\ \text{N/m}^2$

Bars: cross-section area $A_2 = 6 \times 10^{-3}\ \text{m}^2$, modulus of elasticity $E_2 = 2 \times 10^{11}\ \text{N/m}^2$

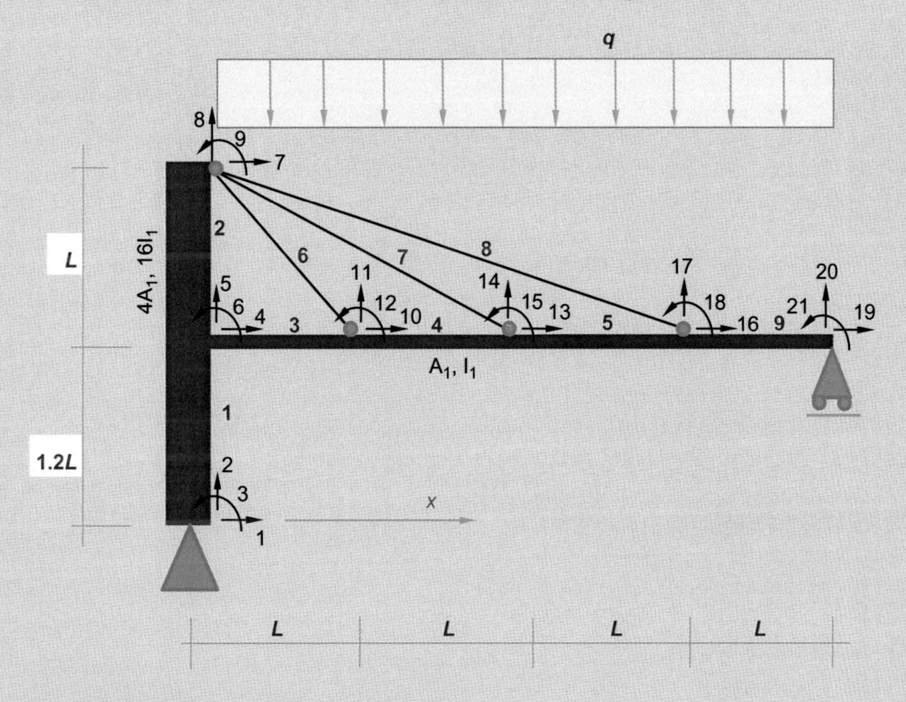

Size of the Stiffness Matrix and the Load Vector

```
f = zeros(21,1);K = zeros(21,21);
```

Boundary Conditions

```
bc = [1 0;2 0;3 0;20 0];
```

Topology Matrix for the Beam Elements

```
Edof1 = [1 1 2 3 4 5 6;2 4 5 6 7 8 9;3 4 5 6 10 11 12;4 10 11 12 13 14 15;5 13 14 15 16 17 18;6 16 17 18 19
20 21];
```

Topology Matrix for the Bar Elements

```
Edof2 = [7 7 8 10 11;8 7 8 13 14;9 7 8 16 17];
```

Continued

EXAMPLE 7.5: ANALYSIS OF A CABLE BRIDGE USING MATLAB/CALFEM—CONT'D

Geometric Data and Elastic Properties

```
L = 15;qx = 0;qy = -5200;
   E1 = 2e11;A1 = 0.36;I1 = 0.0432;ep11 = [E1 A1 I1];ep12 = [E1 4*A1 16*I1];
   E2 = 2e11;A2 = 0.36;ep2 = [E2 A2];
```

Coordinates of the Nodes

```
ex1 = [0 0];ex2 = [0 0];ex3 = [0 L];ex4 = [L 2*L];ex5 = [2*L 3*L];ex6 = [3*L 4*L];ex7 = [0 L];
ex8 = [0 2*L];ex9 = [0 3*L];
   ey1 = [0 1.2*L];ey2 = [1.2*L 2.2*L];ey3 = [1.2*L 1.2*L];ey4 = [1.2*L 1.2*L];ey5 =
[1.2*L 1.2*L];ey6 = [1.2*L 1.2*L];ey7 = [2.2*L 1.2*L];ey8 = [2.2*L 1.2*L];ey9 = [2.2*L
1.2*L];
```

Uniformly Distributed Loads of the Beam Elements in Their Local Coordinate System

```
eq1 = [0 0];eq2 = [0 0];eq3 = [0 qy];eq4 = [0 qy];eq5 = [0 qy];eq6 = [0 qy];
```

Derivation of the Beam Element Matrices

```
[Ke1,fe1] = beam2e(ex1,ey1,ep12,eq1);
   [Ke2,fe2] = beam2e(ex2,ey2,ep12,eq2);
   [Ke3,fe3] = beam2e(ex3,ey3,ep11,eq3);
   [Ke4,fe4] = beam2e(ex4,ey4,ep11,eq4);
   [Ke5,fe5] = beam2e(ex5,ey5,ep11,eq5);
   [Ke6,fe6] = beam2e(ex6,ey6,ep11,eq6);
```

Assembly of the Beam Element Matrices

```
[K,f] = assem(Edof1(1,:),K,Ke1,f,fe1);
   [K,f] = assem(Edof1(2,:),K,Ke2,f,fe2);
   [K,f] = assem(Edof1(3,:),K,Ke3,f,fe3);
   [K,f] = assem(Edof1(4,:),K,Ke4,f,fe4);
   [K,f] = assem(Edof1(5,:),K,Ke5,f,fe5);
   [K,f] = assem(Edof1(6,:),K,Ke6,f,fe6);
```

Derivation of the Bar Element Matrices

```
Ke7 = bar2e(ex7,ey7,ep2);
   Ke8 = bar2e(ex8,ey8,ep2);
   Ke9 = bar2e(ex9,ey9,ep2);
```

Assembly of the Bar Element Matrices

```
K = assem(Edof2(1,:),K,Ke7);
   K = assem(Edof2(2,:),K,Ke8);
   K = assem(Edof2(3,:),K,Ke9);
```

Computation of the Displacement Field [a] and the Support Reactions [r]

```
[a,r] = solveq(K,f,bc);
```

a =	r =
	1.0e+06 *
0	
0	-0.0000
0	0.1957
0.0028	2.3796
-0.0000	-0.0000
-0.0003	0.0000
0.0086	0.0000
-0.0000	0.0000
-0.0004	0.0000
0.0028	0
-0.0059	0.0000
-0.0004	-0.0000
0.0027	0.0000
-0.0122	0.0000
-0.0005	-0.0000
0.0028	-0.0000
-0.0167	0.0000
0.0004	-0.0000
0.0028	0.0000
0	0
0.0015	0.1163
	0.0000

Extraction of the Element Nodal Displacements According to the Topology Matrix

(a) For beam elements

```
Ed1 = extract(Edof1,a);
```

(b) For bar elements

```
Ed2 = extract(Edof2,a);
```

Drawing of the Undeformed Elements

```
figure(1)
   plotpar = [2 1 0];
   eldraw2(ex1,ey1,plotpar);
   eldraw2(ex2,ey2,plotpar);
   eldraw2(ex3,ey3,plotpar);
   eldraw2(ex4,ey4,plotpar);
```

Continued

EXAMPLE 7.5: ANALYSIS OF A CABLE BRIDGE USING MATLAB/CALFEM—CONT'D

```
eldraw2(ex5,ey5,plotpar);
eldraw2(ex6,ey6,plotpar);
eldraw2(ex7,ey7,plotpar);
eldraw2(ex8,ey8,plotpar);
eldraw2(ex9,ey9,plotpar);
```

Drawing of the Deformed Elements

(a) For beam elements

```
sfac = scalfact2(ex6,ey6,Ed1(6,:),0.1);
plotpar = [1 2 1];
eldisp2(ex1,ey1,Ed1(1,:),plotpar,sfac);
eldisp2(ex2,ey2,Ed1(2,:),plotpar,sfac);
eldisp2(ex3,ey3,Ed1(3,:),plotpar,sfac);
eldisp2(ex4,ey4,Ed1(4,:),plotpar,sfac);
eldisp2(ex5,ey5,Ed1(5,:),plotpar,sfac);
eldisp2(ex6,ey6,Ed1(6,:),plotpar,sfac);
```

(b) For bar elements

```
eldisp2(ex7,ey7,Ed2(1,:),plotpar,sfac);
eldisp2(ex8,ey8,Ed2(2,:),plotpar,sfac);
eldisp2(ex9,ey9,Ed2(3,:),plotpar,sfac);
axis([-10 70 -10 40]);
```

Computation of the Distribution of Axial Forces-Shear Forces-Bending Moments

```
es1 = beam2s(ex1,ey1,ep12,Ed1(1,:),eq1,20);
es2 = beam2s(ex2,ey2,ep12,Ed1(2,:),eq2,20);
es3 = beam2s(ex3,ey3,ep11,Ed1(3,:),eq3,20);
es4 = beam2s(ex4,ey4,ep11,Ed1(4,:),eq4,20);
es5 = beam2s(ex5,ey5,ep11,Ed1(5,:),eq5,20);
es6 = beam2s(ex6,ey6,ep11,Ed1(6,:),eq6,20);
```

Drawing of the Axial Forces

```
figure(2)
plotpar = [2 1];
sfac = scalfact2(ex5,ey5,es5(:,1),1);
eldia2(ex1,ey1,es1(:,1),plotpar,sfac);
eldia2(ex2,ey2,es2(:,1),plotpar,sfac);
eldia2(ex3,ey3,es3(:,1),plotpar,sfac);
eldia2(ex4,ey4,es4(:,1),plotpar,sfac);
eldia2(ex5,ey5,es5(:,1),plotpar,sfac);
eldia2(ex6,ey6,es6(:,1),plotpar,sfac);
axis([-10 70 -10 40]);
```

Drawing of the Shear Forces

```
figure(3)
   sfac = scalfact2(ex3,ey3,es3(:,2),0.6);
   eldia2(ex1,ey1,es1(:,2),plotpar,sfac);
   eldia2(ex2,ey2,es2(:,2),plotpar,sfac);
   eldia2(ex3,ey3,es3(:,2),plotpar,sfac);
   eldia2(ex4,ey4,es4(:,2),plotpar,sfac);
   eldia2(ex5,ey5,es5(:,2),plotpar,sfac);
   eldia2(ex6,ey6,es6(:,2),plotpar,sfac);
   axis([-10 70 -10 50]);
```

Drawing of the Bending Moments

```
figure(4)
   sfac = scalfact2(ex6,ey6,es6(:,3),4);
   eldia2(ex1,ey1,es1(:,3),plotpar,sfac);
   eldia2(ex2,ey2,es2(:,3),plotpar,sfac);
   eldia2(ex3,ey3,es3(:,3),plotpar,sfac);
   eldia2(ex4,ey4,es4(:,3),plotpar,sfac);
   eldia2(ex5,ey5,es5(:,3),plotpar,sfac);
   eldia2(ex6,ey6,es6(:,3),plotpar,sfac);
   axis([-10 70 -50 40]);
```

Results

40
30
20
10
0
−10
−10 0 10 20 30 40 50 60 70

Continued

EXAMPLE 7.5: ANALYSIS OF A CABLE BRIDGE USING MATLAB/CALFEM—CONT'D

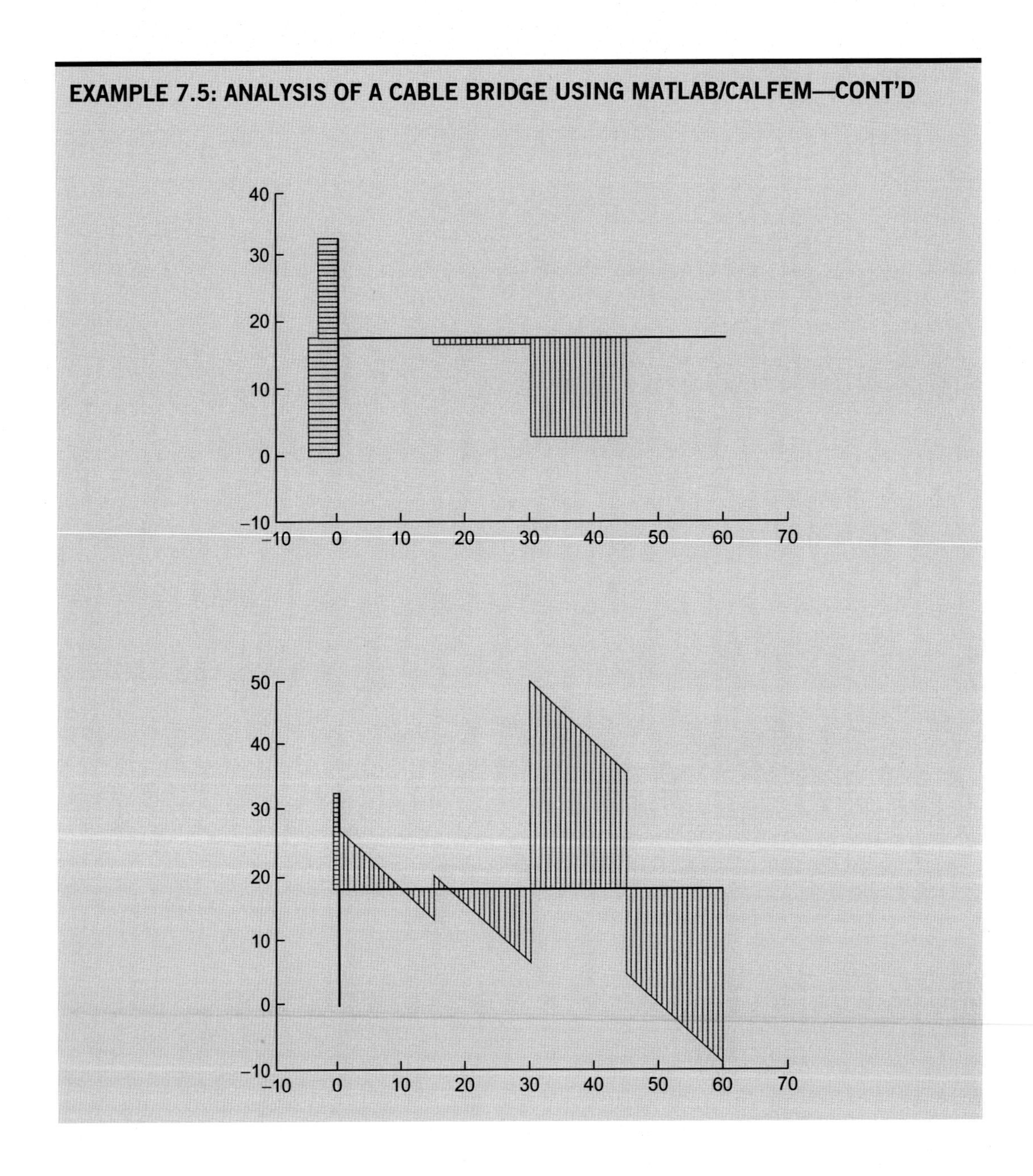

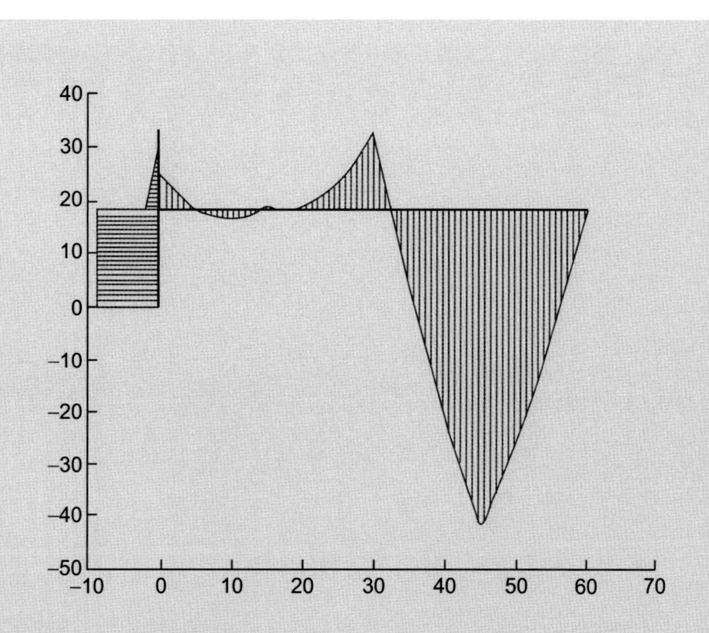

The MATLAB/CALFEM Computer Code

```
% ——— BRIDGE ———
   f = zeros(21,1);K = zeros(21,21);
   bc = [1 0;2 0;3 0;20 0];
   Edof1 = [1 1 2 3 4 5 6;2 4 5 6 7 8 9;3 4 5 6 10 11 12;4 10 11 12 13 14 15;5 13 14 15 16 17 18;6 16 17
18 19 20 21];
   Edof2 = [7 7 8 10 11;8 7 8 13 14;9 7 8 16 17];
   L = 15;qx = 0;qy = -5200;
   E1 = 2e11;A1 = 0.36;I1 = 0.0432;ep11 = [E1 A1 I1];ep12 = [E1 A1 I1];
   E2 = 2e11;A2 = 0.36;ep2 = [E2 A2];
   ex1 = [0 0];ex2 = [0 0];ex3 = [0 L];ex4 = [L 2*L];ex5 = [2*L 3*L];ex6 = [3*L 4*L];ex7 = [0
L];ex8 = [0 2*L];ex9 = [0 3*L];
   ey1 = [0 1.2*L];ey2 = [1.2*L 2.2*L];ey3 = [1.2*L 1.2*L];ey4 = [1.2*L 1.2*L];ey5 = [1.2*L
1.2*L];ey6 = [1.2*L 1.2*L];ey7 = [2.2*L 1.2*L];ey8 = [2.2*L 1.2*L];ey9 = [2.2*L 1.2*L];
   eq1 = [0 0];eq2 = [0 0];eq3 = [0 qy];eq4 = [0 qy];eq5 = [0 qy];eq6 = [0 qy];
   [Ke1,fe1] = beam2e(ex1,ey1,ep12,eq1);
   [Ke2,fe2] = beam2e(ex2,ey2,ep12,eq2);
   [Ke3,fe3] = beam2e(ex3,ey3,ep11,eq3);
   [Ke4,fe4] = beam2e(ex4,ey4,ep11,eq4);
   [Ke5,fe5] = beam2e(ex5,ey5,ep11,eq5);
   [Ke6,fe6] = beam2e(ex6,ey6,ep11,eq6);
```

Continued

EXAMPLE 7.5: ANALYSIS OF A CABLE BRIDGE USING MATLAB/CALFEM—CONT'D

```
[K,f] = assem(Edof1(1,:),K,Ke1,f,fe1);
[K,f] = assem(Edof1(2,:),K,Ke2,f,fe2);
[K,f] = assem(Edof1(3,:),K,Ke3,f,fe3);
[K,f] = assem(Edof1(4,:),K,Ke4,f,fe4);
[K,f] = assem(Edof1(5,:),K,Ke5,f,fe5);
[K,f] = assem(Edof1(6,:),K,Ke6,f,fe6);
Ke7 = bar2e(ex7,ey7,ep2);
Ke8 = bar2e(ex8,ey8,ep2);
Ke9 = bar2e(ex9,ey9,ep2);
K = assem(Edof2(1,:),K,Ke7);
K = assem(Edof2(2,:),K,Ke8);
K = assem(Edof2(3,:),K,Ke9);
[a,r] = solveq(K,f,bc)
Ed1 = extract(Edof1,a);
Ed2 = extract(Edof2,a);
figure(1)
plotpar = [2 1 0];
eldraw2(ex1,ey1,plotpar);
eldraw2(ex2,ey2,plotpar);
eldraw2(ex3,ey3,plotpar);
eldraw2(ex4,ey4,plotpar);
eldraw2(ex5,ey5,plotpar);
eldraw2(ex6,ey6,plotpar);
eldraw2(ex7,ey7,plotpar);
eldraw2(ex8,ey8,plotpar);
eldraw2(ex9,ey9,plotpar);
sfac = scalfact2(ex6,ey6,Ed1(6,:),0.1);
plotpar = [1 2 1];
eldisp2(ex1,ey1,Ed1(1,:),plotpar,sfac);
eldisp2(ex2,ey2,Ed1(2,:),plotpar,sfac);
eldisp2(ex3,ey3,Ed1(3,:),plotpar,sfac);
eldisp2(ex4,ey4,Ed1(4,:),plotpar,sfac);
eldisp2(ex5,ey5,Ed1(5,:),plotpar,sfac);
eldisp2(ex6,ey6,Ed1(6,:),plotpar,sfac);
eldisp2(ex7,ey7,Ed2(1,:),plotpar,sfac);
eldisp2(ex8,ey8,Ed2(2,:),plotpar,sfac);
eldisp2(ex9,ey9,Ed2(3,:),plotpar,sfac);
axis([-10 70 -10 40]);
% —— computation of the distribution of axial forces-shear forces-bending moments
——
```

```
es1 = beam2s(ex1,ey1,ep12,Ed1(1,:),eq1,20);
es2 = beam2s(ex2,ey2,ep12,Ed1(2,:),eq2,20);
es3 = beam2s(ex3,ey3,ep11,Ed1(3,:),eq3,20);
es4 = beam2s(ex4,ey4,ep11,Ed1(4,:),eq4,20);
es5 = beam2s(ex5,ey5,ep11,Ed1(5,:),eq5,20);
es6 = beam2s(ex6,ey6,ep11,Ed1(6,:),eq6,20);
% —— Drawing of axial forces ——
figure(2)
plotpar = [2 1];
sfac = scalfact2(ex5,ey5,es5(:,1),1);
eldia2(ex1,ey1,es1(:,1),plotpar,sfac);
eldia2(ex2,ey2,es2(:,1),plotpar,sfac);
eldia2(ex3,ey3,es3(:,1),plotpar,sfac);
eldia2(ex4,ey4,es4(:,1),plotpar,sfac);
eldia2(ex5,ey5,es5(:,1),plotpar,sfac);
eldia2(ex6,ey6,es6(:,1),plotpar,sfac);
axis([-10 70 -10 40]);
% —— Drawing of shear forces ——
figure(3)
sfac = scalfact2(ex3,ey3,es3(:,2),0.6);
eldia2(ex1,ey1,es1(:,2),plotpar,sfac);
eldia2(ex2,ey2,es2(:,2),plotpar,sfac);
eldia2(ex3,ey3,es3(:,2),plotpar,sfac);
eldia2(ex4,ey4,es4(:,2),plotpar,sfac);
eldia2(ex5,ey5,es5(:,2),plotpar,sfac);
eldia2(ex6,ey6,es6(:,2),plotpar,sfac);
axis([-10 70 -10 50]);
% —— Drawing of bending moments ——
figure(4)
sfac = scalfact2(ex6,ey6,es6(:,3),4);
eldia2(ex1,ey1,es1(:,3),plotpar,sfac);
eldia2(ex2,ey2,es2(:,3),plotpar,sfac);
eldia2(ex3,ey3,es3(:,3),plotpar,sfac);
eldia2(ex4,ey4,es4(:,3),plotpar,sfac);
eldia2(ex5,ey5,es5(:,3),plotpar,sfac);
eldia2(ex6,ey6,es6(:,3),plotpar,sfac);
axis([-10 70 -50 40]);
```

REFERENCES

[1] Hartmann F, Katz C. Structural analysis with finite elements. Berlin: Springer; 2007.
[2] Cook RD, Malkus DS, Plesha ME, Witt RJ. Concepts and applications of finite element analysis. Hoboken: John Wiley & Sons; 2002.
[3] Logan DL. A first course in the finite element method. Boston: Cengage Learning; 2012.
[4] Bhatti MA. Fundamental finite element analysis and applications. Hoboken: John Wiley & Sons; 2005.
[5] Fish J, Belytschko T. A first course in finite elements. Hoboken: John Wiley & Sons; 2007.
[6] Austrell P-E, Dahlblom O, Lindemann J, Olsson A, Olsson K-G, Persson K, et al. CALFEM—a finite element toolbox, Version 3.4., Division of Structural Mechanics, Lund University; 2004.
[7] ANSYS, User e-manual, Version 13.
[8] Melosh RJ. Structural engineering analysis by Finite Elements. Upper Saddle River, NJ: Prentice-Hall; 1990.
[9] Shames IH, Dym CL. Energy and Finite Element methods in structural mechanics. New York: Hemisphere Publishing Corporation; 1985.

CHAPTER

THE PRINCIPLE OF MINIMUM POTENTIAL ENERGY FOR ONE-DIMENSIONAL ELEMENTS

8

In Chapters 1–7, the element equations for springs, bars, and beams was derived using the direct equilibrium method. However, this method is not suitable for analyzing solids that are more complex. For such cases, methods based on the variational principle are more effective. Among variational principles, the principle of minimum potential energy (MPE) is the basis for most finite element method applications in mechanical and structural engineering problems. In order to understand this principle, it is initially applied on one-dimensional elements. It should be noted that the principle of MPE is only applicable to conservative (e.g., elastic) systems.

8.1 THE BASIC CONCEPT

From our knowledge of physics, we know that a system is conservative if work done by its internal and external forces (i.e., forces acting on its surface, volume, or nodes) depends only on the initial and final displaced configuration. Therefore, the change of the total potential energy Π_p of a conservative system is independent on the deformation history from the initial to the final configuration. According to this definition, elastic solids and structures are conservative systems.

The total potential energy includes the strain energy U of the stresses or internal forces causing elastic deformation plus the potential energy W possessed by the external loads (body forces, nodal forces, surface tractions), by virtue having the capacity to do work if they move through a distance. Therefore:

$$\Pi_\mathrm{p} = U + W \tag{8.1}$$

According to the principle of stationary potential energy, among all possible states of an elastic system, those that satisfy the equilibrium equations yield stationary potential energy with respect to small possible displacements. If the stationary potential energy has a relative minimum, the equilibrium state of the system is stable.

Taking into account the above principle, the equilibrium of a conservative system of many degrees of freedom $(d_1, d_2, \ldots, d_n)$ prevails when

$$\frac{\partial \Pi_\mathrm{p}}{\partial d_\mathrm{i}} = 0, \quad i = 1, 2, \ldots, n \tag{8.2}$$

The above equation is the mathematical expression of the MPE principle.

Essentials of the Finite Element Method. http://dx.doi.org/10.1016/B978-0-12-802386-0.00008-6

8.2 APPLICATION OF THE MPE PRINCIPLE ON SYSTEMS OF SPRING ELEMENTS

Let us take the example of the spring of Figure 8.1. The nodal forces r_{1x}, r_{2x} cause nodal displacements d_{1x}, d_{2x}. Therefore, the deformation of the spring is $(d_{2x} - d_{1x})$. As it is already known from the physics, the elastic energy of the spring is as follows:

$$U = \frac{1}{2}k(d_{2x} - d_{1x})^2 \tag{8.3}$$

In contrast, the work done by the external forces r_{1x}, r_{2x} is as follows:

$$W = -(r_{1x}d_{1x} + r_{2x}d_{2x}) \tag{8.4}$$

The forces r_{1x}, r_{2x} are always acting at their full value (their work is independent of the elastic behavior of the spring). Their movement through the corresponding displacements d_{1x}, d_{2x} is doing work in at a quantity of $r_{1x}d_{1x}$ and $r_{2x}d_{2x}$, respectively, thereby losing potential of equal amount (hence the negative sign in Equation 8.4).

According to Equation (8.1), the total potential of the spring demonstrated in Figure 8.1 is as follows:

$$\Pi_{\mathrm{p}} = \frac{1}{2}k(d_{2x} - d_{1x})^2 - (r_{1x}d_{1x} + r_{2x}d_{2x}) \tag{8.5}$$

Therefore, application of the principle of MPE (Equation 8.2) yields

$$\begin{cases} \dfrac{\partial \Pi_{\mathrm{p}}}{\partial d_{1x}} = 0 \\ \dfrac{\partial \Pi_{\mathrm{p}}}{\partial d_{2x}} = 0 \end{cases} \tag{8.6}$$

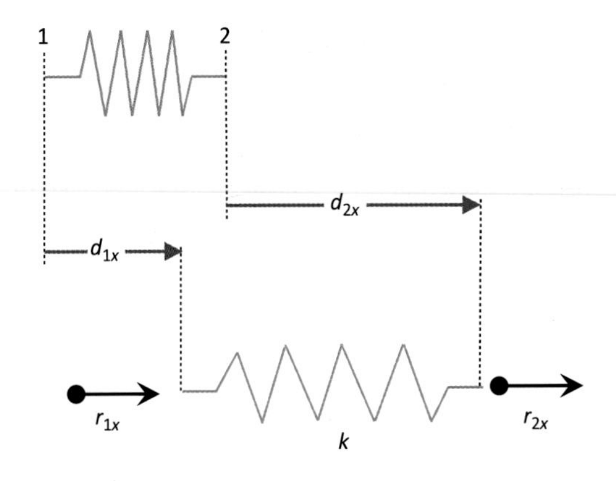

FIGURE 8.1

A single elastic spring subjected to axial forces on its ends.

or

$$\begin{cases} k(d_{1x}-d_{2x})-r_{1x}=0 \\ k(-d_{1x}+d_{2x})-r_{2x}=0 \end{cases} \tag{8.7}$$

The above system of equations can now be written in a matrix form providing the spring element equation:

$$\begin{Bmatrix} r_{1x} \\ r_{2x} \end{Bmatrix} = \begin{bmatrix} k & -k \\ -k & k \end{bmatrix} \begin{Bmatrix} d_{1x} \\ d_{2x} \end{Bmatrix} \tag{8.8}$$

The above equation is the same as Equation (3.7) derived by the use of the principle of direct equilibrium.

Let us now apply Equation (8.2) in a system with more degrees of freedom. In order to do this, we recall the structural system of figure 3.3, which is composed of three spring elements. The total potential for this system is as follows:

$$\begin{aligned} \Pi_p = {} & \frac{1}{2}k_1(d_{2x}-d_{1x})^2 + \frac{1}{2}k_2(d_{3x}-d_{2x})^2 + \frac{1}{2}k_3(d_{4x}-d_{3x})^2 \\ & -(R_{1x}d_{1x}+R_{2x}d_{2x}+R_{3x}d_{3x}+R_{4x}d_{4x}) \end{aligned} \tag{8.9}$$

Taking into account the above equation, Equation (8.2) can be written:

$$\begin{cases} \partial\Pi_p/\partial d_{1x}=0 \\ \partial\Pi_p/\partial d_{2x}=0 \\ \partial\Pi_p/\partial d_{3x}=0 \\ \partial\Pi_p/\partial d_{4x}=0 \end{cases} \tag{8.10}$$

After some simple algebraic operations, the above algebraic system yields

$$\begin{cases} k_1d_{1x}-k_1d_{2x}-R_{1x}=0 \\ -k_1d_{1x}+(k_1+k_2)d_{2x}-k_2d_{3x}-R_{2x}=0 \\ -k_2d_{2x}+(k_2+k_3)d_{3x}-k_3d_{4x}-R_{3x}=0 \\ -k_3d_{3x}+k_3d_{4x}-R_{4x}=0 \end{cases} \tag{8.11}$$

The above system can be written in the following matrix from:

$$\begin{Bmatrix} R_{1x} \\ R_{2x} \\ R_{3x} \\ R_{4x} \end{Bmatrix} = \begin{bmatrix} k_1 & -k_1 & 0 & 0 \\ -k_1 & k_1+k_2 & -k_2 & 0 \\ 0 & -k_2 & k_2+k_3 & -k_3 \\ 0 & 0 & -k_3 & k_3 \end{bmatrix} \begin{Bmatrix} d_{1x} \\ d_{2x} \\ d_{3x} \\ d_{4x} \end{Bmatrix} \tag{8.12}$$

Equation (8.12) is same as Equation (3.15) derived by the use of the direct equilibrium principle.

8.3 APPLICATION OF THE MPE PRINCIPLE ON SYSTEMS OF BAR ELEMENTS

Following similar procedure as in Section 8.2, we assume the bar of Figure 8.2. The nodal forces r_{1x}, r_{2x} cause nodal displacements d_{1x}, d_{2x}.

Therefore, the axial deformation of the bar is

$$\Delta L = (d_{2x}-d_{1x}) \tag{8.13}$$

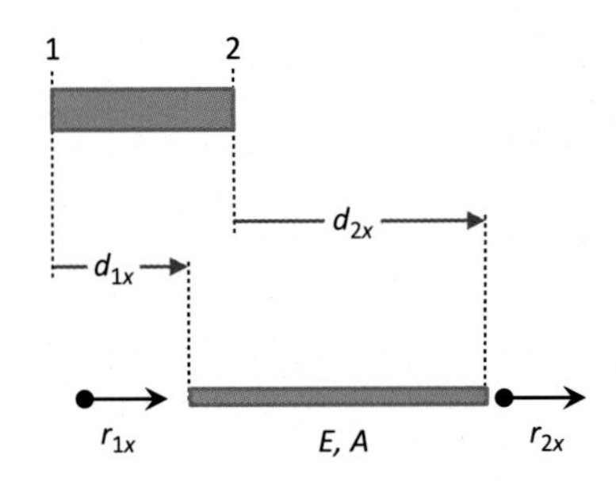

FIGURE 8.2

An elastic bar subjected to axial forces on its ends.

This deformation yields strain

$$\varepsilon = \frac{\Delta L}{L} \tag{8.14}$$

or

$$\varepsilon = \frac{d_{2x} - d_{1x}}{L} \tag{8.15}$$

According to the Hooke's law, the corresponding stress is

$$\sigma = E\varepsilon \tag{8.16}$$

or

$$\sigma = E\frac{d_{2x} - d_{1x}}{L} \tag{8.17}$$

As it is already known from the mechanics of solids, the accumulated strain energy of a bar is as follows:

$$U = \frac{1}{2}\sigma\varepsilon V \tag{8.18}$$

where V is the volume of the bar

$$V = AL \tag{8.19}$$

Taking into account Equations (8.15) and (8.17), Equation (8.18) yields

$$U = \frac{EA}{2L}(d_{2x} - d_{1x})^2 \tag{8.20}$$

The work done by the external forces is as follows:

$$W = -(r_{1x}d_{1x} + r_{2x}d_{2x}) \tag{8.21}$$

Therefore, taking into account Equations (8.20) and (8.21), the total potential energy of the bar is as follows:

$$\Pi_p = \frac{EA}{2L}(d_{2x} - d_{1x})^2 - (r_{1x}d_{1x} + r_{2x}d_{2x}) \tag{8.22}$$

Using the above formula, the principle of the MPE

$$\begin{cases} \dfrac{\partial \Pi_p}{\partial d_{1x}} = 0 \\ \dfrac{\partial \Pi_p}{\partial d_{2x}} = 0 \end{cases} \tag{8.23}$$

yields

$$\begin{cases} \dfrac{EA}{L}(d_{1x} - d_{2x}) - r_{1x} = 0 \\ \dfrac{EA}{L}(-d_{1x} + d_{2x}) - r_{2x} = 0 \end{cases} \tag{8.24}$$

or in a matrix form

$$\begin{Bmatrix} r_{1x} \\ r_{2x} \end{Bmatrix} = \begin{bmatrix} EA/L & -EA/L \\ -EA/L & EA/L \end{bmatrix} \begin{Bmatrix} d_{1x} \\ d_{2x} \end{Bmatrix} \tag{8.25}$$

Equation (8.25) is the same as Equation (4.35) derived by the principle of direct equilibrium.

Let us now apply Equation (8.2) to a system of bars with more degrees of freedom. To do this, we recall the structural system of figure 4.4, which is composed of three bar elements. The total potential of this system is:

$$\begin{aligned} \Pi_p = {} & \frac{1}{2}\frac{E_1A_1}{L_1}(d_{2x} - d_{1x})^2 + \frac{1}{2}\frac{E_2A_2}{L_2}(d_{3x} - d_{2x})^2 + \frac{1}{2}\frac{E_3A_3}{L_3}(d_{4x} - d_{3x})^2 \\ & -(R_{1x}d_{1x} + R_{2x}d_{2x} + R_{3x}d_{3x} + R_{4x}d_{4x}) \end{aligned} \tag{8.26}$$

Therefore, the principle of the MPE

$$\begin{cases} \partial \Pi_p / \partial d_{1x} = 0 \\ \partial \Pi_p / \partial d_{2x} = 0 \\ \partial \Pi_p / \partial d_{3x} = 0 \\ \partial \Pi_p / \partial d_{4x} = 0 \end{cases} \tag{8.27}$$

using Equation (8.26) yields the following matrix equation:

$$\begin{Bmatrix} R_{1x} \\ R_{2x} \\ R_{3x} \\ R_{4x} \end{Bmatrix} = \begin{bmatrix} E_1A_1/L_1 & -E_1A_1/L_1 & 0 & 0 \\ -E_1A_1/L_1 & E_1A_1/L_1 + E_2A_2/L_2 & -E_2A_2/L_2 & 0 \\ 0 & -E_2A_2/L_2 & E_2A_2/L_2 + E_3A_3/L_3 & -E_3A_3/L_3 \\ 0 & 0 & -E_3A_3/L_3 & E_3A_3/L_3 \end{bmatrix} \begin{Bmatrix} d_{1x} \\ d_{2x} \\ d_{3x} \\ d_{4x} \end{Bmatrix} \tag{8.28}$$

The above equation is the same as Equation (4.43) derived using the direct equilibrium principle.

8.4 APPLICATION OF THE MPE PRINCIPLE ON TRUSSES

Trusses are structural systems composed of bars. Since any truss member 1-2 aligned with its local coordinate axis $\bar{x}$ is often inclined with respect to the global coordinate system x-y, the change of its length should be expressed in terms of the nodal displacements d_{1x}, d_{1y} and d_{2x}, d_{2y} with respect to the system x-y. Taking into account Equations (5.3) and (5.4), the change of the length is as follows:

$$\Delta L = d_{2\bar{x}} - d_{1\bar{x}} = \left(Cd_{2x} + Sd_{2y}\right) - \left(Cd_{1x} + Sd_{1y}\right) \tag{8.29}$$

or

$$\Delta L = [-C \;\; -S \;\; C \;\; -S]\begin{Bmatrix} d_{1x} \\ d_{1y} \\ d_{2x} \\ d_{2y} \end{Bmatrix} \tag{8.30}$$

Therefore, the total potential of a truss containing n-bars and k-nodes can be expressed by the following formula:

$$\Pi_{\mathrm{p}} = \sum_{i=1}^{n} \frac{1}{2}\frac{E_i A_i}{L_i}(\Delta L_i)^2 - \sum_{j=1}^{k}\left(R_{jx}d_{jx} + R_{jy}d_{jy}\right) \tag{8.31}$$

Therefore, the corresponding system of algebraic equations yielding the structure's equation is as follows:

$$\left\{\frac{\partial}{\partial d_{1x}} \;\; \frac{\partial}{\partial d_{1y}} \;\; \frac{\partial}{\partial d_{2x}} \;\; \frac{\partial}{\partial d_{2y}} \;\; \cdots \;\; \frac{\partial}{\partial d_{kx}} \;\; \frac{\partial}{\partial d_{ky}}\right\}^{\mathrm{T}} \Pi_{\mathrm{p}} = 0 \tag{8.32}$$

8.5 APPLICATION OF THE MPE PRINCIPLE ON BEAMS

We recall from Chapter 6 that the types of external loads acting on beams are:

1. transverse varying loads q;
2. transverse concentrated nodal forces r_{iy};
3. nodal bending moments m_i.

The consequence of the above loads are transverse deflections $u(x)$ and slopes $\vartheta(x)$ taking place in any point x of the beam.

Therefore, for a beam element of length L, the work of the external loads is as follows:

$$W = -\int_0^L q(x)u(x)\mathrm{d}x - \sum_{i=1}^{2}\left(r_{iy}d_{iy} + m_i\vartheta_i\right) \tag{8.33}$$

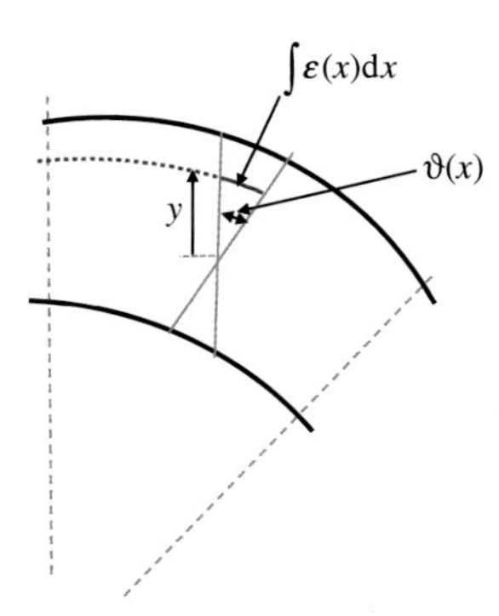

FIGURE 8.3

Deformation of a beam under bending.

where d_{iy} and ϑ_i are the transverse nodal displacement and the slope of the node i, respectively. Taking into account Equation (6.13), the above equation can be expressed in terms of nodal displacements

$$W = -\int_0^L q(x)\{[N]\{d\}\}^{\mathrm{T}}\mathrm{d}x - \{d\}^{\mathrm{T}}[r]^{\mathrm{T}} \tag{8.34}$$

where

$$[N] = [N_1 \ \ N_2 \ \ N_3 \ \ N_4],\ \{d\}^{\mathrm{T}} = \{d_{1x} \ \ \vartheta_1 \ \ d_{2x} \ \ \vartheta_2\},\ [r]^{\mathrm{T}} = [r_{1y} \ \ m_1 \ \ r_{2y} \ \ m_2]^{\mathrm{T}}$$

As it is already known from the mechanics of solids, during bending the slope of each cross-section of a beam (Figure 8.3) is as follows:

$$\vartheta(x) = \frac{\mathrm{d}u(x)}{\mathrm{d}x} \tag{8.35}$$

Therefore, the geometric condition

$$\vartheta(x) = -\frac{\int \varepsilon(x)\mathrm{d}x}{y} \tag{8.36}$$

can now be written

$$\frac{\mathrm{d}u(x)}{\mathrm{d}x} = -\frac{\int \varepsilon(x)\mathrm{d}x}{y} \tag{8.37}$$

The above equation yields

$$\varepsilon(x) = -y\frac{\mathrm{d}^2u(x)}{\mathrm{d}x^2} \tag{8.38}$$

Taking into account Equation (6.13), Equation (8.38) can correlate the strain of each point x of the beam to the nodal displacements

$$\{\varepsilon(x)\} = -y\frac{\mathrm{d}^2}{\mathrm{d}x^2}[N_1 \ N_2 \ N_3 \ N_4]\begin{Bmatrix} d_{1y} \\ \vartheta_1 \\ d_{2y} \\ \vartheta_2 \end{Bmatrix} \tag{8.39}$$

where N_1, N_2, N_3, and N_4 are shape functions given in Equations (6.14)–(6.17).

After some simple algebraic operations, Equation (8.39) yields

$$\{\varepsilon(x)\} = -y[B]\{d\} \tag{8.40}$$

where

$$\{d\} = \{d_{1x} \ \vartheta_1 \ d_{2x} \ \vartheta_2\}^{\mathrm{T}} \tag{8.41}$$

and

$$[B] = \frac{1}{L^3}\left[(12x - 6L) \ (6xL - 4L^2) \ (-12x + 6L) \ (6xL - 2L)\right] \tag{8.42}$$

Since the stress-strain relationship is given by Hooke's law $\sigma(x) = E\varepsilon(x)$, taking into account Equation (8.40), it can be written

$$\{\sigma(x)\} = -Ey[B]\{d\} \tag{8.43}$$

Therefore, the strain energy U of a beam under bending can be correlated to the nodal displacements using Equations (8.40) and (8.43):

$$U = \int_V \frac{1}{2}\{\sigma\}^{\mathrm{T}}\{\varepsilon\}\mathrm{d}V \tag{8.44}$$

where V is the volume of the beam. Since $\mathrm{d}V = \mathrm{d}A\mathrm{d}x$, the above equation can be written

$$U = \frac{1}{2}\int_A\int_L \{\sigma\}^{\mathrm{T}}\{\varepsilon\}\mathrm{d}A\mathrm{d}x \tag{8.45}$$

Taking into account Equations (8.40) and (8.43), Equation (8.45) yields

$$U = \frac{E}{2}\int_A\int_L \{[B]\{d\}\}^{\mathrm{T}}[B]\{d\}y^2\mathrm{d}A\mathrm{d}x \tag{8.46}$$

Taking into account the following property of the linear algebra

$$\{[B]\{d\}\}^{\mathrm{T}} = \{d\}^{\mathrm{T}}[B]^{\mathrm{T}} \tag{8.47}$$

Equation (8.46) can now be written

$$U = \frac{E}{2}\int_A\int_L \{d\}^{\mathrm{T}}[B]^{\mathrm{T}}[B]\{d\}y^2\mathrm{d}A\mathrm{d}x \tag{8.48}$$

Since the quantity

$$I = \int_A y^2 \mathrm{d}A \tag{8.49}$$

represents the moment of inertia, Equation (8.48) yields

$$U = \frac{EI}{2} \int_L \{d\}^{\mathrm{T}} [B]^{\mathrm{T}} [B] \{d\} \mathrm{d}x \tag{8.50}$$

Therefore, taking into account Equations (8.34) and (8.50), the total potential energy for a beam element takes now the form:

$$\Pi_{\mathrm{p}} = \frac{EI}{2} \int_0^L \{d\}^{\mathrm{T}} [B]^{\mathrm{T}} [B] \{d\} \mathrm{d}x - \int_0^L q(x) \{d\}^{\mathrm{T}} [N]^{\mathrm{T}} \mathrm{d}x - \{d\}^{\mathrm{T}} \{r\} \tag{8.51}$$

Applying the MPE principle to the above equation, the following formula can be obtained

$$\frac{\partial \Pi_{\mathrm{p}}}{\partial \{d\}} = 0 \tag{8.52}$$

yielding

$$\left(EI \int_0^L [B]^{\mathrm{T}} [B] \mathrm{d}x \right) \{d\} - \int_0^L [N]^{\mathrm{T}} q(x) \mathrm{d}x - \{r\} = 0 \tag{8.53}$$

Let

$$\{f\} = \int_0^L [N]^{\mathrm{T}} q(x) \mathrm{d}x + \{r\} \tag{8.54}$$

Then, Equation (8.53) takes the form:

$$\left(EI \int_0^L [B]^{\mathrm{T}} [B] \mathrm{d}x \right) \{d\} = \{f\} \tag{8.55}$$

Comparing the above equation with Equation (6.32), it can be concluded that the stiffness matrix of the beam element is as follows:

$$[k] = \left(EI \int_0^L [B]^{\mathrm{T}} [B] \mathrm{d}x \right) \tag{8.56}$$

Taking into account Equation (8.42), the above equation yields

$$[k] = \frac{EI}{L^3} \begin{bmatrix} 12 & 6L & -12 & 6L \\ 6L & 4L^2 & -6L & 2L^2 \\ -12 & -6L & 12 & -6L \\ 6L & 2L^2 & -6L & 4L^2 \end{bmatrix} \tag{8.57}$$

REFERENCES

[1] Wunderlich W, Pilkey WD. Mechanics of structures, variational and computational methods. Boca Raton: CRC Press; 2003.
[2] Logan DL. A first course in the finite element method. Boston MA: Cengage Learning; 2012.
[3] Fish J, Belytschko T. A first course in the finite elements. New York: Wiley; 2007.
[4] Bhatti MA. Fundamental finite element analysis and applications. Hoboken: John Wiley & Sons; 2005.
[5] Hartmann F, Katz C. Structural analysis with finite elements. Heidelberg: Springer; 2007.

CHAPTER

FROM "ISOTROPIC" TO "ORTHOTROPIC" PLANE ELEMENTS: ELASTICITY EQUATIONS FOR TWO-DIMENSIONAL SOLIDS

9

9.1 THE GENERALIZED HOOKE'S LAW

The stress-strain relation adopted in Chapters 1–8 for stiffness matrix derivation for one-dimensional elements was the following familiar form of Hooke's law:

$$\sigma = E\varepsilon \tag{9.1}$$

However, in multiaxial loading cases (Figure 9.1) where the mechanical behavior of solids is characterized by more complicated stress and strain states, for example,

$$\{\bar{\sigma}\} = \left\{\sigma_{\bar{x}}\, \sigma_{\bar{y}}\, \sigma_{\bar{z}}\, \tau_{\bar{x}\bar{y}}\, \tau_{\bar{y}\bar{z}}\, \tau_{\bar{z}\bar{x}}\right\}^{\mathrm{T}} \tag{9.2}$$

$$\{\bar{\varepsilon}\} = \left\{\varepsilon_{\bar{x}}\, \varepsilon_{\bar{y}}\, \varepsilon_{\bar{z}}\, \gamma_{\bar{x}\bar{y}}\, \gamma_{\bar{y}\bar{z}}\, \gamma_{\bar{z}\bar{x}}\right\}^{\mathrm{T}} \tag{9.3}$$

Hooke's law has the following generalized form:

$$\begin{Bmatrix} \sigma_{\bar{x}} \\ \sigma_{\bar{y}} \\ \sigma_{\bar{z}} \\ \tau_{\bar{x}\bar{y}} \\ \tau_{\bar{y}\bar{z}} \\ \tau_{\bar{z}\bar{x}} \end{Bmatrix} = \begin{bmatrix} C_{11} & C_{12} & C_{13} & C_{14} & C_{15} & C_{16} \\ C_{21} & C_{22} & C_{23} & C_{24} & C_{25} & C_{26} \\ C_{31} & C_{32} & C_{33} & C_{34} & C_{35} & C_{36} \\ C_{41} & C_{42} & C_{43} & C_{44} & C_{45} & C_{46} \\ C_{51} & C_{52} & C_{53} & C_{54} & C_{55} & C_{56} \\ C_{61} & C_{62} & C_{63} & C_{64} & C_{65} & C_{66} \end{bmatrix} \begin{Bmatrix} \varepsilon_{\bar{x}} \\ \varepsilon_{\bar{y}} \\ \varepsilon_{\bar{z}} \\ \gamma_{\bar{x}\bar{y}} \\ \gamma_{\bar{y}\bar{z}} \\ \gamma_{\bar{z}\bar{x}} \end{Bmatrix} \tag{9.4}$$

or

$$\{\bar{\sigma}\} = [E]\{\bar{\varepsilon}\} \tag{9.5}$$

where the 36 constants C_{ij} are called elastic constants and must be experimentally derived for each material.

Fortunately, usually the elastic properties of engineering materials are symmetric about all coordinate planes. Therefore, the number of constants C_{ij} is reduced to nine and the material is called orthotropic. In this case, the generalized Hooke's law is expressed as follows:

Essentials of the Finite Element Method. http://dx.doi.org/10.1016/B978-0-12-802386-0.00009-8

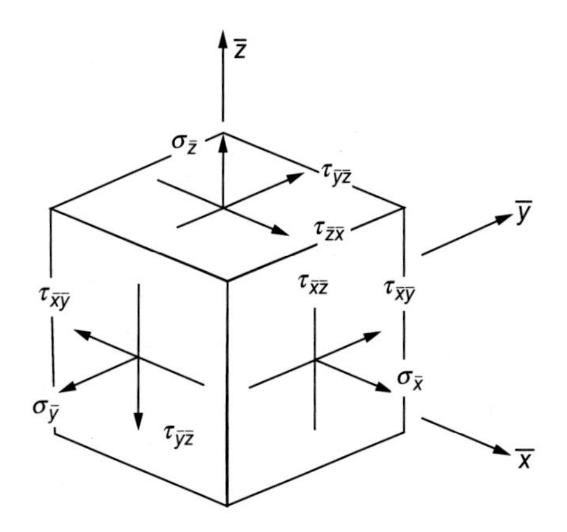

FIGURE 9.1

Stresses acting on the system $\bar{x}-\bar{y}-\bar{z}$.

$$\begin{Bmatrix} \sigma_{\bar{x}} \\ \sigma_{\bar{y}} \\ \sigma_{\bar{z}} \\ \tau_{\bar{x}\bar{y}} \\ \tau_{\bar{y}\bar{z}} \\ \tau_{\bar{z}\bar{x}} \end{Bmatrix} = \begin{bmatrix} C_{11} & C_{12} & C_{13} & 0 & 0 & 0 \\ C_{21} & C_{22} & C_{23} & 0 & 0 & 0 \\ C_{31} & C_{32} & C_{33} & 0 & 0 & 0 \\ 0 & 0 & 0 & C_{44} & 0 & 0 \\ 0 & 0 & 0 & 0 & C_{55} & 0 \\ 0 & 0 & 0 & 0 & 0 & C_{66} \end{bmatrix} \begin{Bmatrix} \varepsilon_{\bar{x}} \\ \varepsilon_{\bar{y}} \\ \varepsilon_{\bar{z}} \\ \gamma_{\bar{x}\bar{y}} \\ \gamma_{\bar{y}\bar{z}} \\ \gamma_{\bar{z}\bar{x}} \end{Bmatrix} \tag{9.6}$$

The above equation can be written in terms of strains:

$$\begin{Bmatrix} \varepsilon_{\bar{x}} \\ \varepsilon_{\bar{y}} \\ \varepsilon_{\bar{z}} \\ \gamma_{\bar{x}\bar{y}} \\ \gamma_{\bar{y}\bar{z}} \\ \gamma_{\bar{z}\bar{x}} \end{Bmatrix} = \begin{bmatrix} S_{11} & S_{12} & S_{13} & 0 & 0 & 0 \\ S_{21} & S_{22} & S_{23} & 0 & 0 & 0 \\ S_{31} & S_{32} & S_{33} & 0 & 0 & 0 \\ 0 & 0 & 0 & S_{44} & 0 & 0 \\ 0 & 0 & 0 & 0 & S_{55} & 0 \\ 0 & 0 & 0 & 0 & 0 & S_{66} \end{bmatrix} \begin{Bmatrix} \sigma_{\bar{x}} \\ \sigma_{\bar{y}} \\ \sigma_{\bar{z}} \\ \tau_{\bar{x}\bar{y}} \\ \tau_{\bar{y}\bar{z}} \\ \tau_{\bar{z}\bar{x}} \end{Bmatrix} \tag{9.7}$$

where

$$S_{11} = \frac{1}{E_{\bar{x}}}, \quad S_{12} = -\frac{\nu_{\bar{y}\bar{x}}}{E_{\bar{y}}}, \quad S_{13} = -\frac{\nu_{\bar{z}\bar{x}}}{E_{\bar{z}}} \tag{9.8}$$

$$S_{21} = -\frac{\nu_{\bar{x}\bar{y}}}{E_{\bar{x}}}, \quad S_{22} = \frac{1}{E_{\bar{y}}}, \quad S_{23} = -\frac{\nu_{\bar{z}\bar{y}}}{E_{\bar{z}}} \tag{9.9}$$

$$S_{31} = -\frac{\nu_{\bar{x}\bar{z}}}{E_{\bar{x}}}, \quad S_{32} = -\frac{\nu_{\bar{y}\bar{z}}}{E_{\bar{y}}}, \quad S_{33} = \frac{1}{E_{\bar{z}}} \tag{9.10}$$

$$S_{44} = \frac{1}{G_{\bar{y}\bar{z}}}, \quad S_{55} = \frac{1}{G_{\bar{x}\bar{z}}}, \quad S_{66} = \frac{1}{G_{\bar{x}\bar{y}}} \tag{9.11}$$

In the above equations, $E_{\bar{x}}, E_{\bar{y}}, E_{\bar{z}}$ are the modulus of elasticity in the directions $\bar{x}, \bar{y}, \bar{z}$, respectively, $G_{\bar{x}\bar{y}}, G_{\bar{y}\bar{z}}, G_{\bar{z}\bar{x}}$ are the shear modulus in the directions $\bar{x}-\bar{y}$, $\bar{y}-\bar{z}$, $\bar{z}-\bar{x}$, and ν_{ij} are the Poisson's ratios given by

$$\nu_{ij} = -\frac{\varepsilon_j}{\varepsilon_i} \tag{9.12}$$

Combining Equations (9.6) and (9.7) it can be concluded that C_{ij} can be written in terms of the above engineering constants S_{ij}. For shorthand notation, Equations (9.6) and (9.7) can be abbreviated by

$$\{\bar{\sigma}\} = [C]\{\bar{\varepsilon}\} \tag{9.13}$$

$$\{\bar{\varepsilon}\} = [S]\{\bar{\sigma}\} \tag{9.14}$$

From Equations (9.13) and (9.14), it is obvious that

$$[C] = [S]^{-1} \tag{9.15}$$

According to Maxwell-Betti Reciprocal Theorem, the following relationships among material properties can be obtained

$$\frac{\nu_{\bar{x}\bar{y}}}{E_{\bar{x}}} = \frac{\nu_{\bar{y}\bar{x}}}{E_{\bar{y}}} \tag{9.16}$$

$$\frac{\nu_{\bar{x}\bar{z}}}{E_{\bar{x}}} = \frac{\nu_{\bar{z}\bar{x}}}{E_{\bar{z}}} \tag{9.17}$$

$$\frac{\nu_{\bar{y}\bar{z}}}{E_{\bar{y}}} = \frac{\nu_{\bar{z}\bar{y}}}{E_{\bar{z}}} \tag{9.18}$$

yielding

$$S_{21} = S_{12} \tag{9.19}$$

$$S_{31} = S_{13} \tag{9.20}$$

$$S_{32} = S_{23} \tag{9.21}$$

Therefore, using Equations (9.15) and (9.19)–(9.21), the members C_{ij} of the matrix C are given by:

$$C_{11} = \frac{S_{22}S_{33} - S_{23}^2}{S} \tag{9.22}$$

$$C_{12} = \frac{S_{13}S_{23} - S_{12}S_{33}}{S} \tag{9.23}$$

$$C_{13} = \frac{S_{12}S_{23} - S_{13}S_{22}}{S} \tag{9.24}$$

$$C_{21} = C_{12} \tag{9.25}$$

$$C_{22} = \frac{S_{33}S_{11} - S_{13}^2}{S} \tag{9.26}$$

$$C_{23} = \frac{S_{12}S_{13} - S_{23}S_{11}}{S} \tag{9.27}$$

$$C_{31} = C_{13} \tag{9.28}$$

$$C_{32} = C_{23} \tag{9.29}$$

$$C_{33} = \frac{S_{11}S_{22} - S_{12}^2}{S} \tag{9.30}$$

$$C_{44} = \frac{1}{S_{44}} \tag{9.31}$$

$$C_{55} = \frac{1}{S_{55}} \tag{9.32}$$

$$C_{66} = \frac{1}{S_{66}} \tag{9.33}$$

where

$$S = S_{11}S_{22}S_{33} - S_{11}S_{23}^2 - S_{22}S_{13}^2 - S_{33}S_{12}^2 + 2S_{12}S_{23}S_{13} \tag{9.34}$$

The matrices $[C]$ and $[S]$ correlating the stress and strains are called stiffness matrix and compliance matrix, respectively.

9.1.1 EFFECTS OF FREE THERMAL STRAINS

Temperature changes in a material element can cause significant stresses. In order to keep the consistency with the definition of stress and strain the term "mechanical strains" will be used in the stress-strain relations. According to this concept, the mechanical strains are the total strains (i.e., the changes in length per unit length of a material element) minus the free thermal strains. Therefore, the stress-strain relations can be written in the form:

$$\begin{Bmatrix} \varepsilon_{\bar{x}} - \alpha_{\bar{x}}\Delta T \\ \varepsilon_{\bar{y}} - \alpha_{\bar{y}}\Delta T \\ \varepsilon_{\bar{z}} - \alpha_{\bar{z}}\Delta T \\ \gamma_{\bar{y}\bar{z}} \\ \gamma_{\bar{x}\bar{z}} \\ \gamma_{\bar{x}\bar{y}} \end{Bmatrix} = \begin{bmatrix} S_{11} & S_{12} & S_{13} & 0 & 0 & 0 \\ S_{12} & S_{22} & S_{23} & 0 & 0 & 0 \\ S_{13} & S_{23} & S_{33} & 0 & 0 & 0 \\ 0 & 0 & 0 & S_{44} & 0 & 0 \\ 0 & 0 & 0 & 0 & S_{55} & 0 \\ 0 & 0 & 0 & 0 & 0 & S_{66} \end{bmatrix} \begin{Bmatrix} \sigma_{\bar{x}} \\ \sigma_{\bar{y}} \\ \sigma_{\bar{z}} \\ \tau_{\bar{y}\bar{z}} \\ \tau_{\bar{x}\bar{z}} \\ \tau_{\bar{x}\bar{y}} \end{Bmatrix} \tag{9.35}$$

or

$$\begin{Bmatrix} \sigma_{\bar{x}} \\ \sigma_{\bar{y}} \\ \sigma_{\bar{z}} \\ \tau_{\bar{y}\bar{z}} \\ \tau_{\bar{x}\bar{z}} \\ \tau_{\bar{x}\bar{y}} \end{Bmatrix} = \begin{bmatrix} C_{11} & C_{12} & C_{13} & 0 & 0 & 0 \\ C_{12} & C_{22} & C_{23} & 0 & 0 & 0 \\ C_{13} & C_{23} & C_{33} & 0 & 0 & 0 \\ 0 & 0 & 0 & C_{44} & 0 & 0 \\ 0 & 0 & 0 & 0 & C_{55} & 0 \\ 0 & 0 & 0 & 0 & 0 & C_{66} \end{bmatrix} \begin{Bmatrix} \varepsilon_{\bar{x}} - \alpha_{\bar{x}}\Delta T \\ \varepsilon_{\bar{y}} - \alpha_{\bar{y}}\Delta T \\ \varepsilon_{\bar{z}} - \alpha_{\bar{z}}\Delta T \\ \gamma_{\bar{y}\bar{z}} \\ \gamma_{\bar{x}\bar{z}} \\ \gamma_{\bar{x}\bar{y}} \end{Bmatrix} \tag{9.36}$$

The mechanical stresses used in above equations are given by:

$$\begin{Bmatrix} \varepsilon_{\bar{x}}^{\text{mech}} \\ \varepsilon_{\bar{y}}^{\text{mech}} \\ \varepsilon_{\bar{z}}^{\text{mech}} \\ \gamma_{\bar{y}\bar{z}}^{\text{mech}} \\ \gamma_{\bar{x}\bar{z}}^{\text{mech}} \\ \gamma_{\bar{x}\bar{y}}^{\text{mech}} \end{Bmatrix} = \begin{Bmatrix} \varepsilon_{\bar{x}} - \alpha_{\bar{x}}\Delta T \\ \varepsilon_{\bar{y}} - \alpha_{\bar{y}}\Delta T \\ \varepsilon_{\bar{z}} - \alpha_{\bar{z}}\Delta T \\ \gamma_{\bar{y}\bar{z}} \\ \gamma_{\bar{x}\bar{z}} \\ \gamma_{\bar{x}\bar{y}} \end{Bmatrix} \tag{9.37}$$

where $\alpha_{\bar{x}}, \alpha_{\bar{y}}, \alpha_{\bar{z}}$ are the thermal expansion coefficients in the directions $\bar{x}, \bar{y}, \bar{z}$, respectively, and ΔT is the temperature change. The above definitions take into account the cases of (a) stress with no thermal effects, (b) thermal effects but no stresses, and (c) thermal effects with stresses. For each of these three cases the corresponding stress-strain relationships can be written in the following form:
Case (a): Stress with no thermal effects

$$\begin{Bmatrix} \varepsilon_{\bar{x}} \\ \varepsilon_{\bar{y}} \\ \varepsilon_{\bar{z}} \\ \gamma_{\bar{y}\bar{z}} \\ \gamma_{\bar{x}\bar{z}} \\ \gamma_{\bar{x}\bar{y}} \end{Bmatrix} = \begin{bmatrix} S_{11} & S_{12} & S_{13} & 0 & 0 & 0 \\ S_{12} & S_{22} & S_{23} & 0 & 0 & 0 \\ S_{13} & S_{23} & S_{33} & 0 & 0 & 0 \\ 0 & 0 & 0 & S_{44} & 0 & 0 \\ 0 & 0 & 0 & 0 & S_{55} & 0 \\ 0 & 0 & 0 & 0 & 0 & S_{66} \end{bmatrix} \begin{Bmatrix} \sigma_{\bar{x}} \\ \sigma_{\bar{y}} \\ \sigma_{\bar{z}} \\ \tau_{\bar{y}\bar{z}} \\ \tau_{\bar{x}\bar{z}} \\ \tau_{\bar{x}\bar{y}} \end{Bmatrix} \tag{9.38}$$

Case (b): Thermal effects of a material element with no constrains on its bounding surfaces

$$\begin{Bmatrix} \varepsilon_{\bar{x}} \\ \varepsilon_{\bar{y}} \\ \varepsilon_{\bar{z}} \\ \gamma_{\bar{y}\bar{z}} \\ \gamma_{\bar{x}\bar{z}} \\ \gamma_{\bar{x}\bar{y}} \end{Bmatrix} = \begin{Bmatrix} 0 \\ 0 \\ 0 \\ 0 \\ 0 \\ 0 \end{Bmatrix} \tag{9.39}$$

Case (c): Stress caused by a temperature change of a fully restrained (against deformation) material element.

$$\begin{Bmatrix} \sigma_{\bar{x}} \\ \sigma_{\bar{y}} \\ \sigma_{\bar{z}} \\ \tau_{\bar{y}\bar{z}} \\ \tau_{\bar{x}\bar{z}} \\ \tau_{\bar{x}\bar{y}} \end{Bmatrix} = \begin{bmatrix} C_{11} & C_{12} & C_{13} & 0 & 0 & 0 \\ C_{12} & C_{22} & C_{23} & 0 & 0 & 0 \\ C_{13} & C_{23} & C_{33} & 0 & 0 & 0 \\ 0 & 0 & 0 & C_{44} & 0 & 0 \\ 0 & 0 & 0 & 0 & C_{55} & 0 \\ 0 & 0 & 0 & 0 & 0 & C_{66} \end{bmatrix} \begin{Bmatrix} -\alpha_{\bar{x}}\Delta T \\ -\alpha_{\bar{y}}\Delta T \\ -\alpha_{\bar{z}}\Delta T \\ 0 \\ 0 \\ 0 \end{Bmatrix} \tag{9.40}$$

9.1.2 EFFECTS OF FREE MOISTURE STRAINS

When certain types of materials (e.g., polymers) are exposed to a liquid, a certain amount of that liquid is absorbed yielding an increase of their weight and expansion. Although the weight gain is negligible, the expansion can be important. The created free moisture strains can be considered to be proportional to the amount of the absorbed liquid. Therefore, in analogy to the coefficient of thermal expansion, the term of coefficient of moisture expansion can be used to describe the free moisture expansion. Using the notations $\beta_{\bar{x}},\beta_{\bar{y}},\beta_{\bar{z}}$ for the moisture expansion coefficients in the directions $\bar{x},\bar{y},\bar{z}$ and ΔM for the change of absorbed moisture, the generalized Hooke's law can be written in the form:

$$\begin{Bmatrix} \varepsilon_{\bar{x}}-\beta_{\bar{x}}\Delta M \\ \varepsilon_{\bar{y}}-\beta_{\bar{y}}\Delta M \\ \varepsilon_{\bar{z}}-\beta_{\bar{z}}\Delta M \\ \gamma_{\bar{y}\bar{z}} \\ \gamma_{\bar{x}\bar{z}} \\ \gamma_{\bar{x}\bar{y}} \end{Bmatrix} = \begin{bmatrix} S_{11} & S_{12} & S_{13} & 0 & 0 & 0 \\ S_{12} & S_{22} & S_{23} & 0 & 0 & 0 \\ S_{13} & S_{23} & S_{33} & 0 & 0 & 0 \\ 0 & 0 & 0 & S_{44} & 0 & 0 \\ 0 & 0 & 0 & 0 & S_{55} & 0 \\ 0 & 0 & 0 & 0 & 0 & S_{66} \end{bmatrix} \begin{Bmatrix} \sigma_{\bar{x}} \\ \sigma_{\bar{y}} \\ \sigma_{\bar{z}} \\ \tau_{\bar{y}\bar{z}} \\ \tau_{\bar{x}\bar{z}} \\ \tau_{\bar{x}\bar{y}} \end{Bmatrix} \tag{9.41}$$

or

$$\begin{Bmatrix} \sigma_{\bar{x}} \\ \sigma_{\bar{y}} \\ \sigma_{\bar{z}} \\ \tau_{\bar{y}\bar{z}} \\ \tau_{\bar{x}\bar{z}} \\ \tau_{\bar{x}\bar{y}} \end{Bmatrix} = \begin{bmatrix} C_{11} & C_{12} & C_{13} & 0 & 0 & 0 \\ C_{12} & C_{22} & C_{23} & 0 & 0 & 0 \\ C_{13} & C_{23} & C_{33} & 0 & 0 & 0 \\ 0 & 0 & 0 & C_{44} & 0 & 0 \\ 0 & 0 & 0 & 0 & C_{55} & 0 \\ 0 & 0 & 0 & 0 & 0 & C_{66} \end{bmatrix} \begin{Bmatrix} \varepsilon_{\bar{x}}-\beta_{\bar{x}}\Delta M \\ \varepsilon_{\bar{y}}-\beta_{\bar{y}}\Delta M \\ \varepsilon_{\bar{z}}-\beta_{\bar{z}}\Delta M \\ \gamma_{\bar{y}\bar{z}} \\ \gamma_{\bar{x}\bar{z}} \\ \gamma_{\bar{x}\bar{y}} \end{Bmatrix} \tag{9.42}$$

The mechanical stresses used in the above equations are given by:

$$\begin{Bmatrix} \varepsilon_{\bar{x}}^{\text{mech}} \\ \varepsilon_{\bar{y}}^{\text{mech}} \\ \varepsilon_{\bar{z}}^{\text{mech}} \\ \gamma_{\bar{y}\bar{z}}^{\text{mech}} \\ \gamma_{\bar{x}\bar{z}}^{\text{mech}} \\ \gamma_{\bar{x}\bar{y}}^{\text{mech}} \end{Bmatrix} = \begin{Bmatrix} \varepsilon_{\bar{x}}-\beta_{\bar{x}}\Delta M \\ \varepsilon_{\bar{y}}-\beta_{\bar{y}}\Delta M \\ \varepsilon_{\bar{z}}-\beta_{\bar{z}}\Delta M \\ \gamma_{\bar{y}\bar{z}} \\ \gamma_{\bar{x}\bar{z}} \\ \gamma_{\bar{x}\bar{y}} \end{Bmatrix} \tag{9.43}$$

In case of interaction of moisture strains with thermal strains, superposition of Equations (9.35) and (9.41) as well as Equations (9.36) and (9.42) yields:

$$\begin{Bmatrix} \varepsilon_{\bar{x}}-\alpha_{\bar{x}}\Delta T-\beta_{\bar{x}}\Delta M \\ \varepsilon_{\bar{y}}-\alpha_{\bar{y}}\Delta T-\beta_{\bar{y}}\Delta M \\ \varepsilon_{\bar{z}}-\alpha_{\bar{z}}\Delta T-\beta_{\bar{z}}\Delta M \\ \gamma_{\bar{y}\bar{z}} \\ \gamma_{\bar{x}\bar{z}} \\ \gamma_{\bar{x}\bar{y}} \end{Bmatrix} = \begin{bmatrix} S_{11} & S_{12} & S_{13} & 0 & 0 & 0 \\ S_{12} & S_{22} & S_{23} & 0 & 0 & 0 \\ S_{13} & S_{23} & S_{33} & 0 & 0 & 0 \\ 0 & 0 & 0 & S_{44} & 0 & 0 \\ 0 & 0 & 0 & 0 & S_{55} & 0 \\ 0 & 0 & 0 & 0 & 0 & S_{66} \end{bmatrix} \begin{Bmatrix} \sigma_{\bar{x}} \\ \sigma_{\bar{y}} \\ \sigma_{\bar{z}} \\ \tau_{\bar{y}\bar{z}} \\ \tau_{\bar{x}\bar{z}} \\ \tau_{\bar{x}\bar{y}} \end{Bmatrix} \tag{9.44}$$

or

$$\begin{Bmatrix} \sigma_{\bar{x}} \\ \sigma_{\bar{y}} \\ \sigma_{\bar{z}} \\ \tau_{\bar{y}\bar{z}} \\ \tau_{\bar{x}\bar{z}} \\ \tau_{\bar{x}\bar{y}} \end{Bmatrix} = \begin{bmatrix} C_{11} & C_{12} & C_{13} & 0 & 0 & 0 \\ C_{12} & C_{22} & C_{23} & 0 & 0 & 0 \\ C_{13} & C_{23} & C_{33} & 0 & 0 & 0 \\ 0 & 0 & 0 & C_{44} & 0 & 0 \\ 0 & 0 & 0 & 0 & C_{55} & 0 \\ 0 & 0 & 0 & 0 & 0 & C_{66} \end{bmatrix} \begin{Bmatrix} \varepsilon_{\bar{x}}-\alpha_{\bar{x}}\Delta T-\beta_{\bar{x}}\Delta M \\ \varepsilon_{\bar{y}}-\alpha_{\bar{y}}\Delta T-\beta_{\bar{y}}\Delta M \\ \varepsilon_{\bar{z}}-\alpha_{\bar{z}}\Delta T-\beta_{\bar{z}}\Delta M \\ \gamma_{\bar{y}\bar{z}} \\ \gamma_{\bar{x}\bar{z}} \\ \gamma_{\bar{x}\bar{y}} \end{Bmatrix} \tag{9.45}$$

9.1.3 PLANE STRESS CONSTITUTIVE RELATIONS

Main characteristic of plates, beams, thin-walled sheets, etc., is that the value of at least one of their characteristic geometric dimensions is much smaller than the two other dimensions. Therefore, three of the six components of stress are much smaller than the other three. For plane structural components (e.g., plates, thin-walled sheets), the in-plane stresses are much larger than the stresses perpendicular to their plane. Usually, for design purposes, the stress components perpendicular to the plane of these structures can be set to zero yielding simplification of many problems. According to the above assumption, in a plane element with the directions shown in Figure 9.1 the stresses $\sigma_{\bar{z}}, \tau_{\bar{x}\bar{z}}, \tau_{\bar{y}\bar{z}}$ can be set to zero yielding the following simplification in the generalized Hooke's law:

$$\begin{Bmatrix} \varepsilon_{\bar{x}} \\ \varepsilon_{\bar{y}} \\ \varepsilon_{\bar{z}} \\ \gamma_{\bar{y}\bar{z}} \\ \gamma_{\bar{x}\bar{z}} \\ \gamma_{\bar{x}\bar{y}} \end{Bmatrix} = \begin{bmatrix} S_{11} & S_{12} & S_{13} & 0 & 0 & 0 \\ S_{12} & S_{22} & S_{23} & 0 & 0 & 0 \\ S_{13} & S_{23} & S_{33} & 0 & 0 & 0 \\ 0 & 0 & 0 & S_{44} & 0 & 0 \\ 0 & 0 & 0 & 0 & S_{55} & 0 \\ 0 & 0 & 0 & 0 & 0 & S_{66} \end{bmatrix} \begin{Bmatrix} \sigma_{\bar{x}} \\ \sigma_{\bar{y}} \\ 0 \\ 0 \\ 0 \\ \tau_{\bar{x}\bar{y}} \end{Bmatrix} \tag{9.46}$$

It is obvious from the above equation that

$$\gamma_{\bar{x}\bar{z}} = 0 \tag{9.47}$$

$$\gamma_{\bar{y}\bar{z}} = 0 \tag{9.48}$$

Equations (9.47) and (9.48) show that the planes $\bar{y}-\bar{z}$ and $\bar{x}-\bar{z}$ can be considered as free of shear strains.

Although the normal strain $\varepsilon_{\bar{z}}$ is not zero, the above described plane stresses assumption yields a simplified Hooke's law involving only $\sigma_{\bar{x}}$, $\sigma_{\bar{y}}$, $\tau_{\bar{x}\bar{y}}$ and $\varepsilon_{\bar{x}}, \varepsilon_{\bar{y}}, \gamma_{\bar{x}\bar{y}}$.

$$\begin{Bmatrix} \varepsilon_{\bar{x}} \\ \varepsilon_{\bar{y}} \\ \gamma_{\bar{x}\bar{y}} \end{Bmatrix} = \begin{bmatrix} S_{11} & S_{12} & 0 \\ S_{21} & S_{22} & 0 \\ 0 & 0 & S_{66} \end{bmatrix} \begin{Bmatrix} \sigma_{\bar{x}} \\ \sigma_{\bar{y}} \\ \tau_{\bar{x}\bar{y}} \end{Bmatrix} \tag{9.49}$$

where

$$S_{11} = \frac{1}{E_{\bar{x}}} \tag{9.50}$$

$$S_{12} = -\frac{\nu_{xy}}{E_{\bar{x}}} = -\frac{\nu_{\bar{y}\bar{x}}}{E_{\bar{y}}} \tag{9.51}$$

$$S_{22} = \frac{1}{E_{\bar{y}}} \tag{9.52}$$

$$S_{66} = \frac{1}{G_{\bar{x}\bar{y}}} \tag{9.53}$$

Inversion of Equation (9.49) yields the corresponding relation between stresses and strains

$$\begin{Bmatrix} \sigma_{\bar{x}} \\ \sigma_{\bar{y}} \\ \tau_{\bar{x}\bar{y}} \end{Bmatrix} = \begin{bmatrix} Q_{11} & Q_{12} & 0 \\ Q_{21} & Q_{22} & 0 \\ 0 & 0 & Q_{66} \end{bmatrix} \begin{Bmatrix} \varepsilon_{\bar{x}} \\ \varepsilon_{\bar{y}} \\ \gamma_{\bar{x}\bar{y}} \end{Bmatrix} \tag{9.54}$$

where the parameters Q_{ij} are given by the following equations

$$Q_{11} = C_{11} - \frac{C_{13}^2}{C_{33}} \tag{9.55}$$

$$Q_{12} = C_{12} - \frac{C_{13}C_{23}}{C_{33}} \tag{9.56}$$

$$Q_{22} = C_{22} - \frac{C_{23}^2}{C_{33}} \tag{9.57}$$

$$Q_{66} = C_{66} \tag{9.58}$$

The above equations, using Equations (9.22)–(9.30), (9.33), (9.34), and (9.8)–(9.11) yield

$$Q_{11} = \frac{E_{\bar{x}}}{1 - \nu_{\bar{x}\bar{y}} \cdot \nu_{\bar{y}\bar{x}}} \tag{9.59}$$

$$Q_{12} = \frac{\nu_{\bar{x}\bar{y}} E_{\bar{y}}}{1 - \nu_{\bar{x}\bar{y}} \cdot \nu_{\bar{y}\bar{x}}} = \frac{\nu_{\bar{y}\bar{x}} E_{\bar{x}}}{1 - \nu_{\bar{x}\bar{y}} \cdot \nu_{\bar{y}\bar{x}}} \tag{9.60}$$

$$Q_{22} = \frac{E_{\bar{y}}}{1 - \nu_{\bar{x}\bar{y}} \cdot \nu_{\bar{y}\bar{x}}} \tag{9.61}$$

$$Q_{66} = G_{\bar{x}\bar{y}} \tag{9.62}$$

Taking into account the plane stresses assumption, Equations (9.44) and (9.45) describing the generalized Hooke's law for the cases of the effects of free thermal and moisture strains can now be written:

$$\begin{Bmatrix} \varepsilon_{\bar{x}} - \alpha_{\bar{x}}\Delta T - \beta_{\bar{x}}\Delta M \\ \varepsilon_{\bar{y}} - \alpha_{\bar{y}}\Delta T - \beta_{\bar{y}}\Delta M \\ \gamma_{\bar{x}\bar{y}} \end{Bmatrix} = \begin{bmatrix} S_{11} & S_{12} & 0 \\ S_{21} & S_{22} & 0 \\ 0 & 0 & S_{66} \end{bmatrix} \begin{Bmatrix} \sigma_{\bar{x}} \\ \sigma_{\bar{y}} \\ \tau_{\bar{x}\bar{y}} \end{Bmatrix} \tag{9.63}$$

and

$$\begin{Bmatrix} \sigma_{\bar{x}} \\ \sigma_{\bar{y}} \\ \tau_{\bar{x}\bar{y}} \end{Bmatrix} = \begin{bmatrix} Q_{11} & Q_{12} & 0 \\ Q_{12} & Q_{22} & 0 \\ 0 & 0 & Q_{66} \end{bmatrix} \begin{Bmatrix} \varepsilon_{\bar{x}} - \alpha_{\bar{x}}\Delta T - \beta_{\bar{x}}\Delta M \\ \varepsilon_{\bar{y}} - \alpha_{\bar{y}}\Delta T - \beta_{\bar{y}}\Delta M \\ \gamma_{\bar{x}\bar{y}} \end{Bmatrix} \tag{9.64}$$

where the mechanical strains are given by

$$\begin{Bmatrix} \varepsilon_{\bar{x}}^{\text{mech}} \\ \varepsilon_{\bar{y}}^{\text{mech}} \\ \gamma_{\bar{x}\bar{y}}^{mech} \end{Bmatrix} = \begin{Bmatrix} \varepsilon_{\bar{x}} - \alpha_{\bar{x}}\Delta T - \beta_{\bar{x}}\Delta M \\ \varepsilon_{\bar{y}} - \alpha_{\bar{y}}\Delta T - \beta_{\bar{y}}\Delta M \\ \gamma_{\bar{x}\bar{y}} \end{Bmatrix} \tag{9.65}$$

9.2 FROM "ISOTROPIC" TO "ORTHOTROPIC" PLANE ELEMENTS

In most finite element treatments the plane elements, that is, plates and sheets, are considered isotropic, that is, having the same material properties in all directions. However, today the advanced materials manufacturers yield products with different material properties in the L and T directions (Figure 9.2). Therefore, the assumption of the equal material properties in all directions is a simplification. Often, the

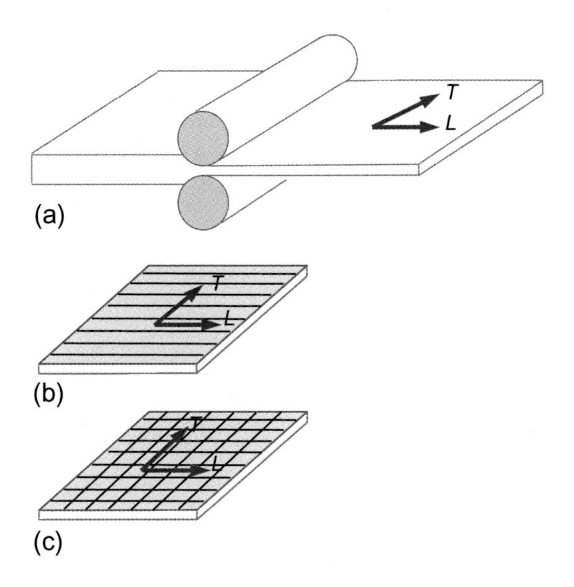

FIGURE 9.2

Orthotropic elasticity of plane structural elements: (a) orthotropic plate, (b) unidirectional fiber-reinforced lamina, (c) bidirectional fiber-reinforced lamina.

material properties are provided in the market for the longitudinal (L) and the transverse (T) direction (Figure 9.2). Therefore, for plane elements, the values E_L, E_T for the modulus of elasticity, G_{LT} for the shear modulus, and ν_{LT} for Poisson's ratio are available. Today, modern materials such as unidirectional or bidirectional fiber-reinforced laminae are orthotropic. For such materials, apart from the elastic constants E, G, ν, the values of thermal and moisture expansion coefficients are also different for the directions of the orthotropy L and T. In the following sections, the already used local coordinate system $\bar{x}-\bar{y}$ will coincide with the coordinate system L-T of the material's orthotropy. That means that we should use the following nomenclature:

Nomenclature for material properties		Nomenclature for stresses	Nomenclature for strains
$E_{\bar{x}} \rightarrow E_L$	$\alpha_{\bar{x}} \rightarrow \alpha_L$	$\sigma_{\bar{x}} \rightarrow \sigma_L$	$\varepsilon_{\bar{x}} \rightarrow \varepsilon_L$
$E_{\bar{y}} \rightarrow E_T$	$\alpha_{\bar{y}} \rightarrow \alpha_T$	$\sigma_{\bar{y}} \rightarrow \sigma_T$	$\varepsilon_{\bar{y}} \rightarrow \varepsilon_T$
$G_{\bar{x}\bar{y}} \rightarrow G_{LT}$	$\beta_{\bar{x}} \rightarrow \beta_L$	$\tau_{\bar{x}\bar{y}} \rightarrow \tau_{LT}$	$\gamma_{\bar{x}\bar{y}} \rightarrow \gamma_{LT}$
$\nu_{\bar{x}\bar{y}} \rightarrow \nu_{LT}$	$\beta_{\bar{y}} \rightarrow \beta_T$		

Generally, the directions of forces in the global coordinate system x-y are not coincided to the directions of materials orthotropy L-T. However, since the material properties are known only for the directions L and T, the stresses and strains in the local system L-T should be defined with respect to the stresses and strains to the global coordinate system x-y. Furthermore, apart from the stresses and strains, the orthotropic material properties should be transformed from the local L-T to the global x-y coordinate system.

9.2.1 COORDINATE TRANSFORMATION OF STRESS AND STRAIN COMPONENTS FOR ORTHOTROPIC TWO-DIMENSIONAL ELEMENTS

Let us now consider the orthotropic plane element shown in Figure 9.3. Its material properties are expressed with respect to the local coordinate system *L-T*. For the case of plane stress, the transformation of the stresses from the system x-y to the system *L-T* is given by the following matrix equation:

$$\begin{Bmatrix} \sigma_L \\ \sigma_T \\ \tau_{LT} \end{Bmatrix} = \begin{bmatrix} \cos^2\vartheta & \sin^2\vartheta & 2\sin\vartheta\cos\vartheta \\ \sin^2\vartheta & \cos^2\vartheta & -2\sin\vartheta\cos\vartheta \\ -\sin\vartheta\cos\vartheta & \sin\vartheta\cos\vartheta & \cos^2\vartheta-\sin^2\vartheta \end{bmatrix} \begin{Bmatrix} \sigma_x \\ \sigma_y \\ \tau_{xy} \end{Bmatrix} \tag{9.66}$$

where ϑ is the angle defining the orientation of the local system with respect to the global one.

Using the notations $m=\cos\vartheta$ and $n=\sin\vartheta$, the above equation is usually written in the following form:

$$\begin{Bmatrix} \sigma_L \\ \sigma_T \\ \tau_{LT} \end{Bmatrix} = [T] \begin{Bmatrix} \sigma_x \\ \sigma_y \\ \tau_{xy} \end{Bmatrix} \tag{9.67}$$

where the transformation matrix [*T*] is given by:

$$[T] = \begin{bmatrix} m^2 & n^2 & 2mn \\ n^2 & m^2 & -2mn \\ -mn & mn & m^2-n^2 \end{bmatrix} \tag{9.68}$$

Using the inversed matrix $[T]^{-1}$

$$[T]^{-1} = \begin{bmatrix} m^2 & n^2 & -2mn \\ n^2 & m^2 & 2mn \\ mn & -mn & m^2-n^2 \end{bmatrix} \tag{9.69}$$

The stresses $\sigma_x, \sigma_y, \tau_{xy}$ can be expressed with respect to $\sigma_L, \sigma_T, \tau_{LT}$:

$$\begin{Bmatrix} \sigma_x \\ \sigma_y \\ \tau_{xy} \end{Bmatrix} = [T]^{-1} \begin{Bmatrix} \sigma_L \\ \sigma_T \\ \tau_{LT} \end{Bmatrix} \tag{9.70}$$

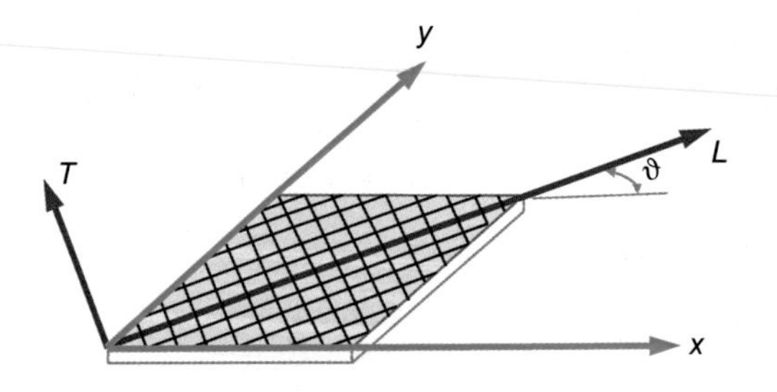

FIGURE 9.3

Demonstration of the local coordinate system *L-T* with respect to the global one *x-y* for an orthotropic two-dimensional element.

9.3 HOOKE'S LAW OF AN ORTHOTROPIC TWO-DIMENSIONAL ELEMENT, WITH RESPECT TO THE GLOBAL COORDINATE SYSTEM

Concerning the strains involved in plane stresses assumption, the following transformation equation can be written as follows:

$$\begin{Bmatrix} \varepsilon_L \\ \varepsilon_T \\ \frac{1}{2}\gamma_{LT} \end{Bmatrix} = [T]\begin{Bmatrix} \varepsilon_x \\ \varepsilon_y \\ \frac{1}{2}\gamma_{xy} \end{Bmatrix} \tag{9.71}$$

Using Equations (9.67) and (9.71), Equation (9.49) yields

$$\begin{Bmatrix} \varepsilon_x \\ \varepsilon_y \\ \gamma_{xy} \end{Bmatrix} = \begin{bmatrix} \bar{S}_{11} & \bar{S}_{12} & \bar{S}_{16} \\ \bar{S}_{12} & \bar{S}_{22} & \bar{S}_{26} \\ \bar{S}_{16} & \bar{S}_{26} & \bar{S}_{66} \end{bmatrix}\begin{Bmatrix} \sigma_x \\ \sigma_y \\ \tau_{xy} \end{Bmatrix} \tag{9.72}$$

where

$$\bar{S}_{11} = S_{11}m^4 + (2S_{12} + S_{66})n^2m^2 + S_{22}n^4 \tag{9.73}$$

$$\bar{S}_{12} = (S_{11} + S_{22} - S_{66})n^2m^2 + S_{12}\left(n^4 + m^4\right) \tag{9.74}$$

$$\bar{S}_{16} = (2S_{11} - 2S_{12} - S_{66})nm^3 - (2S_{22} - 2S_{12} - S_{66})n^3m \tag{9.75}$$

$$\bar{S}_{22} = S_{11}n^4 + (2S_{12} + S_{66})n^2m^2 + S_{22}m^4 \tag{9.76}$$

$$\bar{S}_{26} = (2S_{11} - 2S_{12} - S_{66})n^3m - (2S_{22} - 2S_{12} - S_{66})nm^3 \tag{9.77}$$

$$\bar{S}_{66} = 2(2S_{11} + 2S_{22} - 4S_{12} - S_{66})n^2m^2 + S_{66}\left(n^4 + m^4\right) \tag{9.78}$$

Taking into account Equations (9.67) and (9.71), Equation (9.54) yields Hooke's law for a two-dimensional (2D) orthotropic element with respect to the global coordinate system *x*-*y*:

$$\begin{Bmatrix} \sigma_x \\ \sigma_y \\ \tau_{xy} \end{Bmatrix} = \begin{bmatrix} \bar{Q}_{11} & \bar{Q}_{12} & \bar{Q}_{16} \\ \bar{Q}_{12} & \bar{Q}_{22} & \bar{Q}_{26} \\ \bar{Q}_{16} & \bar{Q}_{26} & \bar{Q}_{66} \end{bmatrix}\begin{Bmatrix} \varepsilon_x \\ \varepsilon_y \\ \gamma_{xy} \end{Bmatrix} \tag{9.79}$$

where

$$\bar{Q}_{11} = Q_{11}m^4 + 2(Q_{12} + 2S_{66})n^2m^2 + Q_{22}n^4 \tag{9.80}$$

$$\bar{Q}_{12} = (Q_{11} + Q_{22} - 4Q_{66})n^2m^2 + Q_{12}\left(n^4 + m^4\right) \tag{9.81}$$

$$\bar{Q}_{16} = (Q_{11} - Q_{12} - 2Q_{66})nm^3 + (Q_{12} - Q_{22} + 2Q_{66})n^3m \tag{9.82}$$

$$\bar{Q}_{22} = Q_{11}n^4 + 2(Q_{12} + 2Q_{66})n^2m^2 + Q_{22}m^4 \tag{9.83}$$

$$\bar{Q}_{26} = (Q_{11} - Q_{12} - 2Q_{66})n^3m + (Q_{12} - Q_{22} + 2Q_{66})nm^3 \tag{9.84}$$

$$\bar{Q}_{66} = (Q_{11} + Q_{22} - 2Q_{12} - 2S_{66})n^2m^2 + Q_{66}\left(n^4 + m^4\right) \tag{9.85}$$

9.4 TRANSFORMATION OF ENGINEERING PROPERTIES

9.4.1 ELASTIC PROPERTIES OF AN ORTHOTROPIC TWO-DIMENSIONAL ELEMENT IN THE GLOBAL COORDINATE SYSTEM

Apart from the engineering properties in the local coordinate system *L-T*, engineering properties can also be defined in the x-y global coordinate system.

Denoting the modulus of elasticity in the x direction by E_x, Hooke's law for this direction can be written in the form:

$$\sigma_x = E_x \varepsilon_x \tag{9.86}$$

For the above situation $\left(\sigma_x \neq 0, \sigma_y = 0, \tau_{xy} = 0\right)$, Equation (9.72) yields

$$\varepsilon_x = \bar{S}_{11} \sigma_x \tag{9.87}$$

Therefore, the combination of Equations (9.86) and (9.87) provides the following definition of the elasticity modulus in the x direction:

$$E_x = \frac{1}{\bar{S}_{11}} \tag{9.88}$$

Taking into account Equations (9.73) and (9.49)–(9.53), the above equation can be written as follows:

$$E_x = \frac{E_L}{m^4 + \left(\frac{E_L}{G_{LT}} - 2\nu_{LT}\right) n^2 m^2 + \frac{E_L}{E_T} n^4} \tag{9.89}$$

Poisson's ratio in x-y direction is defined by the ratio of the contraction strain ε_y in the y direction over the extensional strain ε_x in the x direction:

$$\nu_{xy} = -\frac{\varepsilon_y}{\varepsilon_x} \tag{9.90}$$

Since (for the case of $\sigma_x \neq 0, \sigma_y = 0, \tau_{xy} = 0$) Equation (9.72) yields

$$\varepsilon_y = \bar{S}_{12} \sigma_x \tag{9.91}$$

Equation (9.90) using Equations (9.87) and (9.91) can be written:

$$\nu_{xy} = -\frac{\bar{S}_{12}}{\bar{S}_{11}} \tag{9.92}$$

Using the definitions of $\bar{S}_{12}$, $\bar{S}_{11}$, the above equation provides the following formula:

$$\nu_{xy} = \frac{\nu_{LT}(n^4 + m^4) - \left(1 + \frac{E_L}{E_T} - \frac{E_L}{G_{LT}}\right) n^2 m^2}{m^4 + \left(\frac{E_L}{G_{LT}} - 2\nu_{LT}\right) n^2 m^2 + \frac{E_L}{E_T} n^4} \tag{9.93}$$

For evaluating the modulus of elasticity in the y direction, the stress situation $\sigma_x = 0, \sigma_y \neq 0, \tau_{xy} = 0$ will now be considered. In that case, the modulus E_y is given by the following formula:

$$E_y = \frac{\sigma_y}{\varepsilon_y} \tag{9.94}$$

Since

$$\varepsilon_y = \bar{S}_{22}\sigma_y \tag{9.95}$$

Equation (9.94) yields

$$E_y = \frac{1}{\bar{S}_{22}} \tag{9.96}$$

Using the definition of $\bar{S}_{22}$ given by Equation (9.76) and the definitions of S_{ij} given by Equations (9.50)–(9.53) and the nomenclature described in Section 9.2, Equation (9.96) can now be written

$$E_y = \frac{E_T}{m^4 + \left(\frac{E_L}{G_{LT}} - 2\nu_{LT}\right)n^2m^2 + \frac{E_T}{E_L}n^4} \tag{9.97}$$

Due to the stress in only the y direction, Poisson's ratio ν_{yx} is

$$\nu_{yx} = -\frac{\varepsilon_x}{\varepsilon_y} \tag{9.98}$$

For the situation $\sigma_x = 0, \sigma_y \neq 0, \tau_{xy} = 0$, Equation (9.72) yields

$$\varepsilon_x = \bar{S}_{12}\sigma_y \tag{9.99}$$

Therefore, using Equations (9.95) and (9.99), Equation (9.98) provides

$$\nu_{yx} = -\frac{\bar{S}_{12}}{\bar{S}_{22}} \tag{9.100}$$

Using the definitions of $\bar{S}_{12}$ and $\bar{S}_{22}$, the above equation can now be written

$$\nu_{yx} = \frac{\nu_{TL}(n^4 + m^4) - \left(1 + \frac{E_T}{E_L} - \frac{E_T}{G_{LT}}\right)n^2m^2}{m^4 + \left(\frac{E_T}{G_{LT}} - 2\nu_{TL}\right)n^2m^2 + \frac{E_T}{E_L}n^4} \tag{9.101}$$

For the derivation of the formula providing the shear modulus G_{xy}, the stress situation $\sigma_x = 0, \sigma_y = 0, \tau_{xy} \neq 0$ will be considered. In that case

$$\gamma_{xy} = \frac{\tau_{xy}}{G_{xy}} \tag{9.102}$$

However, for the above stress situation Equation (9.72) provides

$$\gamma_{xy} = \bar{S}_{66}\tau_{xy} \tag{9.103}$$

Therefore, from Equations (9.102) and (9.103) we obtain

$$G_{xy} = \frac{1}{\bar{S}_{66}} \tag{9.104}$$

Using the definition of $\bar{S}_{66}$ the above equation yields

$$G_{xy} = \frac{G_{LT}}{n^4 + m^4 + 2\left(2\frac{G_{LT}}{E_L}(1 + 2\nu_{LT}) + 2\frac{G_{LT}}{E_T} - 1\right)n^2m^2} \tag{9.105}$$

Schematic representation of the variation with fiber orientation ϑ of the elastic properties in the global coordinate system is shown in Figure 9.4a–d. These figures indicate that: (i) the modulus E_x becomes greatest when the two coordinate systems are coincided in (i.e., $\vartheta = 0$) and decreases rapidly with ϑ; (ii) the modulus E_y has its smallest value at $\vartheta = 0$ and increases rapidly as ϑ approaches the value $\pm 90°$; (iii) the shear modulus G_{xy} takes its maximum value at $\vartheta = \pm 45^o$ and its minimum occurs at $\vartheta = 0$; (iv) the maximum value of ν_{xy} occurs in the area $0 < \vartheta < 45°$ and the minimum at $\vartheta = 90$; (v) the Poisson's ratio ν_{yx} has a minimum at $\vartheta = 0$ while its maximum appears in the area $45° < \vartheta < 90°$.

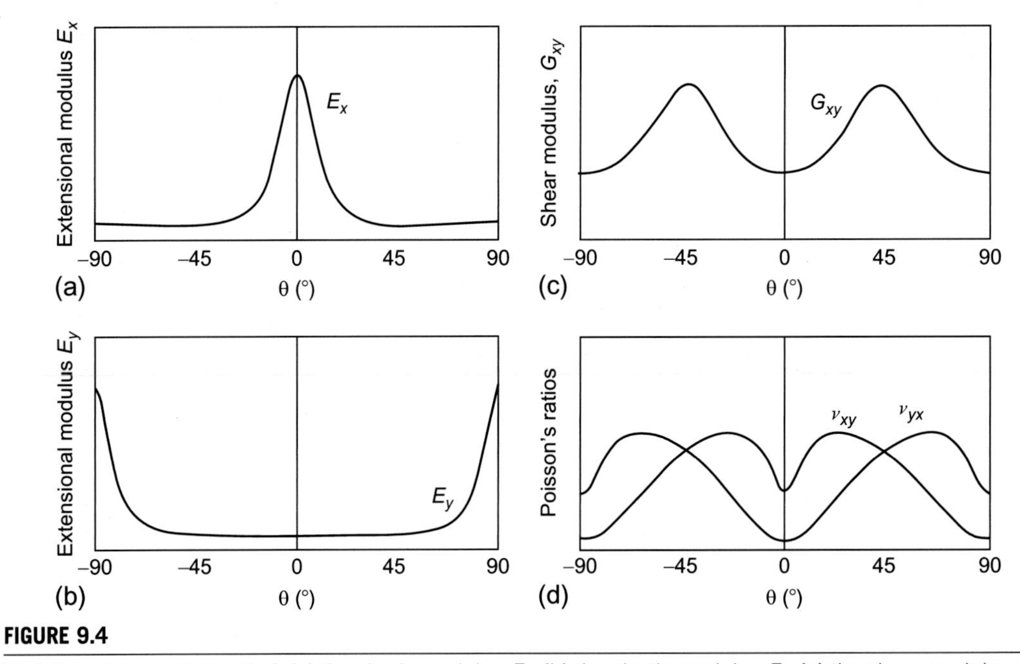

FIGURE 9.4

Variation with orientation ϑ of: (a) the elastic modulus E_x, (b) the elastic modulus E_y, (c) the shear modulus G_{xy}, (d) the Poisson's ratios ν_{xy} and ν_{yx}.

9.4.2 FREE THERMAL AND FREE MOISTURE STRAINS IN GLOBAL COORDINATE SYSTEM

(a) Transformation of thermal and moisture expansion coefficients

Following the concept described in previous paragraphs, the relations of stress and strain components in the x-y-z system including the effects of free thermal and moisture strains will be obtained. To this end, the thermal and moisture expansion coefficients in the global coordinate system will be correlated initially with the thermal and moisture expansion coefficients in local orthotropic system. Inversing Equation (9.71) the following formula can be obtained:

$$\begin{Bmatrix} \varepsilon_x \\ \varepsilon_y \\ \frac{1}{2}\gamma_{xy} \end{Bmatrix} = [T]^{-1} \begin{Bmatrix} \varepsilon_L \\ \varepsilon_T \\ \frac{1}{2}\gamma_{LT} \end{Bmatrix} \tag{9.106}$$

Since in the case of free thermal strains, it is

$$\varepsilon_L = \alpha_L \Delta T \tag{9.107}$$

$$\varepsilon_T = \alpha_T \Delta T \tag{9.108}$$

$$\gamma_{LT} = 0 \tag{9.109}$$

then, Equation (9.106) yields

$$\begin{Bmatrix} \varepsilon_x \\ \varepsilon_y \\ \frac{1}{2}\gamma_{xy} \end{Bmatrix} = [T]^{-1} \begin{Bmatrix} \alpha_L \\ \alpha_T \\ 0 \end{Bmatrix} \Delta T \tag{9.110}$$

Considering the following definitions of free thermal strains in the global coordinate system

$$\varepsilon_x = \alpha_x \Delta T \tag{9.111}$$

$$\varepsilon_y = \alpha_y \Delta T \tag{9.112}$$

$$\gamma_{xy} = \alpha_{xy} \Delta T \tag{9.113}$$

Equation (9.110) can now be written:

$$\begin{Bmatrix} \alpha_x \\ \alpha_y \\ \frac{1}{2}\alpha_{xy} \end{Bmatrix} = [T]^{-1} \begin{Bmatrix} \alpha_L \\ \alpha_T \\ 0 \end{Bmatrix} \tag{9.114}$$

Using Equation (9.69), the above equation provides the correlation of the thermal expansion coefficients in the global coordinate system versus the thermal expansion coefficients in principal coordinate system:

$$\alpha_x = \alpha_L \cos^2\vartheta + \alpha_T \sin^2\vartheta \tag{9.115}$$

$$\alpha_y = \alpha_L \sin^2\vartheta + \alpha_T \cos^2\vartheta \tag{9.116}$$

$$\alpha_{xy} = 2(\alpha_L - \alpha_T)\cos\vartheta \sin\vartheta \tag{9.117}$$

Schematic representation of the variation of the above coefficients with angle ϑ is shown in Figure 9.5.

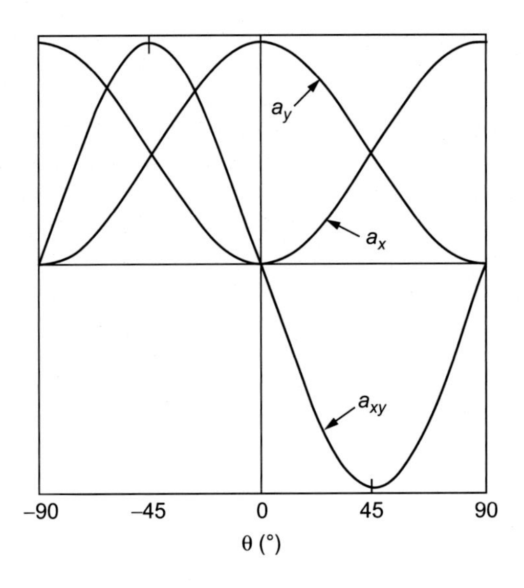

FIGURE 9.5

Schematic representation of the variation of thermal expansion coefficients in global coordinate system versus angle ϑ.

Following the same concept, the coefficients of moisture expansion in global coordinate system are given by the following equations:

$$\begin{Bmatrix} \beta_x \\ \beta_y \\ \frac{1}{2}\beta_{xy} \end{Bmatrix} = [T]^{-1} \begin{Bmatrix} \beta_L \\ \beta_T \\ 0 \end{Bmatrix} \tag{9.118}$$

yielding

$$\beta_x = \beta_L \cos^2\vartheta + \beta_T \sin^2\vartheta \tag{9.119}$$

$$\beta_y = \beta_L \sin^2\vartheta + \beta_T \cos^2\vartheta \tag{9.120}$$

$$\beta_{xy} = 2(\beta_L - \beta_T)\cos\vartheta \sin\vartheta \tag{9.121}$$

We recall that β_L, β_T are the moisture expansion coefficients in orthotropic coordinate system L-T.

(b) Transformation of free thermal and moisture strains

Using the concept of mechanical strains given by Equation (9.63) and the inversed forms of Equations (9.114) and (9.118), that is,

$$\begin{Bmatrix} \alpha_L \\ \alpha_T \\ 0 \end{Bmatrix} = [T] \begin{Bmatrix} \alpha_x \\ \alpha_y \\ \frac{1}{2}\alpha_{xy} \end{Bmatrix} \tag{9.122}$$

and

$$\begin{Bmatrix} \beta_L \\ \beta_T \\ 0 \end{Bmatrix} = [T] \begin{Bmatrix} \beta_x \\ \beta_y \\ \frac{1}{2}\beta_{xy} \end{Bmatrix} \tag{9.123}$$

the following matrix equation can be obtained:

$$\begin{Bmatrix} \varepsilon_L \\ \varepsilon_T \\ \gamma_{LT} \end{Bmatrix} = -[T] \begin{Bmatrix} \alpha_x \\ \alpha_y \\ \frac{1}{2}\alpha_{xy} \end{Bmatrix} \Delta T - [T] \begin{Bmatrix} \beta_x \\ \beta_y \\ \frac{1}{2}\beta_{xy} \end{Bmatrix} = \begin{bmatrix} S_{11} & S_{12} & 0 \\ S_{12} & S_{22} & 0 \\ 0 & 0 & \frac{1}{2}S_{66} \end{bmatrix} \begin{Bmatrix} \sigma_L \\ \sigma_T \\ \tau_{LT} \end{Bmatrix} \tag{9.124}$$

Taking into account Equations (9.66) and (9.71), the above equation yields

$$\begin{Bmatrix} \varepsilon_x - \alpha_x \Delta T - \beta_x \Delta M \\ \varepsilon_y - \alpha_y \Delta T - \beta_y \Delta M \\ \gamma_{xy} - \alpha_{xy} \Delta T - \beta_{xy} \Delta M \end{Bmatrix} = \begin{bmatrix} \bar{S}_{11} & \bar{S}_{12} & \bar{S}_{16} \\ \bar{S}_{12} & \bar{S}_{22} & \bar{S}_{26} \\ \bar{S}_{16} & \bar{S}_{26} & \bar{S}_{66} \end{bmatrix} \begin{Bmatrix} \sigma_x \\ \sigma_y \\ \tau_{xy} \end{Bmatrix} \tag{9.125}$$

The inverse of the above equation provides the stress components $\sigma_x, \sigma_y, \tau_{xy}$ versus the mechanical strains

$$\begin{Bmatrix} \sigma_x \\ \sigma_y \\ \tau_{xy} \end{Bmatrix} = \begin{bmatrix} \bar{Q}_{11} & \bar{Q}_{12} & \bar{Q}_{16} \\ \bar{Q}_{12} & \bar{Q}_{22} & \bar{Q}_{26} \\ \bar{Q}_{16} & \bar{Q}_{26} & \bar{Q}_{66} \end{bmatrix} \begin{Bmatrix} \varepsilon_x - \alpha_x \Delta T - \beta_x \Delta M \\ \varepsilon_y - \alpha_y \Delta T - \beta_y \Delta M \\ \gamma_{xy} - \alpha_{xy} \Delta T - \beta_{xy} \Delta M \end{Bmatrix} \tag{9.126}$$

The above equation represents the general form of Hooke's law containing thermal and moisture effects of an orthotropic 2D element, with respect to the global coordinate system.

9.5 ELASTICITY EQUATIONS FOR ISOTROPIC SOLIDS

9.5.1 GENERALIZED HOOKE'S LAW FOR ISOTROPIC SOLIDS

Materials having mechanical properties symmetric about every plane are called "isotropic". Taking into account the above definition, only two independent mechanical properties are needed to describe their elastic behavior; Young's modulus of elasticity E and Poisson's ratio ν. The generalized Hooke's law for an isotropic material is

$$\begin{Bmatrix} \sigma_x \\ \sigma_y \\ \sigma_z \\ \tau_{xy} \\ \tau_{yz} \\ \tau_{zx} \end{Bmatrix} = \frac{E}{(1+\nu)(1-2\nu)} \begin{bmatrix} (1-\nu) & \nu & \nu & 0 & 0 & 0 \\ \nu & (1-\nu) & \nu & 0 & 0 & 0 \\ \nu & \nu & (1-\nu) & 0 & 0 & 0 \\ 0 & 0 & 0 & (1-2\nu)/2 & 0 & 0 \\ 0 & 0 & 0 & 0 & (1-2\nu)/2 & 0 \\ 0 & 0 & 0 & 0 & 0 & (1-2\nu)/2 \end{bmatrix} \begin{Bmatrix} \varepsilon_x \\ \varepsilon_y \\ \varepsilon_z \\ \gamma_{xy} \\ \gamma_{yz} \\ \gamma_{zx} \end{Bmatrix} \tag{9.127}$$

The above formula can be written in terms of strains

$$\begin{Bmatrix} \varepsilon_x \\ \varepsilon_y \\ \varepsilon_z \\ \gamma_{xy} \\ \gamma_{yz} \\ \gamma_{zx} \end{Bmatrix} = \begin{bmatrix} 1/E & -\nu/E & -\nu/E & 0 & 0 & 0 \\ -\nu/E & 1/E & -\nu/E & 0 & 0 & 0 \\ -\nu/E & -\nu/E & 1/E & 0 & 0 & 0 \\ 0 & 0 & 0 & 1/G & 0 & 0 \\ 0 & 0 & 0 & 0 & 1/G & 0 \\ 0 & 0 & 0 & 0 & 0 & 1/G \end{bmatrix} \begin{Bmatrix} \sigma_x \\ \sigma_y \\ \sigma_z \\ \tau_{xy} \\ \tau_{yz} \\ \tau_{zx} \end{Bmatrix} \tag{9.128}$$

where G is the shear modulus given by

$$G = \frac{E}{2(1+\nu)} \tag{9.129}$$

Equation (9.128) can be written in abbreviated form as

$$\{\varepsilon\} = [E]^{-1}\{\sigma\} \tag{9.130}$$

where

$$[E]^{-1} = \begin{bmatrix} 1/E & -\nu/E & -\nu/E & 0 & 0 & 0 \\ -\nu/E & 1/E & -\nu/E & 0 & 0 & 0 \\ -\nu/E & -\nu/E & 1/E & 0 & 0 & 0 \\ 0 & 0 & 0 & 1/G & 0 & 0 \\ 0 & 0 & 0 & 0 & 1/G & 0 \\ 0 & 0 & 0 & 0 & 0 & 1/G \end{bmatrix} \tag{9.131}$$

In case of plane stress conditions, the stresses $\sigma_z, \tau_{xz}, \tau_{yz}$ should be set to zero. Therefore, Equation (9.128) yields

$$\gamma_{xz} = 0 \tag{9.132}$$

$$\gamma_{yz} = 0 \tag{9.133}$$

Using the above assumptions, Equation (9.128) can be written in the following form:

$$\begin{Bmatrix} \varepsilon_x \\ \varepsilon_y \\ \gamma_{xy} \end{Bmatrix} = [E]^{-1} \begin{Bmatrix} \sigma_x \\ \sigma_y \\ \tau_{xy} \end{Bmatrix} \tag{9.134}$$

and the matrix $[E]^{-1}$ is simplified to

$$[E]^{-1} = \begin{bmatrix} 1/E & -\nu/E & 0 \\ -\nu/E & 1/E & 0 \\ 0 & 0 & 1/G \end{bmatrix} \tag{9.135}$$

Inversion of Equation (9.134) yields the corresponding relation between stresses and strains

$$\begin{Bmatrix} \sigma_x \\ \sigma_y \\ \sigma_{xy} \end{Bmatrix} = [E] \begin{Bmatrix} \varepsilon_x \\ \varepsilon_y \\ \varepsilon_{xy} \end{Bmatrix} \tag{9.136}$$

where

$$[E] = \begin{bmatrix} E/(1-\nu^2) & \nu E/(1-\nu^2) & 0 \\ \nu E/(1-\nu^2) & E/(1-\nu^2) & 0 \\ 0 & 0 & G \end{bmatrix} \tag{9.137}$$

9.5.2 CORRELATION OF STRAINS WITH DISPLACEMENTS

It is known from the Theory of Elasticity that small strains of a three-dimensional solid can be correlated to the displacements through the following formulae:

$$\left.\begin{aligned} \varepsilon_x &= \frac{\partial u}{\partial x}, \quad \gamma_{xy} = \frac{\partial u}{\partial y} + \frac{\partial v}{\partial x} \\ \varepsilon_y &= \frac{\partial v}{\partial y}, \quad \gamma_{yz} = \frac{\partial v}{\partial z} + \frac{\partial w}{\partial y} \\ \varepsilon_z &= \frac{\partial w}{\partial z}, \quad \gamma_{xz} = \frac{\partial v}{\partial z} + \frac{\partial w}{\partial x} \end{aligned}\right\} \tag{9.138}$$

In the above equations, u, v, and w are the displacements of a material element in the x, y, and z directions of the coordinate axes, respectively.

9.5.3 CORRELATION OF STRESSES WITH DISPLACEMENTS

Combining the generalized Hooke's law given by Equation (9.127) with Equation (9.138), the following relations correlating the stresses to displacements can be obtained:

$$\sigma_x = \frac{E}{(1+\nu)(1-2\nu)}\left[(1-\nu)\frac{\partial u}{\partial x} + \nu\frac{\partial v}{\partial y} + \nu\frac{\partial w}{\partial z}\right] \tag{9.139}$$

$$\sigma_y = \frac{E}{(1+\nu)(1-2\nu)}\left[\nu\frac{\partial u}{\partial x} + (1-\nu)\frac{\partial v}{\partial y} + \nu\frac{\partial w}{\partial z}\right] \tag{9.140}$$

$$\sigma_z = \frac{E}{(1+\nu)(1-2\nu)}\left[\nu\frac{\partial u}{\partial x} + \nu\frac{\partial v}{\partial y} + (1-\nu)\frac{\partial w}{\partial z}\right] \tag{9.141}$$

$$\tau_{xy} = G\left(\frac{\partial u}{\partial y} + \frac{\partial v}{\partial x}\right) \tag{9.142}$$

$$\tau_{yz} = G\left(\frac{\partial v}{\partial z} + \frac{\partial w}{\partial y}\right) \tag{9.143}$$

$$\tau_{zx} = G\left(\frac{\partial u}{\partial z} + \frac{\partial w}{\partial x}\right) \tag{9.144}$$

9.5.4 DIFFERENTIAL EQUATIONS OF EQUILIBRIUM

It is well known from the Theory of Elasticity that a consequence of the equilibrium of a material element subjected to stresses $\sigma_x, \sigma_y, \sigma_z, \tau_{xy}, \tau_{yz}, \tau_{xz}$ and body forces per unit volume (e.g., inertia forces) p_x, p_y, p_z are the following differential equations of equilibrium:

$$\frac{\partial \sigma_x}{\partial x}+\frac{\partial \tau_{xy}}{\partial y}+\frac{\partial \tau_{xz}}{\partial z}+p_x=0 \tag{9.145}$$

$$\frac{\partial \tau_{xy}}{\partial x}+\frac{\partial \sigma_y}{\partial y}+\frac{\partial \tau_{yz}}{\partial z}+p_y=0 \tag{9.146}$$

$$\frac{\partial \tau_{zx}}{\partial x}+\frac{\partial \tau_{zy}}{\partial y}+\frac{\partial \sigma_z}{\partial z}+p_z=0 \tag{9.147}$$

9.5.5 DIFFERENTIAL EQUATIONS IN TERMS OF DISPLACEMENTS

Combining the equations of equilibrium (Equations 9.145–9.147) with Equations (9.139)–(9.144), the following governing differential equations in terms of displacements can be obtained:

$$G\left[\frac{\partial^2}{\partial x^2}+\frac{\partial^2}{\partial y^2}+\frac{\partial^2}{\partial z^2}\right]u+\frac{G}{1-2\nu}\frac{\partial}{\partial x}\left[\frac{\partial u}{\partial x}+\frac{\partial \upsilon}{\partial y}+\frac{\partial w}{\partial z}\right]+p_x=0 \tag{9.148}$$

$$G\left[\frac{\partial^2}{\partial x^2}+\frac{\partial^2}{\partial y^2}+\frac{\partial^2}{\partial z^2}\right]\upsilon+\frac{G}{1-2\nu}\frac{\partial}{\partial y}\left[\frac{\partial u}{\partial x}+\frac{\partial \upsilon}{\partial y}+\frac{\partial w}{\partial z}\right]+p_y=0 \tag{9.149}$$

$$G\left[\frac{\partial^2}{\partial x^2}+\frac{\partial^2}{\partial y^2}+\frac{\partial^2}{\partial z^2}\right]w+\frac{G}{1-2\nu}\frac{\partial}{\partial z}\left[\frac{\partial u}{\partial x}+\frac{\partial \upsilon}{\partial y}+\frac{\partial w}{\partial z}\right]+p_z=0 \tag{9.150}$$

9.5.6 THE TOTAL POTENTIAL ENERGY

We recall from Chapter 8 that the total potential energy for a solid is given by the expression

$$\Pi_{\mathrm{p}}=U+W \tag{9.151}$$

where U is the strain energy and W is the work done by the external forces.

Defining the stresses and strains by the following vectors:

$$\{\varepsilon\}=\left\{\varepsilon_x\, \varepsilon_y\, \varepsilon_z\, \gamma_{xy} \gamma_{yz}\, \gamma_{zx}\right\}^{\mathrm{T}} \tag{9.152}$$

$$\{\sigma\}=\left\{\sigma_x\, \sigma_y\, \sigma_z \tau_{xy} \tau_{yz}\, \tau_{zx}\right\}^{\mathrm{T}} \tag{9.153}$$

the expression providing the strain energy of the elastic solid is:

$$U=\frac{1}{2}\int\{\varepsilon\}^{\mathrm{T}}\{\sigma\}\mathrm{d}V \tag{9.154}$$

The work W of the external forces is the summation of three parts

$$W = W_s + W_c + W_b \tag{9.155}$$

where W_s is the work done by the surface forces, W_c is the work done by the concentrated forces, and W_b is the work done by the body forces.

If q_x, q_y, and q_z are the components of the surface distributed forces acting on a surface S, then, the work of the surface forces is given by the following relation:

$$W_s = \int_S \left(q_x u + q_y v + q_z w\right) \mathrm{d}S \tag{9.156}$$

The work of the concentrated forces is expressed as the summation of the components of the forces multiplied by the displacement components at the points of the force actions:

$$W_c = \sum \left(F_{xi} u_i + F_{yi} v_i + F_{zi} w_i\right) \tag{9.157}$$

The work done by the body forces is calculated by an integral over the volume of the solid

$$W_b = \iiint \left[B_x u + B_y v + B_z w\right] \mathrm{d}x\mathrm{d}y\mathrm{d}z \tag{9.158}$$

where B_x, B_y, and B_z are the components of the body forces per unit volume in the x, y, and z directions, respectively.

REFERENCES

[1] Hyer MW. Stress analysis of fiber-reinforced composite materials. Lancaster, PA: Destech Publications; 2009.
[2] Pavlou DG. Composite materials in piping applications. Lancaster, PA: Destech Publications; 2013.
[3] Rand O, Rovenski V. Analytical methods in anisotropic elasticity. Boston: Birkhäuser; 2005.
[4] Hwu C. Anisotropic elastic plates. New York: Springer; 2010.
[5] Ting TCT. Anisotropic elasticity. New York: Oxford University Press; 1996.
[6] Timoshenko SP, Goodier JN. Theory of elasticity. Singapore: McGraw-Hill; 1970.
[7] Pilkey WD. Formulas for stress, strain, and structural matrices. Hoboken: John Wiley & Sons; 2005.

CHAPTER

THE PRINCIPLE OF MINIMUM POTENTIAL ENERGY FOR TWO-DIMENSIONAL AND THREE-DIMENSIONAL ELEMENTS

10

10.1 INTERPOLATION AND SHAPE FUNCTIONS

Chapter 6 described how the element equation of beams can be derived by using either interpolation functions or the governing differential equation. However, for two-dimensional (2D) elements, the governing differential equations are much more complicated, therefore, for such cases, the use of interpolation functions is more effective. In order for this methodology to be easier to understand, it is going to be initially described for the case of one-dimensional (1D) elements where the use of a single independent variable x is sufficient.

Let us denote $\varphi(x)$ the field quantities (e.g., displacements, strains, or stresses) to be interpolated by $\varphi(x)$. The above function can be written in the following polynomial form:

$$\varphi(x)=\begin{bmatrix}1 & x & x^2 & \cdots & x^n\end{bmatrix}\begin{Bmatrix}a_o\\a_1\\a_2\\\vdots\\a_n\end{Bmatrix} \tag{10.1}$$

where $a_0, a_1, a_2, \ldots, a_n$ are constants. The main target of this procedure is the correlation of the function $\varphi(x)$ to the nodal values $\varphi_1, \varphi_2, \varphi_3, \ldots$. To this end, Equation (10.1) should be applied to every pair of nodal data $(x_1,\varphi_1),(x_2,\varphi_2),(x_3,\varphi_3),\ldots$ yielding

$$\begin{Bmatrix}\varphi_1\\\varphi_2\\\varphi_3\\\vdots\\\varphi_n\end{Bmatrix}=\begin{bmatrix}1 & x_1 & x_1^2 & x_1^3 & \cdots & x_1^n\\1 & x_2 & x_2^2 & x_2^3 & \cdots & x_2^n\\1 & x_3 & x_3^2 & x_3^3 & \cdots & x_3^n\\\vdots & \vdots & \vdots & \vdots & \vdots & \vdots\\1 & x_n & x_n^2 & x_n^3 & \cdots & x_n^n\end{bmatrix}\begin{Bmatrix}a_o\\a_1\\a_2\\\vdots\\a_n\end{Bmatrix} \tag{10.2}$$

or in an abbreviated form

$$\begin{Bmatrix}\varphi_1\\\varphi_2\\\varphi_3\\\vdots\\\varphi_n\end{Bmatrix}=[\Xi]\begin{Bmatrix}a_o\\a_1\\a_2\\\vdots\\a_n\end{Bmatrix} \tag{10.3}$$

Essentials of the Finite Element Method. http://dx.doi.org/10.1016/B978-0-12-802386-0.00010-4

From the above equation, the vector of constants can be derived as:

$$\begin{Bmatrix} a_o \\ a_1 \\ a_2 \\ \vdots \\ a_n \end{Bmatrix} = [\Xi]^{-1} \begin{Bmatrix} \varphi_1 \\ \varphi_2 \\ \varphi_3 \\ \vdots \\ \varphi_n \end{Bmatrix} \tag{10.4}$$

Using the above equation, Equation (10.1) can now be written in the following form:

$$\varphi(x) = \begin{bmatrix} 1 & x & x^2 & \cdots & x^n \end{bmatrix} [\Xi]^{-1} \begin{Bmatrix} \varphi_1 \\ \varphi_2 \\ \varphi_3 \\ \vdots \end{Bmatrix} \tag{10.5}$$

The function $\varphi(x)$ is now correlated to the nodal values $\varphi_1, \varphi_2, \varphi_3, \ldots$

Let us assume a 1D element of length L containing two nodes. Therefore, the nodal data are $(x_1, \varphi_1) = (0, \varphi_1)$ and $(x_2, \varphi_2) = (L, \varphi_2)$. If we will choose a linear interpolation function

$$\varphi(x) = [1 \;\; x] \begin{Bmatrix} a_0 \\ a_1 \end{Bmatrix} \tag{10.6}$$

the matrix $[\Xi]$ has the form

$$[\Xi] = \begin{bmatrix} 1 & 0 \\ 1 & L \end{bmatrix} \tag{10.7}$$

Inversion of the above matrix yields

$$[\Xi]^{-1} = \frac{1}{L} \begin{bmatrix} L & 0 \\ -1 & 1 \end{bmatrix} \tag{10.8}$$

Using the above equation, Equation (10.5) can now be written as:

$$\varphi(x) = [1 \;\; x] \frac{1}{L} \begin{bmatrix} L & 0 \\ -1 & 1 \end{bmatrix} \begin{Bmatrix} \varphi_1 \\ \varphi_2 \end{Bmatrix} \tag{10.9}$$

yielding

$$\varphi(x) = \begin{bmatrix} \dfrac{L-x}{L} & \dfrac{x}{L} \end{bmatrix} \begin{Bmatrix} \varphi_1 \\ \varphi_2 \end{Bmatrix} \tag{10.10}$$

or in an abbreviated form

$$\varphi(x) = [N]\{\varphi_e\} \tag{10.11}$$

where

$$[N] = [N_1, N_2] \tag{10.12}$$

$$N_1 = \frac{L-x}{L} \tag{10.13}$$

$$N_2 = \frac{x}{L} \tag{10.14}$$

$$\{\varphi_e\} = \{\varphi_1 \;\; \varphi_2\}^{\mathrm{T}} \tag{10.15}$$

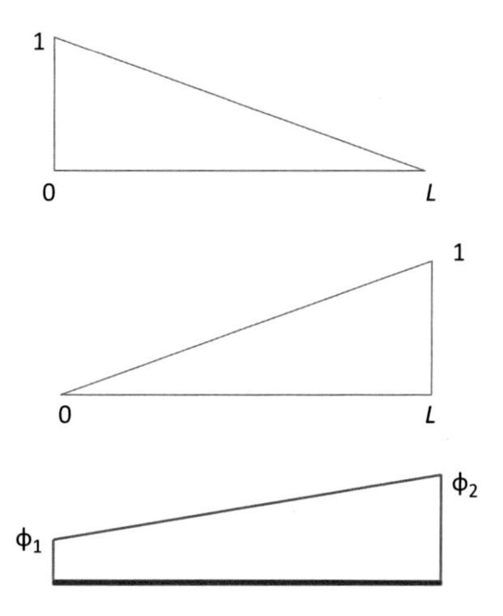

FIGURE 10.1

Graphical representation of the shape functions N_1, N_2 for linear interpolation.

Graphical representation of the shape functions N_I, N_2 is demonstrated in Figure 10.1. The above procedure can be generalized for elements containing more nodes, and for nonlinear interpolation functions. If we will choose an element containing three nodes and a quadratic interpolation, the points $(x_1, \varphi_1), (x_2, \varphi_2), (x_3, \varphi_3)$ will be interpolated by the following formula:

$$\varphi(x) = \begin{bmatrix} 1 & x & x^2 \end{bmatrix} [\Xi]^{-1} \begin{Bmatrix} \varphi_1 \\ \varphi_2 \\ \varphi_3 \end{Bmatrix} \tag{10.16}$$

where

$$[\Xi] = \begin{bmatrix} 1 & x_1 & x_1^2 \\ 1 & x_2 & x_2^2 \\ 1 & x_3 & x_3^2 \end{bmatrix} \tag{10.17}$$

Combining Equations (10.16) and (10.17), the following formula for $\varphi(x)$ can be derived:

$$\varphi(x) = [N_1 \;\; N_2 \;\; N_3] \begin{Bmatrix} \varphi_1 \\ \varphi_2 \\ \varphi_3 \end{Bmatrix} \tag{10.18}$$

where

$$N_1 = \frac{(x_2 - x)(x_3 - x)}{(x_2 - x_1)(x_3 - x_1)} \tag{10.19}$$

$$N_2 = \frac{(x_1 - x)(x_3 - x)}{(x_1 - x_2)(x_3 - x_2)} \tag{10.20}$$

$$N_3 = \frac{(x_1 - x)(x_2 - x)}{(x_1 - x_3)(x_2 - x_3)} \tag{10.21}$$

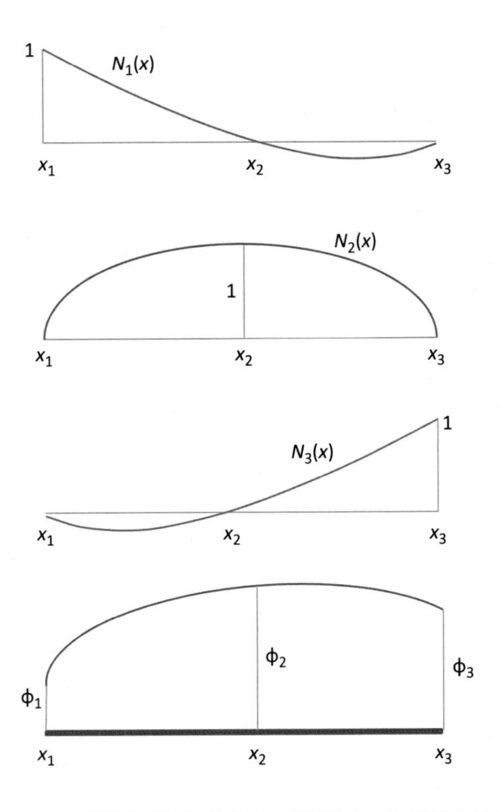

FIGURE 10.2

Graphical representation of the shape functions N_1, N_2, and N_3 for quadratic interpolation.

The above shape functions are demonstrated graphically in Figure 10.2.

It should be noted that even though the function $\varphi(x)$ varies smoothly within an element, the transmission between elements may not be smooth. In order to guarantee smooth variation of $\varphi(x)$ between elements, in addition to the nodal values of φ, its derivatives on the nodes should also be interpolated. The degree of continuity of $\varphi(x)$ in a finite element mesh is characterized (Cook RD et al.) by the symbol C^{m}. Depending on the values of m, we can say that a 1D field is C^0 continuous if φ is continuous but $\mathrm{d}\varphi/\mathrm{d}x$ is not. Similarly, a field is C^1 continuous when the functions φ and $\mathrm{d}\varphi/\mathrm{d}x$ are continuous but the function $\mathrm{d}^2\varphi/\mathrm{d}x^2$ is not. Figure 10.3 demonstrates a C^0 and a C^1 interpolation function.

In addition to displacements, the slopes of a beam or a plate should also be interelement continuous. Then, C^1 element types are usually used to model structures subjected to bending. For plane and solid models, the use of C^0 element types is sufficient. As an example of a C^1 interpolation, the deflection $f(x)$ of a beam of length L under bending will be assumed. In order to achieve smooth variation of the slope $\mathrm{d}f(x)/\mathrm{d}x$ between elements, the following interpolation function:

$$f(x) = \begin{bmatrix} 1 & x & x^2 & x^3 \end{bmatrix} \begin{Bmatrix} a_0 \\ a_1 \\ a_2 \\ a_3 \end{Bmatrix} \tag{10.22}$$

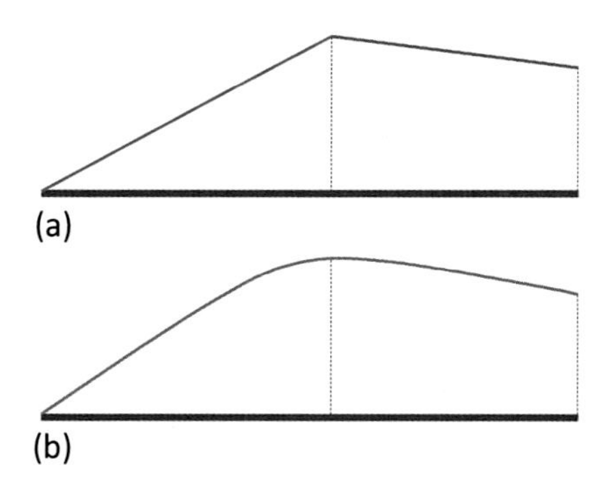

FIGURE 10.3

(a) C^0 continuous function and (b) C^1 continuous function.

and its derivative

$$f'(x) = \begin{bmatrix} 0 & 1 & 2x & 3x^2 \end{bmatrix} \begin{Bmatrix} a_0 \\ a_1 \\ a_2 \\ a_3 \end{Bmatrix} \tag{10.23}$$

will be applied to the nodes 1 and 2, yielding

$$\begin{Bmatrix} f(0) \\ f'(0) \\ f(L) \\ f'(L) \end{Bmatrix} = \begin{bmatrix} 1 & 0 & 0 & 0 \\ 0 & 1 & 0 & 0 \\ 1 & L & L^2 & L^3 \\ 0 & L & 2L & 3L^2 \end{bmatrix} \begin{Bmatrix} a_0 \\ a_1 \\ a_2 \\ a_3 \end{Bmatrix} \tag{10.24}$$

Therefore,

$$[\Xi] = \begin{bmatrix} 1 & 0 & 0 & 0 \\ 0 & 1 & 0 & 0 \\ 1 & L & L^2 & L^3 \\ 0 & L & 2L & 3L^2 \end{bmatrix} \tag{10.25}$$

and

$$f(x) = \begin{bmatrix} 1 & x & x^2 & x^3 \end{bmatrix} [\Xi]^{-1} \begin{Bmatrix} f(0) \\ f'(0) \\ f(L) \\ f'(L) \end{Bmatrix} \tag{10.26}$$

or in an abbreviated form

$$f(x) = [N_1 \;\; N_2 \;\; N_3 \;\; N_4] \begin{Bmatrix} f(0) \\ f'(0) \\ f(L) \\ f'(L) \end{Bmatrix} \tag{10.27}$$

where

$$[N_1 \;\; N_2 \;\; N_3 \;\; N_4] = \begin{bmatrix} 1 & x & x^2 & x^3 \end{bmatrix} [\Xi]^{-1} \tag{10.28}$$

After some simple algebraic operations, the shape functions can be expressed as:

$$N_1 = 1 - \frac{3x^2}{L^2} + \frac{2x^3}{L^3} \tag{10.29}$$

$$N_2 = x - \frac{2x^2}{L} + \frac{x^3}{L^2} \tag{10.30}$$

$$N_3 = \frac{3x^2}{L^2} - \frac{2x^3}{L^3} \tag{10.31}$$

$$N_4 = -\frac{x^2}{L} + \frac{x^3}{L^2} \tag{10.32}$$

For 2D and three-dimensional (3D) elements, the single variable x is not sufficient to describe the field. Therefore, for 2D elements, the indented variables x, y and their products should be used, and for 3D element three variables x, y, z and their combinations are also needed.

Such interpolation functions will be derived for common types of elements, such as linear triangular, quadratic triangular, bilinear rectangular, tetrahedral solid, and rectangular solid, and for plate bending elements.

10.1.1 LINEAR TRIANGULAR ELEMENTS (OR CST ELEMENTS)

Let us consider the triangular element demonstrated in Figure 10.4.

The field functions are the horizontal and vertical displacements $u(x,y)$ and $v(x,y)$, respectively. For these functions, the following linear formulae will be used:

$$u(x,y) = [1 \ \ x \ \ y]\begin{Bmatrix} a_1 \\ a_2 \\ a_3 \end{Bmatrix} \tag{10.33}$$

$$v(x,y) = [1 \ \ x \ \ y]\begin{Bmatrix} a_4 \\ a_5 \\ a_6 \end{Bmatrix} \tag{10.34}$$

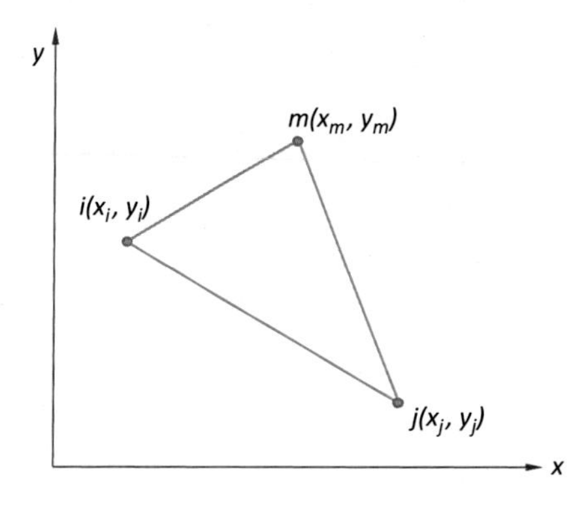

FIGURE 10.4

Coordinate system for a triangular element.

The above interpolation functions will now be applied to the three nodes i, j, and m of the triangular element yielding:

$$\begin{Bmatrix} u_i \\ u_j \\ u_m \end{Bmatrix} = \begin{bmatrix} 1 & x_i & y_i \\ 1 & x_j & y_j \\ 1 & x_m & y_m \end{bmatrix} \begin{Bmatrix} a_1 \\ a_2 \\ a_3 \end{Bmatrix} \tag{10.35}$$

$$\begin{Bmatrix} v_i \\ v_j \\ v_m \end{Bmatrix} = \begin{bmatrix} 1 & x_i & y_i \\ 1 & x_j & y_j \\ 1 & x_m & y_m \end{bmatrix} \begin{Bmatrix} a_4 \\ a_5 \\ a_6 \end{Bmatrix} \tag{10.36}$$

From the above equations, the constants $a_1, a_2, \ldots, a_6$ can now be obtained

$$\begin{Bmatrix} a_1 \\ a_2 \\ a_3 \end{Bmatrix} = [\Xi]^{-1} \begin{Bmatrix} u_i \\ u_j \\ u_m \end{Bmatrix} \tag{10.37}$$

$$\begin{Bmatrix} a_4 \\ a_5 \\ a_6 \end{Bmatrix} = [\Xi]^{-1} \begin{Bmatrix} v_i \\ v_j \\ v_m \end{Bmatrix} \tag{10.38}$$

where

$$[\Xi]^{-1} = \begin{bmatrix} 1 & x_i & y_i \\ 1 & x_j & y_j \\ 1 & x_m & y_m \end{bmatrix}^{-1} = \frac{1}{2A} \begin{bmatrix} a_i & a_j & a_m \\ \beta_i & \beta_j & \beta_m \\ \gamma_i & \gamma_j & \gamma_m \end{bmatrix} \tag{10.39}$$

In the above equation, the parameters $a_i, a_j, a_m, \beta_i, \beta_j, \beta_m, \gamma_i, \gamma_j, \gamma_m$ and the area A of the triangle are given by the following equations:

$$a_i = x_j y_m - y_j x_m \tag{10.40}$$

$$a_j = x_m y_i - y_m x_i \tag{10.41}$$

$$a_m = x_i y_j - y_i x_j \tag{10.42}$$

$$\beta_i = y_j - y_m \tag{10.43}$$

$$\beta_j = y_m - y_i \tag{10.44}$$

$$\beta_m = y_i - y_j \tag{10.45}$$

$$\gamma_i = x_m - x_j \tag{10.46}$$

$$\gamma_j = x_i - x_m \tag{10.47}$$

$$\gamma_m = x_j - x_i \tag{10.48}$$

$$A = \frac{1}{2}\left[x_i\left(y_j - y_m\right) + x_j\left(y_m - y_i\right) + x_m\left(y_i - y_j\right)\right] \tag{10.49}$$

Taking into account Equations (10.37) and (10.38), Equations (10.33) and (10.34) can now be written:

$$u(x,y)=[1 \ \ x \ \ y][\Xi]^{-1}\begin{Bmatrix} u_i \\ u_j \\ u_m \end{Bmatrix} \tag{10.50}$$

$$\upsilon(x,y)=[1 \ \ x \ \ y][\Xi]^{-1}\begin{Bmatrix} \upsilon_i \\ \upsilon_j \\ \upsilon_m \end{Bmatrix} \tag{10.51}$$

Using Equations (10.39)–(10.40), after some algebraic operations, Equations (10.50) and (10.51) yield

$$u(x,y)=[N_i \ \ N_j \ \ N_m]\begin{Bmatrix} u_i \\ u_j \\ u_m \end{Bmatrix} \tag{10.52}$$

$$\upsilon(x,y)=[N_i \ \ N_j \ \ N_m]\begin{Bmatrix} \upsilon_i \\ \upsilon_j \\ \upsilon_m \end{Bmatrix} \tag{10.53}$$

where

$$N_i=\frac{1}{2A}(a_i+\beta_i x+\gamma_i y) \tag{10.54}$$

$$N_j=\frac{1}{2A}(a_j+\beta_j x+\gamma_j y) \tag{10.55}$$

$$N_m=\frac{1}{2A}(a_m+\beta_m x+\gamma_m y) \tag{10.56}$$

The set of Equations (10.52) and (10.53) can also be written in the following matrix form

$$\{\Psi\}=\begin{Bmatrix} u(x,y) \\ \upsilon(x,y) \end{Bmatrix}=\begin{bmatrix} N_i & 0 & N_j & 0 & N_m & 0 \\ 0 & N_i & 0 & N_j & 0 & N_m \end{bmatrix}\begin{Bmatrix} u_i \\ \upsilon_i \\ u_j \\ \upsilon_j \\ u_m \\ \upsilon_m \end{Bmatrix} \tag{10.57}$$

or in an abbreviated form

$$\{\Psi\}=[N]\{d\} \tag{10.58}$$

10.1.2 QUADRATIC TRIANGULAR ELEMENTS (OR LST ELEMENTS)

Let us now consider the triangular element shown in Figure 10.5. This element, apart from the vertex nodes, has side nodes. The field functions are again the horizontal and vertical displacements $u(x,y)$ and $\upsilon(x,y)$, respectively. For these functions, the following quadratic formulae will be used:

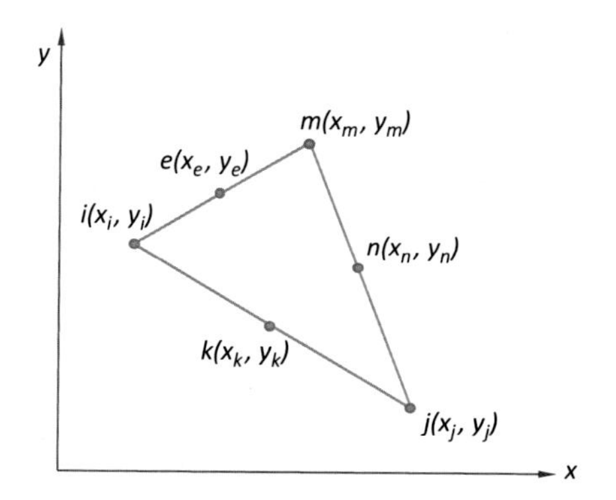

FIGURE 10.5

Coordinate system for a quadratic triangular element.

$$u(x,y) = \begin{bmatrix} 1 & x & y & x^2 & xy & y^2 \end{bmatrix} \begin{Bmatrix} a_1 \\ a_2 \\ a_3 \\ a_4 \\ a_5 \\ a_6 \end{Bmatrix} \quad (10.59)$$

$$v(x,y) = \begin{bmatrix} 1 & x & y & x^2 & xy & y^2 \end{bmatrix} \begin{Bmatrix} a_7 \\ a_8 \\ a_9 \\ a_{10} \\ a_{11} \\ a_{12} \end{Bmatrix} \quad (10.60)$$

The above interpolation functions should be applied to the six nodes i, k, j, n, m, and e of the triangular element. Denoting by $[\Xi]$ the following matrix:

$$[\Xi] = \begin{bmatrix} 1 & x_i & y_i & x_i^2 & x_i y_i & y_i^2 \\ 1 & x_k & y_k & x_k^2 & x_k y_k & y_k^2 \\ 1 & x_j & y_j & x_j^2 & x_j y_j & y_j^2 \\ 1 & x_n & y_n & x_n^2 & x_n y_n & y_n^2 \\ 1 & x_m & y_m & x_m^2 & x_m y_m & y_m^2 \\ 1 & x_e & y_e & x_e^2 & x_e y_e & y_e^2 \end{bmatrix} \quad (10.61)$$

the field functions $u(x,y)$ and $v(x,y)$ given by Equations (10.59) and (10.60) can be written:

$$u(x,y) = \begin{bmatrix} 1 & x & y & x^2 & xy & y^2 \end{bmatrix} [\Xi]^{-1} \begin{Bmatrix} u_i \\ u_k \\ u_j \\ u_n \\ u_m \\ u_e \end{Bmatrix} \quad (10.62)$$

$$v(x,y) = \begin{bmatrix} 1 & x & y & x^2 & xy & y^2 \end{bmatrix} [\Xi]^{-1} \begin{Bmatrix} v_i \\ v_k \\ v_j \\ v_n \\ v_m \\ v_e \end{Bmatrix} \tag{10.63}$$

Denoted by

$$[N_i \;\; N_k \;\; N_j \;\; N_n \;\; N_m \;\; N_e] = \begin{bmatrix} 1 & x & y & x^2 & xy & y^2 \end{bmatrix} [\Xi]^{-1} \tag{10.64}$$

Equations (10.62) and (10.63) can be written in an abbreviated form as follows:

$$u(x,y) = [N_i \;\; N_k \;\; N_j \;\; N_n \;\; N_m \;\; N_e] \begin{Bmatrix} u_i \\ u_k \\ u_j \\ u_n \\ u_m \\ u_e \end{Bmatrix} \tag{10.65}$$

$$v(x,y) = [N_i \;\; N_k \;\; N_j \;\; N_n \;\; N_m \;\; N_e] \begin{Bmatrix} v_i \\ v_k \\ v_j \\ v_n \\ v_m \\ v_e \end{Bmatrix} \tag{10.66}$$

In order to apply the principle of minimum potential energy, it is more convenient to write the above set of equations in the following form:

$$\{\Psi\} = \begin{Bmatrix} u(x,y) \\ v(x,y) \end{Bmatrix} = \begin{bmatrix} N_i & 0 & N_k & 0 & N_j & 0 & N_n & 0 & N_m & 0 & N_e & 0 \\ 0 & N_i & 0 & N_k & 0 & N_j & 0 & N_n & 0 & N_m & 0 & N_e \end{bmatrix} \begin{Bmatrix} u_i \\ v_i \\ u_k \\ v_k \\ u_j \\ v_j \\ u_n \\ v_n \\ u_m \\ v_m \\ u_e \\ v_e \end{Bmatrix} \tag{10.67}$$

or

$$[\Psi] = [N]\{d\} \tag{10.68}$$

10.1.3 BILINEAR RECTANGULAR ELEMENTS (OR Q4 ELEMENTS)

In most membrane problems, it is more practical to use the four-node rectangular plane stress element shown in Figure 10.6.

The field functions for this element, that is, the horizontal and vertical displacements $u(x,y)$ and $\upsilon(x,y)$, respectively, have the following form:

$$u(x,y) = [1 \;\; x \;\; y \;\; xy]\begin{Bmatrix} a_1 \\ a_2 \\ a_3 \\ a_4 \end{Bmatrix} \tag{10.69}$$

$$\upsilon(x,y) = [1 \;\; x \;\; y \;\; xy]\begin{Bmatrix} a_5 \\ a_6 \\ a_7 \\ a_8 \end{Bmatrix} \tag{10.70}$$

Application of the above function to the four nodes i, j, m, and k of the rectangular element yields the following $[\Xi]$ matrix:

$$[\Xi] = \begin{bmatrix} 1 & x_i & y_i & x_i y_i \\ 1 & x_j & y_j & x_j y_j \\ 1 & x_m & y_m & x_m y_m \\ 1 & x_k & y_k & x_k y_k \end{bmatrix} \tag{10.71}$$

Therefore, the field functions of Equations (10.69) and (10.70) can be written in the following form:

$$u(x,y) = [1 \;\; x \;\; y \;\; xy][\Xi]^{-1}\begin{Bmatrix} u_i \\ u_j \\ u_m \\ u_k \end{Bmatrix} \tag{10.72}$$

$$\upsilon(x,y) = [1 \;\; x \;\; y \;\; xy][\Xi]^{-1}\begin{Bmatrix} \upsilon_i \\ \upsilon_j \\ \upsilon_m \\ \upsilon_k \end{Bmatrix} \tag{10.73}$$

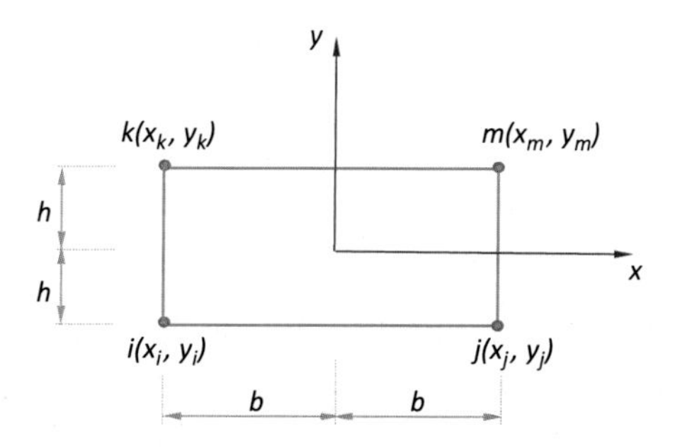

FIGURE 10.6

Coordinate system for a bilinear rectangular element.

or

$$u(x,y)=[N_i\ N_j\ N_m\ N_k]\begin{Bmatrix} u_i \\ u_j \\ u_m \\ u_k \end{Bmatrix} \tag{10.74}$$

$$v(x,y)=[N_i\ N_j\ N_m\ N_k]\begin{Bmatrix} v_i \\ v_j \\ v_m \\ v_k \end{Bmatrix} \tag{10.75}$$

where

$$[N_i\ N_j\ N_m\ N_k]=[1\ x\ y\ xy][\Xi]^{-1} \tag{10.76}$$

Performing the required algebraic operations, the above equation yields:

$$N_1=\frac{(b-x)(h-y)}{4bh} \tag{10.77}$$

$$N_2=\frac{(b+x)(h-y)}{4bh} \tag{10.78}$$

$$N_3=\frac{(b+x)(h+y)}{4bh} \tag{10.79}$$

$$N_4=\frac{(b-x)(h+y)}{4bh} \tag{10.80}$$

Using the following notations

$$\{d\}=\{u_i\ v_i\ u_j\ v_j\ u_m\ v_m\ u_k\ v_k\}^{\mathrm{T}} \tag{10.81}$$

$$[N]=\begin{bmatrix} N_i & 0 & N_j & 0 & N_m & 0 & N_k & 0 \\ 0 & N_i & 0 & N_j & 0 & N_m & 0 & N_k \end{bmatrix} \tag{10.82}$$

$$\{\Psi\}=\begin{Bmatrix} u(x,y) \\ v(x,y) \end{Bmatrix} \tag{10.83}$$

Equations (10.74) and (10.75) can be written in the following form:

$$[\Psi]=[N]\{d\} \tag{10.84}$$

10.1.4 TETRAHEDRAL SOLID ELEMENTS

The simplest 3D solid element is the tetrahedral one element shown in Figure 10.7. Each node now has three degrees of freedom x, y, z. Let us denote the field functions (i.e., the displacements) with u, v, and w in the x, y, and z directions, respectively. Therefore, the number of the degrees of freedom is (four nodes) $\times$ (three directions) $=12$. The displacement functions for this element are

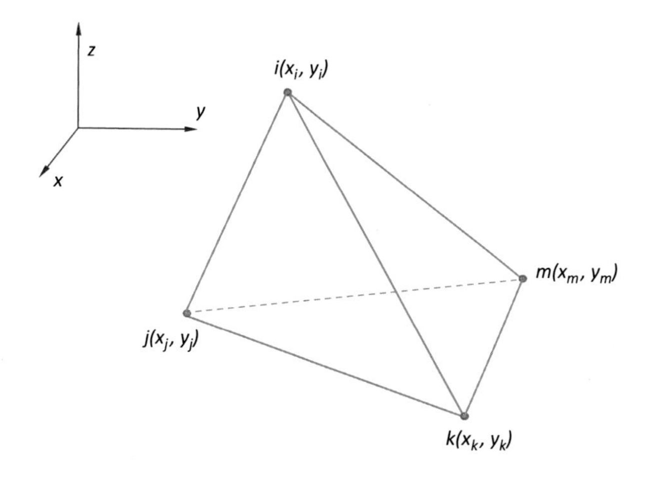

FIGURE 10.7

Tetrahedral solid element.

$$u(x,y,z)=[1\ \ x\ \ y\ \ z]\begin{Bmatrix} a_1 \\ a_2 \\ a_3 \\ a_4 \end{Bmatrix} \tag{10.85}$$

$$v(x,y,z)=[1\ \ x\ \ y\ \ z]\begin{Bmatrix} a_5 \\ a_6 \\ a_7 \\ a_8 \end{Bmatrix} \tag{10.86}$$

$$w(x,y,z)=[1\ \ x\ \ y\ \ z]\begin{Bmatrix} a_9 \\ a_{10} \\ a_{11} \\ a_{12} \end{Bmatrix} \tag{10.87}$$

If we apply the above field equations to the four nodes of the element, the following $[\Xi]$ matrix will be derived:

$$[\Xi]=\begin{bmatrix} 1 & x_i & y_i & z_i \\ 1 & x_j & y_j & z_j \\ 1 & x_m & y_m & z_m \\ 1 & x_k & y_k & z_k \end{bmatrix} \tag{10.88}$$

Now, the field displacements can be correlated to the nodal displacements by the following formulae:

$$u(x,y,z)=[1\ \ x\ \ y\ \ z][\Xi]^{-1}\begin{Bmatrix} u_i \\ u_j \\ u_m \\ u_k \end{Bmatrix} \tag{10.89}$$

$$v(x,y,z)=[1\ \ x\ \ y\ \ z][\Xi]^{-1}\begin{Bmatrix} v_i \\ v_j \\ v_m \\ v_k \end{Bmatrix} \tag{10.90}$$

$$w(x,y,z)=[1\ \ x\ \ y\ \ z][\Xi]^{-1}\begin{Bmatrix} w_i \\ w_j \\ w_m \\ w_k \end{Bmatrix} \tag{10.91}$$

The above three field functions can be written in the following form:

$$\{\Psi\}=\begin{Bmatrix} u(x,y,z) \\ v(x,y,z) \\ w(x,y,z) \end{Bmatrix}=\begin{bmatrix} N_i & 0 & 0 & N_j & 0 & 0 & N_m & 0 & 0 & N_k & 0 & 0 \\ 0 & N_i & 0 & 0 & N_j & 0 & 0 & N_m & 0 & 0 & N_k & 0 \\ 0 & 0 & N_i & 0 & 0 & N_j & 0 & 0 & N_m & 0 & 0 & N_k \end{bmatrix}\begin{Bmatrix} u_i \\ v_i \\ w_i \\ u_j \\ v_j \\ w_j \\ u_m \\ v_m \\ w_m \\ u_k \\ v_k \\ w_k \end{Bmatrix} \tag{10.92}$$

or in an abbreviated form

$$\{\Psi\}=[N]\{d\} \tag{10.93}$$

where the shape functions N_i, N_j, N_m, N_k can be obtained by the following equations:

$$[N_i\ \ N_j\ \ N_m\ \ N_k]=[1\ \ x\ \ y\ \ z][\Xi]^{-1} \tag{10.94}$$

Performing the required algebraic operations, the above equation yields

$$N_i=\frac{H_i+Y_ix+\Phi_iy+\Omega_iz}{M} \tag{10.95}$$

$$N_j=\frac{H_j+Y_jx+\Phi_jy+\Omega_jz}{M} \tag{10.96}$$

$$N_m=\frac{H_m+Y_mx+\Phi_my+\Omega_mz}{M} \tag{10.97}$$

$$N_k=\frac{H_k+Y_kx+\Phi_ky+\Omega_kz}{M} \tag{10.98}$$

where

$$H_i=\det\begin{bmatrix} x_j & y_j & z_j \\ x_m & y_m & z_m \\ x_k & y_k & z_k \end{bmatrix} \tag{10.99}$$

$$Y_i = -\det\begin{bmatrix} 1 & y_j & z_j \\ 1 & y_m & z_m \\ 1 & y_k & z_k \end{bmatrix} \tag{10.100}$$

$$\Phi_i = \det\begin{bmatrix} 1 & x_j & z_j \\ 1 & x_m & z_m \\ 1 & x_k & z_k \end{bmatrix} \tag{10.101}$$

$$\Omega_i = -\det\begin{bmatrix} 1 & x_j & y_j \\ 1 & x_m & z_m \\ 1 & x_k & z_k \end{bmatrix} \tag{10.102}$$

$$H_j = -\det\begin{bmatrix} x_i & y_i & z_i \\ x_m & y_m & z_m \\ x_k & y_k & z_k \end{bmatrix} \tag{10.103}$$

$$Y_j = \det\begin{bmatrix} 1 & y_i & z_i \\ 1 & y_m & z_m \\ 1 & y_k & z_k \end{bmatrix} \tag{10.104}$$

$$\Phi_j = -\det\begin{bmatrix} 1 & x_i & z_i \\ 1 & x_m & z_m \\ 1 & x_k & z_k \end{bmatrix} \tag{10.105}$$

$$\Omega_j = \det\begin{bmatrix} 1 & x_i & y_i \\ 1 & x_m & y_m \\ 1 & x_k & y_k \end{bmatrix} \tag{10.106}$$

$$H_m = \det\begin{bmatrix} x_i & y_i & z_i \\ x_j & y_j & z_j \\ x_k & y_k & z_k \end{bmatrix} \tag{10.107}$$

$$Y_m = -\det\begin{bmatrix} 1 & y_i & z_i \\ 1 & y_j & z_j \\ 1 & y_k & z_k \end{bmatrix} \tag{10.108}$$

$$\Phi_m = \det\begin{bmatrix} 1 & x_i & z_i \\ 1 & x_j & z_j \\ 1 & x_k & z_k \end{bmatrix} \tag{10.109}$$

$$\Omega_m = -\det\begin{bmatrix} 1 & x_i & y_i \\ 1 & x_j & z_j \\ 1 & x_k & z_k \end{bmatrix} \tag{10.110}$$

$$H_k = -\det\begin{bmatrix} x_i & y_i & z_i \\ x_j & y_j & z_j \\ x_m & y_m & z_m \end{bmatrix} \quad (10.111)$$

$$Y_k = \det\begin{bmatrix} 1 & y_i & z_i \\ 1 & y_j & z_j \\ 1 & y_m & z_m \end{bmatrix} \quad (10.112)$$

$$\Phi_k = -\det\begin{bmatrix} 1 & x_i & z_i \\ 1 & x_j & z_j \\ 1 & x_m & z_m \end{bmatrix} \quad (10.113)$$

$$\Omega_k = \det\begin{bmatrix} 1 & x_i & y_i \\ 1 & x_j & y_j \\ 1 & x_m & y_m \end{bmatrix} \quad (10.114)$$

$$M = \det\begin{bmatrix} 1 & x_i & y_i & z_i \\ 1 & x_j & y_j & z_j \\ 1 & x_m & y_m & z_m \\ 1 & x_k & y_k & z_k \end{bmatrix} \quad (10.115)$$

10.1.5 EIGHT-NODE RECTANGULAR SOLID ELEMENTS

The most common type among the solid elements is the "brick" element. Figure 10.8 demonstrates the simplest brick element containing eight nodes, and, therefore, $8 \times 3 = 24$ degrees of freedom.

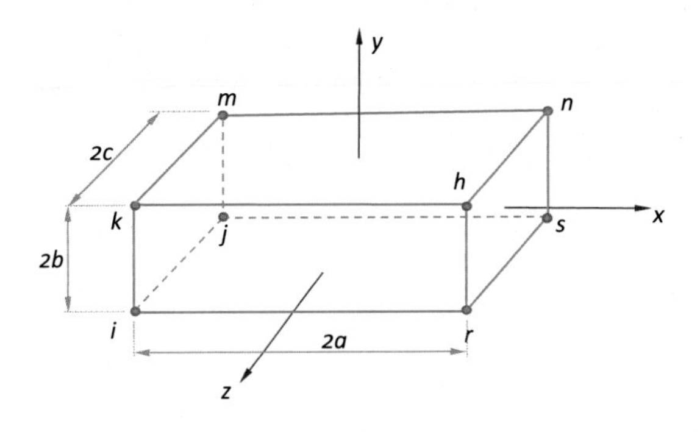

FIGURE 10.8

Eight-nodes brick element.

The displacement functions for this element are:

$$u(x,y,z) = [1 \;\; x \;\; y \;\; z \;\; xy \;\; yz \;\; zx \;\; xyz] \begin{Bmatrix} a_1 \\ a_2 \\ a_3 \\ a_4 \\ a_5 \\ a_6 \\ a_7 \\ a_8 \end{Bmatrix} \tag{10.116}$$

$$v(x,y,z) = [1 \;\; x \;\; y \;\; z \;\; xy \;\; yz \;\; zx \;\; xyz] \begin{Bmatrix} a_9 \\ a_{10} \\ a_{11} \\ a_{12} \\ a_{13} \\ a_{14} \\ a_{15} \\ a_{16} \end{Bmatrix} \tag{10.117}$$

$$w(x,y,z) = [1 \;\; x \;\; y \;\; z \;\; xy \;\; yz \;\; zx \;\; xyz] \begin{Bmatrix} a_{17} \\ a_{18} \\ a_{19} \\ a_{20} \\ a_{21} \\ a_{22} \\ a_{23} \\ a_{24} \end{Bmatrix} \tag{10.118}$$

Application of the above functions to the eight nodes yields the following $[\Xi]$ matrix:

$$[\Xi] = \begin{bmatrix} 1 & x_i & y_i & z_i & x_iy_i & y_iz_i & z_ix_i & x_iy_iz_i \\ 1 & x_j & y_j & z_j & x_jy_j & y_jz_j & z_jx_j & x_jy_jz_j \\ 1 & x_m & y_m & z_m & x_my_m & y_mz_m & z_mx_m & x_my_mz_m \\ 1 & x_k & y_k & z_k & x_ky_k & y_kz_k & z_kx_k & x_ky_kz_k \\ 1 & x_r & y_r & z_r & x_ry_r & y_rz_r & z_rx_r & x_ry_rz_r \\ 1 & x_s & y_s & z_s & x_sy_s & y_sz_s & z_sx_s & x_sy_sz_s \\ 1 & x_n & y_n & z_n & x_ny_n & y_nz_n & z_nx_n & x_ny_nz_n \\ 1 & x_h & y_h & z_h & x_hy_h & y_hz_h & z_hx_h & x_hy_hz_h \end{bmatrix} \tag{10.119}$$

Taking into account the above equation, the shape functions can be calculated as follows:

$$[N_i \;\; N_j \;\; N_m \;\; N_k \;\; N_r \;\; N_s \;\; N_n \;\; N_h] = [1 \;\; x \;\; y \;\; z \;\; xy \;\; yz \;\; zx \;\; xyz][\Xi]^{-1} \tag{10.120}$$

and the matrix of displacements can be derived by the following equation

$$\{\Psi\} = \begin{Bmatrix} u(x,y,z) \\ v(x,y,z) \\ w(x,y,z) \end{Bmatrix} = \left[[A_i]\ [A_j]\ [A_m]\ [A_k]\ [A_r]\ [A_s]\ [A_n]\ [A_h]\right] \begin{Bmatrix} \{d_i\} \\ \{d_j\} \\ \{d_m\} \\ \{d_k\} \\ \{d_r\} \\ \{d_s\} \\ \{d_n\} \\ \{d_k\} \end{Bmatrix} \tag{10.121}$$

where

$$[A_\varphi] = \begin{bmatrix} N_\varphi & 0 & 0 \\ 0 & N_\varphi & 0 \\ 0 & 0 & N_\varphi \end{bmatrix} \quad \text{for } \varphi = i,j,m,k,r,s,n,h \tag{10.122}$$

$$\{d_\varphi\} = \begin{Bmatrix} u_\varphi \\ v_\varphi \\ w_\varphi \end{Bmatrix} \quad \text{for } \varphi = i,j,m,k,r,s,n,h \tag{10.123}$$

Equation (10.121) can be written in the following well-known abbreviated form:

$$\{\Psi\} = [N]\{d\} \tag{10.124}$$

10.1.6 PLATE BENDING ELEMENTS

Even though plates are 3D solids, however, since their thickness is very small, bending is the dominant type of deformation. Therefore, brick-type elements are not suitable for analyzing plates because the use of a very large number of brick elements with very small thickness yields finite element (FE) models with far too many degrees of freedom. To overcome this shortcoming, special elements based on Kirchhoff and Mindlin bending theories have been developed. Since the aim of the present book is to exhibit the FE modeling techniques, and, therefore, the focus on plate theories is beyond its purposes, only the classical Kirchhoff thin plate theory that prohibits transverse shear deformation will be adopted. Figure 10.9 shows a thin plate element with four nodes.

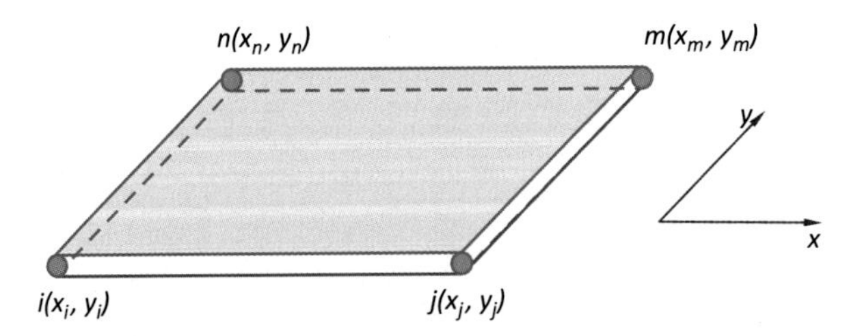

FIGURE 10.9

Four-node thin plate element.

For each node there are three degrees of freedom; the displacement w, the slope θ_x, and the slope θ_y. Since the number of degrees of freedom is $3 \times 4 = 12$, the displacement function is defined by a 12-term polynomial defined as

$$w(x,y) = \begin{bmatrix} 1 & x & y & x^2 & xy & y^2 & x^3 & x^2y & xy^2 & y^3 & x^3y & xy^3 \end{bmatrix} \begin{Bmatrix} a_1 \\ a_2 \\ a_3 \\ a_4 \\ a_5 \\ a_6 \\ a_7 \\ a_8 \\ a_9 \\ a_{10} \\ a_{11} \\ a_{12} \end{Bmatrix} \tag{10.125}$$

Therefore, taking into account the above equation, the slopes

$$\theta_x(x,y) = \frac{\partial w(x,y)}{\partial y} \tag{10.126}$$

$$\theta_y(x,y) = -\frac{\partial w(x,y)}{\partial x} \tag{10.127}$$

yields

$$\theta_x(x,y) = \begin{bmatrix} 0 & 0 & 1 & 0 & x & 2y & 0 & x^2 & 2xy & 3y^2 & x^3 & 3xy^2 \end{bmatrix} \begin{Bmatrix} a_1 \\ a_2 \\ a_3 \\ \vdots \\ a_{12} \end{Bmatrix} \tag{10.128}$$

$$\theta_y(x,y) = \begin{bmatrix} 0 & -1 & 0 & -2x & -y & 0 & -3x^2 & -2xy & -y^2 & 0 & -3x^2y & -y^3 \end{bmatrix} \begin{Bmatrix} a_1 \\ a_2 \\ a_3 \\ \vdots \\ a_{12} \end{Bmatrix} \tag{10.129}$$

The set of Equations (10.125), (10.128), and (10.129) can be written in the following abbreviated form:

$$\{\Psi\} = [X]\{a_e\} \tag{10.130}$$

where

$$\{\Psi\} = \begin{Bmatrix} w(x,y) \\ \theta_x(x,y) \\ \theta_y(x,y) \end{Bmatrix} \tag{10.131}$$

$$\{a_e\} = \begin{Bmatrix} a_1 \\ a_2 \\ a_3 \\ \vdots \\ a_{12} \end{Bmatrix} \tag{10.132}$$

$$[X] = \begin{bmatrix} 1 & x & y & x^2 & xy & y^2 & x^3 & x^2y & xy^2 & y^3 & x^3y & xy^3 \\ 0 & 0 & 1 & 0 & x & 2y & 0 & x^2 & 2xy & 3y^2 & x^3 & 3xy^2 \\ 0 & -1 & 0 & -2x & -y & 0 & -3x^2 & -2xy & -y^2 & 0 & -3x^2y & -y^3 \end{bmatrix} \tag{10.133}$$

Equation (10.130) can be applied to the nodal coordinates of the four nodes yielding the following equation

$$\{d\} = [\Lambda]\{a_e\} \tag{10.134}$$

where

$$\{d\} = \begin{Bmatrix} \{d_i\} \\ \{d_j\} \\ \{d_m\} \\ \{d_n\} \end{Bmatrix} \tag{10.135}$$

$$[\Lambda] = \begin{bmatrix} [M_i] \\ [M_j] \\ [M_m] \\ [M_n] \end{bmatrix} \tag{10.136}$$

In the above equations, the submatrices $\{d_i\}, \{d_j\}, \{d_m\}, \{d_n\}$ and $[M_i], [M_j], [M_m], [M_n]$ are given by the following equations:

$$\{d_i\} = \begin{Bmatrix} w_i \\ \theta_{xi} \\ \theta_{yi} \end{Bmatrix} \tag{10.137}$$

$$\{d_j\} = \begin{Bmatrix} w_j \\ \theta_{xj} \\ \theta_{yj} \end{Bmatrix} \tag{10.138}$$

$$\{d_m\} = \begin{Bmatrix} w_m \\ \theta_{xm} \\ \theta_{ym} \end{Bmatrix} \tag{10.139}$$

$$\{d_n\} = \begin{Bmatrix} w_n \\ \theta_{xn} \\ \theta_{yn} \end{Bmatrix} \tag{10.140}$$

$$[M_i] = \begin{bmatrix} 1 & x_i & y_i & x_i^2 & x_iy_i & y_i^2 & x_i^3 & x_i^2y_i & x_iy_i^2 & y_i^3 & x_i^3y_i & x_iy_i^3 \\ 0 & 0 & 1 & 0 & x_i & 2y_i & 0 & x_i^2 & 2x_iy_i & 3y_i^2 & x_i^3 & 3x_iy_i^2 \\ 0 & -1 & 0 & -2x_i & -y_i & 0 & -3x_i^2 & -2x_iy_i & -y_i^2 & 0 & -3x_i^2y_i & -y_i^3 \end{bmatrix} \tag{10.141}$$

$$[M_j]=\begin{bmatrix}1 & x_j & y_j & x_j^2 & x_jy_j & y_j^2 & x_j^3 & x_j^2y_j & x_jy_j^2 & y_j^3 & x_j^3y_j & x_jy_j^3\\ 0 & 0 & 1 & 0 & x_j & 2y_j & 0 & x_j^2 & 2x_jy_j & 3y_j^2 & x_j^3 & 3x_jy_j^2\\ 0 & -1 & 0 & -2x_j & -y_j & 0 & -3x_j^2 & -2x_jy_j & -y_j^2 & 0 & -3x_j^2y_j & -y_j^3\end{bmatrix} \quad (10.142)$$

$$[M_m]=\begin{bmatrix}1 & x_m & y_m & x_m^2 & x_my_m & y_m^2 & x_m^3 & x_m^2y_m & x_my_m^2 & y_m^3 & x_m^3y_m & x_my_m^3\\ 0 & 0 & 1 & 0 & x_m & 2y_m & 0 & x_m^2 & 2x_my_m & 3y_m^2 & x_m^3 & 3x_my_m^2\\ 0 & -1 & 0 & -2x_m & -y_m & 0 & -3x_m^2 & -2x_my_m & -y_m^2 & 0 & -3x_m^2y_m & -y_m^3\end{bmatrix} \quad (10.143)$$

$$[M_n]=\begin{bmatrix}1 & x_n & y_n & x_n^2 & x_ny_n & y_n^2 & x_n^3 & x_n^2y_n & x_ny_n^2 & y_n^3 & x_n^3y_n & x_ny_n^3\\ 0 & 0 & 1 & 0 & x_n & 2y_n & 0 & x_n^2 & 2x_ny_n & 3y_n^2 & x_n^3 & 3x_ny_n^2\\ 0 & -1 & 0 & -2x_n & -y_n & 0 & -3x_n^2 & -2x_ny_n & -y_n^2 & 0 & -3x_n^2y_n & -y_n^3\end{bmatrix} \quad (10.144)$$

Therefore, the parameters $\{a_e\}$ can be obtained by the following equation

$$\{a_e\}=[\Lambda]^{-1}\{d\} \quad (10.145)$$

Using the above equation, Equation (10.130) can now be written:

$$\{\Psi\}=[X][\Lambda]^{-1}\{d\} \quad (10.146)$$

or

$$\{\Psi\}=[N]\{d\} \quad (10.147)$$

where the matrix $[N]$ containing the shape functions is

$$[N]=[X][\Lambda]^{-1} \quad (10.148)$$

Apart from the displacement matrix $\{\psi\}$, and the curvature matrix $\{\kappa\}$ can also be correlated with the nodal displacements $\{d\}$:

$$\{\kappa\}=\begin{Bmatrix}\kappa_x\\ \kappa_y\\ \kappa_{xy}\end{Bmatrix}=\begin{Bmatrix}-\partial^2 w/\partial x^2\\ -\partial^2 w/\partial y^2\\ -2\partial^2 w/\partial x\partial y\end{Bmatrix} \quad (10.149)$$

Using Equation (10.125), the above equation yields

$$\{\kappa\}=[Y]\{a_e\} \quad (10.150)$$

where

$$[Y]=\begin{bmatrix}0 & 0 & 0 & -2 & 0 & 0 & -6 & -2y & 0 & 0 & -6xy & 0\\ 0 & 0 & 0 & 0 & 0 & -2 & 0 & 0 & -2x & -6y & 0 & 6xy\\ 0 & 0 & 0 & 0 & -2 & 0 & 0 & -4 & -4y & 0 & -6x^2 & -6y^2\end{bmatrix} \quad (10.151)$$

Taking into account Equation (10.145), Equation (10.150) can now be written as

$$\{\kappa\}=[Y][\Lambda]^{-1}\{d\} \quad (10.152)$$

or

$$\{\kappa\}=[B]\{d\} \quad (10.153)$$

where

$$[B]=[Y][\Lambda]^{-1} \quad (10.154)$$

10.2 ISOPARAMETRIC ELEMENTS

10.2.1 DEFINITION OF ISOPARAMETRIC ELEMENTS

As it is already known, the strain energy of a solid element is given by the relation

$$U=\int_{V}\frac{1}{2}\{\sigma\}^{\mathrm{T}}\{\varepsilon\}\mathrm{d}V \tag{10.155}$$

In order to apply the MPE principle for deriving the stiffness matrix, the field functions $\{\varepsilon\}$ and $\{\sigma\}=[E]\{\varepsilon\}$ in the above equation must be correlated with the nodal displacements $\{d\}$. This target was the subject of Section 10.1. However, the followed concept in that section was restricted to elements with rectangular shapes.

Since it is impractical to model complicated geometries by using only rectangular elements, the aim of the present section is to present the concept of the isoparametric elements, that is, nonrectangular elements that can even have curved sides. By using isoparametric elements, modeling of structures with curved edges, as well as grading from a coarse mesh to fine one, is possible.

Isoparametric elements use auxiliary coordinates, usually denoted $\xi-\eta-\zeta$ in three dimensions, or $\xi-\eta$ in two dimensions, in order to simulate a physical element with a reference element that is a cube or square. In the present section, emphasis will be given to plane elements. Therefore, the procedure for the correlation of the displacement field $\{u \;\; v\}$ of a popular nonrectangular plane element to its nodal displacements will be presented.

An element is isoparametric if the matrix $[N]$ correlating the field functions $\{u \;\; v \;\; w\}$ to the nodal functions $\{d\}$ is the same with the matrix $[N^*]$ correlating the coordinate variables $\{x \;\; y \;\; z\}$ with the nodal coordinates $\{c\}$, that is,

$$\left.\begin{array}{l}\{u \;\; v \;\; w\}^{\mathrm{T}}=[N]\{d\}\\ \{x \;\; y \;\; z\}^{\mathrm{T}}=[N^*]\{c\}\end{array}\right\}\;\text{the element is isoparametric if } [N]=[N^*]$$

10.2.2 LAGRANGE POLYNOMIALS

The interpolation functions N_1, N_2, N_3 presented in Equations (10.19)–(10.21) or the functions N_1, N_2, N_3, N_4 presented in Equations (10.77)–(10.80) can be considered as particular cases of the Lagrange polynomials given by

$$N_k=\prod_{\substack{m=1\\k\neq m}}^{n}\frac{x_m-x}{x_m-x_k}=\frac{(x_1-x)(x_2-x)\cdots[x_k-x]\cdots(x_n-x)}{(x_1-x_k)(x_2-x_k)\cdots[x_k-x_k]\cdots(x_n-x_k)} \tag{10.156}$$

where n is the number of nodes. Usually the number of polynomials $N_1, N_2,\ldots, N_k$ should be equal to the number of nodes. In Equation (10.156), the quantities in the brackets $[x_k-x], [x_k-x_k]$ should be omitted. For 2D elements (see Figure 10.6), the interpolation functions N_k ($k=1, 2, \ldots, n$) depends on two variables x, y. If we will use the following notations:

$$N_{k,x}=\prod_{\substack{m=1\\k\neq m}}^{n}\frac{x_m-x}{x_m-x_k} \tag{10.157}$$

$$N_{k,y}=\prod_{\substack{m=1\\k\neq m}}^{n}\frac{y_m-y}{y_m-y_k} \tag{10.158}$$

the corresponding interpolation functions are given by the formula:

$$N_k = N_{k,x} N_{k,y} \tag{10.159}$$

In Equations (10.157) and (10.158) n is the number of nodes in each side of the plane element. Since in each line $x = -b, x = b, y = -b, y = b$ of the element shown in Figure 10.6 we have two nodes, the value of n is $n = 2$.

Taking into account Figure 10.6, if we set in Equations (10.157) and (10.158)

$$\begin{aligned} x_1 &= -b, \quad x_2 = b, \quad x_3 = b, \quad x_4 = -b \\ y_1 &= -h, \quad y_2 = -h, \quad y_3 = h, \quad y_4 = h \end{aligned} \tag{10.160}$$

the following expressions can be obtained:

$$N_1 = N_{1,x} N_{1,y} = \frac{(b-x)(h-y)}{4ah} \tag{10.161}$$

$$N_2 = N_{2,x} N_{2,y} = \frac{(b+x)(h-y)}{4ah} \tag{10.162}$$

$$N_3 = N_{3,x} N_{3,y} = \frac{(b+x)(h+y)}{4ah} \tag{10.163}$$

$$N_4 = N_{4,x} N_{4,y} = \frac{(b-x)(h+y)}{4ah} \tag{10.164}$$

10.2.3 THE BILINEAR QUADRILATERAL ELEMENT

Let us assume the four-node, nonrectangular element of Figure 10.10a. This is a typical example for describing the concept of using isoparametric formulation to model nonrectangular elements.

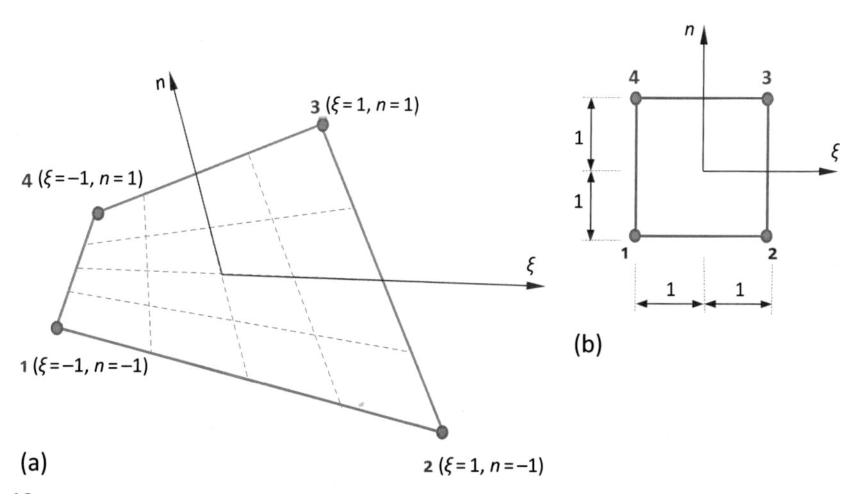

FIGURE 10.10

(a) Physical plane element and (b) mapped plane element.

We recall Equation (10.84) correlating the displacement field $\{\Psi\}=\{u(x,y)\ \ v(x,y)\}^{\mathrm{T}}$ to the nodal displacements $\{d\}=\{u_1\ \ v_1\ \ u_2\ \ v_2\ \ u_3\ \ v_3\ \ u_4\ \ v_4\}^{\mathrm{T}}$:

$$\begin{Bmatrix} u(x,y) \\ v(x,y) \end{Bmatrix} = \begin{bmatrix} N_1 & 0 & N_2 & 0 & N_3 & 0 & N_4 & 0 \\ 0 & N_1 & 0 & N_2 & 0 & N_3 & 0 & N_4 \end{bmatrix} \{d\} \tag{10.165}$$

According to the definition of the isoparametric elements, the coordinates $\{x,y\}$ should be correlated to the nodal coordinates

$$\{c\}=\{x_1\ \ y_1\ \ x_2\ \ y_2\ \ x_3\ \ y_3\ \ x_4\ \ y_4\}^{\mathrm{T}} \tag{10.166}$$

by the following equation:

$$\begin{Bmatrix} x \\ y \end{Bmatrix} = \begin{bmatrix} N_1 & 0 & N_2 & 0 & N_3 & 0 & N_4 & 0 \\ 0 & N_1 & 0 & N_2 & 0 & N_3 & 0 & N_4 \end{bmatrix} \{c\} \tag{10.167}$$

where the polynomials N_1, N_2, N_3, N_4 are given by Equations (10.77)–(10.80) or by Equations (10.161)–(10.164). Setting to these equations the parameters corresponding to the mapped element shown in Figure 10.10b, that is, $x \to \xi, y \to \eta, a=1, b=1$, the following formulae can be obtained:

$$N_1 = \frac{(1-\xi)(1-\eta)}{4} \tag{10.168}$$

$$N_2 = \frac{(1+\xi)(1-\eta)}{4} \tag{10.169}$$

$$N_3 = \frac{(1+\xi)(1+\eta)}{4} \tag{10.170}$$

$$N_4 = \frac{(1-\xi)(1+\eta)}{4} \tag{10.171}$$

Since, according to Equation (9.138), the strains $\varepsilon_x \varepsilon_y \gamma_{xy}$ are correlated to the derivatives of the displacements $\partial u/\partial x, \partial v/\partial y, \partial u/\partial y, \partial v/\partial x$

$$\{\varepsilon\} = \begin{Bmatrix} \varepsilon_x \\ \varepsilon_y \\ \gamma_{xy} \end{Bmatrix} = \begin{bmatrix} 1 & 0 & 0 & 0 \\ 0 & 0 & 0 & 1 \\ 0 & 1 & 1 & 0 \end{bmatrix} \begin{Bmatrix} \partial u/\partial x \\ \partial u/\partial y \\ \partial v/\partial x \\ \partial v/\partial y \end{Bmatrix} \tag{10.172}$$

then, Equation (10.165) should be expressed in terms of the derivatives $\partial N_i/\partial x, \partial N_i/\partial y (i=1,2,3,4)$. However, according to Equations (10.168)–(10.171), the functions $N_i (i=1,2,3,4)$ depend on ξ, η. Therefore, in order to perform the derivatives $\partial N_i(\xi,\eta)/\partial x, \partial N_i(\xi,\eta)/\partial y (i=1,2,3,4)$, the following properties can be used

$$\frac{\partial N_i}{\partial \xi} = \frac{\partial N_i}{\partial x}\frac{\partial x}{\partial \xi} + \frac{\partial N_i}{\partial y}\frac{\partial y}{\partial \xi} \tag{10.173}$$

$$\frac{\partial N_i}{\partial \eta} = \frac{\partial N_i}{\partial x}\frac{\partial x}{\partial \eta} + \frac{\partial N_i}{\partial y}\frac{\partial y}{\partial \eta} \tag{10.174}$$

or

$$\begin{Bmatrix} \partial N_i/\partial \xi \\ \partial N_i/\partial \eta \end{Bmatrix} = \begin{bmatrix} \partial x/\partial \xi & \partial y/\partial \xi \\ \partial x/\partial \eta & \partial y/\partial \eta \end{bmatrix} \begin{Bmatrix} \partial N_i/\partial x \\ \partial N_i/\partial y \end{Bmatrix} \tag{10.175}$$

In the above equation the matrix

$$[J] = \begin{bmatrix} \partial x/\partial \xi & \partial y/\partial \xi \\ \partial x/\partial \eta & \partial y/\partial \eta \end{bmatrix} \tag{10.176}$$

is called Jacobian matrix.

Taking into account Equation (10.175), the following expression can be obtained:

$$\begin{Bmatrix} \partial N_i/\partial x \\ \partial N_i/\partial y \end{Bmatrix} = [J]^{-1} \begin{Bmatrix} \partial N_i/\partial \xi \\ \partial N_i/\partial \eta \end{Bmatrix} \tag{10.177}$$

Using Equation (10.167), the Jacobian matrix can now be written:

$$\begin{bmatrix} \partial x/\partial \xi & \partial y/\partial \xi \\ \partial x/\partial \eta & \partial y/\partial \eta \end{bmatrix} = \begin{bmatrix} \sum \frac{\partial N_i}{\partial \xi} x_i & \sum \frac{\partial N_i}{\partial \xi} y_i \\ \sum \frac{\partial N_i}{\partial \eta} x_i & \sum \frac{\partial N_i}{\partial \eta} y_i \end{bmatrix} \tag{10.178}$$

Taking into account Equations (10.168)–(10.171), the above equation can be written:

$$[J] = \begin{bmatrix} J_{11} & J_{12} \\ J_{21} & J_{22} \end{bmatrix} \tag{10.179}$$

where

$$J_{11} = \frac{\partial N_1}{\partial \xi} x_1 + \frac{\partial N_2}{\partial \xi} x_2 + \frac{\partial N_3}{\partial \xi} x_3 + \frac{\partial N_4}{\partial \xi} x_4 \tag{10.180}$$

$$J_{12} = \frac{\partial N_1}{\partial \xi} y_1 + \frac{\partial N_2}{\partial \xi} y_2 + \frac{\partial N_3}{\partial \xi} y_3 + \frac{\partial N_4}{\partial \xi} y_4 \tag{10.181}$$

$$J_{21} = \frac{\partial N_1}{\partial \eta} x_1 + \frac{\partial N_2}{\partial \eta} x_2 + \frac{\partial N_3}{\partial \eta} x_3 + \frac{\partial N_4}{\partial \eta} x_4 \tag{10.182}$$

$$J_{22} = \frac{\partial N_1}{\partial \eta} y_1 + \frac{\partial N_2}{\partial \eta} y_2 + \frac{\partial N_3}{\partial \eta} y_3 + \frac{\partial N_4}{\partial \eta} y_4 \tag{10.183}$$

As it can be shown by Equation (10.177), the derivatives of any field function g can now be derived by the following equation:

$$\begin{Bmatrix} \partial g/\partial x \\ \partial g/\partial y \end{Bmatrix} = [J]^{-1} \begin{Bmatrix} \partial g/\partial \xi \\ \partial g/\partial \eta \end{Bmatrix} \tag{10.184}$$

where

$$[J]^{-1} = \begin{bmatrix} I_{11} & I_{12} \\ I_{21} & I_{22} \end{bmatrix} = \frac{1}{\det[J]} \begin{bmatrix} J_{22} & -J_{12} \\ -J_{21} & J_{11} \end{bmatrix} \tag{10.185}$$

Setting $g = u$ or $g = v$ into Equation (10.184), Equation (10.172) can now be written as

$$\{\varepsilon\} = \begin{bmatrix} 1 & 0 & 0 & 0 \\ 0 & 0 & 0 & 1 \\ 0 & 1 & 1 & 0 \end{bmatrix} \begin{bmatrix} I_{11} & I_{12} & 0 & 0 \\ I_{21} & I_{22} & 0 & 0 \\ 0 & 0 & I_{11} & I_{12} \\ 0 & 0 & I_{21} & I_{22} \end{bmatrix} \begin{Bmatrix} \partial u/\partial\xi \\ \partial u/\partial\eta \\ \partial v/\partial\xi \\ \partial v/\partial\eta \end{Bmatrix} \tag{10.186}$$

In the above equation, the matrix $\{\partial u/\partial\xi \;\; \partial u/\partial\eta \;\; \partial v/\partial\xi \;\; \partial v/\partial\eta\}^{\mathrm{T}}$ can be derived using Equation (10.165). Therefore:

$$\begin{Bmatrix} \partial u/\partial\xi \\ \partial u/\partial\eta \\ \partial v/\partial\xi \\ \partial v/\partial\eta \end{Bmatrix} = \begin{bmatrix} \frac{\partial N_1}{\partial\xi} & 0 & \frac{\partial N_2}{\partial\xi} & 0 & \frac{\partial N_3}{\partial\xi} & 0 & \frac{\partial N_4}{\partial\xi} & 0 \\ \frac{\partial N_1}{\partial\eta} & 0 & \frac{\partial N_2}{\partial\eta} & 0 & \frac{\partial N_3}{\partial\eta} & 0 & \frac{\partial N_4}{\partial\eta} & 0 \\ 0 & \frac{\partial N_1}{\partial\xi} & 0 & \frac{\partial N_2}{\partial\xi} & 0 & \frac{\partial N_3}{\partial\xi} & 0 & \frac{\partial N_4}{\partial\xi} \\ 0 & \frac{\partial N_1}{\partial\eta} & 0 & \frac{\partial N_2}{\partial\eta} & 0 & \frac{\partial N_3}{\partial\eta} & 0 & \frac{\partial N_4}{\partial\eta} \end{bmatrix} \begin{Bmatrix} u_1 \\ v_1 \\ u_2 \\ v_2 \\ u_3 \\ v_3 \\ u_4 \\ v_4 \end{Bmatrix} \tag{10.187}$$

Combining Equations (10.186) and (10.187), the matrix $\{\varepsilon\}$ for the nonrectangular element shown in Figure 10.10a can be correlated to the nodal displacements $\{d\}$ by the following equation:

$$\{\varepsilon\} = [B]\{d\} \tag{10.188}$$

where

$$[B] = [H_1][H_2][H_3] \tag{10.189}$$

$$[H_1] = \begin{bmatrix} 1 & 0 & 0 & 0 \\ 0 & 0 & 0 & 1 \\ 0 & 1 & 1 & 0 \end{bmatrix} \tag{10.190}$$

$$[H_2] = \begin{bmatrix} I_{11} & I_{12} & 0 & 0 \\ I_{21} & I_{22} & 0 & 0 \\ 0 & 0 & I_{11} & I_{12} \\ 0 & 0 & I_{21} & I_{22} \end{bmatrix} \tag{10.191}$$

$$[H_3] = \begin{bmatrix} \partial N_1/\partial\xi & 0 & \partial N_2/\partial\xi & 0 & \partial N_3/\partial\xi & 0 & \partial N_4/\partial\xi & 0 \\ \partial N_1/\partial\eta & 0 & \partial N_2/\partial\eta & 0 & \partial N_3/\partial\eta & 0 & \partial N_4/\partial\eta & 0 \\ 0 & \partial N_1/\partial\xi & 0 & \partial N_2/\partial\xi & 0 & \partial N_3/\partial\xi & 0 & \partial N_4/\partial\xi \\ 0 & \partial N_1/\partial\eta & 0 & \partial N_2/\partial\eta & 0 & \partial N_3/\partial\eta & 0 & \partial N_4/\partial\eta \end{bmatrix} \tag{10.192}$$

Apart from the bilinear quadrilateral element, a similar procedure can be followed for $[B]$ matrix derivation of other types of isoparametric elements. However, the research in FEM is on-going and special elements and new techniques for a variety of engineering problems are available in publications.

10.3 DERIVATION OF STIFFNESS MATRICES

In Section 10.1, matrix equations of the type $\{u(x,y)\ v(x,y)\}^{\mathrm{T}} = [N]\{d\}$ or $\{u(x,y,z)\ v(x,y,z)\ w(x,y,z)\}^{\mathrm{T}} = [N]\{d\}$ correlating the field displacements to the nodal displacements for the linear triangular, quadratic triangular, bilinear rectangular, tetrahedral solid, rectangular solid, and plate bending elements have been presented. The above equations will now be used for the correlation of the strain and stress fields, $\{\varepsilon\}$ and $\{\sigma\}$, respectively, with the nodal displacements $\{d\}$, in order to derive the corresponding stiffness matrices using the MPE principle.

10.3.1 THE LINEAR TRIANGULAR ELEMENT (OR CST ELEMENT)

We recall the linear triangular element shown in Figure 10.4. As it is known from the mechanics of solids, the matrix $\{\varepsilon\}$ is correlated to the field displacements $\{u(x,y)\ v(x,y)\}$ by the following equation:

$$\{\varepsilon\} = \begin{Bmatrix} \varepsilon_x \\ \varepsilon_y \\ \gamma_{xy} \end{Bmatrix} = \begin{Bmatrix} \partial u/\partial x \\ \partial v/\partial y \\ \partial u/\partial y + \partial v/\partial x \end{Bmatrix} \tag{10.193}$$

Combining the above equation with Equation (10.57), the following expression can be derived

$$\begin{Bmatrix} \varepsilon_x \\ \varepsilon_y \\ \gamma_{xy} \end{Bmatrix} = \frac{1}{2A} \begin{bmatrix} \beta_i & 0 & \beta_j & 0 & \beta_m & 0 \\ 0 & \gamma_i & 0 & \gamma_j & 0 & \gamma_m \\ \gamma_i & \beta_i & \gamma_j & \beta_j & \gamma_m & \beta_m \end{bmatrix} \begin{Bmatrix} u_i \\ v_i \\ u_j \\ v_j \\ u_m \\ v_m \end{Bmatrix} \tag{10.194}$$

where $\beta_i, \beta_j, \beta_m, \gamma_i, \gamma_j, \gamma_m$ can be obtained from Equations (10.43)–(10.48). Equation (10.194) can be written in the following abbreviated form:

$$\{\varepsilon\} = [B]\{d\} \tag{10.195}$$

where

$$[B] = \frac{1}{2A} \begin{bmatrix} \beta_i & 0 & \beta_j & 0 & \beta_m & 0 \\ 0 & \gamma_i & 0 & \gamma_j & 0 & \gamma_m \\ \gamma_i & \beta_i & \gamma_j & \beta_j & \gamma_m & \beta_m \end{bmatrix} \tag{10.196}$$

However, according to Hooke's law for an isotropic plate in plane stress conditions, the stress $\{\sigma\}$ is correlated to $\{\varepsilon\}$ by Equation (9.136)

$$\{\sigma\} = \begin{Bmatrix} \sigma_x \\ \sigma_y \\ \sigma_{xy} \end{Bmatrix} = [E]\{\varepsilon\} \tag{10.197}$$

where the matrix $[E]$ is already known from Equation (9.137). Combining Equations (10.195) and (10.197), it can be written:

$$\{\sigma\} = [E][B]\{d\} \tag{10.198}$$

Therefore, using Equations (10.195) and (10.198), the strain energy U given by Equation (9.154) now takes the form

$$U = \frac{1}{2}\int_V \{d\}^{\mathrm{T}}[B]^{\mathrm{T}}[E][B]\{d\}\mathrm{d}V \tag{10.199}$$

Since (a) the surface forces acting on an element are usually simulated by concentrated forces acting on the nodes, and (b) we can neglect the inertia forces (body forces) because the problem is static, the work of the external forces is given by Equation (9.157). The last equation can be written in the following abbreviated form:

$$W_c = -\{d\}^{\mathrm{T}}\{f\} \tag{10.200}$$

where

$$\{d\}^{\mathrm{T}} = \{u_i \;\; v_i \;\; u_j \;\; v_j \;\; u_m \;\; v_m\} \tag{10.201}$$

and

$$\{f\} = \{f_{xi} \;\; f_{yi} \;\; f_{xj} \;\; f_{yj} \;\; f_{xm} \;\; f_{ym}\}^{\mathrm{T}} \tag{10.202}$$

Therefore, combining Equations (9.151), (10.199), and (10.200), the total potential energy can now be written as:

$$\Pi_p = \frac{1}{2}\int_V \{d\}^{\mathrm{T}}[B]^{\mathrm{T}}[E][B]\{d\}\mathrm{d}V - \{d\}^{\mathrm{T}}\{f\} \tag{10.203}$$

Applying the principle of the MPE, the above equation yields

$$\frac{\partial \Pi_p}{\partial\{d\}} = 0 \tag{10.204}$$

or

$$\left[\int_V [B]^{\mathrm{T}}[E][B]\mathrm{d}V\right]\{d\} - \{f\} = 0 \tag{10.205}$$

For an element of constant thickness t, the above equation can be written as

$$\left[t\iint [B]^{\mathrm{T}}[E][B]\mathrm{d}x\mathrm{d}y\right]\{d\} = \{f\} \tag{10.206}$$

From the above equation, it can be concluded that the stiffness matrix $\{k\}$ of the linear triangular element (or CST element) can be derived by the following equation:

$$[k] = \left[t\iint [B]^{\mathrm{T}}[E][B]\mathrm{d}x\mathrm{d}y\right] \tag{10.207}$$

However, the matrices $[E]$ and $[B]$ are independent of x or y. Therefore, the above equation can now be written in the following form:

$$[k] = tA[B]^{\mathrm{T}}[E][B] \tag{10.208}$$

where A is the area of the triangle given by Equation (10.49).

10.3.2 THE QUADRATIC TRIANGULAR ELEMENT (OR LST ELEMENT)

Following a similar procedure, the stiffness matrix for the quadratic triangular element shown in Figure 10.5 is given by the following well-known equation:

$$[k] = tA[B]^{\mathrm{T}}[E][B] \tag{10.209}$$

In the above equation, t is the thickness and A the area of the triangular element, $[E]$ is given from Equation (9.137), and the matrix $[B]$ can be derived from the combination of Equations (10.193) and (10.67).

10.3.3 THE BILINEAR RECTANGULAR ELEMENT (OR Q4 ELEMENT)

As it has already been shown, the stiffness matrix is given again by the following general equation

$$[k] = t\int_{-h}^{h}\int_{-b}^{b}[B]^{\mathrm{T}}[E][B]\mathrm{d}x\mathrm{d}y \tag{10.210}$$

For the case of the bilinear rectangular element demonstrated in Figure 10.6, the matrix $[E]$ is given in Equation (9.137) and the matrix $[B]$ can be derived from the combination of Equations (10.84) and (10.193).

10.3.4 THE TETRAHEDRAL SOLID ELEMENT

Even though the stiffness matrix for the tetrahedral solid element shown in Figure 10.7 is given in the following already-known type of equation,

$$[k] = \left[\iint_V [B]^{\mathrm{T}}[E][B]\mathrm{d}V\right] \tag{10.211}$$

the matrix $[E]$ should now be obtained by the inversion of Equation (9.131), which is valid for 3D isotropic solids.

The matrix $[B]$ can be obtained from the combination of Equation (9.138) with Equation (10.93).

10.3.5 EIGHT-NODE RECTANGULAR SOLID ELEMENT

For the eight-node rectangular solid element shown in Figure 10.8, the stiffness matrix is given again from Equation (10.211). The matrix $[E]$ can be obtained from the inversion of Equation (9.131) and the matrix $[B]$ can be derived from the combination of Equation (9.138) with Equation (10.124).

10.3.6 PLATE BENDING ELEMENT

For the plate bending element shown in Figure 10.9, the stiffness matrix is given from the following formula:

$$[k] = \frac{t^3}{12}\iint_A [B]^{\mathrm{T}}[E][B]\mathrm{d}x\mathrm{d}y \tag{10.212}$$

In the above formula, the matrix $[E]$ can be obtained from Equation (9.137) and the matrix $[B]$ is given in Equation (10.154).

10.3.7 ISOPARAMETRIC FORMULATION

When we use isoparametric elements, the matrix $[B]$ is a function of the variables $\xi, \ \eta$. In that case, the stiffness matrix for 2D elements given by Equation (10.210) has to be transformed with respect to the variables $\xi, \ \eta$. Therefore, for 2D isoparametric elements the following equation should be used:

$$[k] = t \iint_A [B]^{\mathrm{T}} [E][B] J \mathrm{d}\xi \mathrm{d}\eta \tag{10.213}$$

We recall that for a bilinear quadrilateral element shown in Figure 10.10a, the matrix $[B]$ is given by Equation (10.189). The matrix $[E]$ contains only elastic constants and can be obtained by Equation (9.137). The parameter J is the Jacobian given by

$$J = \det[J] \tag{10.214}$$

where $[J]$ is given in Equation (10.179). It should be noted that the calculation of the integral of Equation (10.213) usually is possible only by numerical methods. Among the numerical methods, the Gauss Quadrature method seems to be the most effective. Since there are many available important commercial software tools (e.g., Mathematica, MatLab, etc.) for numerical integration, the focus of such methods is beyond the scope of this book.

EXAMPLE 10.1

Determine the nodal displacements and stresses of the following structural part.

Data

$$E = 200 \text{ GPa}$$
$$\nu = 0.3$$
$$a = 70 \text{ cm}$$
$$b = 35 \text{ cm}$$
$$t = 2 \text{ cm}$$
$$q_x = 240 \text{MPa}$$
$$q_y = 200 \text{MPa}$$

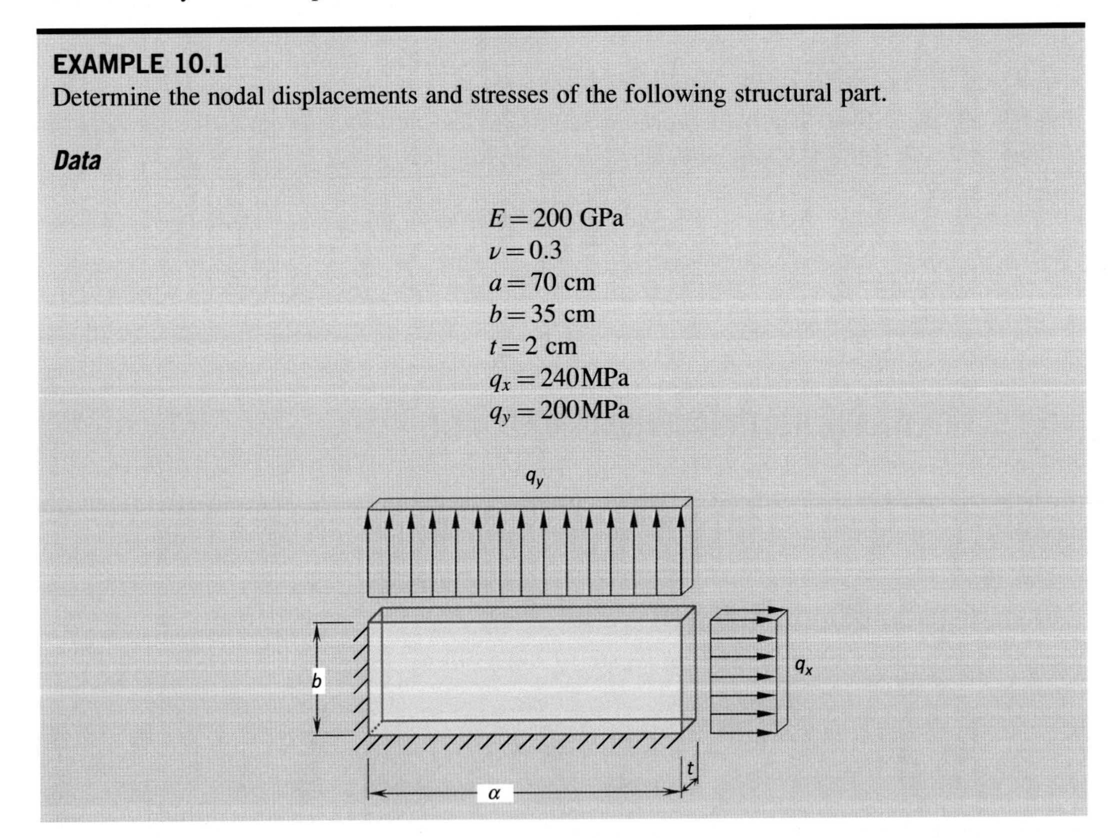

Solution

The above structure is discretized into two linear triangular elements (or CST elements). The first step for the solution of the above problem is the transformation of the variable loads q_x, q_y into nodal concentrated ones.

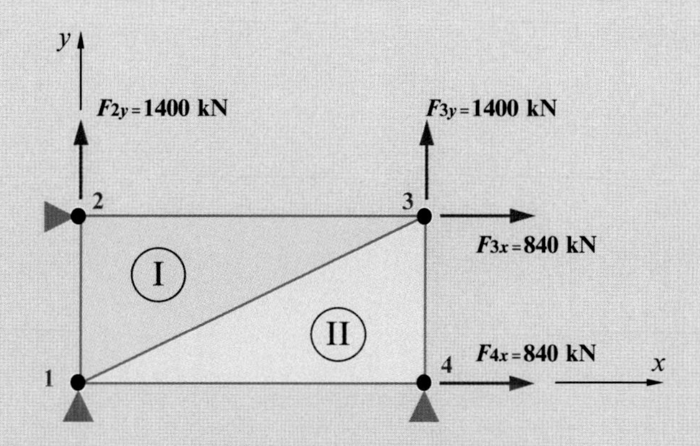

According to the following equilibrium equations, the concentrated loads are:

$$F_{3x} = F_{4x} = \frac{q_x bt}{2} = 840\text{kN} \tag{e 10.1-1}$$

$$F_{2y} = F_{3y} = \frac{q_y at}{2} = 1400 \text{ kN} \tag{e 10.1-2}$$

The stiffness matrix for each element is given by the following equation (10.208):

$$[k] = tA[B]^{\mathrm{T}}[E][B] \tag{e 10.1-3}$$

where A is the area of each triangle, and $[E]$ and $[B]$ are given by the following equations:

$$[E] = \begin{bmatrix} E/(1-\nu^2) & \nu E/(1-\nu^2) & 0 \\ \nu E/(1-\nu^2) & E/(1-\nu^2) & 0 \\ 0 & 0 & G \end{bmatrix} = \begin{bmatrix} 21.978 & 6.593 & 0 \\ 6.593 & 21.978 & 0 \\ 0 & 0 & 7.69 \end{bmatrix} \times 10^{10}$$

$$[B] = \frac{1}{2A}\begin{bmatrix} \beta_i & 0 & \beta_j & 0 & \beta_m & 0 \\ 0 & \gamma_i & 0 & \gamma_j & 0 & \gamma_m \\ \gamma_i & \beta_i & \gamma_j & \beta_j & \gamma_m & \beta_m \end{bmatrix}$$

where the parameters $\beta_i, \beta_j, \beta_m, \gamma_i, \gamma_j, \gamma_m$ can be obtained from Equations (10.43)–(10.48):

$$\beta_i = y_j - y_m, \quad \gamma_i = x_m - x_j$$

$$\beta_j = y_m - y_i, \quad \gamma_j = x_i - x_m$$

$$\beta_m = y_i - y_j, \quad \gamma_m = x_j - x_i$$

Continued

EXAMPLE 10.1—CONT'D

Stiffness Matrix for Element I

Since the nodes of each element should be labeled in a counter-clockwise manner, the following nomenclature should be used:

$$i=1,\ j=3,\ m=2$$

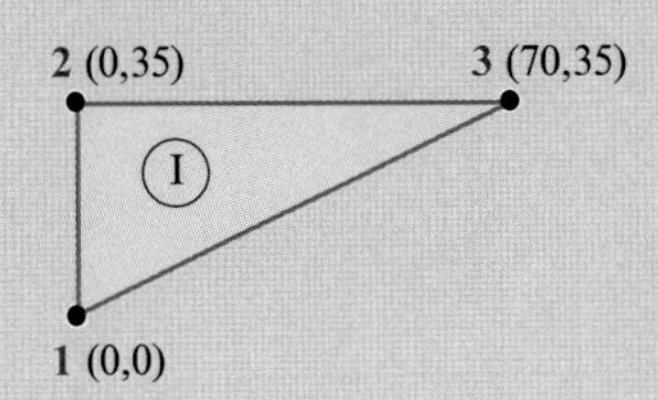

Therefore:

$$\beta_i = y_3 - y_2 = (35-35)\times 10^{-2} = 0,\ \ \gamma_i = x_2 - x_3 = (0-70)\times 10^{-2} = -70\times 10^{-2}$$

$$\beta_j = y_2 - y_1 = (35-0)\times 10^{-2} = 35\times 10^{-2},\ \ \gamma_j = x_1 - x_2 = (0-0)\times 10^{-2} = 0$$

$$\beta_m = y_1 - y_3 = (0-35)\times 10^{-2} = -35\times 10^{-2},\ \ \gamma_m = x_3 - x_1 = (70-0)\times 10^{-2} = 70\times 10^{-2}$$

and

$$A = \frac{1}{2}ba = 0.1225\ \text{m}^2$$

Taking into account the above values, the following result for the matrix $[B]$ for the element I, denoted by $[B_I]$, can be derived:

$$[B_I] = \begin{bmatrix} 0 & 0 & 1.42857 & 0 & -1.42857 & 0 \\ 0 & -2.85714 & 0 & 0 & 0 & 2.85714 \\ -2.85714 & 0 & 0 & 1.42857 & 2.85714 & -1.42857 \end{bmatrix} \quad \text{(e 10.1-4)}$$

Therefore, using Equations (e 10.1-3), (9.137), and (10.196), the following results for the stiffness matrix of the element I can be obtained:

$$\left[kI_{ij}\right] = \begin{bmatrix} 28.5714 & 0 & 0 & -14.2857 & -28.5714 & 14.2857 \\ 0 & 43.956 & -6.59341 & 0 & 6.59341 & -43.956 \\ 0 & -6.59341 & 10.989 & 0 & -10.989 & 6.59341 \\ -14.2857 & 0 & 0 & 7.14286 & 14.2857 & -7.14286 \\ -28.5714 & 6.59341 & -10.989 & 14.2857 & 39.5604 & -20.8791 \\ 14.2857 & -43.956 & 6.59341 & -7.14286 & -20.8791 & 51.0989 \end{bmatrix} \times 10^8$$

(e 10.1-5)

Using the above results, the element equation can be written as:

$$\begin{bmatrix} kI_{11} & kI_{12} & kI_{13} & kI_{14} & kI_{15} & kI_{16} \\ kI_{21} & kI_{22} & kI_{23} & kI_{24} & kI_{25} & kI_{26} \\ kI_{31} & kI_{32} & kI_{33} & kI_{34} & kI_{35} & kI_{36} \\ kI_{41} & kI_{42} & kI_{43} & kI_{44} & kI_{45} & kI_{46} \\ kI_{51} & kI_{52} & kI_{53} & kI_{54} & kI_{55} & kI_{56} \\ kI_{61} & kI_{62} & kI_{63} & kI_{64} & kI_{65} & kI_{66} \end{bmatrix} \begin{Bmatrix} d_{1x} \\ d_{1y} \\ d_{3x} \\ d_{3y} \\ d_{2x} \\ d_{2y} \end{Bmatrix} = \begin{Bmatrix} f_{1x} \\ f_{1y} \\ f_{3x} \\ f_{3y} \\ f_{2x} \\ f_{2y} \end{Bmatrix} \qquad \text{(e 10.1-6)}$$

The above element equation should be expanded to the global coordinate system. To achieve this target, the columns of the above matrix should be rearranged in order to comply with the order of the global coordinate system:

$$\begin{bmatrix} kI_{11} & kI_{12} & kI_{15} & kI_{16} & kI_{13} & kI_{14} & 0 & 0 \\ kI_{21} & kI_{22} & kI_{25} & kI_{26} & kI_{23} & kI_{24} & 0 & 0 \\ kI_{31} & kI_{32} & kI_{35} & kI_{36} & kI_{33} & kI_{34} & 0 & 0 \\ kI_{41} & kI_{42} & kI_{45} & kI_{46} & kI_{43} & kI_{44} & 0 & 0 \\ kI_{51} & kI_{52} & kI_{55} & kI_{56} & kI_{53} & kI_{54} & 0 & 0 \\ kI_{61} & kI_{62} & kI_{65} & kI_{66} & kI_{63} & kI_{64} & 0 & 0 \\ 0 & 0 & 0 & 0 & 0 & 0 & 0 & 0 \\ 0 & 0 & 0 & 0 & 0 & 0 & 0 & 0 \end{bmatrix} \begin{Bmatrix} d_{1x} \\ d_{1y} \\ d_{2x} \\ d_{2y} \\ d_{3x} \\ d_{3y} \\ d_{4x} \\ d_{4y} \end{Bmatrix} = \begin{Bmatrix} f_{1x} \\ f_{1y} \\ f_{2x} \\ f_{2y} \\ f_{3x} \\ f_{3y} \\ f_{4x} \\ f_{4y} \end{Bmatrix} \qquad \text{(e 10.1-7)}$$

Stiffness Matrix for Element II

Again, the nodes of each element should be labeled in a counter-clockwise manner. Therefore, the following nomenclature should be used:

$$i=1, \; j=4, \; m=3$$

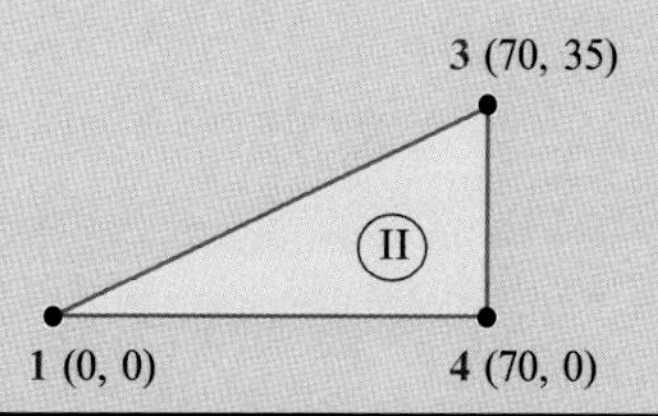

Continued

EXAMPLE 10.1—CONT'D

Therefore:

$$\beta_i = y_4 - y_3 = (0-35)\times 10^{-2} = -35\times 10^{-2},\quad \gamma_i = x_3 - x_4 = (70-70)\times 10^{-2} = 0$$

$$\beta_j = y_3 - y_1 = (35-0)\times 10^{-2} = 35\times 10^{-2},\quad \gamma_j = x_1 - x_3 = (0-70)\times 10^{-2} = -70\times 10^{-2}$$

$$\beta_m = y_1 - y_4 = (0-0) = 0,\quad \gamma_m = x_4 - x_1 = (70-0)\times 10^{-2} = 70\times 10^{-2}$$

Taking into account the above values, as well as the value $A = 0.1225\ \mathrm{m}^2$, the following results for the matrix $[B_{\mathrm{II}}]$ for the element II can be obtained:

$$[B_{\mathrm{II}}] = \begin{bmatrix} -1.42857 & 0 & 1.42857 & 0 & 0 & 0 \\ 0 & 0 & 0 & -2.85714 & 0 & 2.85714 \\ 0 & -1.42857 & -2.85714 & 1.42857 & 2.85714 & 0 \end{bmatrix} \qquad \text{(e 10.1-8)}$$

Then, using Equations (e 10.1-3), (9.137), and (10.196), the following results for the stiffness matrix of the element II can be obtained:

$$[k\mathrm{II}_{ij}] = \begin{bmatrix} 10.989 & 0 & -10.898 & 6.5934 & 0 & -6.5934 \\ 0 & 7.1428 & 14.2857 & -7.1428 & -14.2857 & 0 \\ -10.989 & 14.2857 & 39.5604 & -20.8791 & -28.5714 & 6.5934 \\ 6.5934 & -7.1428 & -20.8791 & 51.0989 & 14.2857 & -43.956 \\ 0 & -14.2857 & -28.5714 & 14.2857 & 28.5714 & 0 \\ -6.5934 & 0 & 6.5934 & -43.956 & 0 & 43.956 \end{bmatrix} \times 10^8 \qquad \text{(e 10.1-8)}$$

Taking into account the nomenclature for the nodes of the element II, the following equation can be derived:

$$\begin{bmatrix} k\mathrm{II}_{11} & k\mathrm{II}_{12} & k\mathrm{II}_{13} & k\mathrm{II}_{14} & k\mathrm{II}_{15} & k\mathrm{II}_{16} \\ k\mathrm{II}_{21} & k\mathrm{II}_{22} & k\mathrm{II}_{23} & k\mathrm{II}_{24} & k\mathrm{II}_{25} & k\mathrm{II}_{26} \\ k\mathrm{II}_{31} & k\mathrm{II}_{32} & k\mathrm{II}_{33} & k\mathrm{II}_{34} & k\mathrm{II}_{35} & k\mathrm{II}_{36} \\ k\mathrm{II}_{41} & k\mathrm{II}_{42} & k\mathrm{II}_{43} & k\mathrm{II}_{44} & k\mathrm{II}_{45} & k\mathrm{II}_{46} \\ k\mathrm{II}_{51} & k\mathrm{II}_{52} & k\mathrm{II}_{53} & k\mathrm{II}_{54} & k\mathrm{II}_{55} & k\mathrm{II}_{56} \\ k\mathrm{II}_{61} & k\mathrm{II}_{62} & k\mathrm{II}_{63} & k\mathrm{II}_{64} & k\mathrm{II}_{65} & k\mathrm{II}_{66} \end{bmatrix} \begin{Bmatrix} d_{1x} \\ d_{1y} \\ d_{4x} \\ d_{4y} \\ d_{3x} \\ d_{3y} \end{Bmatrix} = \begin{Bmatrix} f_{1x} \\ f_{1y} \\ f_{4x} \\ f_{4y} \\ f_{3x} \\ f_{3y} \end{Bmatrix} \qquad \text{(e 10.1-9)}$$

The above element equation will now be expanded to the global coordinate system. It should be noticed that the columns and rows of the above matrix should be rearranged in order to comply with the order of the global coordinate system:

$$\begin{bmatrix} kII_{11} & kII_{12} & 0 & 0 & kII_{15} & kII_{16} & kII_{13} & kII_{14} \\ kII_{21} & kII_{22} & 0 & 0 & kII_{25} & kII_{26} & kII_{23} & kII_{24} \\ 0 & 0 & 0 & 0 & 0 & 0 & 0 & 0 \\ 0 & 0 & 0 & 0 & 0 & 0 & 0 & 0 \\ kII_{51} & kII_{52} & 0 & 0 & kII_{55} & kII_{56} & kII_{53} & kII_{54} \\ kII_{61} & kII_{62} & 0 & 0 & kII_{65} & kII_{66} & kII_{63} & kII_{64} \\ kII_{31} & kII_{32} & 0 & 0 & kII_{35} & kII_{36} & kII_{33} & kII_{34} \\ kII_{41} & kII_{42} & 0 & 0 & kII_{45} & kII_{46} & kII_{43} & kII_{44} \end{bmatrix} \begin{Bmatrix} d_{1x} \\ d_{1y} \\ d_{2x} \\ d_{2y} \\ d_{3x} \\ d_{3y} \\ d_{4x} \\ d_{4y} \end{Bmatrix} = \begin{Bmatrix} f_{1x} \\ f_{1y} \\ f_{2x} \\ f_{2y} \\ f_{3x} \\ f_{3y} \\ f_{4x} \\ f_{4y} \end{Bmatrix} \quad \text{(e 10.1-10)}$$

The Global Matrix for the Structure

Superposition of Equations (e 10.1-7) and (e 10.1-10) yields the following structure equation:

$$\begin{bmatrix} kI_{11}+kII_{11} & kI_{12}+kII_{12} & kI_{15} & kI_{16} & kI_{13}+kII_{16} & kI_{14}+kII_{16} & kII_{13} & kII_{14} \\ kI_{21}+kII_{21} & kI_{22}+kII_{22} & kI_{25} & kI_{26} & kI_{23}+kII_{25} & kI_{24}+kII_{26} & kII_{23} & kII_{24} \\ kI_{31} & kI_{32} & kI_{35} & kI_{36} & kI_{33} & kI_{34} & 0 & 0 \\ kI_{41} & kI_{42} & kI_{45} & kI_{46} & kI_{43} & kI_{44} & 0 & 0 \\ kI_{51}+kII_{51} & kI_{52}+kII_{52} & kI_{55} & kI_{56} & kI_{53}+kII_{55} & kI_{54}+kII_{56} & kII_{53} & kII_{54} \\ kI_{61}+kII_{61} & kI_{62}+kII_{62} & kI_{65} & kI_{66} & kI_{63}+kII_{65} & kI_{64}+kII_{66} & kII_{63} & kII_{64} \\ kII_{31} & kII_{32} & 0 & 0 & kII_{35} & kII_{36} & kII_{33} & kII_{34} \\ kII_{41} & kII_{42} & 0 & 0 & kII_{45} & kII_{46} & kII_{43} & kII_{44} \end{bmatrix} \begin{Bmatrix} d_{1x} \\ d_{1y} \\ d_{2x} \\ d_{2y} \\ d_{3x} \\ d_{3y} \\ d_{4x} \\ d_{4y} \end{Bmatrix} = \begin{Bmatrix} F_{1x} \\ F_{1y} \\ F_{2x} \\ F_{2y} \\ F_{3x} \\ F_{3y} \\ F_{4x} \\ F_{4y} \end{Bmatrix} \quad \text{(e10.1-11)}$$

or in abbreviated form

$$[K]\{d\} = \{F\} \quad \text{(e 10.1-12)}$$

Boundary Conditions

The FE model of the structure has the following eight boundary conditions:

(a) Boundary conditions regarding nodal displacements

(1) $d_{1x} = 0$
(2) $d_{1y} = 0$
(3) $d_{2x} = 0$
(4) $d_{2y} = 0$
(5) $d_{3x} = 0$
(6) $d_{3y} = 0$

Continued

EXAMPLE 10.1—CONT'D

(b) Boundary conditions regarding nodal forces

$$(7)\ F_{3x} = 840$$

$$(8)\ F_{3y} = 1400$$

It should be noticed that since nodes 2 and 4 are supported, the forces acting on these nodes do not affect the displacements of the structure. Therefore, the external forces on nodes 2 and 4 will not be taken into account in the list of boundary conditions.

The above boundary conditions can be written in the following matrix formats:

$$\begin{bmatrix} 1 & 0 & 0 & 0 & 0 & 0 & 0 & 0 \\ 0 & 1 & 0 & 0 & 0 & 0 & 0 & 0 \\ 0 & 0 & 1 & 0 & 0 & 0 & 0 & 0 \\ 0 & 0 & 0 & 1 & 0 & 0 & 0 & 0 \\ 0 & 0 & 0 & 0 & 0 & 0 & 1 & 0 \\ 0 & 0 & 0 & 0 & 0 & 0 & 0 & 1 \\ 0 & 0 & 0 & 0 & 0 & 0 & 0 & 0 \\ 0 & 0 & 0 & 0 & 0 & 0 & 0 & 0 \end{bmatrix} \begin{Bmatrix} d_{1x} \\ d_{1y} \\ d_{2x} \\ d_{2y} \\ d_{3x} \\ d_{3y} \\ d_{4x} \\ d_{4y} \end{Bmatrix} = \begin{Bmatrix} 0 \\ 0 \\ 0 \\ 0 \\ 0 \\ 0 \\ 0 \\ 0 \end{Bmatrix} \quad \text{(e 10.1-13)}$$

or in an abbreviated form:

$$[BCd]\{d\} = \{DO\} \quad \text{(e 10.1-14)}$$

and

$$\begin{bmatrix} 0 & 0 & 0 & 0 & 0 & 0 & 0 & 0 \\ 0 & 0 & 0 & 0 & 0 & 0 & 0 & 0 \\ 0 & 0 & 0 & 0 & 0 & 0 & 0 & 0 \\ 0 & 0 & 0 & 0 & 0 & 0 & 0 & 0 \\ 0 & 0 & 0 & 0 & 0 & 0 & 0 & 0 \\ 0 & 0 & 0 & 0 & 0 & 0 & 0 & 0 \\ 0 & 0 & 0 & 0 & 1 & 0 & 0 & 0 \\ 0 & 0 & 0 & 0 & 0 & 1 & 0 & 0 \end{bmatrix} \begin{Bmatrix} F_{1x} \\ F_{1y} \\ F_{2x} \\ F_{2y} \\ F_{3x} \\ F_{3y} \\ F_{4x} \\ F_{4y} \end{Bmatrix} = \begin{Bmatrix} 0 \\ 0 \\ 0 \\ 0 \\ 0 \\ 0 \\ 840 \\ 1400 \end{Bmatrix} \quad \text{(e 10.1-15)}$$

or in abbreviated form:

$$[BCR]\{f\} = \{RO\} \quad \text{(e 10.1-16)}$$

where

$$\{RO\} = \{0\ 0\ 0\ 0\ 0\ 0\ 840\ 1400\}^{T} \quad \text{(e 10.1-17)}$$

System of Algebraic Equations and Results

Equations (e 10.1-12), (e 10.1-14), and (e 10.1-16) can be combined to compose the following 16×16 system of algebraic equations:

$$\begin{bmatrix} [K] & [-I] \\ [BCd] & [BCR] \end{bmatrix} \begin{Bmatrix} \{d\} \\ \{F\} \end{Bmatrix} = \begin{Bmatrix} \{O\} \\ \{DO+RO\} \end{Bmatrix} \quad \text{(e 10.1-18)}$$

The solution of the above system yields the nodal displacements

$$\{d\} = \begin{Bmatrix} d_{1x} \\ d_{1y} \\ d_{2x} \\ d_{2y} \\ d_{3x} \\ d_{3y} \\ d_{4x} \\ d_{4y} \end{Bmatrix} = \begin{Bmatrix} 0 \\ 0 \\ 0 \\ 0 \\ 1.97498 \times 10^{-7} \\ 3.44926 \times 10^{-7} \\ 0 \\ 0 \end{Bmatrix} \text{m} \quad \text{(e 10.1-19)}$$

and the nodal reactions

$$\{F\} = \begin{Bmatrix} F_{1x} \\ F_{1y} \\ F_{2x} \\ F_{2y} \\ F_{3x} \\ F_{3y} \\ F_{4x} \\ F_{4y} \end{Bmatrix} = \begin{Bmatrix} -720.175 \\ -412.358 \\ 217.031 \\ 246.376 \\ 840 \\ 1400 \\ -336.856 \\ -1234.02 \end{Bmatrix} \text{kN} \quad \text{(e 10.1-20)}$$

In order to check whereas the above solution is correct, we can apply the results $\{F\}$ to the following equilibrium equations:

$$\sum F_x = F_{1x} + F_{2x} + F_{3x} + F_{4x} = -720.175 + 217.031 + 840 - 336.856 = 0$$

$$\sum F_y = F_{1y} + F_{2y} + F_{3y} + F_{4y} = -412.358 + 246.376 + 1400 - 1234.02 = 0$$

Since the members of the matrix $\{F\}$ confirm the equilibrium equations, the accuracy of the solution can be concluded.

Continued

EXAMPLE 10.1—CONT'D

Stresses of Each Element

The stresses of each element of the structure can now be determined by the following equations:

Stresses for element I

$$\begin{Bmatrix} \sigma_x \\ \sigma_y \\ \sigma_{xy} \end{Bmatrix} = [E][B_I]\begin{Bmatrix} d_{1x} \\ d_{1y} \\ d_{3x} \\ d_{3y} \\ d_{2x} \\ d_{2y} \end{Bmatrix} \tag{e 10.1-21}$$

yielding

$$\begin{Bmatrix} \sigma_x \\ \sigma_y \\ \sigma_{xy} \end{Bmatrix} = \begin{Bmatrix} 62{,}008.7 \\ 18{,}602.6 \\ 70{,}393.0 \end{Bmatrix} \text{kN/m}^2 \tag{e 10.2-22}$$

Stresses for element II

$$\begin{Bmatrix} \sigma_x \\ \sigma_y \\ \sigma_{xy} \end{Bmatrix} = [E][B_{II}]\begin{Bmatrix} d_{1x} \\ d_{1y} \\ d_{4x} \\ d_{4y} \\ d_{3x} \\ d_{3y} \end{Bmatrix} \tag{e 10.1-23}$$

yielding

$$\begin{Bmatrix} \sigma_x \\ \sigma_y \\ \sigma_{xy} \end{Bmatrix} = \begin{Bmatrix} 64{,}978.2 \\ 216{,}594.0 \\ 80{,}611.4 \end{Bmatrix} \text{kN/m}^2 \tag{e 10.1-24}$$

EXAMPLE 10.2

Determine the stiffness matrix for the following element.

Data

$$r_1 = 0.5\text{m}$$

$$r_2 = 1.0\text{m}$$

$$t = 1.5\text{cm} \quad \text{(thickness)}$$

$$E = 200\text{GPa}$$

$$\nu = 0.3$$

$$\alpha = \pi/4$$

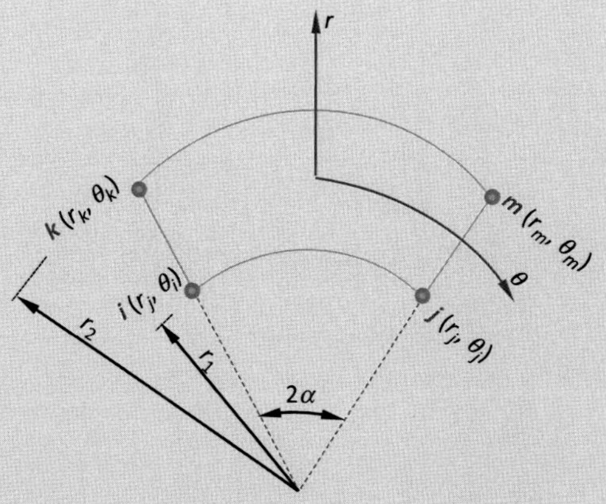

Solution

This problem will be formulated in polar coordinates (r, θ). Since the number of nodes is 4, the following interpolation functions will be adopted:

$$u(r,\theta) = [1 \;\; r \;\; \theta \;\; r\theta]\begin{Bmatrix} a_1 \\ a_2 \\ a_3 \\ a_4 \end{Bmatrix} \quad \text{(e 10.2-1)}$$

$$v(r,\theta) = [1 \;\; r \;\; \theta \;\; r\theta]\begin{Bmatrix} a_5 \\ a_6 \\ a_7 \\ a_8 \end{Bmatrix} \quad \text{(e 10.2-2)}$$

The above interpolation functions will now be applied to the four nodes $i(r_i,\theta_i), j(r_j,\theta_j), m(r_m,\theta_m), k(r_k,\theta_k)$ of the element yielding:

$$\begin{Bmatrix} u_i \\ u_j \\ u_m \\ u_k \end{Bmatrix} = \begin{bmatrix} 1 & r_i & \theta_i \\ 1 & r_j & \theta_j \\ 1 & r_m & \theta_m \\ 1 & r_k & \theta_k \end{bmatrix}\begin{Bmatrix} a_1 \\ a_2 \\ a_3 \\ a_4 \end{Bmatrix} \quad \text{(e 10.2-3)}$$

$$\begin{Bmatrix} v_i \\ v_j \\ v_m \\ v_k \end{Bmatrix} = \begin{bmatrix} 1 & r_i & \theta_i \\ 1 & r_j & \theta_j \\ 1 & r_m & \theta_m \\ 1 & r_k & \theta_k \end{bmatrix}\begin{Bmatrix} a_5 \\ a_6 \\ a_7 \\ a_8 \end{Bmatrix} \quad \text{(e 10.2-4)}$$

Continued

EXAMPLE 10.2—CONT'D

Taking into account the above equations, the constants $a_1, a_2, \ldots, a_8$ can be derived by the following formulae:

$$\begin{Bmatrix} a_1 \\ a_2 \\ a_3 \\ a_4 \end{Bmatrix} = [\Xi]^{-1} \begin{Bmatrix} u_i \\ u_j \\ u_m \\ u_k \end{Bmatrix} \quad \text{(e 10.2-5)}$$

$$\begin{Bmatrix} a_5 \\ a_6 \\ a_7 \\ a_8 \end{Bmatrix} = [\Xi]^{-1} \begin{Bmatrix} v_i \\ v_j \\ v_m \\ v_k \end{Bmatrix} \quad \text{(e 10.2-6)}$$

where

$$[\Xi] = \begin{bmatrix} 1 & r_i & \theta_i \\ 1 & r_j & \theta_j \\ 1 & r_m & \theta_m \\ 1 & r_k & \theta_k \end{bmatrix} \quad \text{(e 10.2-7)}$$

Using Equations (e 10.2-5) and (e 10.2-6), Equation (e 10.2-1), Equation (e 10.2-2) can now be written as:

$$u(r,\theta) = [1 \ \ r \ \ \theta \ \ r\theta][\Xi]^{-1} \begin{Bmatrix} u_i \\ u_j \\ u_m \\ u_k \end{Bmatrix} \quad \text{(e 10.2-8)}$$

$$v(r,\theta) = [1 \ \ r \ \ \theta \ \ r\theta][\Xi]^{-1} \begin{Bmatrix} v_i \\ v_j \\ v_m \\ v_k \end{Bmatrix} \quad \text{(e 10.2-9)}$$

or

$$u(r,\theta) = [N_i \ \ N_j \ \ N_m \ \ N_k] \begin{Bmatrix} u_i \\ u_j \\ u_m \\ u_k \end{Bmatrix} \quad \text{(e 10.2-10)}$$

$$v(r,\theta) = [N_i \ \ N_j \ \ N_m \ \ N_k] \begin{Bmatrix} v_i \\ v_j \\ v_m \\ v_k \end{Bmatrix} \quad \text{(e 10.2-11)}$$

where

$$[N_i \; N_j \; N_m \; N_k] = [1 \; r \; \theta \; r\theta][\Xi]^{-1} \tag{e 10.2-12}$$

Since $r_i = r_1$, $\theta_i = -a$, $r_j = r_1$, $\theta_j = a$, $r_m = r_2$, $\theta_m = a$, $r_k = r_2$, $\theta_k = -a$, after some standard algebraic operations, Equation (e 10.2-12) yields

$$N_1 = \frac{(r - r_2)(\theta - a)}{2a(r_2 - r_1)} \tag{e 10.2-13}$$

$$N_2 = -\frac{(r - r_2)(\theta + a)}{2a(r_2 - r_1)} \tag{e 10.2-14}$$

$$N_3 = \frac{(r - r_1)(\theta + a)}{2a(r_2 - r_1)} \tag{e 10.2-15}$$

$$N_4 = -\frac{(r - r_1)(\theta - a)}{2a(r_2 - r_1)} \tag{e 10.2-16}$$

It is known from the mechanics of solids that the strains $\varepsilon_r, \varepsilon_\theta, \varepsilon_{r\theta}$ are correlated to the displacements $u(r,\theta), v(r,\theta)$ by the following equation:

$$\begin{Bmatrix} \varepsilon_r \\ \varepsilon_\theta \\ \varepsilon_{r\theta} \end{Bmatrix} = \begin{bmatrix} \dfrac{\partial}{\partial r} & 0 \\ \dfrac{1}{r} & \dfrac{1}{r}\dfrac{\partial}{\partial \theta} \\ \dfrac{1}{r}\dfrac{\partial}{\partial \theta} & \dfrac{\partial}{\partial r} - \dfrac{1}{r} \end{bmatrix} \begin{Bmatrix} u(r,\theta) \\ v(r,\theta) \end{Bmatrix} \tag{e 10.2-17}$$

Taking into account Equations (e 9.2-10) and (e 9.2-11), the above equation yields

$$\begin{Bmatrix} \varepsilon_r \\ \varepsilon_\theta \\ \varepsilon_{r\theta} \end{Bmatrix} = [B] \begin{Bmatrix} u_i \\ v_i \\ u_j \\ v_j \\ u_m \\ v_m \\ u_k \\ v_k \end{Bmatrix} \tag{e 10.2-18}$$

where

$$[B] = \begin{bmatrix} \dfrac{\partial N_1}{\partial r} & 0 & \dfrac{\partial N_2}{\partial r} & 0 & \dfrac{\partial N_3}{\partial r} & 0 & \dfrac{\partial N_4}{\partial r} & 0 \\ \dfrac{N_1}{r} & \dfrac{1}{r}\dfrac{\partial N_1}{\partial \theta} & \dfrac{N_2}{r} & \dfrac{1}{r}\dfrac{\partial N_2}{\partial \theta} & \dfrac{N_3}{r} & \dfrac{1}{r}\dfrac{\partial N_3}{\partial \theta} & \dfrac{N_4}{r} & \dfrac{1}{r}\dfrac{\partial N_4}{\partial \theta} \\ \dfrac{1}{r}\dfrac{\partial N_1}{\partial \theta} & \dfrac{\partial N_1}{\partial r} - \dfrac{N_1}{r} & \dfrac{1}{r}\dfrac{\partial N_2}{\partial \theta} & \dfrac{\partial N_2}{\partial r} - \dfrac{N_2}{r} & \dfrac{1}{r}\dfrac{\partial N_3}{\partial \theta} & \dfrac{\partial N_3}{r} - \dfrac{N_3}{r} & \dfrac{1}{r}\dfrac{\partial N_4}{\partial \theta} & \dfrac{\partial N_4}{\partial r} - \dfrac{N_4}{r} \end{bmatrix} \tag{e 10.2-19}$$

Continued

EXAMPLE 10.2—CONT'D

Performing the required algebraic operations, the above matrix takes the form:

$$[B] = \begin{bmatrix} \beta_{11} & \beta_{12} & \beta_{13} & \beta_{14} & \beta_{15} & \beta_{16} & \beta_{17} & \beta_{18} \\ \beta_{21} & \beta_{22} & \beta_{23} & \beta_{24} & \beta_{25} & \beta_{26} & \beta_{27} & \beta_{28} \\ \beta_{31} & \beta_{32} & \beta_{33} & \beta_{34} & \beta_{35} & \beta_{36} & \beta_{37} & \beta_{38} \end{bmatrix} \qquad \text{(e 10.2-20)}$$

where

$$\beta_{11} = \frac{(a-\theta)}{2a(r_1 - r_2)} \qquad \text{(e 10.2-21)}$$

$$\beta_{12} = 0 \qquad \text{(e 10.2-22)}$$

$$\beta_{13} = \frac{(a+\theta)}{2a(r_1 - r_2)} \qquad \text{(e 10.2-23)}$$

$$\beta_{14} = 0 \qquad \text{(e 10.2-24)}$$

$$\beta_{15} = -\frac{(a+\theta)}{2a(r_1 - r_2)} \qquad \text{(e 10.2-25)}$$

$$\beta_{16} = 0 \qquad \text{(e 10.2-26)}$$

$$\beta_{17} = -\frac{(a-\theta)}{2a(r_1 - r_2)} \qquad \text{(e 10.2-27)}$$

$$\beta_{18} = 0 \qquad \text{(e 10.2-28)}$$

$$\beta_{21} = \frac{(r-r_2)(a-\theta)}{2ar(r_1 - r_2)} \qquad \text{(e 10.2-29)}$$

$$\beta_{22} = -\frac{(r-r_2)}{2ar(r_1 - r_2)} \qquad \text{(e 10.2-30)}$$

$$\beta_{23} = \frac{(r-r_2)(a+\theta)}{2ar(r_1 - r_2)} \qquad \text{(e 10.2-31)}$$

$$\beta_{24} = \frac{(r-r_2)}{2ar(r_1 - r_2)} \qquad \text{(e 10.2-32)}$$

$$\beta_{25} = -\frac{(r-r_1)(a+\theta)}{2ar(r_1-r_2)} \quad \text{(e 10.2-33)}$$

$$\beta_{26} = -\frac{(r-r_1)}{2ar(r_1-r_2)} \quad \text{(e 10.2-34)}$$

$$\beta_{27} = -\frac{(r-r_1)(a-\theta)}{2ar(r_1-r_2)} \quad \text{(e 10.2-35)}$$

$$\beta_{28} = \frac{(r-r_1)}{2ar(r_1-r_2)} \quad \text{(e 10.2-36)}$$

$$\beta_{31} = -\frac{(r-r_2)}{2ar(r_1-r_2)} \quad \text{(e 10.2-37)}$$

$$\beta_{32} = \frac{r_2(a-\theta)}{2ar(r_1-r_2)} \quad \text{(e 10.2-38)}$$

$$\beta_{33} = \frac{(r-r_2)}{2ar(r_1-r_2)} \quad \text{(e 10.2-39)}$$

$$\beta_{34} = \frac{r_2(a+\theta)}{2ar(r_1-r_2)} \quad \text{(e 10.2-40)}$$

$$\beta_{35} = -\frac{(r-r_1)}{2ar(r_1-r_2)} \quad \text{(e 10.2-41)}$$

$$\beta_{36} = -\frac{r_1(a+\theta)}{2ar(r_1-r_2)} \quad \text{(e 10.2-42)}$$

$$\beta_{37} = \frac{(r-r_1)}{2ar(r_1-r_2)} \quad \text{(e 10.2-43)}$$

$$\beta_{38} = -\frac{r_1(a-\theta)}{2ar(r_1-r_2)} \quad \text{(e 10.2-44)}$$

Therefore, the stiffness matrix of the above element can be derived from the following equation:

$$[k] = t\int_{r_1}^{r_2}\int_{-a}^{a} [B]^{\mathrm{T}}[E][B]\mathrm{d}\theta\mathrm{d}r \quad \text{(e 10.2-45)}$$

where the matrix $[E]$ can be obtained from Equation (9.137). Taking into account Equation (e 10.2-20) and the numerical data of the problem, and then performing the required integration, the above equation yields:

Continued

EXAMPLE 10.2—CONT'D

$$[k]=10^8\times\begin{bmatrix} 37.54977603 & 3.5570839558 & 13.876502140 & 10.28590037 & -16.905893377 & -4.195340026 & -30.38948220 & -2.726152141 \\ 3.5570839558 & 33.36902898 & -10.28590037 & 2.0624671042 & -11.322880433 & -9.310527511 & -2.609473792 & -8.405220534 \\ 13.876502140 & -10.28590037 & 37.54977603 & -3.5570839558 & -30.38948220 & 2.726152141 & -16.905893377 & 4.195340026 \\ 10.28590037 & 2.0624671042 & -3.5570839558 & 33.36902898 & 2.609473792 & -8.405220534 & 11.322880433 & -9.310527511 \\ -16.905893377 & -11.322880433 & -30.38948220 & 2.609473792 & 48.036720610 & 9.415487872 & 21.56916736 & -5.058784551 \\ -4.195340026 & -9.310527511 & 2.726152141 & -8.405220534 & 9.415487872 & 10.77926514 & 5.058784551 & -1.92139112 \\ -30.38948220 & -2.609473792 & -16.905893377 & 11.322880433 & 21.56916736 & 5.058784551 & 48.036720610 & -9.415487872 \\ -2.726152141 & -8.405220534 & 4.195340026 & -9.310527511 & -5.058784551 & -1.92139112 & -9.415487872 & 10.77926514 \end{bmatrix}$$

EXAMPLE 10.3

Using the stiffness matrix of the Example 10.2, calculate the stress field $\sigma_r(r,\theta)$, $\sigma_\theta(r,\theta)$, $\sigma_{r\theta}(r,\theta)$ for the following structure:

Data

$$q = 6.0\text{MPa}$$
$$r_1 = 0.5\text{m}$$
$$r_2 = 1.0\text{m}$$
$$t = 1.5\text{cm}(\text{thickness})$$
$$E = 200\text{GPa}$$
$$\nu = 0.3$$
$$2\alpha = \pi/2$$

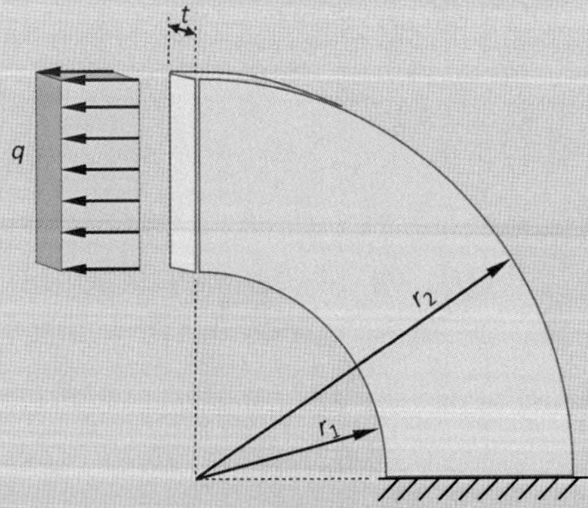

Solution

First step for the solution of the above problem is the transformation of the variable loading q into an equivalent system of nodal forces. The total force is:

$$F = qt(r_2 - r_1) = (6\times10^6)\times(1.5\times10^{-2})\times(1.0-0.5) = 45{,}000\text{N}$$

The above force should be distributed into two equal forces $F_{1\theta}$ and $F_{4\theta}$ acting on the corresponding nodes 1 and 4 as shown in the following model.

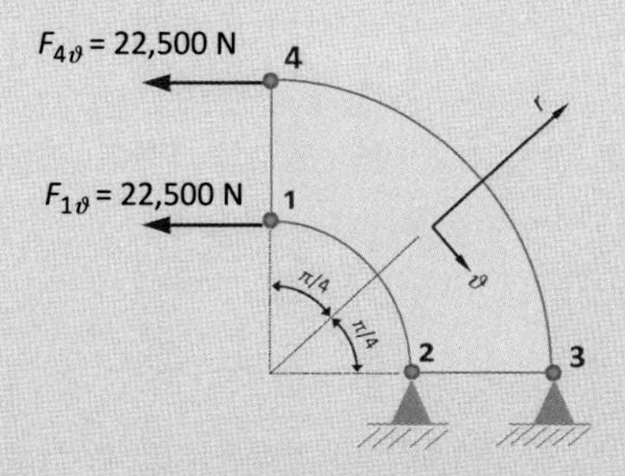

The stiffness matrix $[k]$ obtained in the Example 10.2 correlates the nodal displacements $\{d\}$ with the nodal forces $\{f\}$, that is:

$$[k]\begin{Bmatrix} u_1 \\ v_1 \\ u_2 \\ v_2 \\ u_3 \\ v_3 \\ u_4 \\ v_4 \end{Bmatrix} = \begin{Bmatrix} f_{r,1} \\ f_{\theta,1} \\ f_{r,2} \\ f_{\theta,2} \\ f_{r,3} \\ f_{\theta,3} \\ f_{r,4} \\ f_{\theta,4} \end{Bmatrix} \qquad \text{(e 10.3-1)}$$

where $u_n, v_n \, (n = 1,2,3,4)$ are the nodal displacements in the r and θ directions, respectively.

Let $[k] = [k_{ij}] \, (i = 1,2,3,\ldots,8$ and $j = 1,2,3,\ldots,8)$. Then Equation (e 10.3-1) can be written in the following form:

$$\begin{bmatrix} k11 & k12 & k13 & k14 & k15 & k16 & k17 & k18 & -1 & 0 & 0 & 0 & 0 & 0 & 0 & 0 \\ k21 & k22 & k23 & k24 & k25 & k26 & k27 & k28 & 0 & -1 & 0 & 0 & 0 & 0 & 0 & 0 \\ k31 & k32 & k33 & k34 & k35 & k36 & k37 & k38 & 0 & 0 & -1 & 0 & 0 & 0 & 0 & 0 \\ k41 & k42 & k43 & k44 & k45 & k46 & k47 & k48 & 0 & 0 & 0 & -1 & 0 & 0 & 0 & 0 \\ k51 & k52 & k53 & k54 & k55 & k56 & k57 & k58 & 0 & 0 & 0 & 0 & -1 & 0 & 0 & 0 \\ k61 & k62 & k63 & k64 & k65 & k66 & k67 & k68 & 0 & 0 & 0 & 0 & 0 & -1 & 0 & 0 \\ k71 & k72 & k73 & k74 & k75 & k76 & k77 & k78 & 0 & 0 & 0 & 0 & 0 & 0 & -1 & 0 \\ k81 & k82 & k83 & k84 & k85 & k86 & k87 & k88 & 0 & 0 & 0 & 0 & 0 & 0 & 0 & -1 \end{bmatrix} \begin{Bmatrix} u_1 \\ v_1 \\ u_2 \\ v_2 \\ u_3 \\ v_3 \\ u_4 \\ v_4 \\ f_{r,1} \\ f_{\theta,1} \\ f_{r,2} \\ f_{\theta,2} \\ f_{r,3} \\ f_{\theta,3} \\ f_{r,4} \\ f_{\theta,4} \end{Bmatrix} = \begin{Bmatrix} 0 \\ 0 \\ 0 \\ 0 \\ 0 \\ 0 \\ 0 \\ 0 \end{Bmatrix} \qquad \text{(e 10.3-2)}$$

Continued

EXAMPLE 10.3—CONT'D

The boundary conditions of the problem are:

$$(1)\ f_{1,r}=0$$

$$(2)\ f_{1,\theta}=-22{,}500$$

$$(3)\ u_2=0$$

$$(4)\ v_2=0$$

$$(5)\ u_3=0$$

$$(6)\ v_3=0$$

$$(7)\ f_{4,r}=0$$

$$(8)\ f_{4,\theta}=-22{,}500$$

The above boundary conditions can be written in the following matrix form:

$$\begin{bmatrix} 0&0&0&0&0&0&0&0&1&0&0&0&0&0&0&0 \\ 0&0&0&0&0&0&0&0&0&1&0&0&0&0&0&0 \\ 0&0&1&0&0&0&0&0&0&0&0&0&0&0&0&0 \\ 0&0&0&1&0&0&0&0&0&0&0&0&0&0&0&0 \\ 0&0&0&0&1&0&0&0&0&0&0&0&0&0&0&0 \\ 0&0&0&0&0&1&0&0&0&0&0&0&0&0&0&0 \\ 0&0&0&0&0&0&0&0&0&0&0&0&0&0&1&0 \\ 0&0&0&0&0&0&0&0&0&0&0&0&0&0&0&1 \end{bmatrix} \begin{Bmatrix} u_1 \\ v_1 \\ u_2 \\ v_2 \\ u_3 \\ v_3 \\ u_4 \\ v_4 \\ f_{r,1} \\ f_{\theta,1} \\ f_{r,2} \\ f_{\theta,2} \\ f_{r,3} \\ f_{\theta,3} \\ f_{r,4} \\ f_{\theta,4} \end{Bmatrix} = \begin{Bmatrix} 0 \\ -22{,}500 \\ 0 \\ 0 \\ 0 \\ 0 \\ 0 \\ -22{,}500 \end{Bmatrix} \qquad \text{(e 10.3-3)}$$

Combining Equations (e 10.3-2) and (e 10.3-3), the following 16×16 system of algebraic equations can be obtained.

$$\begin{bmatrix} k11 & k12 & k13 & k14 & k15 & k16 & k17 & k18 & -1 & 0 & 0 & 0 & 0 & 0 & 0 & 0 \\ k21 & k22 & k23 & k24 & k25 & k26 & k27 & k28 & 0 & -1 & 0 & 0 & 0 & 0 & 0 & 0 \\ k31 & k32 & k33 & k34 & k35 & k36 & k37 & k38 & 0 & 0 & -1 & 0 & 0 & 0 & 0 & 0 \\ k41 & k42 & k43 & k44 & k45 & k46 & k47 & k48 & 0 & 0 & 0 & -1 & 0 & 0 & 0 & 0 \\ k51 & k52 & k53 & k54 & k55 & k56 & k57 & k58 & 0 & 0 & 0 & 0 & -1 & 0 & 0 & 0 \\ k61 & k62 & k63 & k64 & k65 & k66 & k67 & k68 & 0 & 0 & 0 & 0 & 0 & -1 & 0 & 0 \\ k71 & k72 & k73 & k74 & k75 & k76 & k77 & k78 & 0 & 0 & 0 & 0 & 0 & 0 & -1 & 0 \\ k81 & k82 & k83 & k84 & k85 & k86 & k87 & k88 & 0 & 0 & 0 & 0 & 0 & 0 & 0 & -1 \\ 0 & 0 & 0 & 0 & 0 & 0 & 0 & 0 & 1 & 0 & 0 & 0 & 0 & 0 & 0 & 0 \\ 0 & 0 & 0 & 0 & 0 & 0 & 0 & 0 & 0 & 1 & 0 & 0 & 0 & 0 & 0 & 0 \\ 0 & 0 & 1 & 0 & 0 & 0 & 0 & 0 & 0 & 0 & 0 & 0 & 0 & 0 & 0 & 0 \\ 0 & 0 & 0 & 1 & 0 & 0 & 0 & 0 & 0 & 0 & 0 & 0 & 0 & 0 & 0 & 0 \\ 0 & 0 & 0 & 0 & 1 & 0 & 0 & 0 & 0 & 0 & 0 & 0 & 0 & 0 & 0 & 0 \\ 0 & 0 & 0 & 0 & 0 & 1 & 0 & 0 & 0 & 0 & 0 & 0 & 0 & 0 & 0 & 0 \\ 0 & 0 & 0 & 0 & 0 & 0 & 0 & 0 & 0 & 0 & 0 & 0 & 0 & 0 & 1 & 0 \\ 0 & 0 & 0 & 0 & 0 & 0 & 0 & 0 & 0 & 0 & 0 & 0 & 0 & 0 & 0 & 1 \end{bmatrix} \begin{Bmatrix} u_1 \\ v_1 \\ u_2 \\ v_2 \\ u_3 \\ v_3 \\ u_4 \\ v_4 \\ f_{r,1} \\ f_{\theta,1} \\ f_{r,2} \\ f_{\theta,2} \\ f_{r,3} \\ f_{\theta,3} \\ f_{r,4} \\ f_{\theta,4} \end{Bmatrix} = \begin{Bmatrix} 0 \\ 0 \\ 0 \\ 0 \\ 0 \\ 0 \\ 0 \\ 0 \\ 0 \\ -22{,}500 \\ 0 \\ 0 \\ 0 \\ 0 \\ 0 \\ -22{,}500 \end{Bmatrix} \quad \text{(e 10.3-4)}$$

Continued

EXAMPLE 10.3—CONT'D

From the solution of the above system, the following results can be obtained for the nodal displacements:

$$\begin{Bmatrix} u_1 \\ v_1 \\ u_2 \\ v_2 \\ u_3 \\ v_3 \\ u_4 \\ v_4 \end{Bmatrix} = \begin{Bmatrix} -0.0000429092 \\ -0.0000304028 \\ 0 \\ 0 \\ 0 \\ 0 \\ -0.0000478555 \\ -0.0000972332 \end{Bmatrix} \qquad \text{(e 10.3-5)}$$

Then, the strain field $\{\varepsilon\} = \{\varepsilon_r(r,\theta) \;\; \varepsilon_\theta(r,\theta) \;\; \varepsilon_{r\theta}(r,\theta)\}$ can be derived from Equation (e 10.2-18), where the matrix $[B]$ is given in Equation (e 10.2-20). Therefore, using the data of the problem and taking into account Equations (e 10.2-18) and (e 10.2-20), the following results can be derived:

$$\begin{Bmatrix} \varepsilon_r(r,\theta) \\ \varepsilon_\theta(r,\theta) \\ \varepsilon_{r\theta}(r,\theta) \end{Bmatrix} = [B] \begin{Bmatrix} u_1 \\ v_1 \\ u_2 \\ v_2 \\ u_3 \\ v_3 \\ u_4 \\ v_4 \end{Bmatrix} = \begin{Bmatrix} (-4.9463 + 6.2978\ \theta) \times 10^{-6} \\ \frac{1}{r}[-42.172 + 80.1448r + \theta(24.168 + 6.29783r)] \times 10^{-6} \\ \left[6.29783 + \frac{5.95416 + 23.1905\ \theta}{r}\right] \times 10^{-6} \end{Bmatrix}$$

Then, using Hooke's law and Equation (9.137), the stress field $\{\sigma\} = \{\sigma_r(r,\theta) \;\; \sigma_\theta(r,\theta) \;\; \sigma_{r\theta}(r,\theta)\}$ will be derived by the following equation:

$$\begin{Bmatrix} \sigma_r(r,\theta) \\ \sigma_\theta(r,\theta) \\ \sigma_{r\theta}(r,\theta) \end{Bmatrix} = [E] \begin{Bmatrix} \varepsilon_r(r,\theta) \\ \varepsilon_\theta(r,\theta) \\ \varepsilon_{r\theta}(r,\theta) \end{Bmatrix} = \begin{Bmatrix} \frac{1}{r}[-3.1235 + 4.82583r + \theta(1.79002 + 1.87995r)] \times 10^{-6} \\ \frac{1}{r}[-9.46515 + 17.6215r + \theta(5.4243 + 1.87995r)] \times 10^{-6} \\ 473521 + \frac{1}{r}[0.447622 + 1.74365\ \theta] \times 10^{-6} \end{Bmatrix}$$

EXAMPLE 10.4: ANALYSIS OF A PLANE STRESS PROBLEM USING ANSYS

Determine the displacement, strain, and stress fields for the following plane stress problem.

Data

$R_1 = 0.5\text{m}$	$t = 1.5\text{cm}$	$E = 200\ \text{GPa}$
$R_2 = 1.0\text{m}$	$q = 7.0\text{MPa}$	$\nu = 0.3$

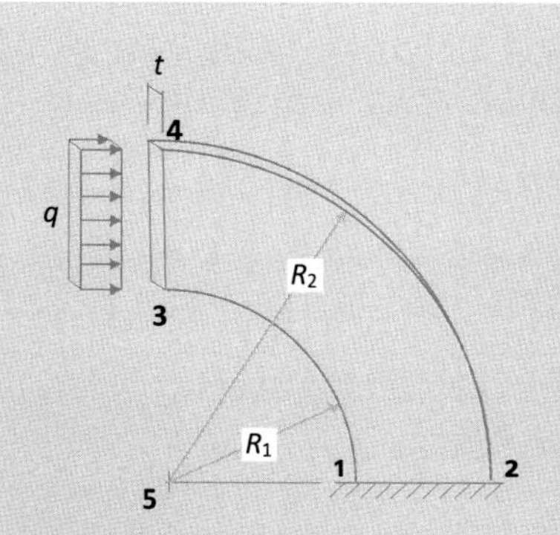

UM: File → Change Directory

Using this command we select the directory to save all files generated for this exercise. Let us choose Directory **F**, and file **ANSYS TUTORIALS**.

INPUT: /UNITS, SI

This command defines the units to SI.

UM: File → Change Job Name

With this command, we specify the job name. Let us choose **Tutorial 8**.

UM: File → Change Title

We can choose again **Tutorial 8**.

MM: Preferences

The following window will appear. We must choose "*Structural*"

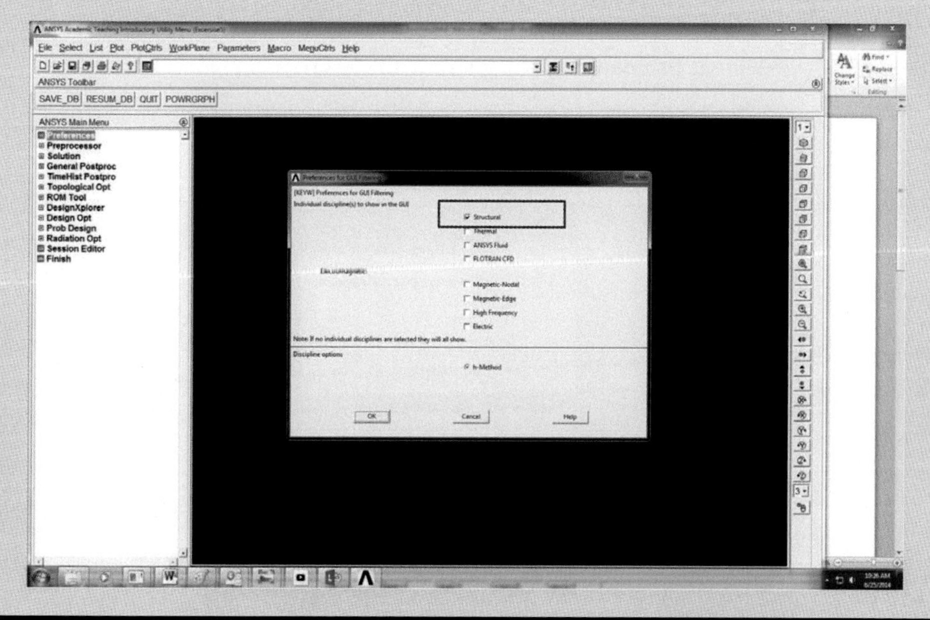

Continued

EXAMPLE 10.4: ANALYSIS OF A PLANE STRESS PROBLEM USING ANSYS—CONT'D

MM: Preprocessor → Element Type → Add/Edit/Delete → Add → Solid → Quad 4 node 182 → OK →

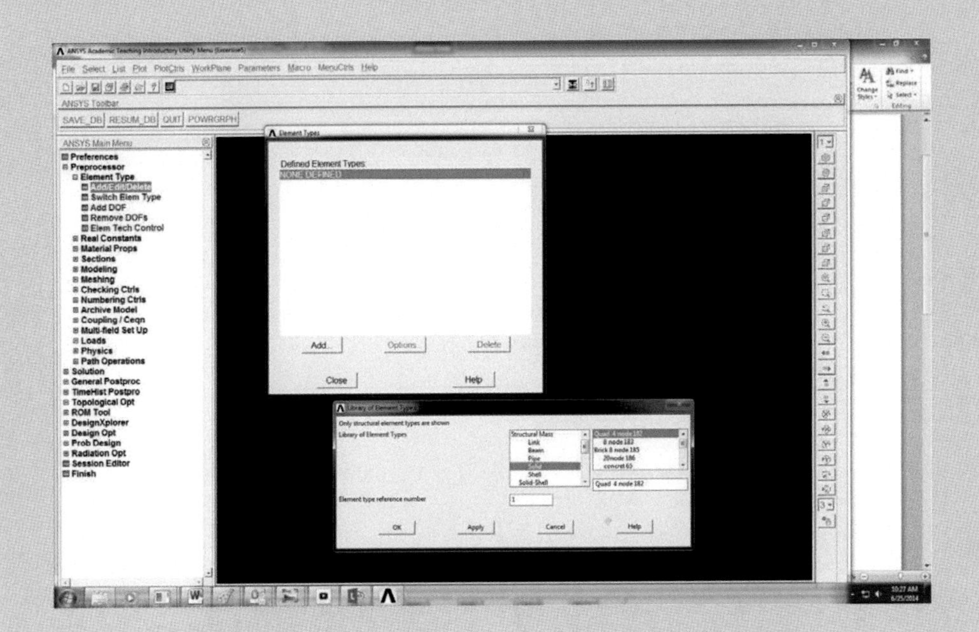

→ Options → (Element behavior) → Plane strs w/thk → OK

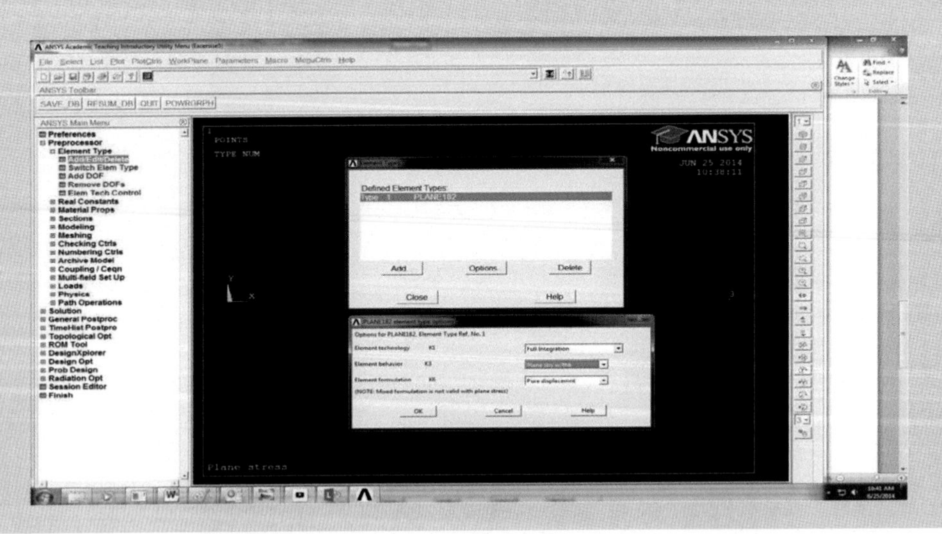

MM: Preprocessor → Real Constants → Add/Edit/Del → Add → OK

The following window will appear. We have to declare the thickness of the structure (thickness 0.015).

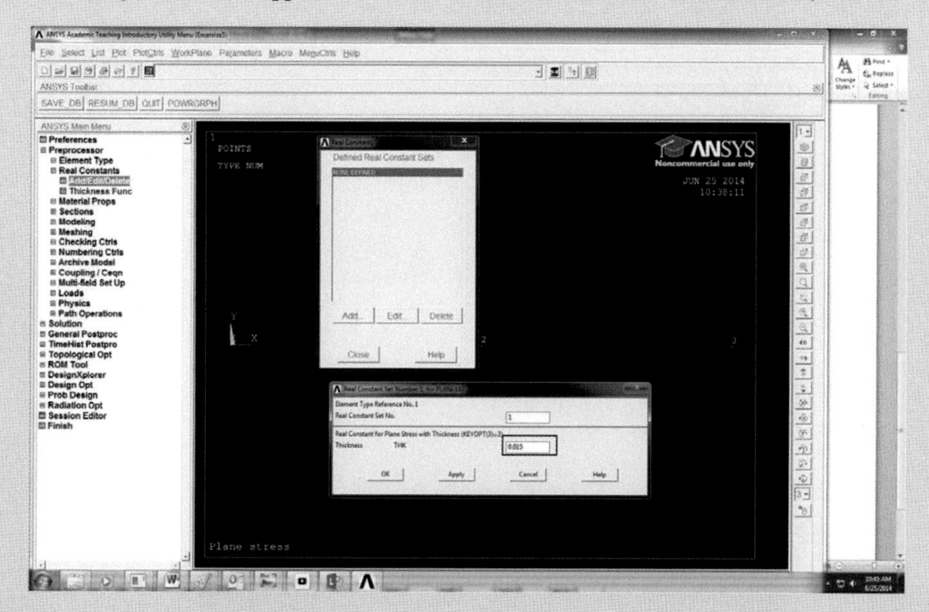

MM: Preprocessor → Material Props → Material Models → Structural → Linear → Elastic → Isotropic

Then, the following window will appear and the values EX = 2e11 and PRXY = 0.3 for the modulus of elasticity and Poisson's ratio should be filled, respectively.

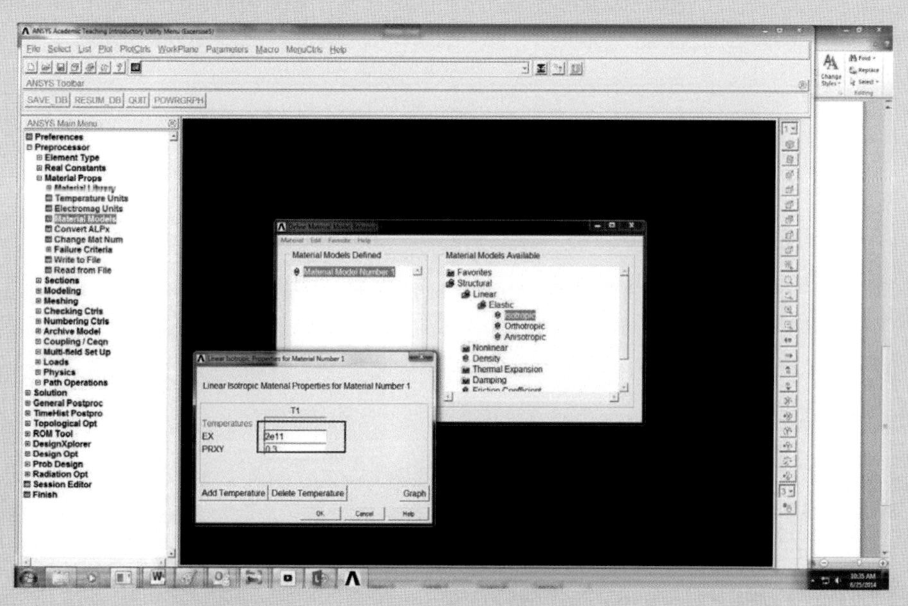

Continued

EXAMPLE 10.4: ANALYSIS OF A PLANE STRESS PROBLEM USING ANSYS—CONT'D

MM: Preprocessor → Modeling → Create → Keypoints → In Active CS

With the above command, the following window will appear. We have to fill the Keypoint numbers and the corresponding coordinates. After filling each form we click the button "**Apply.**" After the completion of the coordinates of the last Keypoint we click the button "**OK.**" Attention: Apart from the four Keypoints on the corners of the solid, we have to specify the Keypoint of the coordinate system (the center of the cycle). Therefore, we should specify five keypoints.

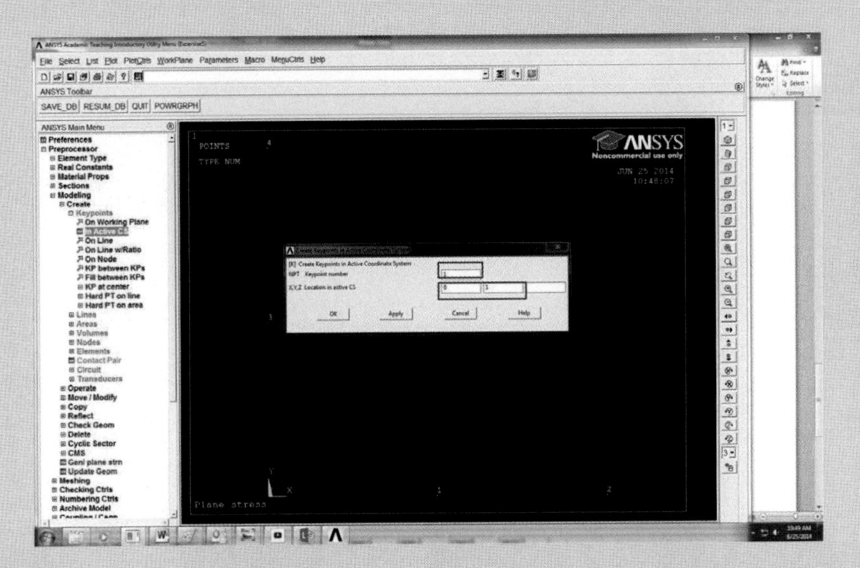

Now we must create the lines surrounding the solid. There are two curved lines, and two straight ones. Let us start the creation of the straight lines:

MM: Preprocessor → Modeling → Create → Lines → In Active CS

After the above command, we have to pick the Keypoints defining the two straight lines, and then to press the button "OK" in the window entitled "Lines in Active Coord." Therefore, the following window will appear.

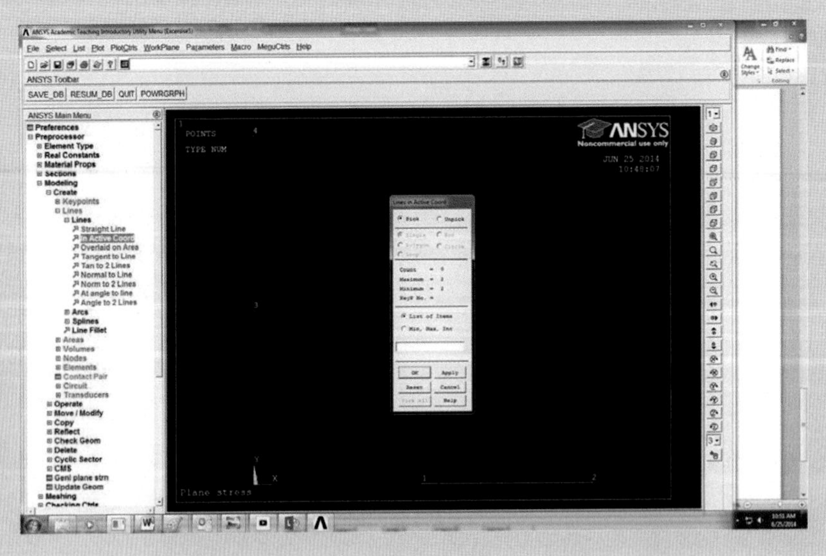

The next step is to create the curved lines (Arcs) using the following commands:
MM: Preprocessor → Modeling → Create → Arcs → By End KPs & Rad
After the above command, we must pick the end points of each arch and press the button "**Apply.**" Then we have to pick the center of the arch, and press the button "**OK.**" Then the following window will appear asking us to fill the radius of the arc, the numbers of its end Keypoints, and the number of the keypoint located at the center of the arc. After filling this form, the button "**OK**" should be pressed.

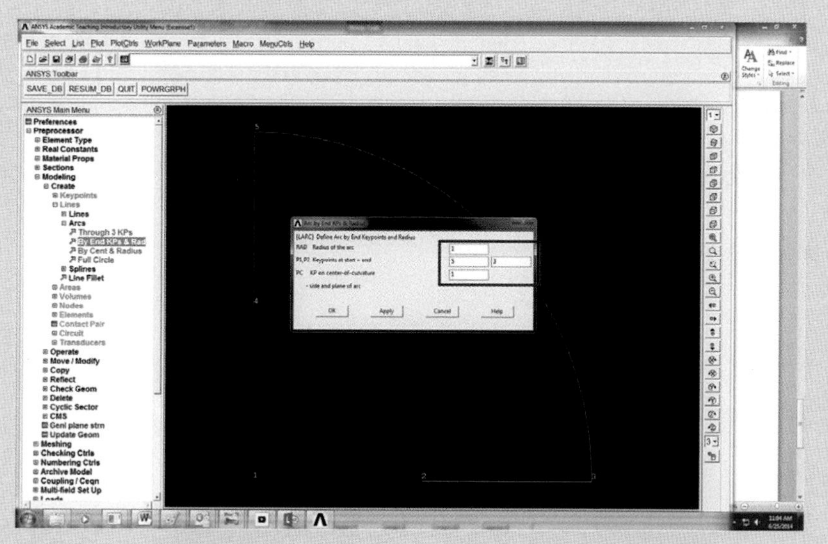

The next step is to create the area of the plane solid by the use of the following commands.
MM: Preprocessor → Modeling → Create → Areas → Arbitrary → By Lines
We pick all lines, and then we press the button "**OK.**" Then the following area will be created:

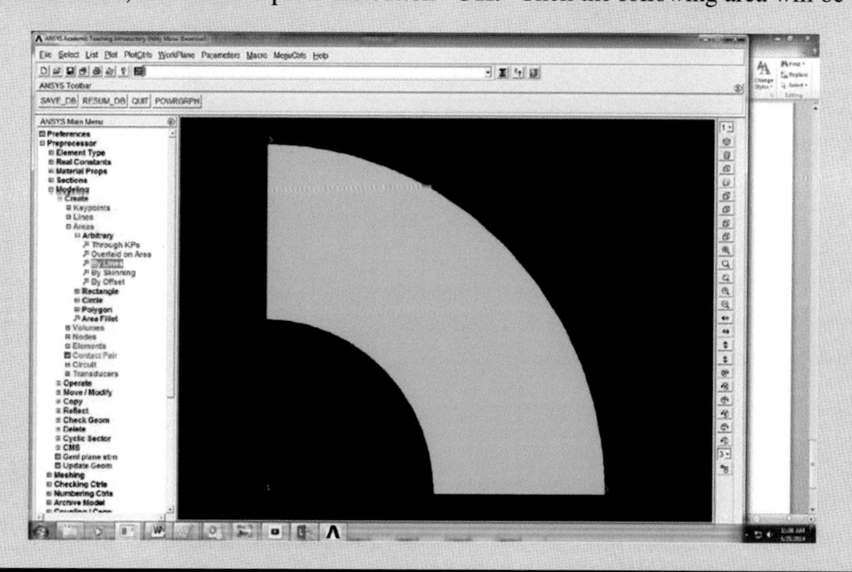

Continued

EXAMPLE 10.4: ANALYSIS OF A PLANE STRESS PROBLEM USING ANSYS—CONT'D

Now it is time to create the mesh of the solid. We can do this using the following commands:

MM: Preprocessor → Meshing → MeshTool

A window entitled "MeshTool" will appear. In this window, we must press the button "**set**" for the Lines. After that, we have to pick the pair of the curved lines and press "Apply.". Then, a window entitled "Element Sizes on Picked Lines" will appear. On this window, we must fill the number of element divisions (NDIV). Let us set NDIV = 20, and press "Apply." After that, we have to pick the pair of the straight lines and press "Apply." Then, a window entitled "Element Sizes on Picked Lines" will appear. On this window, we must fill the number of element divisions (NDIV). Let us set NDIV = 10, and press "OK."

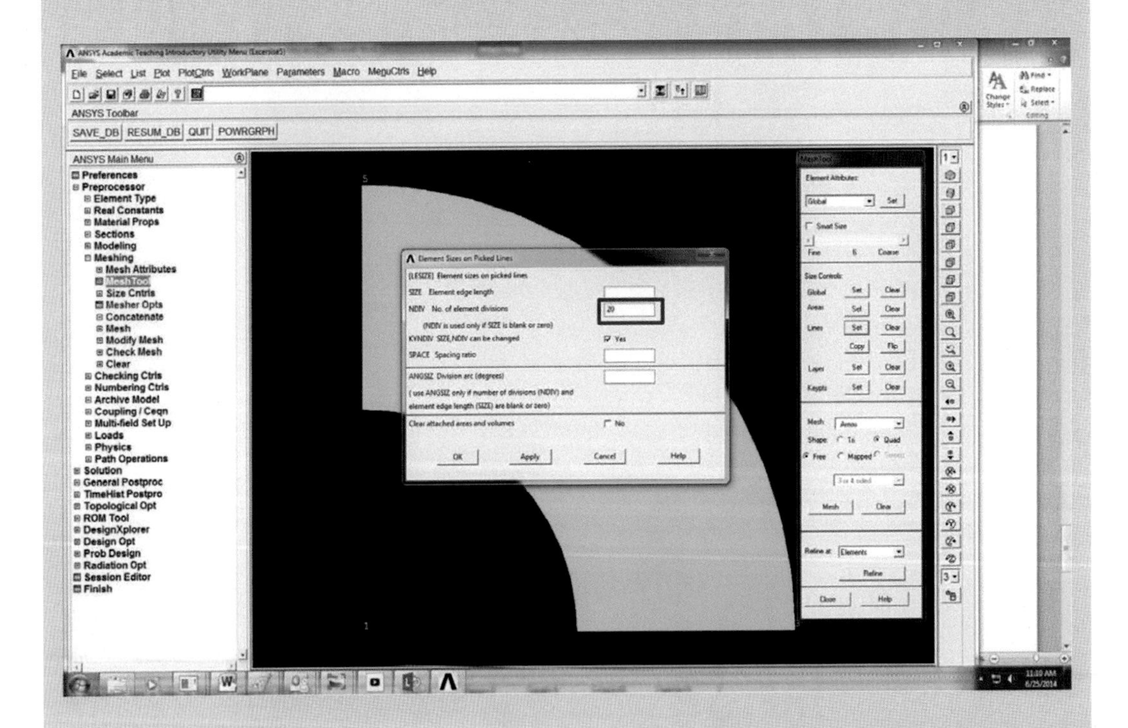

After finishing the above task we must go again to the mesh tool using the commands

MM: Preprocessor → Meshing → MeshTool

Then, we must press the button "Mesh."

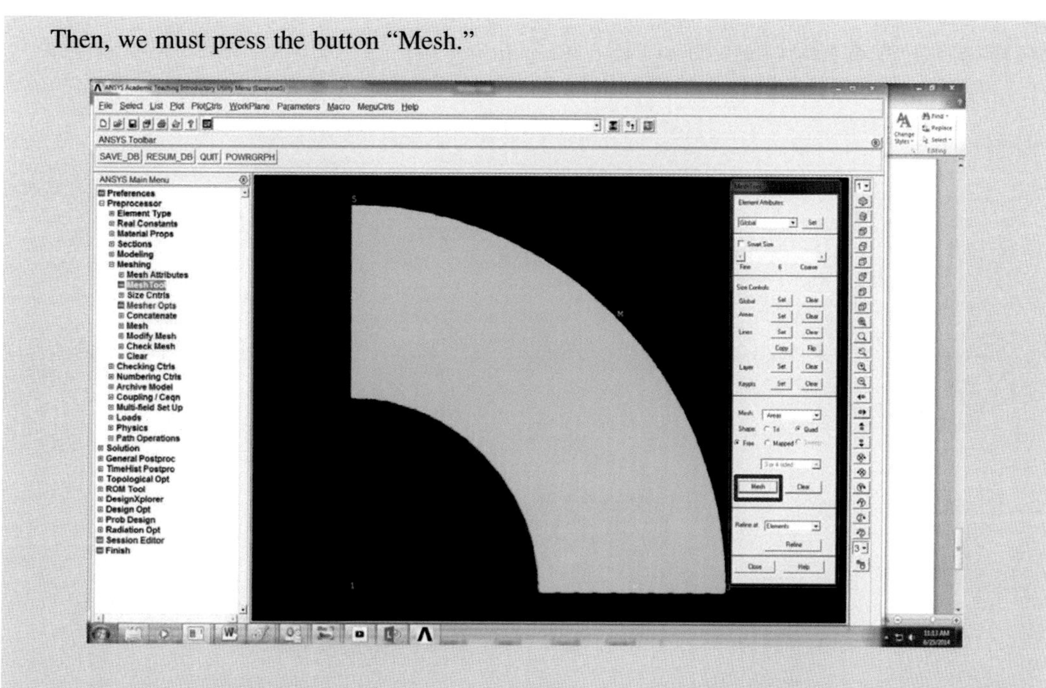

Next we must click on the area of the solid, and press the button "OK" on the window entitled "Mesh Areas." Then the following mesh will be created:

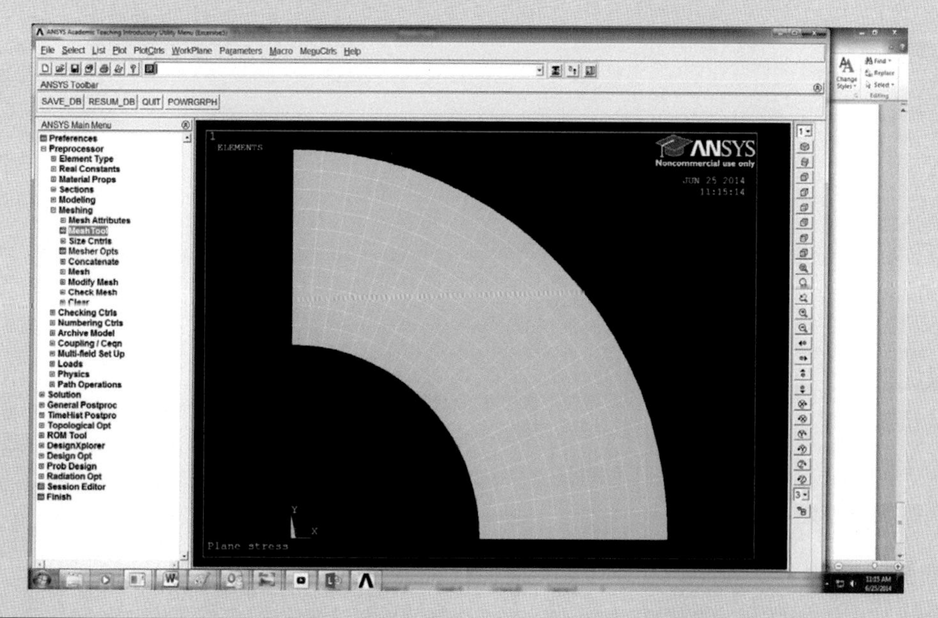

Continued

EXAMPLE 10.4: ANALYSIS OF A PLANE STRESS PROBLEM USING ANSYS—CONT'D

Now it is time to perform the solution by the use of the following commands:

MM: Solution → Analysis Type → New Analysis → Static → OK

Next step is to define the boundary conditions on the supports by the following commands:

MM: Solution → Define Loads → Apply → Structural → Displacements → On Lines

Then, we must click on the line 1–2 to specify the support conditions, and press the button "**Apply.**" In the window entitled "Apply U,ROT on lines" we must select "**All DOF**" and fill in the box "Displacement value" = 0. Then press "**OK.**"

After that, we must specify the pressure acting on the line 3–4 by the following commands:

MM: Solution → Define Loads → Apply → Structural → Pressure → On Lines

Now we have to click on the line 3–4 and press "OK" in the window entitled "Apply PRES on Lines." A new window will appear, asking for the value of pressure. In the box "VALUE Load PRES value" we fill the value 7e6, and press "**OK.**"

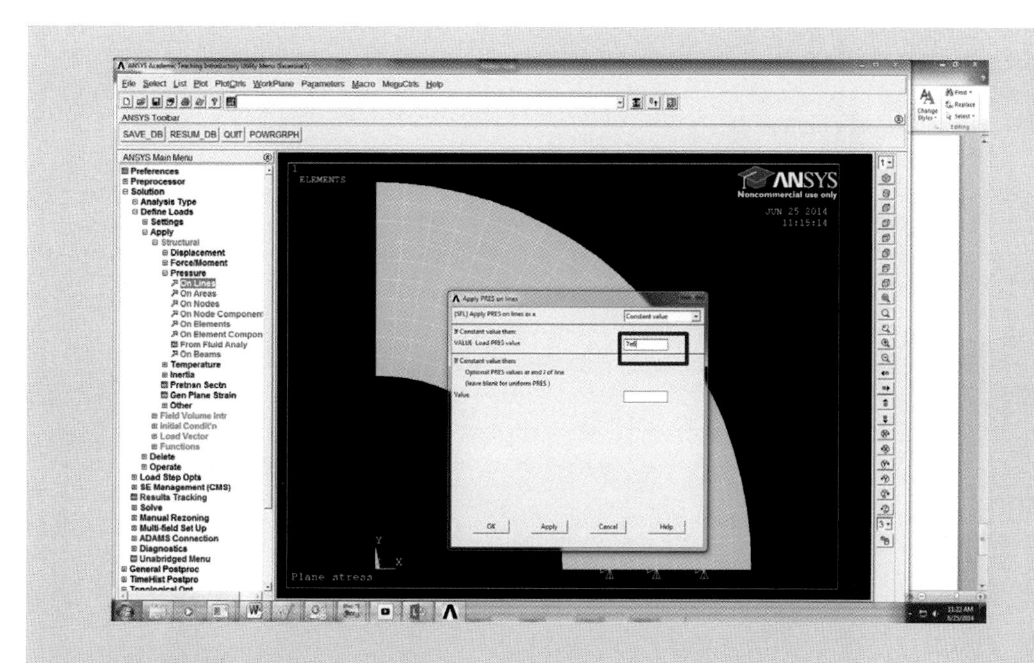

The next step is to solve the model using the following commands:

MM: Solution → Solve → Current LS → OK

A message that the solution is done will appear:

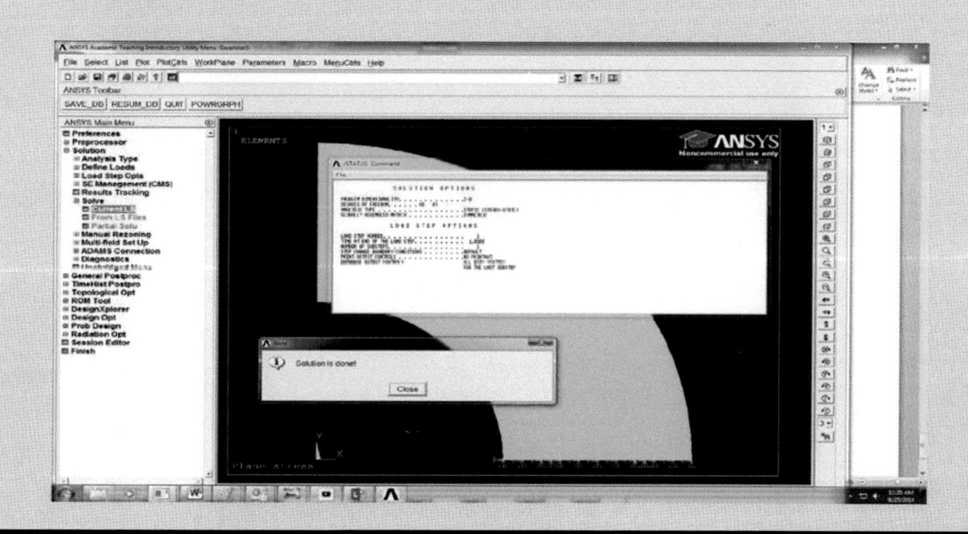

Continued

EXAMPLE 10.4: ANALYSIS OF A PLANE STRESS PROBLEM USING ANSYS—CONT'D

Now it is time to demonstrate the results. The post processor of the program must read the results using the following commands:

MM: General Postproc → Read Results → First set

And then, it can plot the deformed structure by the following commands:

MM: General Postproc → Plot Results → Deformed Shape

We can select "Def + undeformed" from the window entitled "Plot Deformed Shape," and then "**OK**," yielding the following graphic:

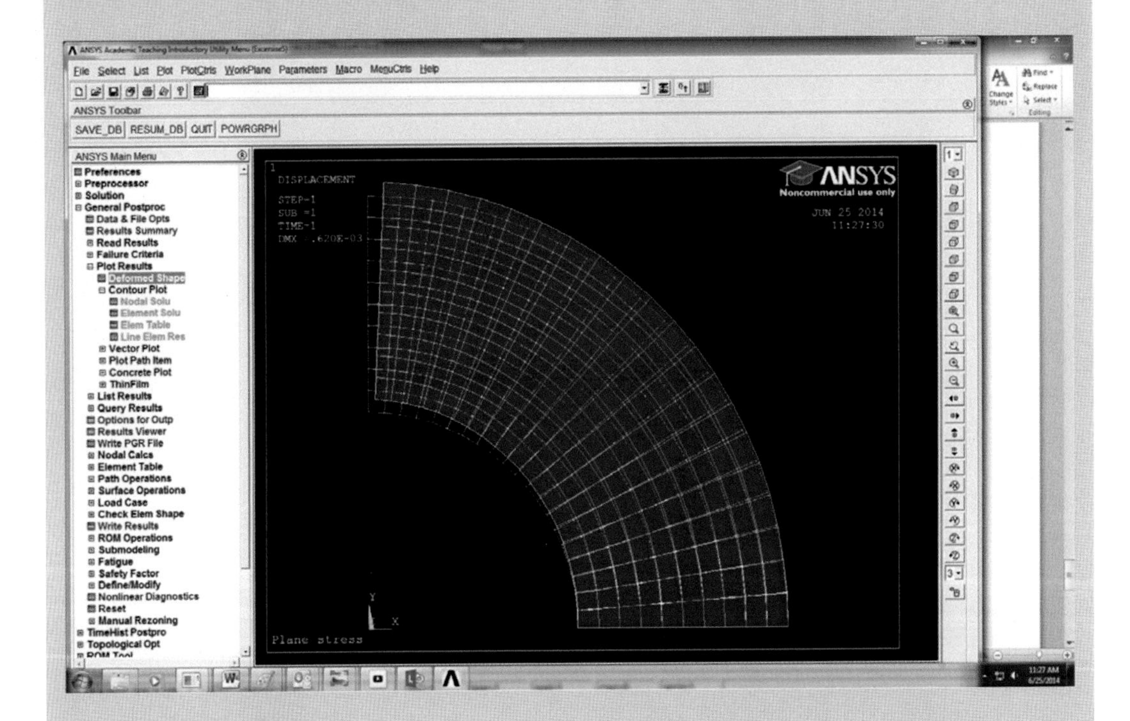

Now we can derive a colored map of the displacements distribution by the commands:

MM: General Postproc → Contour Plot → Nodal solution

In the new window entitled "Item to be contoured" we can select, for example, "**DOD Solution,**" and then "**X-Component of displacement,**" or "**Y-Component of displacement**." The resulted graphic for the Y-component of displacements is the following:

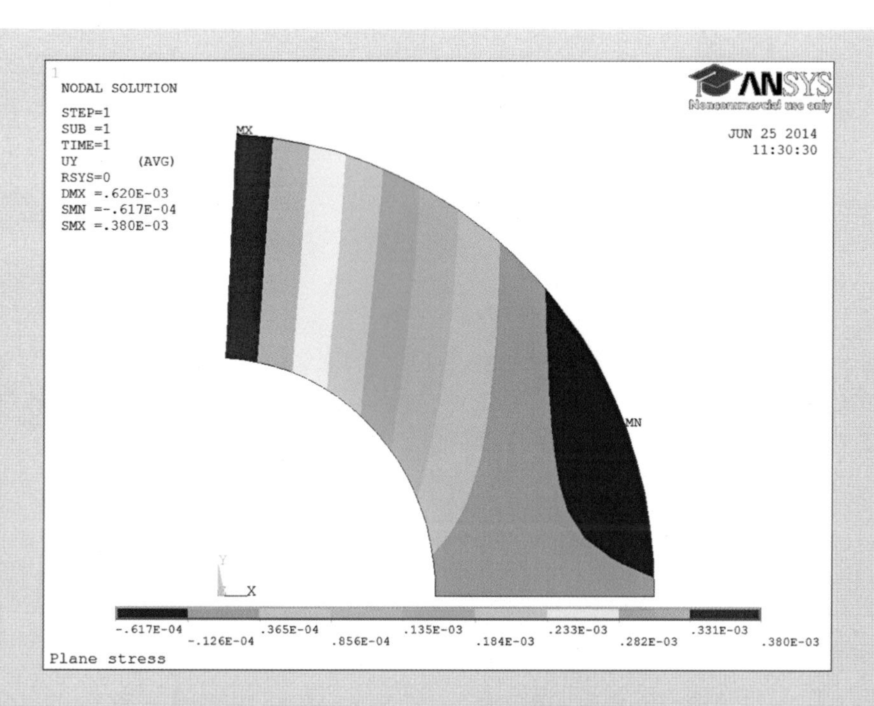

Now we can derive a colored map of strains by the following commands:
MM: General Postproc → Contour Plot → Element solution
In the new window entitled "Item to be contoured" we can select
Element Solution → Total Mechanical Strain → Total Mechanical Strain intensity → OK
according to the following window:

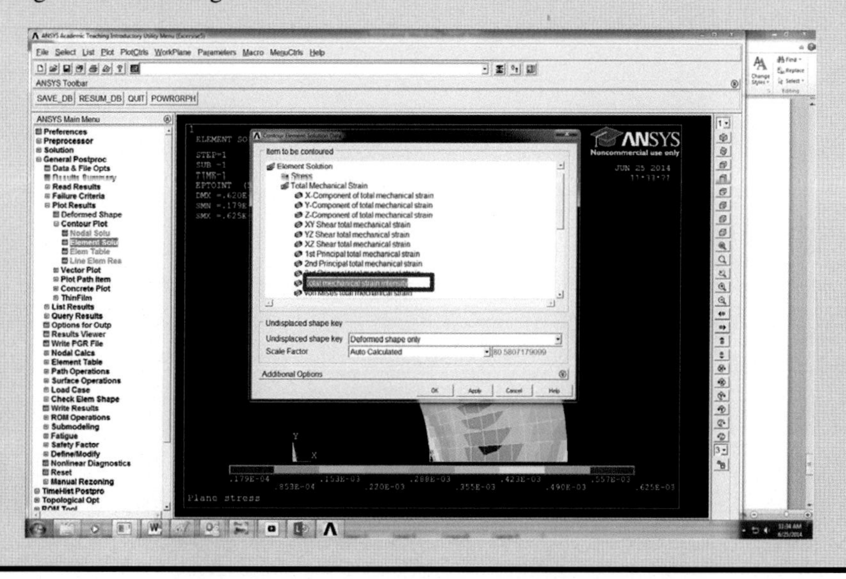

Continued

EXAMPLE 10.4: ANALYSIS OF A PLANE STRESS PROBLEM USING ANSYS—CONT'D

Then the following graphic will be obtained:

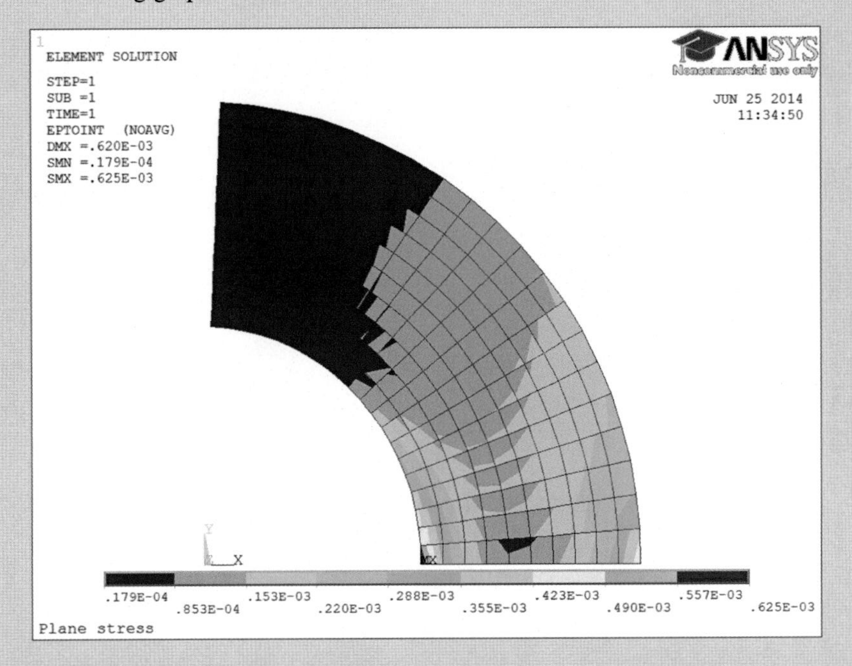

Following a similar procedure, we can derive the graphical representation of the stress distribution as well:

MM: General Postproc → Contour Plot → Element solution

In the new window entitled "Item to be contoured," we can select

Element Solution → Stress → Stress intensity → OK

according to the following window:

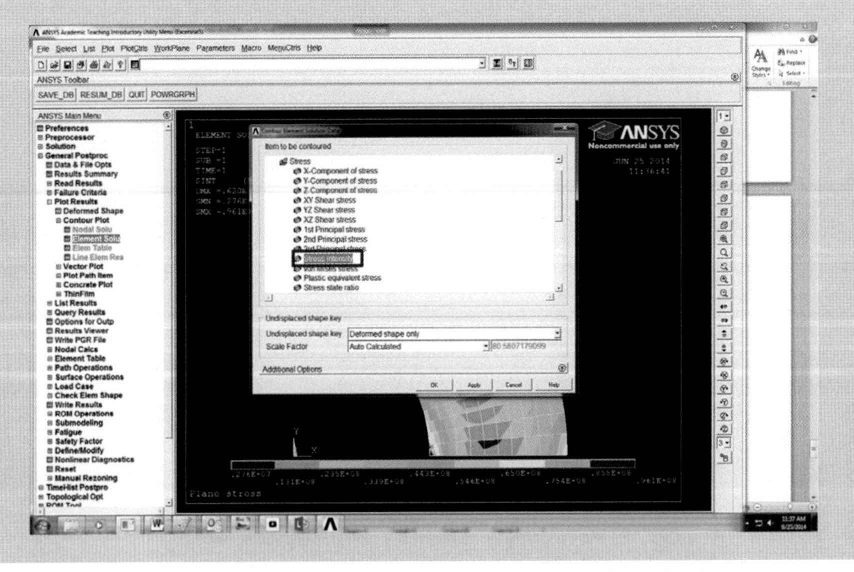

yielding the following result:

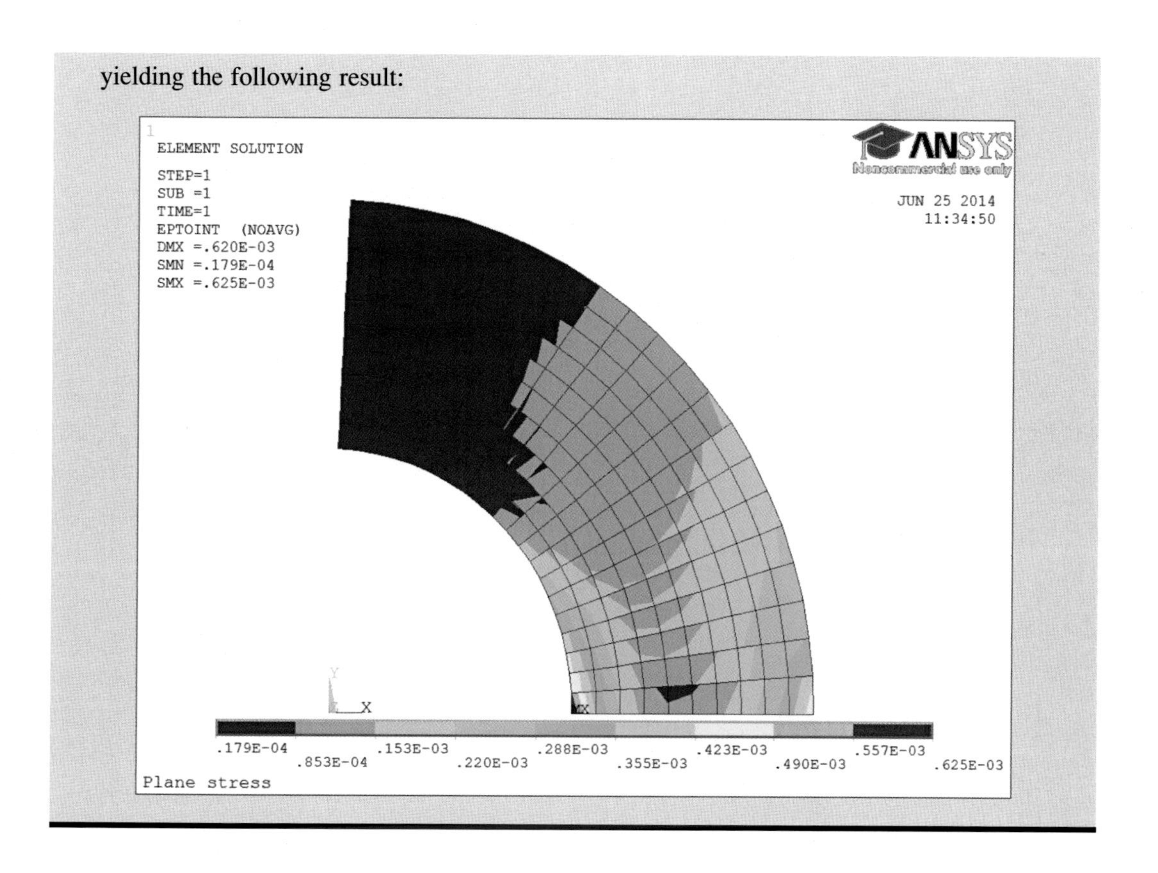

REFERENCES

[1] Cook RD, Malkus DS, Plesha ME, Witt RJ. Concepts and applications of finite element analysis. Hoboken: John Wiley & Sons; 2002.

[2] Shames IH, Dym CL. Energy and finite element methods in structural mechanics. New York: Hemisphere Publishing Corporation; 1985.

[3] Alawadhi EM. Finite element simulations using ANSYS. Boca Raton: CRC Press; 2010.

[4] Moaveni S. Finite element analysis, theory and application with ANSYS. Upper Saddle River, NJ: Pearson; 2008.

[5] Logan DL. A first course in the finite element method. Boston, MA: Cengage Learning; 2012.

[6] Wunderlich W, Pilkey W. Mechanics of structures—variational and computational methods. Boca Raton: CRC Press; 2003.

[7] Hartmann F, Katz C. Structural analysis with finite elements. Berlin: Springer; 2007.

[8] Bhatti MA. Fundamental finite element analysis and applications. Hoboken: John Wiley & Sons; 2005.

[9] Fish J, Belytschko T. A first course in finite elements. New York: John Wiley & Sons; 2007.

[10] Austrell P-E, Dahlblom O, Lindemann J, Olsson A, Olsson K-G, Persson K, et al., CALFEM—a finite element toolbox, Version 3.4., Division of Structural Mechanics, Lund University; 2004.

[11] ANSYS, User e-manual, Version 13.

[12] Melosh RJ. Structural engineering analysis by finite elements. Upper Saddle River, NJ: Prentice-Hall; 1990.

CHAPTER

STRUCTURAL DYNAMICS

11

11.1 THE DYNAMIC EQUATION

In the previous chapters, we discussed methods of finite element analysis of structural systems subjected to static loads, that is, loads keeping constant with respect to time.

For time-varying loads (dynamic loads), the mass distribution of the structure yields inertia forces (dynamic effects). Therefore, the derived matrix equations of the previous chapters should be modified in order to take into account the mass distribution. The influence of the mass on the dynamic equilibrium of forces will be initially discussed with a simple example (mass-spring system).

According to the Newton's second law of motion, the equilibrium equation of the mass shown in Figure 11.1 is given by the following equation:

$$ku + m\ddot{u} = F(t) \tag{11.1}$$

where $\ddot{u}$ is the second derivative of the displacement in the x-direction with respect to time

$$\ddot{u} = \frac{\mathrm{d}^2 u}{\mathrm{d}t^2} \tag{11.2}$$

The free vibration of the mass can be described by Equation (11.1) by setting $F(t) = 0$. In this case, Equation (11.1) can be written as

$$\ddot{u} + \omega^2 u = 0 \tag{11.3}$$

where the parameter

$$\omega^2 = \frac{k}{m} \tag{11.4}$$

is called natural circular frequency of the free vibration.

For more complex structural systems, Equation (11.1) is written in a matrix format

$$[k]\{d\} + [m]\{\ddot{d}\} = \{F\} \tag{11.5}$$

In the above equation, the vectors $\{d\}, \{\ddot{d}\}, \{F\}$ are time dependent. The matrix $[k]$ is the well-known stiffness matrix (see previous chapters). The matrix $[m]$ is called mass matrix. In the subsequent subchapters, mass matrices will be derived for several types of structural members.

Essentials of the Finite Element Method. http://dx.doi.org/10.1016/B978-0-12-802386-0.00011-6

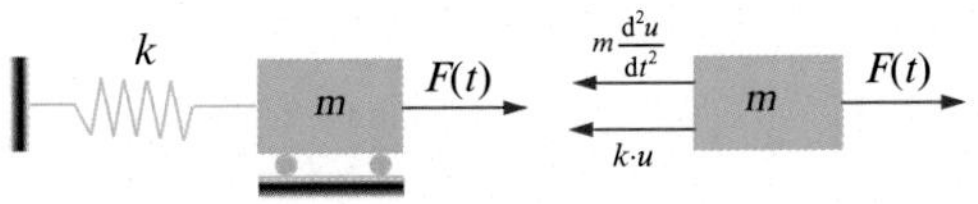

FIGURE 11.1

Dynamic response of a mass-spring system subjected to time-dependent force $F(t)$.

11.2 MASS MATRIX

11.2.1 BAR ELEMENT

Let us now consider the bar element with density ρ shown in Figure 11.2. The mass m of this element is uniformly distributed along its length L.

A simple mechanical model of a weightless bar of same length containing concentrated masses of magnitude

$$M=\frac{1}{2}\rho AL \tag{11.6}$$

on its ends can be used to approximate the dynamic response of the bar under the action of the dynamic nodal forces $r_{1x}(t), r_{2x}(t)$.

Taking into account that the direction of the inertia forces, $M\ddot{d}_{1x}, M\ddot{d}_{2x}$ is always opposite to the direction of motion, the equilibrium Equations (4.31) and (4.32) of the two pieces of the bar (Figure 11.3) should now be modified as:

$$r_{1x}(t)+F-M\ddot{d}_{1x}=0 \tag{11.7}$$

$$-F-M\ddot{d}_{2x}+r_{2x}(t)=0 \tag{11.8}$$

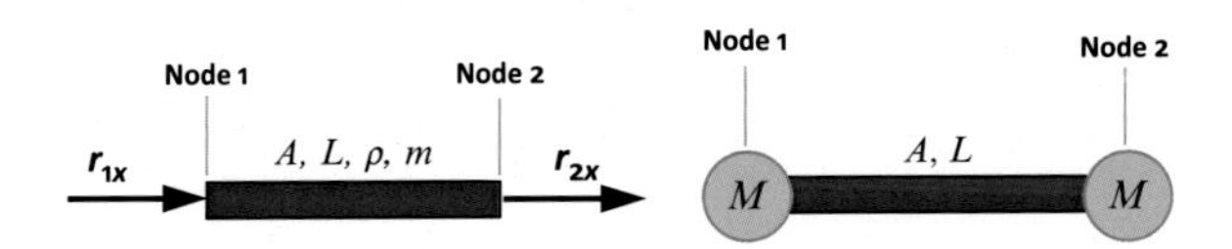

FIGURE 11.2

Bar element.

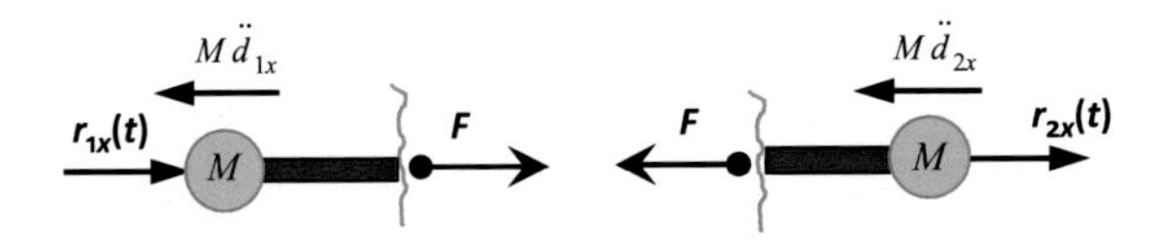

FIGURE 11.3

Equilibrium of the two pieces of the bar.

The above equations can be written in the following matrix form:

$$\begin{Bmatrix} r_{1x}(t) \\ r_{2x}(t) \end{Bmatrix} - \begin{bmatrix} M & 0 \\ 0 & M \end{bmatrix} \begin{Bmatrix} \ddot{d}_{1x} \\ \ddot{d}_{2x} \end{Bmatrix} = \begin{Bmatrix} -F \\ F \end{Bmatrix} \tag{11.9}$$

Taking into account Equation (4.29), the above equation yields

$$\begin{Bmatrix} r_{1x}(t) \\ r_{2x}(t) \end{Bmatrix} - \begin{bmatrix} M & 0 \\ 0 & M \end{bmatrix} \begin{Bmatrix} \ddot{d}_{1x} \\ \ddot{d}_{2x} \end{Bmatrix} = \begin{bmatrix} EA/L & -EA/L \\ -EA/L & EA/L \end{bmatrix} \begin{Bmatrix} d_{1x} \\ d_{2x} \end{Bmatrix} \tag{11.10}$$

or in an abbreviated form

$$[k]\{d\} + [m]\{\ddot{d}\} = \{r\} \tag{11.11}$$

In the above equation, $[k]$ is the well-known stiffness matrix given by Equation (4.36) and $[m]$ is the mass matrix

$$[m] = \begin{bmatrix} M & 0 \\ 0 & M \end{bmatrix} = \frac{1}{2}\rho AL \begin{bmatrix} 1 & 0 \\ 0 & 1 \end{bmatrix} \tag{11.12}$$

Since the above expression of the mass matrix is based on the assumption that the nodal masses M have been obtained by lumping equally at the two nodes of the total bar's mass $m = \rho AL$, it is called lumped-mass matrix. However, an alternative more accurate matrix simulation of the bar's mass can be derived by using D'Alambert's principle. According to this principle, the distribution of the inertia force $-\rho\ddot{d}(x)$ within the body can be used to derive the mass matrix $[m]$ using the shape functions $N_1(x)$, $N_2(x)$ given by Equations (4.12) and (4.13):

$$[m]\begin{Bmatrix} \ddot{d}_{1x} \\ \ddot{d}_{2x} \end{Bmatrix} = \int_V \begin{Bmatrix} N_1(x) \\ N_2(x) \end{Bmatrix} \rho\ddot{d}(x)\mathrm{d}V \tag{11.13}$$

Taking into account Equation (4.11), it can be written:

$$\ddot{d}(x) = [N_1(x) \;\; N_2(x)]\begin{Bmatrix} \ddot{d}_{1x} \\ \ddot{d}_{2x} \end{Bmatrix} \tag{11.14}$$

Then, Equation (11.13) yields

$$[m]\begin{Bmatrix} \ddot{d}_{1x} \\ \ddot{d}_{2x} \end{Bmatrix} = \rho\int_V \begin{Bmatrix} N_1(x) \\ N_2(x) \end{Bmatrix} [N_1(x) \;\; N_2(x)]\begin{Bmatrix} \ddot{d}_{1x} \\ \ddot{d}_{2x} \end{Bmatrix}\mathrm{d}V \tag{11.15}$$

Since the nodal accelerations $\ddot{d}_{1x}, \ddot{d}_{2x}$ are independent from the variable x, the above equation can be written as:

$$[m] = \rho\int_V \begin{Bmatrix} N_1(x) \\ N_2(x) \end{Bmatrix} [N_1(x) \;\; N_2(x)]\mathrm{d}V \tag{11.16}$$

For bars with uniform cross-section, the value of the volume element is

$$\mathrm{d}V = A\mathrm{d}x \tag{11.17}$$

In contrast, we recall from Equations (4.12) and (4.13) the following expressions for the shape functions:

$$N_1(x) = 1 - \frac{x}{L}, \quad N_2(x) = \frac{x}{L} \tag{11.18}$$

Taking into account Equations (11.17) and (11.18), Equation (11.16) yields

$$[m] = \rho A \int_0^L \begin{Bmatrix} (1-x/L) \\ (x/L) \end{Bmatrix} [(1-x/L) \ \ (x/L)] \mathrm{d}x \tag{11.19}$$

or

$$[m] = \frac{\rho AL}{6} \begin{bmatrix} 2 & 1 \\ 1 & 2 \end{bmatrix} \tag{11.20}$$

The mass matrix given by the above equation is called consistent-mass matrix.

11.2.2 TWO-DIMENSIONAL TRUSS ELEMENT

As has already been mentioned in Chapter 5, in any two-dimensional (2D) truss element, two coordinate systems exist: the local system $\bar{x}, \bar{y}$, which is aligned to the element, and the global system x, y, which is the system of coordinates of the whole structure and it is common for all truss members (Figure 11.4).

Let us now recall the element equation (Equation 11.11) and express it with respect to the nomenclature of the local coordinate system:

$$[\bar{k}] \begin{Bmatrix} d_{1\bar{x}} \\ d_{2\bar{x}} \end{Bmatrix} + [\bar{m}] \begin{Bmatrix} \ddot{d}_{1\bar{x}} \\ \ddot{d}_{2\bar{x}} \end{Bmatrix} = \begin{Bmatrix} r_{1\bar{x}} \\ r_{2\bar{x}} \end{Bmatrix} \tag{11.21}$$

In the above equation, the matrices $[\bar{k}], [\bar{m}]$ are given by the following known formulae:

$$[\bar{k}] = \begin{bmatrix} EA/L & -EA/L \\ -EA/L & EA/L \end{bmatrix} \tag{11.22}$$

and

$$[\bar{m}] = \frac{1}{2} \rho AL \begin{bmatrix} 1 & 0 \\ 0 & 1 \end{bmatrix} \quad \text{lumped} - \text{mass matrix} \tag{11.23}$$

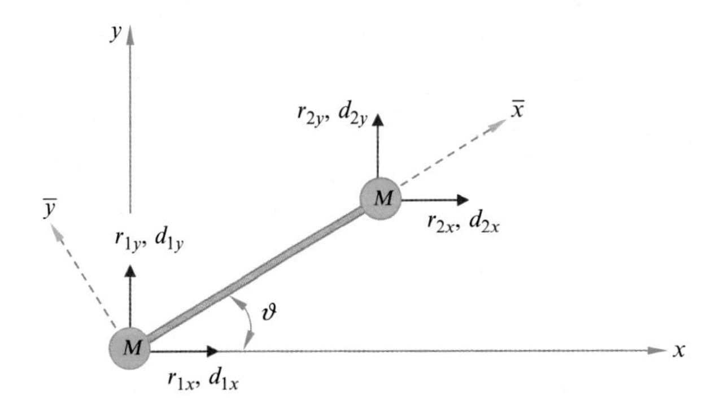

FIGURE 11.4

Local and global coordinate systems for a 2D truss element.

or

$$[\bar{m}] = \frac{1}{6}\rho AL\begin{bmatrix} 2 & 1 \\ 1 & 2 \end{bmatrix} \quad \text{consistent} - \text{mass matrix} \tag{11.24}$$

Our target now is to transform Equation (11.21) with respect to the global coordinate system x-y. Taking into account Equations (5.5) and (5.6), the following equations can be written:

$$\begin{Bmatrix} d_{1\bar{x}} \\ d_{2\bar{x}} \end{Bmatrix} = [\bar{T}]\begin{Bmatrix} d_{1x} \\ d_{1y} \\ d_{2x} \\ d_{2y} \end{Bmatrix} \tag{11.25}$$

$$\begin{Bmatrix} \ddot{d}_{1\bar{x}} \\ \ddot{d}_{2\bar{x}} \end{Bmatrix} = [\bar{T}]\begin{Bmatrix} \ddot{d}_{1x} \\ \ddot{d}_{1y} \\ \ddot{d}_{2x} \\ \ddot{d}_{2y} \end{Bmatrix} \tag{11.26}$$

$$\begin{Bmatrix} r_{1\bar{x}} \\ r_{2\bar{x}} \end{Bmatrix} = [\bar{T}]\begin{Bmatrix} r_{1x} \\ r_{1y} \\ r_{2x} \\ r_{2y} \end{Bmatrix} \tag{11.27}$$

where

$$[\bar{T}] = \begin{bmatrix} C & S & 0 & 0 \\ 0 & 0 & C & S \end{bmatrix} \tag{11.28}$$

and $C = \cos(\theta), S = \sin(\theta)$.

Using the above equations, Equation (11.21) yields

$$[\bar{k}][\bar{T}]\begin{Bmatrix} d_{1x} \\ d_{1y} \\ d_{2x} \\ d_{2y} \end{Bmatrix} + [\bar{m}][\bar{T}]\begin{Bmatrix} \ddot{d}_{1x} \\ \ddot{d}_{1y} \\ \ddot{d}_{2x} \\ \ddot{d}_{2y} \end{Bmatrix} = [\bar{T}]\begin{Bmatrix} r_{1x} \\ r_{1y} \\ r_{2x} \\ r_{2y} \end{Bmatrix} \tag{11.29}$$

To correlate the global coordinates we must invert the matrix $[\bar{T}]$. However, this is not possible because $[\bar{T}]$ is not a square matrix. Therefore, Equations (11.25)–(11.27) and (5.1) should be expressed in the following form:

$$\begin{Bmatrix} d_{1\bar{x}} \\ 0 \\ d_{2\bar{x}} \\ 0 \end{Bmatrix} = \begin{bmatrix} C & S & 0 & 0 \\ 0 & 0 & 0 & 0 \\ 0 & 0 & C & S \\ 0 & 0 & 0 & 0 \end{bmatrix}\begin{Bmatrix} d_{1x} \\ d_{1y} \\ d_{2x} \\ d_{2y} \end{Bmatrix} \tag{11.30}$$

$$\begin{Bmatrix} \ddot{d}_{1\bar{x}} \\ 0 \\ \ddot{d}_{2\bar{x}} \\ 0 \end{Bmatrix} = \begin{bmatrix} C & S & 0 & 0 \\ 0 & 0 & 0 & 0 \\ 0 & 0 & C & S \\ 0 & 0 & 0 & 0 \end{bmatrix}\begin{Bmatrix} \ddot{d}_{1x} \\ \ddot{d}_{1y} \\ \ddot{d}_{2x} \\ \ddot{d}_{2y} \end{Bmatrix} \tag{11.31}$$

$$\begin{Bmatrix} r_{1\bar{x}} \\ 0 \\ r_{2\bar{x}} \\ 0 \end{Bmatrix} = \begin{bmatrix} C & S & 0 & 0 \\ 0 & 0 & 0 & 0 \\ 0 & 0 & C & S \\ 0 & 0 & 0 & 0 \end{bmatrix} \begin{Bmatrix} r_{1x} \\ r_{1y} \\ r_{2x} \\ r_{2y} \end{Bmatrix} \tag{11.32}$$

$$\begin{Bmatrix} r_{1\bar{x}} \\ 0 \\ r_{2\bar{x}} \\ 0 \end{Bmatrix} = \begin{bmatrix} EA/L & 0 & -EA/L & 0 \\ 0 & 0 & 0 & 0 \\ -EA/L & 0 & EA/L & 0 \\ 0 & 0 & 0 & 0 \end{bmatrix} \begin{Bmatrix} d_{1\bar{x}} \\ 0 \\ d_{2\bar{x}} \\ 0 \end{Bmatrix} \tag{11.33}$$

Following the same concept, the member $[m][\ddot{d}]$ of Equation (11.21), which is

$$\frac{1}{2}\rho AL \begin{bmatrix} 1 & 0 \\ 0 & 1 \end{bmatrix} \begin{Bmatrix} \ddot{d}_{1\bar{x}} \\ \ddot{d}_{2\bar{x}} \end{Bmatrix}$$

for the lumped-mass matrix, can be written in the following alternative form:

$$\frac{1}{2}\rho AL \begin{bmatrix} 1 & 0 & 0 & 0 \\ 0 & 0 & 0 & 0 \\ 0 & 0 & 1 & 0 \\ 0 & 0 & 0 & 0 \end{bmatrix} \begin{Bmatrix} \ddot{d}_{1\bar{x}} \\ 0 \\ \ddot{d}_{2\bar{x}} \\ 0 \end{Bmatrix}$$

Therefore, the following modified lumped-mass matrix is created:

$$[m^*] = \frac{1}{2}\rho AL \begin{bmatrix} 1 & 0 & 0 & 0 \\ 0 & 0 & 0 & 0 \\ 0 & 0 & 1 & 0 \\ 0 & 0 & 0 & 0 \end{bmatrix} \quad \text{lumped} - \text{mass matrix} \tag{11.34}$$

Similarly, the modified consistent-mass matrix is

$$[m^*] = \frac{1}{6}\rho AL \begin{bmatrix} 2 & 0 & 1 & 0 \\ 0 & 0 & 0 & 0 \\ 1 & 0 & 2 & 0 \\ 0 & 0 & 0 & 0 \end{bmatrix} \quad \text{consistent} - \text{mass matrix} \tag{11.35}$$

According to Equation (11.33), the modified stiffness matrix is:

$$[k^*] = \begin{bmatrix} EA/L & 0 & -EA/L & 0 \\ 0 & 0 & 0 & 0 \\ -EA/L & 0 & EA/L & 0 \\ 0 & 0 & 0 & 0 \end{bmatrix} \tag{11.36}$$

Then Equation (11.21) can be expressed in the following equivalent format:

$$[k^*] \begin{Bmatrix} d_{1\bar{x}} \\ 0 \\ d_{2\bar{x}} \\ 0 \end{Bmatrix} + [m^*] \begin{Bmatrix} \ddot{d}_{1\bar{x}} \\ 0 \\ \ddot{d}_{2\bar{x}} \\ 0 \end{Bmatrix} = \begin{Bmatrix} r_{1\bar{x}} \\ 0 \\ r_{2\bar{x}} \\ 0 \end{Bmatrix} \tag{11.37}$$

Taking into account Equations (11.30)–(11.32) and using the following notation for the transformation matrix

$$[T]=\begin{bmatrix} C & S & 0 & 0 \\ 0 & 0 & 0 & 0 \\ 0 & 0 & C & S \\ 0 & 0 & 0 & 0 \end{bmatrix} \tag{11.38}$$

Equation (11.37) yields

$$[k^*][T]\begin{Bmatrix} d_{1x} \\ d_{1y} \\ d_{2x} \\ d_{2y} \end{Bmatrix} + [m^*][T]\begin{Bmatrix} \ddot{d}_{1x} \\ \ddot{d}_{1y} \\ \ddot{d}_{2x} \\ \ddot{d}_{2y} \end{Bmatrix} = [T]\begin{Bmatrix} r_{1x} \\ r_{1y} \\ r_{2x} \\ r_{2y} \end{Bmatrix} \tag{11.39}$$

It is known from the matrix algebra that the transformation matrix is orthogonal. Taking into account the property $[T]^{-1}=[T]^{\mathrm{T}}$ of the orthogonal matrices, the above equation can be written:

$$[T]^T[k^*][T]\begin{Bmatrix} d_{1x} \\ d_{1y} \\ d_{2x} \\ d_{2y} \end{Bmatrix} + [T]^T[m^*][T]\begin{Bmatrix} \ddot{d}_{1x} \\ \ddot{d}_{1y} \\ \ddot{d}_{2x} \\ \ddot{d}_{2y} \end{Bmatrix} = \begin{Bmatrix} r_{1x} \\ r_{1y} \\ r_{2x} \\ r_{2y} \end{Bmatrix} \tag{11.40}$$

or

$$[k]\begin{Bmatrix} d_{1x} \\ d_{1y} \\ d_{2x} \\ d_{2y} \end{Bmatrix} + [m]\begin{Bmatrix} \ddot{d}_{1x} \\ \ddot{d}_{1y} \\ \ddot{d}_{2x} \\ \ddot{d}_{2y} \end{Bmatrix} = \begin{Bmatrix} r_{1x} \\ r_{1y} \\ r_{2x} \\ r_{2y} \end{Bmatrix} \tag{11.41}$$

where

$$[k]=\frac{AE}{L}\begin{bmatrix} C^2 & CS & -C^2 & -CS \\ & S^2 & -CS & -S^2 \\ \mathit{symmetric} & & C^2 & CS \\ & & & S^2 \end{bmatrix} \tag{11.42}$$

$$[m]=\frac{1}{2}\rho AL\cdot[T]^T\cdot\begin{bmatrix} 1 & 0 & 0 & 0 \\ 0 & 0 & 0 & 0 \\ 0 & 0 & 1 & 0 \\ 0 & 0 & 0 & 0 \end{bmatrix}\cdot[T] \quad \begin{matrix} \textit{lumped}-\textit{mass matrix} \\ \textit{in global coordinates } x-y \end{matrix} \tag{11.43}$$

$$[m]=\frac{1}{6}\rho AL\cdot[T]^T\cdot\begin{bmatrix} 2 & 0 & 1 & 0 \\ 0 & 0 & 0 & 0 \\ 1 & 0 & 2 & 0 \\ 0 & 0 & 0 & 0 \end{bmatrix}\cdot[T] \quad \begin{matrix} \textit{consistent}-\textit{mass matrix} \\ \textit{in global coordinates } x-y \end{matrix} \tag{11.44}$$

11.2.3 THREE-DIMENSIONAL TRUSS ELEMENT

The derivation of the mass matrix for a three-dimensional (3D) bar element (Figure 11.5) can be based on the same concept as the 2D bar element.

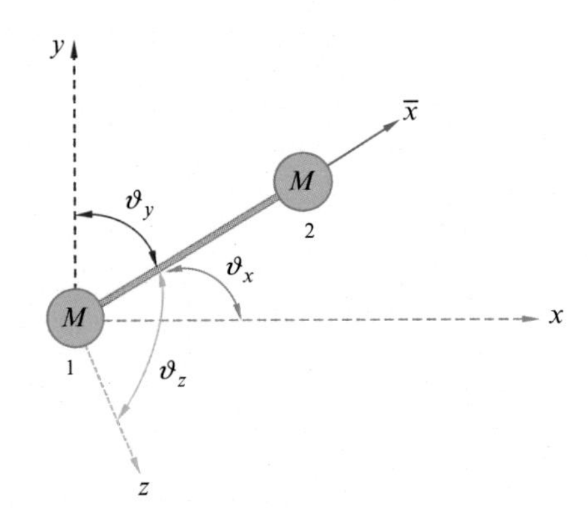

FIGURE 11.5

Local and global coordinate systems for a 3D bar element.

For the 3D element, Equation (11.37) has now the form:

$$[k^*]\begin{Bmatrix} d_{1\bar{x}} \\ 0 \\ 0 \\ d_{2\bar{x}} \\ 0 \\ 0 \end{Bmatrix} + [m^*]\begin{Bmatrix} \ddot{d}_{1\bar{x}} \\ 0 \\ 0 \\ \ddot{d}_{2\bar{x}} \\ 0 \\ 0 \end{Bmatrix} = \begin{Bmatrix} r_{1\bar{x}} \\ 0 \\ 0 \\ r_{2\bar{x}} \\ 0 \\ 0 \end{Bmatrix} \tag{11.45}$$

where

$$[k^*] = \begin{bmatrix} EA/L & 0 & 0 & -EA/L & 0 & 0 \\ 0 & 0 & 0 & 0 & 0 & 0 \\ 0 & 0 & 0 & 0 & 0 & 0 \\ -EA/L & 0 & 0 & EA/L & 0 & 0 \\ 0 & 0 & 0 & 0 & 0 & 0 \\ 0 & 0 & 0 & 0 & 0 & 0 \end{bmatrix} \tag{11.46}$$

$$[m^*] = \frac{\rho AL}{2}\begin{bmatrix} 1 & 0 & 0 & 0 & 0 & 0 \\ 0 & 0 & 0 & 0 & 0 & 0 \\ 0 & 0 & 0 & 0 & 0 & 0 \\ 0 & 0 & 0 & 1 & 0 & 0 \\ 0 & 0 & 0 & 0 & 0 & 0 \\ 0 & 0 & 0 & 0 & 0 & 0 \end{bmatrix} \quad \text{lumped} - \text{mass matrix} \tag{11.47}$$

$$[m^*] = \frac{\rho AL}{6}\begin{bmatrix} 2 & 0 & 0 & 1 & 0 & 0 \\ 0 & 0 & 0 & 0 & 0 & 0 \\ 0 & 0 & 0 & 0 & 0 & 0 \\ 1 & 0 & 0 & 2 & 0 & 0 \\ 0 & 0 & 0 & 0 & 0 & 0 \\ 0 & 0 & 0 & 0 & 0 & 0 \end{bmatrix} \quad \text{consistent} - \text{mass matrix} \tag{11.48}$$

In contrast, Equations (5.17) and (5.18) can be expressed as:

$$\begin{Bmatrix} d_{1\bar{x}} \\ 0 \\ 0 \\ d_{2\bar{x}} \\ 0 \\ 0 \end{Bmatrix} = \begin{bmatrix} C_x & C_y & C_z & 0 & 0 & 0 \\ 0 & 0 & 0 & 0 & 0 & 0 \\ 0 & 0 & 0 & 0 & 0 & 0 \\ 0 & 0 & 0 & C_x & C_y & C_z \\ 0 & 0 & 0 & 0 & 0 & 0 \\ 0 & 0 & 0 & 0 & 0 & 0 \end{bmatrix} \begin{Bmatrix} d_{1x} \\ d_{1y} \\ d_{1z} \\ d_{2x} \\ d_{2y} \\ d_{2z} \end{Bmatrix} \tag{11.49}$$

$$\begin{Bmatrix} \ddot{d}_{1\bar{x}} \\ 0 \\ 0 \\ \ddot{d}_{2\bar{x}} \\ 0 \\ 0 \end{Bmatrix} = \begin{bmatrix} C_x & C_y & C_z & 0 & 0 & 0 \\ 0 & 0 & 0 & 0 & 0 & 0 \\ 0 & 0 & 0 & 0 & 0 & 0 \\ 0 & 0 & 0 & C_x & C_y & C_z \\ 0 & 0 & 0 & 0 & 0 & 0 \\ 0 & 0 & 0 & 0 & 0 & 0 \end{bmatrix} \begin{Bmatrix} \ddot{d}_{1x} \\ \ddot{d}_{1y} \\ \ddot{d}_{1z} \\ \ddot{d}_{2x} \\ \ddot{d}_{2y} \\ \ddot{d}_{2z} \end{Bmatrix} \tag{11.50}$$

$$\begin{Bmatrix} r_{1\bar{x}} \\ 0 \\ 0 \\ r_{2\bar{x}} \\ 0 \\ 0 \end{Bmatrix} = \begin{bmatrix} C_x & C_y & C_z & 0 & 0 & 0 \\ 0 & 0 & 0 & 0 & 0 & 0 \\ 0 & 0 & 0 & 0 & 0 & 0 \\ 0 & 0 & 0 & C_x & C_y & C_z \\ 0 & 0 & 0 & 0 & 0 & 0 \\ 0 & 0 & 0 & 0 & 0 & 0 \end{bmatrix} \begin{Bmatrix} r_{1x} \\ r_{1y} \\ r_{1z} \\ r_{2x} \\ r_{2y} \\ r_{2z} \end{Bmatrix} \tag{11.51}$$

where $C_x = \cos\theta_x, C_y = \cos\theta_y, C_z = \cos\theta_z$.

According to the following notation for the transformation matrix:

$$[T] = \begin{bmatrix} C_x & C_y & C_z & 0 & 0 & 0 \\ 0 & 0 & 0 & 0 & 0 & 0 \\ 0 & 0 & 0 & 0 & 0 & 0 \\ 0 & 0 & 0 & C_x & C_y & C_z \\ 0 & 0 & 0 & 0 & 0 & 0 \\ 0 & 0 & 0 & 0 & 0 & 0 \end{bmatrix} \tag{11.52}$$

and taking into account Equations (11.46)–(11.51), Equation (11.45) can be written as:

$$[k^*][T]\begin{Bmatrix} d_{1x} \\ d_{1y} \\ d_{1z} \\ d_{2x} \\ d_{2y} \\ d_{2z} \end{Bmatrix} + [m^*][T]\begin{Bmatrix} \ddot{d}_{1x} \\ \ddot{d}_{1y} \\ \ddot{d}_{1z} \\ \ddot{d}_{2x} \\ \ddot{d}_{2y} \\ \ddot{d}_{2z} \end{Bmatrix} = [T]\begin{Bmatrix} r_{1x} \\ r_{1y} \\ r_{1z} \\ r_{2x} \\ r_{2y} \\ r_{2z} \end{Bmatrix} \tag{11.53}$$

Therefore, multiplication of both sides of the above equation with $[T]^{\mathrm{T}}$ yields

$$[k]\begin{Bmatrix} d_{1x} \\ d_{1y} \\ d_{1z} \\ d_{2x} \\ d_{2y} \\ d_{2z} \end{Bmatrix} + [m]\begin{Bmatrix} \ddot{d}_{1x} \\ \ddot{d}_{1y} \\ \ddot{d}_{1z} \\ \ddot{d}_{2x} \\ \ddot{d}_{2y} \\ \ddot{d}_{2z} \end{Bmatrix} = \begin{Bmatrix} r_{1x} \\ r_{1y} \\ r_{1z} \\ r_{2x} \\ r_{2y} \\ r_{2z} \end{Bmatrix} \tag{11.54}$$

where

$$[k]=[T]^{\mathrm{T}}[k^*][T]=\frac{EA}{L}\begin{bmatrix} C_x^2 & C_xC_y & C_xC_z & -C_x^2 & -C_xC_y & -C_xC_z \\ & C_y^2 & C_yC_z & -C_xC_y & C_y^2 & -C_yC_z \\ & & C_z^2 & -C_xC_z & -C_yC_z & -C_z^2 \\ & & & C_x^2 & C_xC_y & C_xC_z \\ & \text{symmetric} & & & C_y^2 & C_yC_z \\ & & & & & C_z^2 \end{bmatrix} \tag{11.55}$$

$$[m]=\frac{\rho AL}{2}\cdot[T]^{\mathrm{T}}\cdot\begin{bmatrix} 1&0&0&0&0&0\\0&0&0&0&0&0\\0&0&0&0&0&0\\0&0&0&1&0&0\\0&0&0&0&0&0\\0&0&0&0&0&0 \end{bmatrix}\cdot[T] \quad \begin{matrix}\text{lumped}-\text{mass matrix}\\ \text{in global coordinates } x-y-z\end{matrix} \tag{11.56}$$

$$[m]=\frac{\rho AL}{6}\cdot[T]^{\mathrm{T}}\cdot\begin{bmatrix} 2&0&0&1&0&0\\0&0&0&0&0&0\\0&0&0&0&0&0\\1&0&0&2&0&0\\0&0&0&0&0&0\\0&0&0&0&0&0 \end{bmatrix}\cdot[T] \quad \begin{matrix}\text{consistent}-\text{mass matrix}\\ \text{in global coordinates } x-y-z\end{matrix} \tag{11.57}$$

11.2.4 TWO-DIMENSIONAL BEAM ELEMENT

As it has been mentioned in Chapter 6, the degrees of freedom of a 2D beam element 1-2 are

$$\{d\}=[u_1\ \ \theta_1\ \ u_2\ \ \theta_2]^{\mathrm{T}} \tag{11.58}$$

Taking into account only the translational inertia forces in the direction of the motions u_1, u_2 (Figure 11.6) the following lamped-mass matrix can be obtained

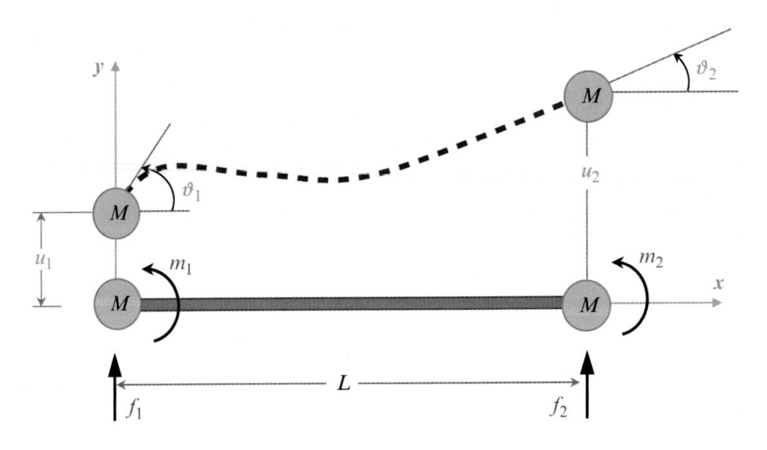

FIGURE 11.6

A 2D beam element.

$$[m]\{\ddot{d}\} = \begin{bmatrix} \rho AL/2 & 0 & 0 & 0 \\ 0 & 0 & 0 & 0 \\ 0 & 0 & \rho AL/2 & 0 \\ 0 & 0 & 0 & 0 \end{bmatrix} \begin{Bmatrix} \ddot{u}_1 \\ \ddot{\theta}_1 \\ \ddot{u}_2 \\ \ddot{\theta}_2 \end{Bmatrix} \tag{11.59}$$

or

$$[m] = \frac{\rho AL}{2} \begin{bmatrix} 1 & 0 & 0 & 0 \\ 0 & 0 & 0 & 0 \\ 0 & 0 & 1 & 0 \\ 0 & 0 & 0 & 0 \end{bmatrix} \tag{11.60}$$

The consistent-mass matrix can be derived by using the shape functions of the beam element (Equations 6.14–6.17) and the already known equation

$$[m] = \rho \int_V \{N\}^T [N] dV = \rho A \int_0^L \begin{Bmatrix} N_1(x) \\ N_2(x) \\ N_3(x) \\ N_4(x) \end{Bmatrix} [N_1(x) \;\; N_2(x) \;\; N_3(x) \;\; N_4(x)] dx \tag{11.61}$$

Then, the consistent-mass matrix for a beam element is

$$[m] = \frac{\rho AL}{420} \begin{bmatrix} 156 & 22L & 54 & -13L \\ 22L & 4L^2 & 13L & -3L^2 \\ 54 & 13L & 156 & -22L \\ -13L & -3L^2 & -22L & 4L^2 \end{bmatrix} \tag{11.62}$$

11.2.5 THREE-DIMENSIONAL BEAM ELEMENT

Following the same concept as for the 2D case and taking into account only the translational inertia forces in the direction of the motions u_{1x}, u_{1y}, u_{1z} and u_{2x}, u_{2y}, u_{2z} (Figure 11.7), the following expression for the lumped-mass matrix can be obtained:

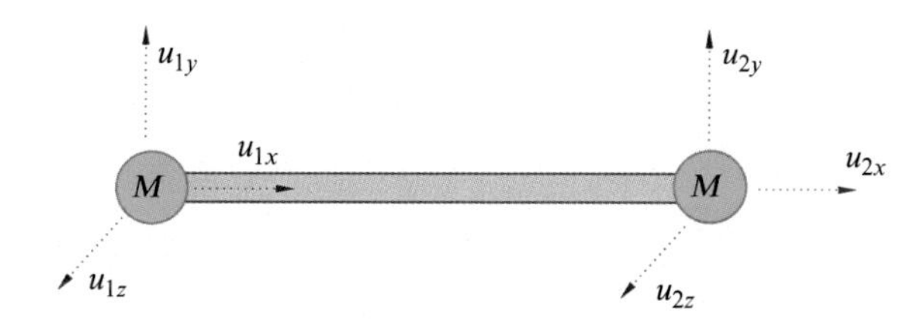

FIGURE 11.7

Translational nodal motions of a beam element in a 3D coordinate system.

$$[m] = \frac{\rho AL}{2}\begin{bmatrix} 1 & 0 & 0 & 0 & 0 & 0 & 0 & 0 & 0 & 0 & 0 & 0 \\ 0 & 1 & 0 & 0 & 0 & 0 & 0 & 0 & 0 & 0 & 0 & 0 \\ 0 & 0 & 1 & 0 & 0 & 0 & 0 & 0 & 0 & 0 & 0 & 0 \\ 0 & 0 & 0 & 0 & 0 & 0 & 0 & 0 & 0 & 0 & 0 & 0 \\ 0 & 0 & 0 & 0 & 0 & 0 & 0 & 0 & 0 & 0 & 0 & 0 \\ 0 & 0 & 0 & 0 & 0 & 0 & 0 & 0 & 0 & 0 & 0 & 0 \\ 0 & 0 & 0 & 0 & 0 & 0 & 1 & 0 & 0 & 0 & 0 & 0 \\ 0 & 0 & 0 & 0 & 0 & 0 & 0 & 1 & 0 & 0 & 0 & 0 \\ 0 & 0 & 0 & 0 & 0 & 0 & 0 & 0 & 1 & 0 & 0 & 0 \\ 0 & 0 & 0 & 0 & 0 & 0 & 0 & 0 & 0 & 0 & 0 & 0 \\ 0 & 0 & 0 & 0 & 0 & 0 & 0 & 0 & 0 & 0 & 0 & 0 \\ 0 & 0 & 0 & 0 & 0 & 0 & 0 & 0 & 0 & 0 & 0 & 0 \end{bmatrix} \tag{11.63}$$

The consistent-mass matrix can be derived by the following known equation:

$$[m] = \rho A \int_0^L \{N\}^{\mathrm{T}}[N]\mathrm{d}x \tag{11.64}$$

Expanding the mass matrices given by Equations (11.20) and (11.62) to the degrees of freedom of a 3D element, the consistent mass matrix can be obtained by superposition of the following equations:

$$[m_x] = \frac{\rho AL}{6}\begin{bmatrix} 2 & 0 & 0 & 0 & 0 & 0 & 1 & 0 & 0 & 0 & 0 & 0 \\ 0 & 0 & 0 & 0 & 0 & 0 & 0 & 0 & 0 & 0 & 0 & 0 \\ 0 & 0 & 0 & 0 & 0 & 0 & 0 & 0 & 0 & 0 & 0 & 0 \\ 0 & 0 & 0 & 0 & 0 & 0 & 0 & 0 & 0 & 0 & 0 & 0 \\ 0 & 0 & 0 & 0 & 0 & 0 & 0 & 0 & 0 & 0 & 0 & 0 \\ 0 & 0 & 0 & 0 & 0 & 0 & 0 & 0 & 0 & 0 & 0 & 0 \\ 1 & 0 & 0 & 0 & 0 & 0 & 2 & 0 & 0 & 0 & 0 & 0 \\ 0 & 0 & 0 & 0 & 0 & 0 & 0 & 0 & 0 & 0 & 0 & 0 \\ 0 & 0 & 0 & 0 & 0 & 0 & 0 & 0 & 0 & 0 & 0 & 0 \\ 0 & 0 & 0 & 0 & 0 & 0 & 0 & 0 & 0 & 0 & 0 & 0 \\ 0 & 0 & 0 & 0 & 0 & 0 & 0 & 0 & 0 & 0 & 0 & 0 \\ 0 & 0 & 0 & 0 & 0 & 0 & 0 & 0 & 0 & 0 & 0 & 0 \end{bmatrix} \tag{11.65}$$

$$[m_{x-y}] = \frac{\rho AL}{420}\begin{bmatrix} 0 & 0 & 0 & 0 & 0 & 0 & 0 & 0 & 0 & 0 & 0 & 0 \\ 0 & 156 & 0 & 0 & 0 & 22L & 0 & 54 & 0 & 0 & 0 & -13L \\ 0 & 0 & 0 & 0 & 0 & 0 & 0 & 0 & 0 & 0 & 0 & 0 \\ 0 & 0 & 0 & 0 & 0 & 0 & 0 & 0 & 0 & 0 & 0 & 0 \\ 0 & 0 & 0 & 0 & 0 & 0 & 0 & 0 & 0 & 0 & 0 & 0 \\ 0 & 22L & 0 & 0 & 0 & 4L^2 & 0 & 13L & 0 & 0 & 0 & -3L^2 \\ 0 & 0 & 0 & 0 & 0 & 0 & 0 & 0 & 0 & 0 & 0 & 0 \\ 0 & 54 & 0 & 0 & 0 & 13L & 0 & 156 & 0 & 0 & 0 & -22L \\ 0 & 0 & 0 & 0 & 0 & 0 & 0 & 0 & 0 & 0 & 0 & 0 \\ 0 & 0 & 0 & 0 & 0 & 0 & 0 & 0 & 0 & 0 & 0 & 0 \\ 0 & 0 & 0 & 0 & 0 & 0 & 0 & 0 & 0 & 0 & 0 & 0 \\ 0 & -13L & 0 & 0 & 0 & -3L^2 & 0 & -22L & 0 & 0 & 0 & 4L^2 \end{bmatrix} \tag{11.66}$$

$$[m_{x-z}] = \frac{\rho AL}{420}\begin{bmatrix} 0 & 0 & 0 & 0 & 0 & 0 & 0 & 0 & 0 & 0 & 0 & 0 \\ 0 & 0 & 0 & 0 & 0 & 0 & 0 & 0 & 0 & 0 & 0 & 0 \\ 0 & 0 & 156 & 0 & 22L & 0 & 0 & 0 & 54 & 0 & -13L & 0 \\ 0 & 0 & 0 & 0 & 0 & 0 & 0 & 0 & 0 & 0 & 0 & 0 \\ 0 & 0 & 22L & 0 & 4L^2 & 0 & 0 & 0 & 13L & 0 & -3L^2 & 0 \\ 0 & 0 & 0 & 0 & 0 & 0 & 0 & 0 & 0 & 0 & 0 & 0 \\ 0 & 0 & 0 & 0 & 0 & 0 & 0 & 0 & 0 & 0 & 0 & 0 \\ 0 & 0 & 0 & 0 & 0 & 0 & 0 & 0 & 0 & 0 & 0 & 0 \\ 0 & 0 & 54 & 0 & 13L & 0 & 0 & 0 & 156 & 0 & -22L & 0 \\ 0 & 0 & 0 & 0 & 0 & 0 & 0 & 0 & 0 & 0 & 0 & 0 \\ 0 & 0 & -13L & 0 & -3L^2 & 0 & 0 & 0 & -22L & 0 & 4L^2 & 0 \\ 0 & 0 & 0 & 0 & 0 & 0 & 0 & 0 & 0 & 0 & 0 & 0 \end{bmatrix} \quad (11.67)$$

Then,

$$[m] = [m_x] + [m_{x-y}] + [m_{x-z}] \quad (11.68)$$

Using the above expression of the mass matrix, the following dynamic equation for a 3D beam element can be used:

$$[k]\begin{Bmatrix} u_{1x} \\ u_{1y} \\ u_{1z} \\ \theta_{1x} \\ \theta_{1y} \\ \theta_{1z} \\ u_{2x} \\ u_{2y} \\ u_{2z} \\ \theta_{2x} \\ \theta_{2y} \\ \theta_{2z} \end{Bmatrix} + [m]\begin{Bmatrix} \ddot{u}_{1x} \\ \ddot{u}_{1y} \\ \ddot{u}_{1z} \\ \ddot{\theta}_{1x} \\ \ddot{\theta}_{1y} \\ \ddot{\theta}_{1z} \\ \ddot{u}_{2x} \\ \ddot{u}_{2y} \\ \ddot{u}_{2z} \\ \ddot{\theta}_{2x} \\ \ddot{\theta}_{2y} \\ \ddot{\theta}_{2z} \end{Bmatrix} = \begin{Bmatrix} f_{1x} \\ f_{1y} \\ f_{1z} \\ m_{1x} \\ m_{1y} \\ m_{1z} \\ f_{2x} \\ f_{2y} \\ f_{2z} \\ m_{2x} \\ m_{2y} \\ m_{2z} \end{Bmatrix} \quad (11.69)$$

In the above equation, the stiffness matrix $[k]$ can be obtained using Equation (6.194).

11.2.6 INCLINED TWO-DIMENSIONAL BEAM ELEMENT (TWO-DIMENSIONAL FRAME ELEMENT)

Following the same concept as the inclined bar element, the consistent-mass matrix for the 2D inclined beam element shown in the Figure 11.8 can be obtained by the transformation with respect to x-y coordinates of the following equation:

$$[\bar{k}]\begin{Bmatrix} d_{1\bar{x}} \\ d_{1\bar{y}} \\ \bar{\varphi}_1 \\ d_{2\bar{x}} \\ d_{2\bar{y}} \\ \bar{\varphi}_2 \end{Bmatrix} + [m^*]\begin{Bmatrix} \ddot{d}_{1\bar{x}} \\ \ddot{d}_{1\bar{y}} \\ \ddot{\bar{\varphi}}_1 \\ \ddot{d}_{2\bar{x}} \\ \ddot{d}_{2\bar{y}} \\ \ddot{\bar{\varphi}}_2 \end{Bmatrix} = \begin{Bmatrix} f_{1\bar{x}} \\ f_{1\bar{y}} \\ \bar{m}_1 \\ f_{2\bar{x}} \\ f_{2\bar{y}} \\ \bar{m}_2 \end{Bmatrix} \quad (11.70)$$

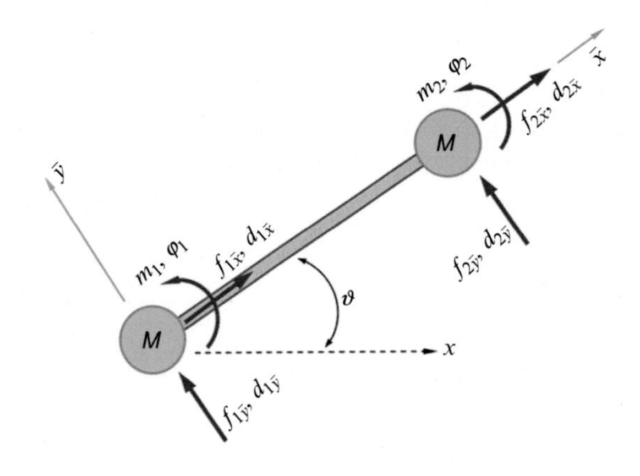

FIGURE 11.8

Local and global coordinate systems for a 2D frame element.

where the stiffness matrix $[\bar{k}]$ is given by Equation (7.17), and the consistent-mass matrix $[m^*]$ can be obtained by the superposition of the expanded matrices (Equations (11.20) and (11.62)):

$$[m^*] = \frac{\rho AL}{6}\begin{bmatrix} 2 & 0 & 0 & 1 & 0 & 0 \\ 0 & 0 & 0 & 0 & 0 & 0 \\ 0 & 0 & 0 & 0 & 0 & 0 \\ 1 & 0 & 0 & 2 & 0 & 0 \\ 0 & 0 & 0 & 0 & 0 & 0 \\ 0 & 0 & 0 & 0 & 0 & 0 \end{bmatrix} + \frac{\rho AL}{420}\begin{bmatrix} 0 & 0 & 0 & 0 & 0 & 0 \\ 0 & 156 & 22L & 0 & 54 & -13L \\ 0 & 22L & 4L^2 & 0 & 13L & -3L^2 \\ 0 & 0 & 0 & 0 & 0 & 0 \\ 0 & 54 & 13L & 0 & 156 & -22L \\ 0 & -13L & -3L^2 & 0 & -22L & 4L^2 \end{bmatrix} \tag{11.71}$$

Now we recall the transformation matrix given by Equation (7.16)

$$[T] = \begin{bmatrix} C & S & 0 & 0 & 0 & 0 \\ -S & C & 0 & 0 & 0 & 0 \\ 0 & 0 & 1 & 0 & 0 & 0 \\ 0 & 0 & 0 & C & S & 0 \\ 0 & 0 & 0 & -S & C & 0 \\ 0 & 0 & 0 & 0 & 0 & 1 \end{bmatrix} \tag{11.72}$$

where C, S depends on the inclination angle θ, that is, $C = \cos\theta, S = \sin\theta$.

Then, using the following equations (see Equations (7.13) and (7.14)):

$$\begin{Bmatrix} d_{1\bar{x}} \\ d_{1\bar{y}} \\ \bar{\varphi}_1 \\ d_{2\bar{x}} \\ d_{2\bar{y}} \\ \bar{\varphi}_2 \end{Bmatrix} = [T]\begin{Bmatrix} d_{1x} \\ d_{1y} \\ \varphi_1 \\ d_{2x} \\ d_{2y} \\ \varphi_2 \end{Bmatrix} \tag{11.73}$$

$$\begin{Bmatrix} f_{1\bar{x}} \\ f_{1\bar{y}} \\ \bar{m}_1 \\ f_{2\bar{x}} \\ f_{2\bar{y}} \\ \bar{m}_2 \end{Bmatrix} = [T]\begin{Bmatrix} f_{1x} \\ f_{1y} \\ m_1 \\ f_{2x} \\ f_{2y} \\ m_2 \end{Bmatrix} \tag{11.74}$$

Equation (11.70) can be written:

$$[\bar{k}][T]\begin{Bmatrix} d_{1x} \\ d_{1y} \\ \varphi_1 \\ d_{2x} \\ d_{2y} \\ \varphi_2 \end{Bmatrix} + [m^*][T]\begin{Bmatrix} \ddot{d}_{1x} \\ \ddot{d}_{1y} \\ \ddot{\varphi}_1 \\ \ddot{d}_{2x} \\ \ddot{d}_{2y} \\ \ddot{\varphi}_2 \end{Bmatrix} = [T]\begin{Bmatrix} f_{1x} \\ f_{1y} \\ m_1 \\ f_{2x} \\ f_{2y} \\ m_2 \end{Bmatrix} \tag{11.75}$$

Since the transformation matrix $[T]$ is orthogonal, multiplying the above equation by $[T]^T$ the following dynamic equation with respect to the global coordinate system x-y can be obtained:

$$[k]\begin{Bmatrix} d_{1x} \\ d_{1y} \\ \varphi_1 \\ d_{2x} \\ d_{2y} \\ \varphi_2 \end{Bmatrix} + [m]\begin{Bmatrix} \ddot{d}_{1x} \\ \ddot{d}_{1y} \\ \ddot{\varphi}_1 \\ \ddot{d}_{2x} \\ \ddot{d}_{2y} \\ \ddot{\varphi}_2 \end{Bmatrix} = \begin{Bmatrix} f_{1x} \\ f_{1y} \\ m_1 \\ f_{2x} \\ f_{2y} \\ m_2 \end{Bmatrix} \tag{11.76}$$

where the obtained stiffness matrix $[k]$ is the same with the one given by Equation (7.25), and the consistent-mass matrix $[m]$ is

$$[m] = [T]^{\mathrm{T}}[m^*][T] \tag{11.77}$$

11.2.7 LINEAR TRIANGULAR ELEMENT (CST ELEMENT)

Using the shape functions given in Equations (10.54)–(10.56), the consistent-mass matrix for the linear triangular element shown in Figure 10.4 can be obtained by the following equation:

$$[m] = \rho t\int_A \begin{bmatrix} N_i & 0 \\ 0 & N_i \\ N_j & 0 \\ 0 & N_j \\ N_m & 0 \\ 0 & N_m \end{bmatrix} \begin{bmatrix} N_i & 0 & N_j & 0 & N_m & 0 \\ 0 & N_i & 0 & N_j & 0 & N_m \end{bmatrix} dA \tag{11.78}$$

After some algebraic manipulations, the above integration yields:

$$[m] = \frac{\rho hA}{12}\begin{bmatrix} 2 & 0 & 1 & 0 & 1 & 0 \\ & 2 & 0 & 1 & 0 & 1 \\ & & 2 & 0 & 1 & 0 \\ & & & 2 & 0 & 1 \\ & \text{symmetric} & & & 2 & 0 \\ & & & & & 2 \end{bmatrix} \tag{11.79}$$

In the above equation, h is the thickness of the triangular element and A is its area.

11.3 SOLUTION METHODOLOGY FOR THE DYNAMIC EQUATION

We recall that the vectors $\{d\},\{\ddot{d}\},\{F\}$ of Equation (11.5) are time dependent. Therefore, in addition to the discretization of the solid into finite elements, a discretization with respect to time is also needed. Even though several methodologies have been proposed for the numerical integration in time, the Central Difference Method (explicit method) and the Newmark-Beta method (implicit method) seem to be convenient for structural analysis under dynamic loads.

11.3.1 CENTRAL DIFFERENCE METHOD

To apply the central difference method the values of the vectors of displacements $\{d(0)\}$ and velocities $\{\dot{d}(0)\}$ at time $t=0$ (initials conditions) as well as a selection of a time increment Δt are initially needed. Suitable values for Δt are the values that are smaller than $2/\omega_{max}$ (ω_{max} is the maximum natural frequency). Particularly for bars, we can choose $\Delta t \leq L\sqrt{\rho/E}$.

The first step of the procedure is the evaluation of the initial acceleration from the dynamic equation (11.5):

$$\{\ddot{d}(0)\} = [m]^{-1}(\{F(0)\} - [k]\{d(0)\}) \tag{11.80}$$

Based on the classical finite difference method (see Numerical Analysis books), the subsequent steps for evaluation of the displacement and acceleration vectors $\{d(i)\},\{\ddot{d}(i)\}$ at any time step i are as follows:

Calculation of {d(−1)}

$$\{d(-1)\} = \{d(0)\} - \Delta t\{\dot{d}(0)\} + \frac{\Delta t^2}{2}\{\ddot{d}(0)\} \tag{11.81}$$

In the above equation, the vectors $\{d(0)\},\{\dot{d}(0)\}$ are given by the initial conditions and the vector $\{\ddot{d}(0)\}$ can be derived by Equation (11.80).

Calculation of {d(1)}

$$[m]\{d(1)\} = \Delta t^2\{F(0)\} + 2([m] - \Delta t^2[k])\{d(0)\} - [m]\{d(-1)\} \tag{11.82}$$

In the above equation, the force vector $\{F(0)\}$ is obtained by the given vector $\{F(t)\}$, the vector $\{d(0)\}$ is obtained by the initial conditions of the problem, and the vector $\{d(-1)\}$ is obtained by Equation (11.81).

Calculation of the Vectors {d(i)}, {$\ddot{d}$(i)} for i > 1

Using the initial values $\{\ddot{d}(0)\}, \{d(-1)\}, \{d(1)\}$ derived in the previous steps, the evaluation of the vectors $\{d(i)\}, \{\ddot{d}(i)\}$ for $i > 1$ can be performed by the following loop:

For $i=1$ *to* n (n: number of time steps)

$$\{d(i+1)\} = [m]^{-1}\left(\Delta t^2 \{F(i)\} + 2([m] - \Delta t^2 [k])\{d(i)\} - [m]\{d(i-1)\}\right) \tag{11.83}$$

$$\{\dot{d}(i)\} = \frac{\{d(i+1)\} - \{d(i-1)\}}{2\Delta t} \tag{11.84}$$

$$\{\ddot{d}(i)\} = [m]^{-1}(\{F(i)\} - [k]\{d(i)\}) \tag{11.85}$$

Next i

11.3.2 NEWMARK-BETA METHOD

Even though the Newmark-Beta method is very old, it has been adopted from the most contemporary commercial FEM programs because of its general versatility. The basic steps of this method are:

Calculation of {$\ddot{d}$(0)}

$$\{\ddot{d}(0)\} = [m]^{-1}(\{F(0)\} - [k]\{d(0)\}) \tag{11.86}$$

In the above equation, the vector $\{d(0)\}$ is given by the initial conditions and the vector $\{F(0)\}$ is obtained by the given vector of loading $\{F(t)\}$.

Calculation of the vectors {d(i)}, {$\ddot{d}$(i)} for i > 1

Knowing the initial values $\{d(0)\}, \{\dot{d}(0)\}$, the load vector $\{F(t)\}$ and the result $\{\ddot{d}(0)\}$ from Equation (11.80), the vectors $\{d(i)\}, \{\ddot{d}(i)\}$ for the subsequent load steps can be obtained by the following loop:

For $i=1$ *to* n (n: number of time steps)

$$[k']=[k]+\frac{1}{\beta\Delta t^2}[m] \tag{11.87}$$

$$\{F'(i+1)\}=\{F(i+1)\}+\frac{1}{\beta\Delta t^2}[m]\left(\{d(i)\}+\Delta t\{\dot{d}(i)\}+\left(\frac{1}{2}-\beta\right)\Delta t^2\{\ddot{d}(i)\}\right) \tag{11.88}$$

$$\{d(i+1)\}=[k']^{-1}\{F'(i+1)\} \tag{11.89}$$

$$\{\ddot{d}(i+1)\}=\frac{1}{\beta\Delta t^2}\left(\{d(i+1)\}-\{d(i)\}-\Delta t\{\dot{d}(i)\}-\Delta t^2\left(\frac{1}{2}-\beta\right)\{\ddot{d}(i)\}\right) \tag{11.90}$$

$$\{\dot{d}(i+1)\}=\{\dot{d}(i)\}+\Delta t\left((1-\gamma)\{\ddot{d}(i)\}+\gamma\{\ddot{d}(i+1)\}\right) \tag{11.91}$$

Next i

In the above equations, the optimum values for the parameters β and γ are $\beta=0.25$ and $\gamma=0.5$. Furthermore, the optimum value of the time step is $\Delta t=0.1\times$ (shortest natural frequency of the structure).

11.4 FREE VIBRATION—NATURAL FREQUENCIES

As it has already been mentioned in the previous subchapter, the optimum time step Δt depends on the shortest natural frequency of the structure. The main steps for the determination of the natural frequencies of structures will be described in the simple case of the bar element free axial vibration.

If we apply instantly an axial force F at the ends of the bar element, that is,

$$F(t)=\begin{cases} F & \text{for } t=0 \\ 0 & \text{for } t>0 \end{cases} \tag{11.92}$$

then, for $t>0$ the bar will vibrate even though the value of the force is $F=0$. Therefore, the equation describing this vibration is:

$$[m]\{\ddot{d}\}+[k]\{d\}=0 \tag{11.93}$$

It should be mentioned that the vectors $\{\ddot{d}\},\{d\}$ are functions of the variables x,t, that is, $\{\ddot{d}(x,t)\},\{d(x,t)\}$. The solution of Equation (11.93) can be given by the following harmonic function

$$\{d(x,t)\}=\{d^*(x)\}e^{i\omega t} \tag{11.94}$$

In the above equation, the parameter ω is called natural frequency and the members of the matrix $\{d^*(x)\}$ are called natural modes.

Substitution of Equation (11.94) into Equation (11.93) yields the following set of algebraic equations:

$$\left([k]-\omega^2[m]\right)\{d^*(x)\}=0 \tag{11.95}$$

It is known from mathematics that the condition for nontrivial solution of the above equation is:

$$\det\left([k]-\omega^2[m]\right)=0 \tag{11.96}$$

Equation (11.96) yields n solutions, that is, $\omega_1,\omega_2,\ldots,\omega_n$ where n is the number of degrees of freedom of the structure. Letting $d^*(L)=1$, the set of algebraic equations (Equation 11.95) yields the modal response of the bar for each natural frequency $\omega_1,\omega_2,\cdots,\omega_n$.

For better understanding of the essence of the modal response of a structure, the above procedures will be implemented in the bar shown in Figure 11.9.

The bar is discretized into two elements having length L and cross-section A. Following the procedure for derivation of the stiffness matrix and mass matrix of structures described in Chapter 4 and in this chapter, respectively, the following local matrices

$$[k_1]=\frac{AE}{L}\begin{bmatrix}1 & -1\\ -1 & 1\end{bmatrix} \tag{11.97}$$

$$[k_2]=\frac{AE}{L}\begin{bmatrix}1 & -1\\ -1 & 1\end{bmatrix} \tag{11.98}$$

$$[m_1]=\frac{\rho AL}{2}\begin{bmatrix}1 & 0\\ 0 & 1\end{bmatrix} \tag{11.99}$$

$$[m_2]=\frac{\rho AL}{2}\begin{bmatrix}1 & 0\\ 0 & 1\end{bmatrix} \tag{11.100}$$

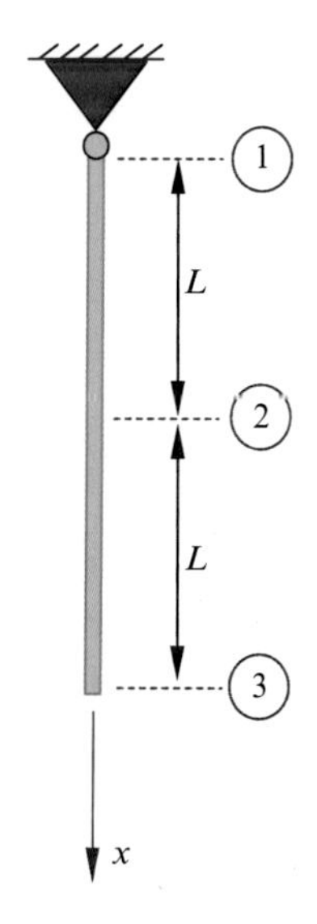

FIGURE 11.9

Finite element model of a bar under free axial vibration.

can be expanded in the global degrees of freedom, yielding:

$$[k_1] = \frac{AE}{L}\begin{bmatrix} 1 & -1 & 0 \\ -1 & 1 & 0 \\ 0 & 0 & 0 \end{bmatrix} \tag{11.101}$$

$$[k_2] = \frac{AE}{L}\begin{bmatrix} 0 & 0 & 0 \\ 0 & 1 & -1 \\ 0 & -1 & 1 \end{bmatrix} \tag{11.102}$$

$$[m_1] = \frac{\rho AL}{2}\begin{bmatrix} 1 & 0 & 0 \\ 0 & 1 & 0 \\ 0 & 0 & 0 \end{bmatrix} \tag{11.103}$$

$$[m_2] = \frac{\rho AL}{2}\begin{bmatrix} 0 & 0 & 0 \\ 0 & 1 & 0 \\ 0 & 0 & 1 \end{bmatrix} \tag{11.104}$$

Then, the stiffness and mass matrix of the structure can be derived by the assembly of the expanded stiffness and mass matrices:

$$[k] = [k_1] + [k_2] \tag{11.105}$$

$$[m] = [m_1] + [m_2] \tag{11.106}$$

Using the above equations, Equation (11.95) can be written:

$$\left(\frac{AE}{L}\begin{bmatrix} 1 & -1 & 0 \\ -1 & 2 & -1 \\ 0 & -1 & 1 \end{bmatrix} - \omega^2\frac{\rho AL}{2}\begin{bmatrix} 1 & 0 & 0 \\ 0 & 2 & 0 \\ 0 & 0 & 1 \end{bmatrix}\right)\begin{Bmatrix} d_{1x}^* \\ d_{2x}^* \\ d_{3x}^* \end{Bmatrix} = \begin{Bmatrix} 0 \\ 0 \\ 0 \end{Bmatrix} \tag{11.107}$$

Taking into account the boundary condition $d_{1x}^* = 0$, the above set of three equations can be reduced into the following set of two equations:

$$\left(\frac{AE}{L}\begin{bmatrix} 2 & -1 \\ -1 & 1 \end{bmatrix} - \omega^2\frac{\rho AL}{2}\begin{bmatrix} 2 & 0 \\ 0 & 1 \end{bmatrix}\right)\begin{Bmatrix} d_{2x}^* \\ d_{3x}^* \end{Bmatrix} = \begin{Bmatrix} 0 \\ 0 \end{Bmatrix} \tag{11.108}$$

The condition of the nontrivial solution of the above equation yields

$$\det\left(\frac{AE}{L}\begin{bmatrix} 2 & -1 \\ -1 & 1 \end{bmatrix} - \omega^2\frac{\rho AL}{2}\begin{bmatrix} 2 & 0 \\ 0 & 1 \end{bmatrix}\right) = 0 \tag{11.109}$$

Equation (11.109) has the following two solutions:

$$\omega_a = \frac{1.85}{L}\sqrt{E/\rho} \tag{11.110}$$

and

$$\omega_b = \frac{0.77}{L}\sqrt{E/\rho} \tag{11.111}$$

Using the above natural frequencies ω_a, ω_b, Equation (11.108) can be written:

(a) For $\omega = \omega_a$:

$$\left(\frac{AE}{L}\begin{bmatrix} 2 & -1 \\ -1 & 1 \end{bmatrix} - \frac{1.85^2}{L^2}\frac{E\rho AL}{\rho}\frac{}{2}\begin{bmatrix} 2 & 0 \\ 0 & 1 \end{bmatrix}\right)\begin{Bmatrix} d_{2x}^* \\ d_{3x}^* \end{Bmatrix} = \begin{Bmatrix} 0 \\ 0 \end{Bmatrix} \tag{11.112}$$

(b) For $\omega = \omega_b$:

$$\left(\frac{AE}{L}\begin{bmatrix} 2 & -1 \\ -1 & 1 \end{bmatrix} - \frac{0.77^2}{L^2}\frac{E\rho AL}{\rho}\frac{}{2}\begin{bmatrix} 2 & 0 \\ 0 & 1 \end{bmatrix}\right)\begin{Bmatrix} d_{2x}^* \\ d_{3x}^* \end{Bmatrix} = \begin{Bmatrix} 0 \\ 0 \end{Bmatrix} \tag{11.113}$$

Letting $d_{3x}^* = 1$, the above two equations yield the following results:

$$d_{2x}^* = \begin{cases} -0.7 \ \ for \ \omega = \omega_a \\ 0.7 \ \ for \ \omega = \omega_b \end{cases}$$

Since the value of the displacement d_{1x}^* is $d_{1x}^* = 0$ (due to the support on the node 1) and taking into account our choice $d_{3x}^* = 1$, the two sets of nodal displacements

$$\begin{Bmatrix} d_{1x}^* \\ d_{2x}^* \\ d_{3x}^* \end{Bmatrix} = \begin{Bmatrix} 0 \\ -0.7 \\ 1 \end{Bmatrix} \ for \ \omega = \omega_a \tag{11.114}$$

$$\begin{Bmatrix} d_{1x}^* \\ d_{2x}^* \\ d_{3x}^* \end{Bmatrix} = \begin{Bmatrix} 0 \\ 0.7 \\ 1 \end{Bmatrix} \ for \ \omega = \omega_b \tag{11.115}$$

can be demonstrated in Figure 11.10, representing the two vibration nodes.

Since the higher frequencies are damped rapidly due to the internal friction of the material (or due to the interaction of the structure with the environment), the lowest natural frequency (and the corresponding mode of vibration) is the most important. Therefore, from an engineering point of view, the first natural frequency:

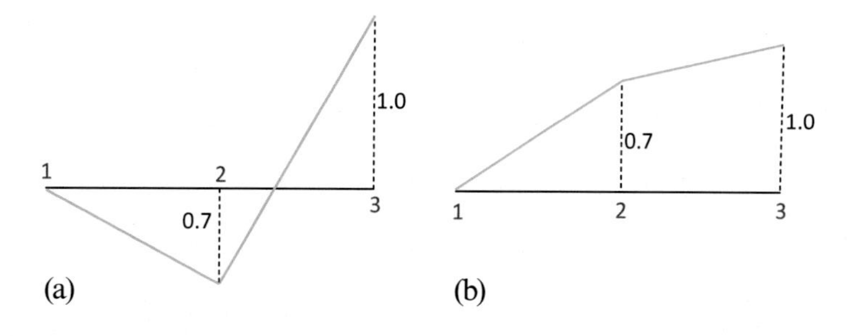

FIGURE 11.10

Modes of a bar under free longitudinal vibration: (a) for $\omega = \omega_a$ and (b) for $\omega = \omega_b$.

$$\omega_1 = \min\{\omega_a, \omega_b\} = \frac{0.77}{L}\sqrt{E/\rho} \tag{11.116}$$

and the corresponding mode of vibration shown in Figure 11.10b are the most important.

EXAMPLE 11.1: DYNAMIC RESPONSE OF A BAR UNDER DYNAMIC LOADING

A bar of length $2L = 6.0\text{m}$, cross-sectional area $A = 1.2\text{cm}^2$, and density $\rho = 8000\text{kg/m}^3$ is fixed supported on its both ends (Figure 11.11a). In the middle of the bar, a dynamic load $F(t)$ is acting (Figure 11.11c). Calculate the dynamic response of the bar.

Solution

The bar is discretized into two finite elements (Figure 11.11b). For the solution of the problem, the central difference time integration method is going to be adopted because it is easier for longhand calculations. Taking into account the procedure described in Section 11.3 the following procedure should be implemented:

Selection of the Time Step Increment

$$\Delta t \leq L\sqrt{\rho/E} \tag{e1}$$

or

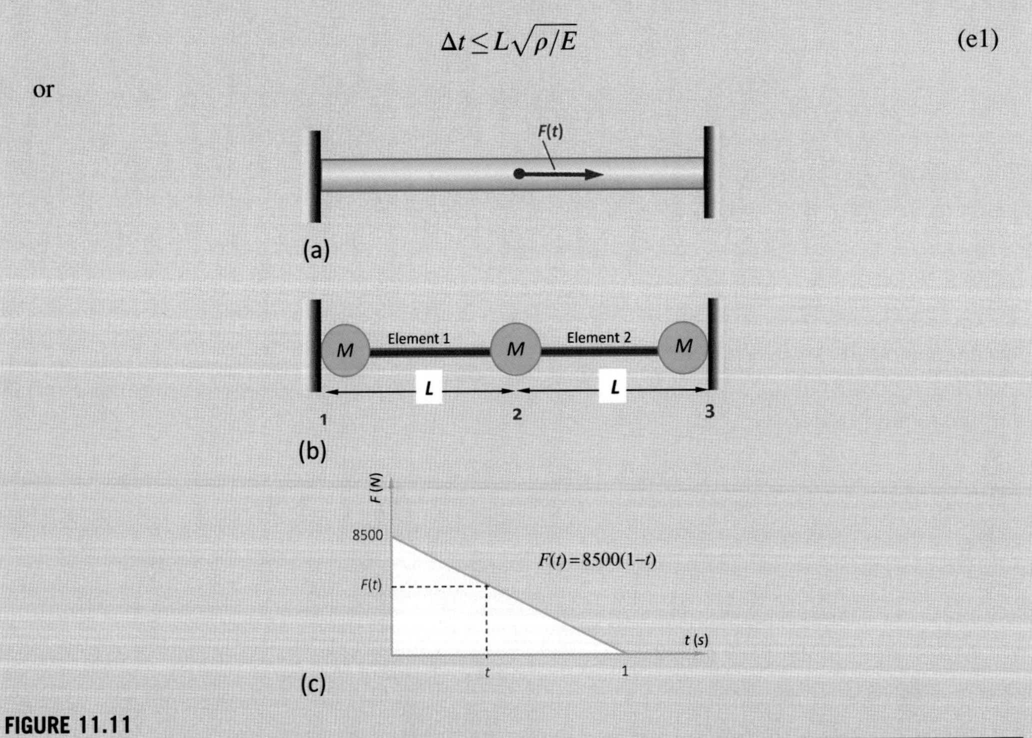

FIGURE 11.11

Fixed supported bar under dynamic loading: (a) the physical model, (b) FEM discretization with lumped-masses, and (c) force-time diagram for the load acting on the node 2.

$$\Delta t \le 3\sqrt{8000/210\times 10^9}$$

or

$$\Delta t \le 0.58\times 10^{-3}s$$

Taking into account the above result, the value

$$\Delta t = 0.5\times 10^{-3}s$$

is adopted.

Local Stiffness Matrices

The local stiffness matrices for the elements 1 and 2 can be derived from the following well-known equations:

$$[k_1] = \frac{AE}{L}\begin{bmatrix} 1 & -1 \\ -1 & 1 \end{bmatrix} \tag{e2}$$

$$[k_2] = \frac{AE}{L}\begin{bmatrix} 1 & -1 \\ -1 & 1 \end{bmatrix} \tag{e3}$$

Expansion of the Local Stiffness Matrices to the Degrees of Freedom of the Structure

$$[k_1] = \frac{AE}{L}\begin{bmatrix} 1 & -1 & 0 \\ -1 & 1 & 0 \\ 0 & 0 & 0 \end{bmatrix} \tag{e4}$$

$$[k_2] = \frac{AE}{L}\begin{bmatrix} 0 & 0 & 0 \\ 0 & 1 & -1 \\ 0 & -1 & 1 \end{bmatrix} \tag{e5}$$

Global Stiffness Matrix of the Structure

$$[k] = [k_1] + [k_2] = \frac{AE}{L}\begin{bmatrix} 1 & -1 & 0 \\ -1 & 2 & -1 \\ 0 & -1 & 1 \end{bmatrix} \tag{e6}$$

Local Mass Matrices

$$[m_1] = \frac{\rho AL}{2}\begin{bmatrix} 1 & 0 \\ 0 & 1 \end{bmatrix} \tag{e7}$$

$$[m_2] = \frac{\rho AL}{2}\begin{bmatrix} 1 & 0 \\ 0 & 1 \end{bmatrix} \tag{e8}$$

Continued

EXAMPLE 11.1: DYNAMIC RESPONSE OF A BAR UNDER DYNAMIC LOADING—CONT'D

Expansion of the Local Mass Matrices to the Degrees of Freedom of the Structure

$$[m_1]=\frac{\rho AL}{2}\begin{bmatrix}1&0&0\\0&1&0\\0&0&0\end{bmatrix} \tag{e9}$$

$$[m_2]=\frac{\rho AL}{2}\begin{bmatrix}0&0&0\\0&1&0\\0&0&1\end{bmatrix} \tag{e10}$$

Global Mass Matrix of the Structure

$$[m]=[m_1]+[m_2]=\frac{\rho AL}{2}\begin{bmatrix}1&0&0\\0&2&0\\0&0&1\end{bmatrix} \tag{e11}$$

Dynamic Equation of the Structure

Taking into account the global stiffness and mass matrices, Equation (11.5) can be written as:

$$\underbrace{\frac{AE}{L}\begin{bmatrix}1&-1&0\\-1&2&-1\\0&-1&1\end{bmatrix}}_{[k]}\begin{Bmatrix}d_1(t)\\d_2(t)\\d_3(t)\end{Bmatrix}+\underbrace{\frac{\rho AL}{2}\begin{bmatrix}1&0&0\\0&2&0\\0&0&1\end{bmatrix}}_{[m]}\begin{Bmatrix}\ddot{d}_1(t)\\\ddot{d}_2(t)\\\ddot{d}_3(t)\end{Bmatrix}=\begin{Bmatrix}F_1(t)\\F_2(t)\\F_3(t)\end{Bmatrix} \tag{e12}$$

The above equation can be expressed in the following abbreviated form:

$$\left[[k]\ \ [m]\ \ [-I_{3x3}]\right]\begin{Bmatrix}d_1(t)\\d_2(t)\\d_3(t)\\\ddot{d}_1(t)\\\ddot{d}_2(t)\\\ddot{d}_3(t)\\F_1(t)\\F_2(t)\\F_3(t)\end{Bmatrix}=\begin{Bmatrix}0\\0\\0\\0\\0\\0\\0\\0\\0\end{Bmatrix} \tag{e13}$$

Initial Conditions

For $t=0$ the initial conditions of the structure are:

Displacements {d(0)}.

$$d_1(0)=0$$
$$d_2(0)=0$$
$$d_3(0)=0$$

The above conditions can be written in the following format:

$$\underbrace{\begin{bmatrix} 1 & 0 & 0 \\ 0 & 1 & 0 \\ 0 & 0 & 1 \\ 0 & 0 & 0 \\ 0 & 0 & 0 \\ 0 & 0 & 0 \end{bmatrix}}_{[BCd(0)]} \underbrace{\begin{Bmatrix} d_1(0) \\ d_2(0) \\ d_3(0) \end{Bmatrix}}_{\{d(0)\}} = \underbrace{\begin{Bmatrix} 0 \\ 0 \\ 0 \\ 0 \\ 0 \\ 0 \end{Bmatrix}}_{\{DO(0)\}} \tag{e14}$$

Equation (e14) can be written in an abbreviated format

$$[BCd(0)]\{d(0)\} = \{DO(0)\}$$

or

$$\begin{bmatrix} [BCd(0)] & [O_{6\times3}] & [O_{6\times3}] \end{bmatrix} \begin{Bmatrix} \{d(0)\} \\ \{\ddot{d}(0)\} \\ \{F(0)\} \end{Bmatrix} = \{DO(0)\} \tag{e15}$$

where

$$\{d(0)\} = [d_1(0) \;\; d_2(0) \;\; d_3(0)]^T \tag{e16}$$

$$\{\ddot{d}(0)\} = [\ddot{d}_1(0) \;\; \ddot{d}_2(0) \;\; \ddot{d}_3(0)]^T \tag{e17}$$

$$\{F(0)\} = [F_1(0) \;\; F_2(0) \;\; F_3(0)]^T \tag{e18}$$

The matrix $[O_{6x3}]$ has six lines and three columns containing zeros.

Accelerations $\{\ddot{d}(0)\}$.

$\ddot{d}_1(0) = 0$ due to the support on the node 1
$\ddot{d}_3(0) = 0$ due to the support on the node 3

The above conditions can be written in the following format:

$$\underbrace{\begin{bmatrix} 0 & 0 & 0 \\ 0 & 0 & 0 \\ 0 & 0 & 0 \\ 1 & 0 & 0 \\ 0 & 0 & 1 \\ 0 & 0 & 0 \end{bmatrix}}_{[BCa(0)]} \underbrace{\begin{Bmatrix} \ddot{d}_1(0) \\ \ddot{d}_2(0) \\ \ddot{d}_3(0) \end{Bmatrix}}_{\{\ddot{d}(0)\}} = \underbrace{\begin{Bmatrix} 0 \\ 0 \\ 0 \\ 0 \\ 0 \\ 0 \end{Bmatrix}}_{\{AO(0)\}} \tag{e19}$$

Continued

EXAMPLE 11.1: DYNAMIC RESPONSE OF A BAR UNDER DYNAMIC LOADING—CONT'D

Equation (e19) can be written in an abbreviated format

$$[BCa(0)]\{\ddot{d}(0)\}=\{AO(0)\}$$

or

$$\begin{bmatrix}[O_{6\times3}] & [BCa(0)] & [O_{6\times3}]\end{bmatrix}\begin{Bmatrix}\{d(0)\}\\ \{\ddot{d}(0)\}\\ \{F(0)\}\end{Bmatrix}=\{AO(0)\} \tag{e20}$$

Forces {F(0)}. According to the force-time diagram shown in Figure 11.11b, the value of the external force acting to the node 2 at $t=0$ is:

$$F_2(0)=8500$$

This condition can be written in the following matrix format:

$$\underbrace{\begin{bmatrix}0&0&0\\0&0&0\\0&0&0\\0&0&0\\0&0&0\\0&1&0\end{bmatrix}}_{[BCR(0)]}\underbrace{\begin{Bmatrix}F_1(0)\\F_2(0)\\F_3(0)\end{Bmatrix}}_{\{F(0)\}}=\underbrace{\begin{Bmatrix}0\\0\\0\\0\\0\\8500\end{Bmatrix}}_{\{RO(0)\}} \tag{e21}$$

Equation (e21) can be written in an abbreviated format

$$[BCR(0)]\{F(0)\}=\{RO(0)\}$$

or

$$\begin{bmatrix}[O_{6\times3}] & [O_{6\times3}] & [BCR(0)]\end{bmatrix}\begin{Bmatrix}\{d(0)\}\\ \{\ddot{d}(0)\}\\ \{F(0)\}\end{Bmatrix}=\{RO(0)\} \tag{e22}$$

Solution for t = 0

Taking into account Equations (e13), (e15), (e20), and (e22) the following matrix equation can be derived:

$$\begin{bmatrix}[k] & [m] & [-I_{3\times3}]\\ [BCd(0)] & [BCa(0)] & [BCR(0)]\end{bmatrix}\begin{Bmatrix}\{d(0)\}\\ \{\ddot{d}(0)\}\\ \{F(0)\}\end{Bmatrix}=\begin{Bmatrix}\{O_{3\times1}\}\\ \{DO(0)\}+\{AO(0)\}+\{RO(0)\}\end{Bmatrix} \tag{e23}$$

The above matrix equation is a complete set of nine equations with nine unknowns, namely $\{d_1(0)\ d_2(0)\ d_3(0)\}$, $\{\ddot{d}_1(0)\ \ddot{d}_2(0)\ \ddot{d}_3(0)\}$, and $\{F_1(0)\ F_2(0)\ F_3(0)\}$. Solution of this system yields

$$\{d(0)\} = \begin{Bmatrix} d_1(0) \\ d_2(0) \\ d_3(0) \end{Bmatrix} = \begin{Bmatrix} 0 \\ 0 \\ 0 \end{Bmatrix} \quad \text{(e24)}$$

$$\{\ddot{d}(0)\} = \begin{Bmatrix} \ddot{d}_1(0) \\ \ddot{d}_2(0) \\ \ddot{d}_3(0) \end{Bmatrix} = \begin{Bmatrix} 0 \\ 2951 \\ 0 \end{Bmatrix} \quad \text{(e25)}$$

$$\{F(0)\} = \begin{Bmatrix} F_1(0) \\ F_2(0) \\ F_3(0) \end{Bmatrix} = \begin{Bmatrix} 0 \\ 8500 \\ 0 \end{Bmatrix} \quad \text{(e26)}$$

Solution for $t = t_1$ $(t_1 = \Delta t)$

Taking into account the initial conditions

$$\begin{Bmatrix} d_1(0) \\ d_2(0) \\ d_3(0) \end{Bmatrix} = \begin{Bmatrix} 0 \\ 0 \\ 0 \end{Bmatrix}$$

$$\begin{Bmatrix} \dot{d}_1(0) \\ \dot{d}_2(0) \\ \dot{d}_3(0) \end{Bmatrix} = \begin{Bmatrix} 0 \\ 0 \\ 0 \end{Bmatrix}$$

and the results given by Equation (e25), we can calculate the displacements $\{d(-1)\}$ by using Equation (11.88):

$$\begin{Bmatrix} d_1(-1) \\ d_2(-1) \\ d_3(-1) \end{Bmatrix} = \begin{Bmatrix} d_1(0) \\ d_2(0) \\ d_3(0) \end{Bmatrix} - \Delta t \begin{Bmatrix} \dot{d}_1(0) \\ \dot{d}_2(0) \\ \dot{d}_3(0) \end{Bmatrix} + \frac{\Delta t^2}{2} \begin{Bmatrix} \ddot{d}_1(0) \\ \ddot{d}_2(0) \\ \ddot{d}_3(0) \end{Bmatrix} \quad \text{(e27)}$$

or

$$\begin{Bmatrix} d_1(-1) \\ d_2(-1) \\ d_3(-1) \end{Bmatrix} = \begin{Bmatrix} 0 \\ 0.000368875 \\ 0 \end{Bmatrix} \quad \text{(e28)}$$

Then, using Equation (11.89) we can calculate the displacements $\{d(1)\}$ corresponding to the first time increment $t_1 = \Delta t$:

$$\begin{Bmatrix} d_1(1) \\ d_2(1) \\ d_3(1) \end{Bmatrix} = [m]^{-1}\Delta t^2 \begin{Bmatrix} F_1(0) \\ F_2(0) \\ F_3(0) \end{Bmatrix} + [m]^{-1}(2[m] - \Delta t^2[k]) \begin{Bmatrix} d_1(0) \\ d_2(0) \\ d_3(0) \end{Bmatrix} - [I] \begin{Bmatrix} d_1(-1) \\ d_2(-1) \\ d_3(-1) \end{Bmatrix} \quad \text{(e29)}$$

Continued

EXAMPLE 11.1: DYNAMIC RESPONSE OF A BAR UNDER DYNAMIC LOADING—CONT'D

or

$$\begin{Bmatrix} d_1(1) \\ d_2(1) \\ d_3(1) \end{Bmatrix} = \begin{Bmatrix} 0 \\ 0.000368972 \\ 0 \end{Bmatrix} \tag{e30}$$

Taking into account the above equation, as well as the boundary conditions (conditions specified by the supports of the bar), the matrices $[BCd(1)]$, $[BCa(1)]$, $[BCR(1)]$ and the vectors $\{DO(1)\}$, $\{AO(1)\}$, $\{RO(1)\}$ can be derived as:

Matrix [BCd(1)] and vector {DO(1)}. Using the results given by Equation (30), the following matrix equation can be obtained:

$$\underbrace{\begin{bmatrix} 1 & 0 & 0 \\ 0 & 1 & 0 \\ 0 & 0 & 1 \\ 0 & 0 & 0 \\ 0 & 0 & 0 \\ 0 & 0 & 0 \end{bmatrix}}_{[BCd(1)]} \underbrace{\begin{Bmatrix} d_1(1) \\ d_2(1) \\ d_3(1) \end{Bmatrix}}_{\{d(1)\}} = \underbrace{\begin{Bmatrix} 0 \\ 0 \\ 0.000368972 \\ 0 \\ 0 \\ 0 \end{Bmatrix}}_{\{DO(1)\}} \tag{e31}$$

The above equation can be written in an abbreviated notation:

$$[BCd(1)]\{d(1)\} = \{DO(1)\} \tag{e32}$$

or

$$[[BCd(1)] \ [O_{6\times3}] \ [O_{6\times3}]] \begin{Bmatrix} \{d(1)\} \\ \{\ddot{d}(1)\} \\ \{F(1)\} \end{Bmatrix} = \{DO(1)\} \tag{e33}$$

Matrix [BCa(1)] and vector {AO(1)}. Since the nodes 1 and 3 are fixed supported, the following boundary conditions are valid:

$$\ddot{d}_1(1) = 0 \text{ due to the support on the node 1}$$
$$\ddot{d}_3(1) = 0 \text{ due to the support on the node 3}$$

The above conditions can be written as follows:

$$\underbrace{\begin{bmatrix} 0 & 0 & 0 \\ 0 & 0 & 0 \\ 0 & 0 & 0 \\ 1 & 0 & 0 \\ 0 & 0 & 1 \\ 0 & 0 & 0 \end{bmatrix}}_{[BCa(1)]} \underbrace{\begin{Bmatrix} \ddot{d}_1(1) \\ \ddot{d}_2(1) \\ \ddot{d}_3(1) \end{Bmatrix}}_{\{\ddot{d}(1)\}} = \underbrace{\begin{Bmatrix} 0 \\ 0 \\ 0 \\ 0 \\ 0 \\ 0 \end{Bmatrix}}_{\{AO(1)\}} \quad \text{(e34)}$$

Equation (e34) can be written in an abbreviated format

$$[BCa(1)]\{\ddot{d}(1)\} = \{AO(1)\} \quad \text{(e35)}$$

or

$$\left[[O_{6\times 3}] \;\; [BCa(1)] \;\; [O_{6\times 3}] \right] \begin{Bmatrix} \{d(1)\} \\ \{\ddot{d}(1)\} \\ \{F(1)\} \end{Bmatrix} = \{AO(1)\} \quad \text{(e36)}$$

Matrix [BCR(1)] and vector {RO(1)}. According to the force-time diagram shown in Figure 11.11b, the value of the external force acting to the node 2 at $t_1 = \Delta t = 0.5 \times 10^{-3} s$ is

$$F(t_1) = F_2(1) = 8500(1 - \Delta t) = 8495.75\text{N}$$

This condition can be written in the following matrix format:

$$\underbrace{\begin{bmatrix} 0 & 0 & 0 \\ 0 & 0 & 0 \\ 0 & 0 & 0 \\ 0 & 0 & 0 \\ 0 & 0 & 0 \\ 0 & 1 & 0 \end{bmatrix}}_{[BCR(1)]} \underbrace{\begin{Bmatrix} F_1(1) \\ F_2(1) \\ F_3(1) \end{Bmatrix}}_{\{F(1)\}} = \underbrace{\begin{Bmatrix} 0 \\ 0 \\ 0 \\ 0 \\ 0 \\ 8495.75 \end{Bmatrix}}_{\{RO(1)\}} \quad \text{(e37)}$$

Continued

EXAMPLE 11.1: DYNAMIC RESPONSE OF A BAR UNDER DYNAMIC LOADING—CONT'D

Equation (e21) can be written in an abbreviated format

$$[BCR(1)]\{F(1)\}=\{RO(1)\} \tag{e38}$$

or

$$\begin{bmatrix}[O_{6x3}] & [O_{6x3}] & [BCR(1)]\end{bmatrix}\begin{Bmatrix}\{d(1)\}\\ \{\ddot{d}(1)\}\\ \{F(1)\}\end{Bmatrix}=\{RO(1)\} \tag{e39}$$

Dynamic Equation for t=t₁

For the first time step $t_1=\Delta t$, Equations (e13), (e33), (e36), and (e39) can compose the following matrix equation:

$$\begin{bmatrix}[k] & [m] & [-I_{3\times3}]\\ [BCd(1)] & [BCa(1)] & [BCR(1)]\end{bmatrix}\begin{Bmatrix}\{d(1)\}\\ \{\ddot{d}(1)\}\\ \{F(1)\}\end{Bmatrix}=\begin{Bmatrix}\{O_{3\times1}\}\\ \{DO(1)\}+\{AO(1)\}+\{RO(1)\}\end{Bmatrix} \tag{e40}$$

Solution of this system yields

$$\{d(1)\}=\begin{Bmatrix}d_1(1)\\ d_2(1)\\ d_3(1)\end{Bmatrix}=\begin{Bmatrix}0\\ 0.000368972\\ 0\end{Bmatrix} \tag{e41}$$

$$\{\ddot{d}(1)\}=\begin{Bmatrix}\ddot{d}_1(1)\\ \ddot{d}_2(1)\\ \ddot{d}_3(1)\end{Bmatrix}=\begin{Bmatrix}0\\ 797.58\\ 0\end{Bmatrix} \tag{e42}$$

$$\{F(1)\}=\begin{Bmatrix}F_1(1)\\ F_2(1)\\ F_3(1)\end{Bmatrix}=\begin{Bmatrix}-3099.36\\ 8495.75\\ -3099.36\end{Bmatrix} \tag{e43}$$

Solution for $t=t_2$ $(t_2=2\Delta t)$

Matrix [BCd(2)] and vector {DO(2)}. Using Equations (e43) and (e41), as well as the initial conditions given by Equation (e14), Equation (11.90) yields the following displacements:

$$\begin{Bmatrix}d_1(2)\\ d_2(2)\\ d_3(2)\end{Bmatrix}=[m]^{-1}\Delta t^2\begin{Bmatrix}F_1(1)\\ F_2(1)\\ F_3(1)\end{Bmatrix}+[m]^{-1}(2[m]-\Delta t^2[k])\begin{Bmatrix}d_1(1)\\ d_2(1)\\ d_3(1)\end{Bmatrix}-[I]\begin{Bmatrix}d_1(0)\\ d_2(0)\\ d_3(0)\end{Bmatrix} \tag{e44}$$

or

$$\begin{Bmatrix}d_1(2)\\ d_2(2)\\ d_3(2)\end{Bmatrix}=\begin{Bmatrix}0\\ 0.000937338\\ 0\end{Bmatrix} \tag{e45}$$

The above equation can now be written as

$$\left[[BCd(2)]\ [O_{6\times3}]\ [O_{6\times3}]\right]\begin{bmatrix}\{d(2)\}\\\{\ddot{d}(2)\}\\\{F(2)\}\end{bmatrix}=\{DO(2)\} \tag{e46}$$

where

$$[BCd(2)]=\begin{bmatrix}1&0&0\\0&1&0\\0&0&1\\0&0&0\\0&0&0\\0&0&0\end{bmatrix} \tag{e47}$$

$$\{DO(2)\}=\begin{Bmatrix}0\\0.000937338\\0\\0\\0\\0\end{Bmatrix} \tag{e48}$$

Matrix [BCα(2)] and vector {AO(2)}. Since the matrix $[BCa(2)]$ and the vector $\{AO(2)\}$ are related to the types of support, there is not a change. Therefore:

$$[BCa(2)]=[BCa(1)] \tag{e49}$$

and

$$\{AO(2)\}=\{AO(1)\} \tag{e50}$$

Then, the following equation can be formulated:

$$\left[[O_{6\times3}]\ [BCa(2)]\ [O_{6\times3}]\right]\begin{bmatrix}\{d(2)\}\\\{\ddot{d}(2)\}\\\{F(2)\}\end{bmatrix}=\{AO(2)\} \tag{e51}$$

Matrix [BCR(2)] and vector {RO(2)}. According to the force-time diagram shown in Figure 11.11b, the value of the external force acting to the node 2 at $t_2=2\Delta t=10^{-3}$ s is:

$$F(t_2)=F_2(2)=8500(1-2\Delta t)=8491.5\text{N}$$

Continued

EXAMPLE 11.1: DYNAMIC RESPONSE OF A BAR UNDER DYNAMIC LOADING—CONT'D

This condition can be written in the following matrix format:

$$\begin{bmatrix}[O_{6\times3}] & [O_{6\times3}] & [BCR(2)]\end{bmatrix}\begin{bmatrix}\{d(2)\}\\ \{\ddot{d}(2)\}\\ \{F(2)\}\end{bmatrix}=\{RO(2)\} \tag{e52}$$

Since the location of the force is unchanged (node 2), then

$$[BCR(2)]=[BCR(1)] \tag{e53}$$

The vector $\{RO(2)\}$ contains the value of the force $F(t_2)=8491.5\text{N}$:

$$\{RO(2)\}=\begin{Bmatrix}0\\0\\0\\0\\0\\8491.5\end{Bmatrix} \tag{e54}$$

Dynamic Equation for $t=t_2$

For the second time step $t_2=2\Delta t$, Equations (e13), (e46), (e51), and (e52) can compose the following dynamic equation:

$$\begin{bmatrix}[k] & [m] & [-I_{3\times3}]\\ [BCd(2)] & [BCa(2)] & [BCR(2)]\end{bmatrix}\begin{Bmatrix}\{d(2)\}\\ \{\ddot{d}(2)\}\\ \{F(2)\}\end{Bmatrix}=\begin{Bmatrix}\{O_{3\times1}\}\\ \{DO(2)\}+\{AO(2)\}+\{RO(2)\}\end{Bmatrix} \tag{e55}$$

Solution of this system yields

$$\{d(2)\}=\begin{Bmatrix}d_1(2)\\d_2(2)\\d_3(2)\end{Bmatrix}=\begin{Bmatrix}0\\0.000937338\\0\end{Bmatrix} \tag{e56}$$

$$\{\ddot{d}(2)\}=\begin{Bmatrix}\ddot{d}_1(2)\\\ddot{d}_2(2)\\\ddot{d}_3(2)\end{Bmatrix}=\begin{Bmatrix}0\\-2519.37\\0\end{Bmatrix} \tag{e57}$$

$$\{F(2)\}=\begin{Bmatrix}F_1(2)\\F_2(2)\\F_3(2)\end{Bmatrix}=\begin{Bmatrix}-7873.64\\8491.5\\-7873.64\end{Bmatrix} \tag{e58}$$

Solution for $t=t_3$ $(t_3=3\Delta t)$

Following the same procedure the matrices $[BCd(3)]$, $\{DO(3)\}$, $[BCa(3)]$, $\{AO(3)\}$, $[BCR(3)]$, $\{RO(3)\}$ composing the dynamic equation can be derived as:

Matrix [BCd(3)] and vector {DO(3)}. For $t=t_3$, Equation (11.83) can be written as:

$$\begin{Bmatrix} d_1(3) \\ d_2(3) \\ d_3(3) \end{Bmatrix} = [m]^{-1}\Delta t^2 \begin{Bmatrix} F_1(2) \\ F_2(2) \\ F_3(2) \end{Bmatrix} + [m]^{-1}\left(2[m]-\Delta t^2[k]\right)\begin{Bmatrix} d_1(2) \\ d_2(2) \\ d_3(2) \end{Bmatrix} - [I]\begin{Bmatrix} d_1(1) \\ d_2(1) \\ d_3(1) \end{Bmatrix} \quad \text{(e59)}$$

Taking into account Equations (e58), (e56), and (e41), the above equation yields:

$$\begin{Bmatrix} d_1(3) \\ d_2(3) \\ d_3(3) \end{Bmatrix} = \begin{Bmatrix} 0 \\ 0.000875862 \\ 0 \end{Bmatrix} \quad \text{(e60)}$$

Then the matrices $[BCd(3)]$ and $\{DO(3)\}$ are

$$[BCd(3)] = [BCd(2)] \quad \text{(e61)}$$

$$\{DO(3)\} = \begin{Bmatrix} 0 \\ 0.000875862 \\ 0 \\ 0 \\ 0 \\ 0 \end{Bmatrix} \quad \text{(e62)}$$

Matrix [BCa(3)] and vector {AO(3)}.

$$[BCa(3)] = [BCa(2)] \quad \text{(e63)}$$

and

$$\{AO(3)\} = \{AO(2)\} \quad \text{(e64)}$$

Matrix [BCR(3)] and vector {RO(3)}. According to the force-time diagram shown in Figure 11.11b, the value of the external force acting to the node 2 at $t_3 = 3\Delta t$ is

$$F(t_3) = F_2(3) = 8500(1-3\Delta t) = 8487.25N$$

Therefore,

$$[BCR(3)] = [BCR(2)] \quad \text{(e65)}$$

and

$$\{RO(3)\} = \begin{Bmatrix} 0 \\ 0 \\ 0 \\ 0 \\ 0 \\ 0 \\ 8487.25 \end{Bmatrix} \quad \text{(e66)}$$

Dynamic Equation for $t=t_3$

$$\begin{bmatrix} [k] & [m] & [-I_{3\times 3}] \\ [BCd(3)] & [BCa(3)] & [BCR(3)] \end{bmatrix} \begin{Bmatrix} \{d(3)\} \\ \{\ddot{d}(3)\} \\ \{F(3)\} \end{Bmatrix} = \begin{Bmatrix} \{O_{3\times 1}\} \\ \{DO(3)\}+\{AO(3)\}+\{RO(3)\} \end{Bmatrix} \quad \text{(e67)}$$

Continued

EXAMPLE 11.1: DYNAMIC RESPONSE OF A BAR UNDER DYNAMIC LOADING—CONT'D

The solution of this system yields

$$\{d(3)\} = \begin{Bmatrix} d_1(3) \\ d_2(3) \\ d_3(3) \end{Bmatrix} = \begin{Bmatrix} 0 \\ 0.000875862 \\ 0 \end{Bmatrix} \tag{e68}$$

$$\{\ddot{d}(3)\} = \begin{Bmatrix} \ddot{d}_1(3) \\ \ddot{d}_2(3) \\ \ddot{d}_3(3) \end{Bmatrix} = \begin{Bmatrix} 0 \\ -2162.23 \\ 0 \end{Bmatrix} \tag{e69}$$

$$\{F(3)\} = \begin{Bmatrix} F_1(3) \\ F_2(3) \\ F_3(3) \end{Bmatrix} = \begin{Bmatrix} -7357.24 \\ 8487.25 \\ -7357.24 \end{Bmatrix} \tag{e70}$$

For the subsequent time steps $t = t_4, t_5, t_6, \ldots$ the same procedure can be repeated.

Sometimes instead of the force-time diagram, the displacement-time diagram on the nodes is known. Furthermore, during structural health monitoring, accelerometers located on the nodes of the structure can provide acceleration-time diagrams. In both cases, the above described dynamic FEM procedure can be followed to derive the dynamic response of the structure.

EXAMPLE 11.2: DYNAMIC RESPONSE OF A FRAME WITH MATLAB/CALFEM

The frame shown in the following figure is subjected to a dynamic loading acting on a point located at the middle of the horizontal beam.

1. Develop a computer code in MATLAB/CALFEM to derive the dynamic response in the horizontal direction of a point located at the middle of the vertical beams an in the vertical direction of the point subjected to dynamic load.
2. Derive the deformed shape of the frame at 10 time segments from $t = 0.1$ s to $t = 1$ s every 0.1 s (snapshots $t = 0.1, 0.2, 0.3, \ldots, 1.0$).

Data

Data	Vertical beams	Horizontal beams
Modulus of elasticity (in Pa)	$E = 3 \times 10^{10}$	$E = 3 \times 10^{10}$
Cross-section area (in m^2)	$Av = 0.103 \times 10^{-2}$	$Ah = 0.0764 \times 10^{-2}$
Moment if inertia (in m^4)	$Iv = 0.171 \times 10^{-5}$	$Ih = 0.0801 \times 10^{-5}$
Material density (in kg/m^3)	rho = 2500	rho = 2500

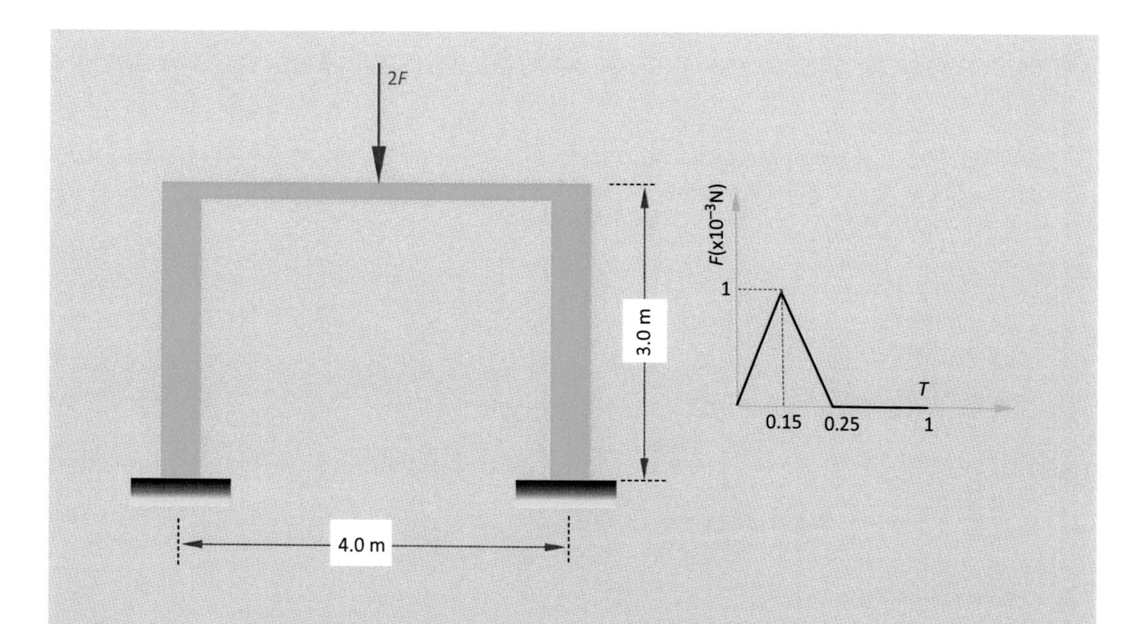

Solution

Since the frame under consideration has a symmetric geometry and since it is subjected to symmetric loading, the following loading model of the frame can be used:

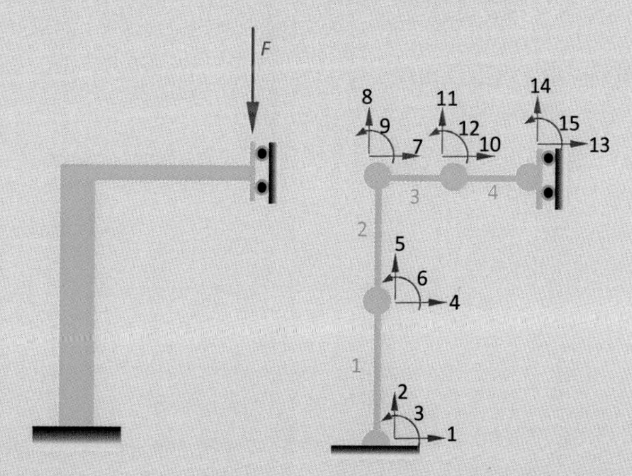

For the above mechanical model, the following MATLAB/CALFEM commands can be used to derive the dynamic response of the structure:

Continued

EXAMPLE 11.2: DYNAMIC RESPONSE OF A FRAME WITH MATLAB/CALFEM—CONT'D

Material Parameters

```
E=3e10; rho=2500;
    Av=0.1030e-2; Iv=0.0171e-4;
    Ah=0.0764e-2; Ih=0.00801e-4;
    epv=[E Av Iv rho*Av]; eph=[E Ah Ih rho*Ah];
```

Topology

```
Edof=[1 1 2 3 4 5 6
    2 4 5 6 7 8 9
    3 7 8 9 10 11 12
    4 10 11 12 13 14 15];
```

Coordinates

```
Coord=[0 0;0 1.5;0 3;1 3;2 3];
```

Degrees of Freedom

```
    Dof=[1 2 3; 4 5 6; 7 8 9; 10 11 12; 13 14 15];
```

Assembly of stiffness and mass matrices

```
K=zeros(15); M=zeros(15);
    [Ex,Ey]=coordxtr(Edof,Coord,Dof,2);
    for i=1:2
    [k,m,c]=beam2d(Ex(i,:),Ey(i,:),epv);
    K=assem(Edof(i,:),K,k); M=assem(Edof(i,:),M,m);
    end
    for i=3:4
    [k,m,c]=beam2d(Ex(i,:),Ey(i,:),eph);
    K=assem(Edof(i,:),K,k); M=assem(Edof(i,:),M,m);
    End
```

Definition of the time step of integration

```
dt=0.005;
```

Definition of the time end of integration

```
T=1;
```

Specification of the dynamic load

% ...Comment: The dynamic load should be modeled in a matrix format. The following matrix G contains the numerical values of the dynamic loading diagram....

```
    G=[0 0; 0.15 1; 0.25 0; T 0];
```

Interpolation function g=g(t) for the dynamic load

% ...Comment: Taking into account the matrix G, the dynamic load can be expressed in an analytic format by using the following interpolation function g=g(t) ....

```
    [t,g]=gfunc(G,dt);
```

Specification of the load vector f

*% ...Comment: The command "f(14,:)=1000*g" means that on the degree of freedom 4 there is action of the time history 1000*g (in Newtons) ....*

```
f=zeros(15, length(g)); f(14,:)=1000*g;
```

Specification of boundary conditions and initial conditions

% ...Comment: The boundary conditions are incorporated in the matrix "bc". The vectors d0, v0 contain the initial values on the 15 degrees of freedom of the displacements and velocities respectively ....

```
bc=[1 0; 2 0; 3 0; 13 0; 15 0];
d0=zeros(15,1); v0=zeros(15,1);
```

Specification of program's output parameters

% ...Comment: The command "ntimes=[0.1:0.1:1]" means that the program is going to derive the snapshots from t=0.1 till t=1 every 0.1 sec. The command "nhist=[4 14]" means that the program is going to derive the response of the structure in the degrees of freedom 4 and 14 ....

```
ntimes=[0.1:0.1:1]; nhist=[4 14];
```

Specification of time integration parameters

% ...Comment: The command "ip=[dt T 0.25 0.5 10 2 ntimes nhist]" means that the time integration procedure is governed by the parameters given in the vector "ip" defined as

ip=[dt T a δ nsnap ndof ntimes nhist]

where a=0.25, δ=0.5 are time integration constants in the Newmark method, nsnap=10 is the number of snapshots will be generated, ndof=2 means the reponse in 2 degrees of freedom ....

```
ip=[dt T 0.25 0.5 10 2 ntimes nhist];
```

Time integration

*% ...Comment: The command "sparse" in order to consume small number of memory spaces. The command "**[Dsnap,D,V,A]=step2(k,[],m,d0,v0,ip,f,bc)**" is used to compute the dynamic solution (displacements D, velocities V, accelerations A) to the set of the second order dynamic differential equation and the associated boundary conditions.*

```
k=sparse(K); m=sparse(M);
[Dsnap,D,V,A]=step2(k,[],m,d0,v0,ip,f,bc);
```

Demonstration of the results

% ...Comment: The command "Edb=extract(Edof,Dsnap(:,i))" extracts elements from the matrix Dsnap(:,i) according to Edof...

```
figure(1), plot(t,D(1,:),'-',t,D(2,:),'–')
figure(2),clf,axis('equal'), hold on, axis off
sfac=25;
title('Snapshots (sec), magnification = 25');
for i=1:5;
Ext=Ex+(i-1)*3; eldraw2(Ext,Ey,[2 3 0]);
Edb=extract(Edof,Dsnap(:,i));
eldisp2(Ext,Ey,Edb,[1 2 2],sfac);
Time=num2str(ntimes(i)); text(3*(i-1)+.5,1.5,Time);
end;
Eyt=Ey-4;
for i=6:10;
Ext=Ex+(i-6)*3; eldraw2(Ext,Eyt,[2 3 0]);
```

Continued

EXAMPLE 11.2: DYNAMIC RESPONSE OF A FRAME WITH MATLAB/CALFEM—CONT'D

```
Edb=extract(Edof,Dsnap(:,i));
eldisp2(Ext,Eyt,Edb,[1 2 2],sfac);
Time=num2str(ntimes(i)); text(3*(i-6)+.5,-2.5,Time);
end;
```

Taking into account the above commands, the MatLab/CalFem code and the corresponding results are the following:

MatLab/CalFem code

```
>> E=3e10; rho=2500;
Av=0.1030e-2; Iv=0.0171e-4;
Ah=0.0764e-2; Ih=0.00801e-4;
epv=[E Av Iv rho*Av]; eph=[E Ah Ih rho*Ah];
Edof=[1 1 2 3 4 5 6
2 4 5 6 7 8 9
3 7 8 9 10 11 12
4 10 11 12 13 14 15];
Coord=[0 0;0 1.5;0 3;1 3;2 3];
Dof=[1 2 3; 4 5 6; 7 8 9; 10 11 12; 13 14 15];
K=zeros(15); M=zeros(15);
[Ex,Ey]=coordxtr(Edof,Coord,Dof,2);
for i=1:2
[k,m,c]=beam2d(Ex(i,:),Ey(i,:),epv);
K=assem(Edof(i,:),K,k); M=assem(Edof(i,:),M,m);
end
for i=3:4
[k,m,c]=beam2d(Ex(i,:),Ey(i,:),eph);
K=assem(Edof(i,:),K,k); M=assem(Edof(i,:),M,m);
end
dt=0.005; T=1;
% ———————— specification of the load ————————
G=[0 0; 0.15 1; 0.25 0; T 0]; [t,g]=gfunc(G,dt);
f=zeros(15, length(g)); f(14,:)=1000*g;
% —— boundary condition, initial condition ————
bc=[1 0; 2 0; 3 0; 13 0; 15 0];
d0=zeros(15,1); v0=zeros(15,1);
% —— output parameters ——————————
ntimes=[0.1:0.1:1]; nhist=[4 14];
% —— time integration parameters ————
ip=[dt T 0.25 0.5 10 2 ntimes nhist];
% —— time integration ————
k=sparse(K); m=sparse(M);
[Dsnap,D,V,A]=step2(k,[],m,d0,v0,ip,f,bc);
```

```
figure(1), plot(t,D(1,:),'-',t,D(2,:),'–')
figure(2),clf,axis('equal'), hold on, axis off
sfac=25;
title('Snapshots (sec), magnification = 25');
for i=1:5;
Ext=Ex+(i-1)*3; eldraw2(Ext,Ey,[2 3 0]);
Edb=extract(Edof,Dsnap(:,i));
eldisp2(Ext,Ey,Edb,[1 2 2],sfac);
Time=num2str(ntimes(i)); text(3*(i-1)+.5,1.5,Time);
end;
Eyt=Ey-4;
for i=6:10;
Ext=Ex+(i-6)*3; eldraw2(Ext,Eyt,[2 3 0]);
Edb=extract(Edof,Dsnap(:,i));
eldisp2(Ext,Eyt,Edb,[1 2 2],sfac);
Time=num2str(ntimes(i)); text(3*(i-6)+.5,-2.5,Time);
end;
>>
```

Results

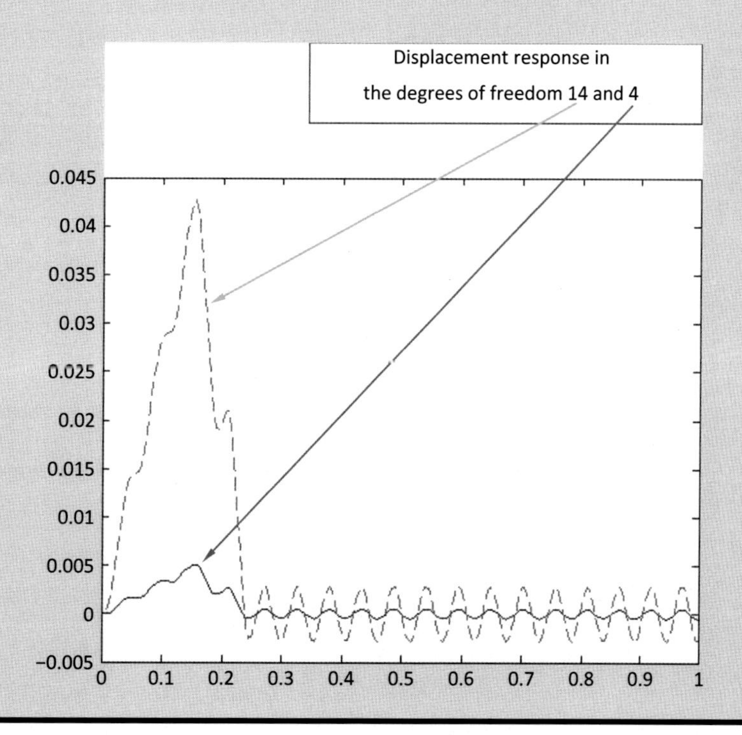

Continued

EXAMPLE 11.2: DYNAMIC RESPONSE OF A FRAME WITH MATLAB/CALFEM—CONT'D

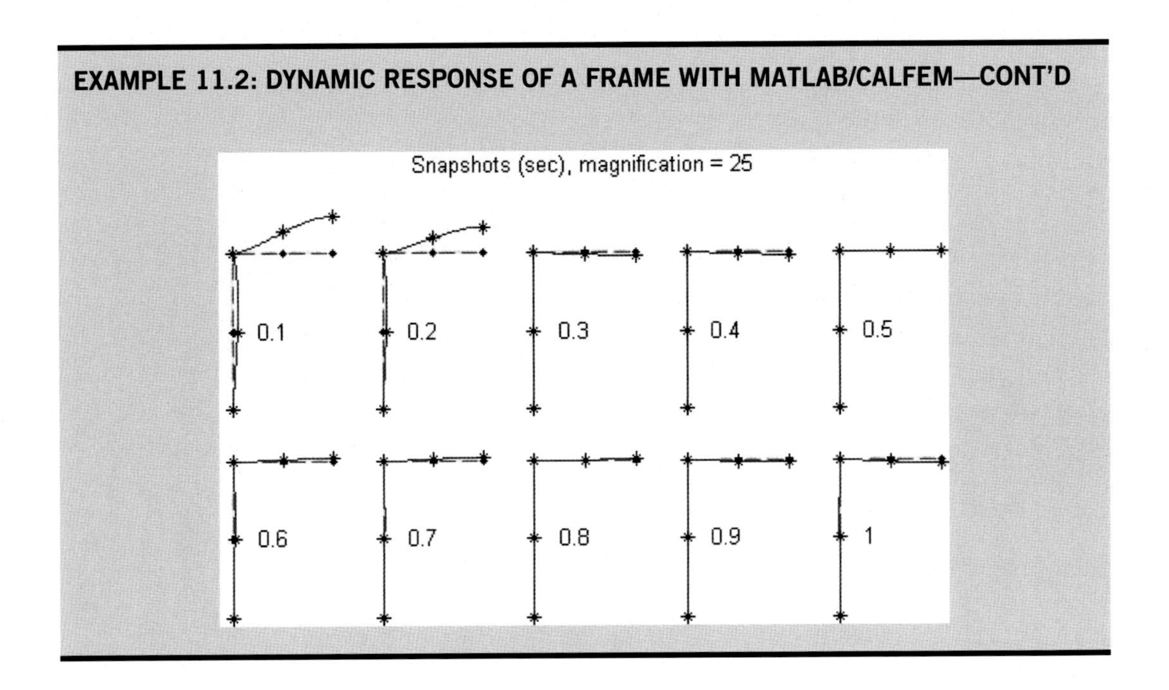

REFERENCES

[1] Wu SR, Gu L. Introduction to the explicit finite element method for nonlinear transient dynamics. Hoboken: Wiley; 2012.

[2] Newmark NM. A method of computation for structural dynamics. J Eng Mech Div ASCE 1959;85 (EM3):67–94.

[3] Logan DL. A first course in the finite element method. Boston, MA: Cengage Learning; 2012.

[4] Austrell P-E, Dahlblom O, Lindemann J, Olsson A, Olsson K-G, Persson K, et al. CALFEM—a finite element toolbox, Version 3.4., Division of Structural Mechanics, Lund University; 2004.

[5] Cook RD, Malkus DS, Plesha ME, Witt RJ. Concepts and applications of finite element analysis. Hoboken: John Wiley & Sons; 2002.

[6] Moaveni S. Finite element analysis. Upper Saddle River, NJ: Pearson Education; 2008.

[7] Wunderlich W, Pilkey W. Mechanics of structures—variational and computational methods. Boca Raton: CRC Press; 2003.

[8] Melosh RJ. Structural engineering analysis by finite elements. Upper Saddle River, NJ: Prentice-Hall; 1990.

CHAPTER

HEAT TRANSFER

12

This chapter seeks to predict the steady-state temperature distribution and heat flow that take place in a material body due to the temperature difference with its environment and/or due to internal energy generation. The three modes of heat transfer are conduction, convection, and radiation (Figure 12.1a).

The heat exchange between the body and its environment takes place through its bounding surface. However, some surface areas can be isolated (Figure 12.1b). In this chapter, we are going to analyze heat transfer problems due to conduction and convection. Radiation heat transfer is a special topic that is beyond the targets of this chapter.

12.1 CONDUCTION HEAT TRANSFER

When a temperature gradient $\partial T/\partial x$ exists in a body, the heat transfer rate per unit area f_x (heat flux in Watt per meter square) is proportional to the temperature gradient in the direction of the heat flow according to the Fourier's law for heat conduction:

$$f_x = -k\frac{\partial T}{\partial x} \tag{12.1}$$

In Equation (12.1), k is a material constant called thermal conductivity. The minus sign in the equation means that heat flows in opposite directions to temperature gradient.

Let us now consider the one-dimensional (1D) thermal equilibrium of a material element of thickness dx and cross-sectional area A (Figure 12.2).

Then the following energy balance can be written as:

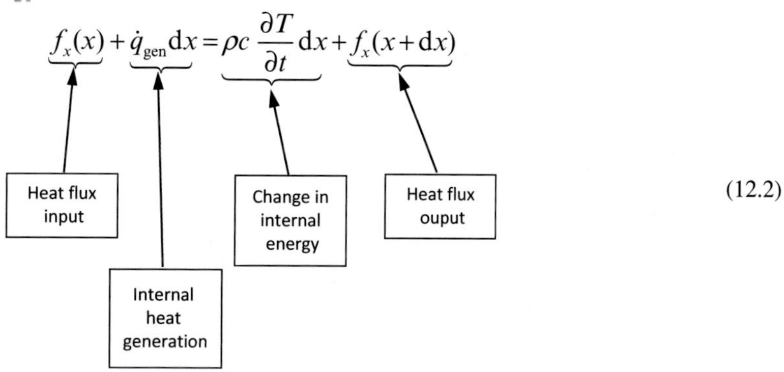

(12.2)

Essentials of the Finite Element Method. http://dx.doi.org/10.1016/B978-0-12-802386-0.00012-8

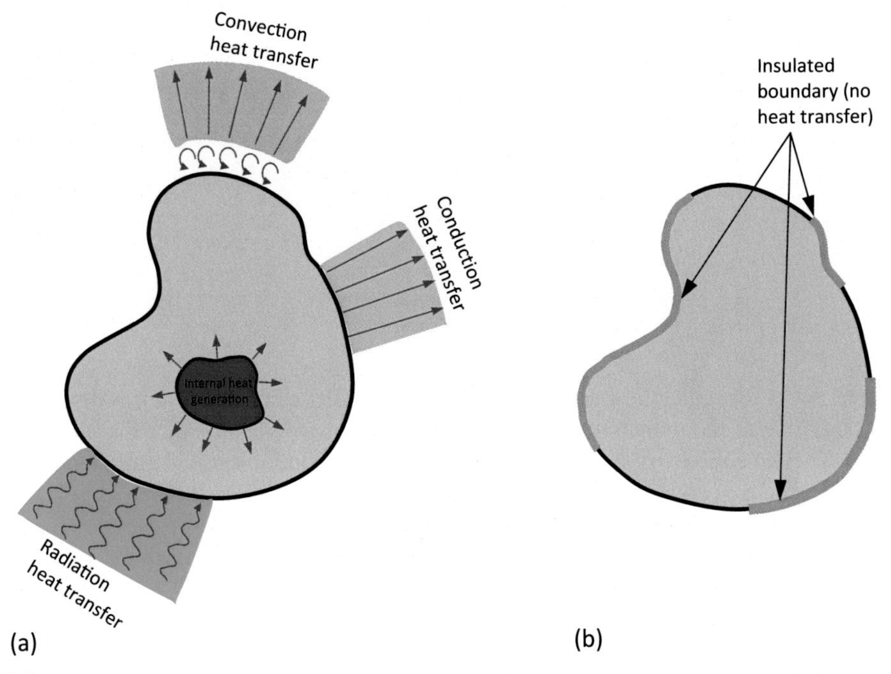

FIGURE 12.1

(a) The three modes of heat transfer and (b) boundary conditions of a body.

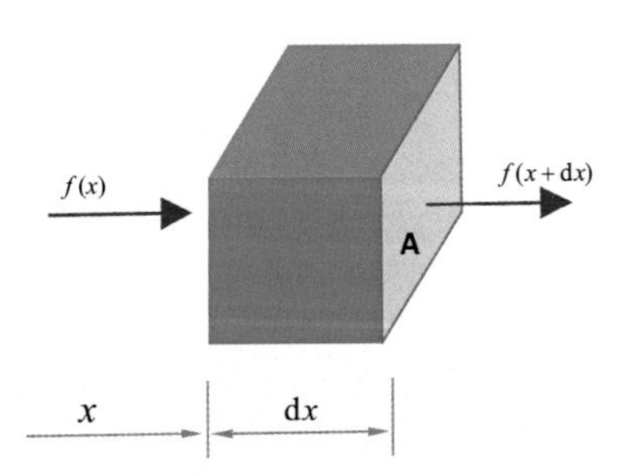

FIGURE 12.2

Thermal equilibrium in one-dimensional element.

The interpretation of the members of Equation (12.2) is the following:

$f_x(x)$	is the heat flux conduction in the left area, given by Equation (12.1).
$\dot{q}_{\text{gen}}$	is the generated heat transfer rate per unit volume (W/m^3) within the element.
$\rho c \frac{\partial T}{\partial t} \mathrm{d}x$	is the increase of the stored energy within the material element. The material parameter c is called specific heat (J/kg K), and the symbols ρ and t denote the material density and time, respectively.
$f_x(x + \mathrm{d}x)$	is the heat conduction out of the right area, given by the following equation:

$$f_x(x+\mathrm{d}x) = -k\frac{\partial T}{\partial x} - \frac{\partial}{\partial x}\left(k\frac{\partial T}{\partial x}\right)\mathrm{d}x \tag{12.3}$$

Combining Equations (12.1)–(12.3), the following equilibrium equation of 1D conduction heat transfer can be obtained:

$$\frac{\partial}{\partial x}\left(k\frac{\partial T}{\partial x}\right) + \dot{q}_{\mathrm{gen}} = \rho c\frac{\partial T}{\partial t} \tag{12.4}$$

In the case of steady state (the temperature does not change with time), the differentiation with respect to time t is equal to zero. Therefore, Equation (12.4) yields

$$\frac{\partial}{\partial x}\left(k\frac{\partial T}{\partial x}\right) + \dot{q}_{\mathrm{gen}} = 0 \tag{12.5}$$

This is the 1D heat conduction equation. Generalization of the above procedure yields the following two-dimensional (2D) and three-dimensional (3D) heat conduction equations in Cartesian coordinates:

2D STEADY-STATE HEAT CONDUCTION EQUATION IN CARTESIAN COORDINATES

$$\frac{\partial}{\partial x}\left(k_x\frac{\partial T}{\partial x} + k_{xy}\frac{\partial T}{\partial y}\right) + \frac{\partial}{\partial y}\left(k_y\frac{\partial T}{\partial y} + k_{xy}\frac{\partial T}{\partial x}\right) + \dot{q}_{\mathrm{gen}} = 0 \tag{12.6}$$

3D STEADY-STATE HEAT CONDUCTION EQUATION IN CARTESIAN COORDINATES (FIGURE 12.3a)

$$\begin{aligned}&\frac{\partial}{\partial x}\left(k_x\frac{\partial T}{\partial x} + k_{xy}\frac{\partial T}{\partial y} + k_{xz}\frac{\partial T}{\partial z}\right) + \frac{\partial}{\partial y}\left(k_y\frac{\partial T}{\partial y} + k_{yx}\frac{\partial T}{\partial x} + k_{yz}\frac{\partial T}{\partial z}\right)\\ &+ \frac{\partial}{\partial z}\left(k_z\frac{\partial T}{\partial z} + k_{zx}\frac{\partial T}{\partial x} + k_{zy}\frac{\partial T}{\partial y}\right) + \dot{q}_{\mathrm{gen}} = 0\end{aligned} \tag{12.7}$$

In the above two equations, k_x, k_y, k_z are the thermal conductivity coefficients in the directions x, y, z, respectively (orthotropic medium). If the medium is isotropic, that is, $k_x = k_y = k_z = k_{xy} = k_{yz} = k_{zx} = k$, then Equations (12.6) and (12.7) can be simplified as:

$$\frac{\partial^2 T}{\partial x^2} + \frac{\partial^2 T}{\partial y^2} + \frac{\dot{q}_{\mathrm{gen}}}{k} = 0 \tag{12.8}$$

$$\frac{\partial^2 T}{\partial x^2} + \frac{\partial^2 T}{\partial y^2} + \frac{\partial^2 T}{\partial z^2} + \frac{\dot{q}_{\mathrm{gen}}}{k} = 0 \tag{12.9}$$

The last equation can also be expressed in cylindrical and spherical coordinates.

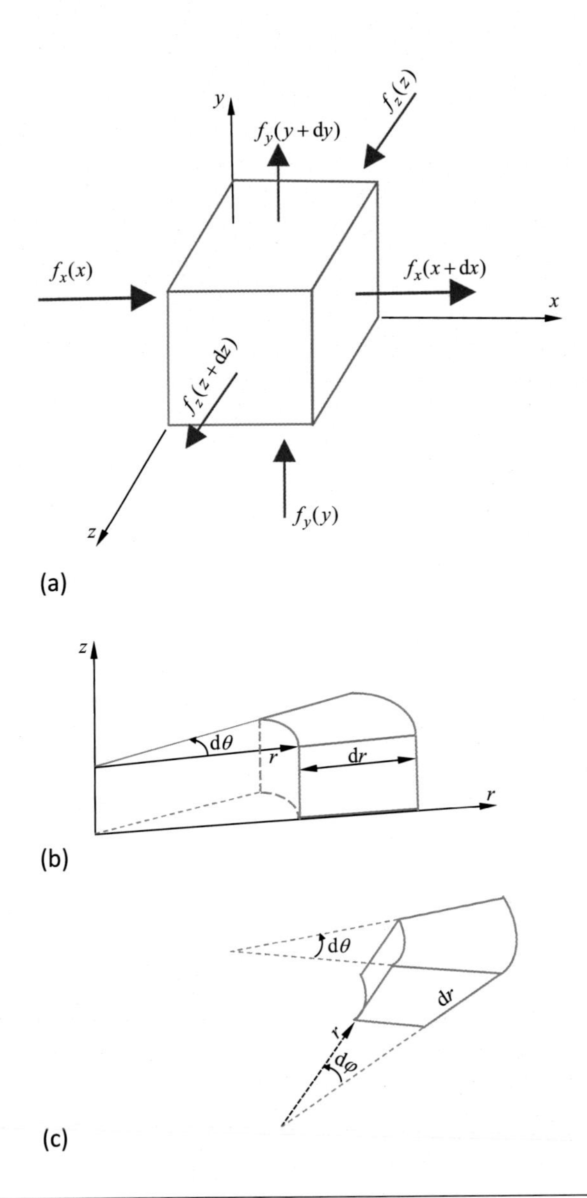

FIGURE 12.3

(a) Cartesian coordinates, (b) cylindrical coordinates, and (c) spherical coordinates.

3D STEADY-STATE HEAT CONDUCTION EQUATION IN CYLINDRICAL COORDINATES (FIGURE 12.3b)

$$\frac{\partial^2 T}{\partial r^2}+\frac{1}{r}\frac{\partial T}{\partial r}+\frac{1}{r^2}\frac{\partial^2 T}{\partial \theta^2}+\frac{\partial^2 T}{\partial z^2}+\frac{\dot{q}_{\text{gen}}}{k}=0 \tag{12.10}$$

3D STEADY-STATE HEAT CONDUCTION EQUATION IN SPHERICAL COORDINATES (FIGURE 12.3c)

$$\frac{1}{r}\frac{\partial^2}{\partial r^2}(rT)+\frac{1}{r^2\sin\varphi}\frac{\partial}{\partial\varphi}\left(\sin\varphi\frac{\partial T}{\partial\varphi}\right)+\frac{1}{r^2\sin^2\varphi}\frac{\partial^2 T}{\partial\theta^2}+\frac{\dot{q}_{\text{gen}}}{k}=0 \tag{12.11}$$

HEAT CONDUCTION OF ORTHOTROPIC MATERIALS

Let us now consider a 2D orthotropic material (Figure 12.4). For such materials we know the values of the thermal conductivity coefficients in the principal directions X and Y, namely, k_X, k_Y. According to Fourier's law, the heat flux into X and Y directions can be obtained by the following equation:

$$\begin{Bmatrix} f_X \\ f_Y \end{Bmatrix}=-\begin{bmatrix} k_X & 0 \\ 0 & k_Y \end{bmatrix}\begin{Bmatrix} \partial T/\partial X \\ \partial T/\partial Y \end{Bmatrix} \tag{12.12}$$

In contrast, by chain rule differentiation, the temperature gradients on the X and Y directions can be expressed versus the gradients on the x and y directions

$$\begin{Bmatrix} \partial T/\partial X \\ \partial T/\partial Y \end{Bmatrix}=[\Theta]\begin{Bmatrix} \partial T/\partial x \\ \partial T/\partial y \end{Bmatrix} \tag{12.13}$$

where

$$[\Theta]=\begin{bmatrix} \partial x/\partial X & \partial y/\partial X \\ \partial x/\partial Y & \partial y/\partial Y \end{bmatrix} \tag{12.14}$$

or

$$[\Theta]=\begin{bmatrix} \cos\vartheta & \sin\vartheta \\ -\sin\vartheta & \cos\vartheta \end{bmatrix} \tag{12.15}$$

It should be noted that the heat flux vector $\{f_x \ f_y\}^{\mathrm{T}}$ can be rotated in the same manner as the displacement vector according to the following well-known equation of the theory of elasticity:

$$\begin{Bmatrix} f_x \\ f_y \end{Bmatrix}=[\Theta]^{\mathrm{T}}\begin{Bmatrix} f_X \\ f_Y \end{Bmatrix} \tag{12.16}$$

Combining Equations (12.12), (12.13), and (12.16) the following formula can be obtained:

$$\begin{Bmatrix} f_x \\ f_y \end{Bmatrix}=-[\Theta]^{\mathrm{T}}\begin{bmatrix} k_X & 0 \\ 0 & k_Y \end{bmatrix}[\Theta]\begin{Bmatrix} \partial T/\partial x \\ \partial T/\partial y \end{Bmatrix} \tag{12.17}$$

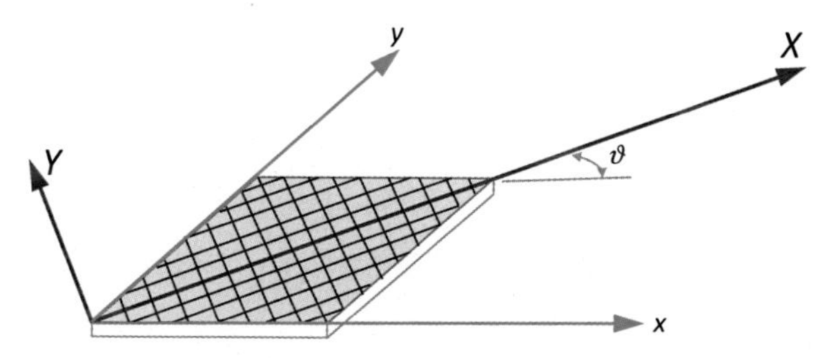

FIGURE 12.4

Principal coordinate system X-Y for a 2D orthotropic material.

or

$$\begin{Bmatrix} f_x \\ f_y \end{Bmatrix} = -\begin{bmatrix} k_x & k_{xy} \\ k_{xy} & k_y \end{bmatrix} \begin{Bmatrix} \partial T/\partial x \\ \partial T/\partial y \end{Bmatrix} \tag{12.18}$$

where

$$\begin{bmatrix} k_x & k_{xy} \\ k_{xy} & k_y \end{bmatrix} = [\Theta]^{\mathrm{T}} \begin{bmatrix} k_X & 0 \\ 0 & k_Y \end{bmatrix} [\Theta] = [\kappa] \tag{12.19}$$

Taking into account Equations (12.18) and (12.19), the 2D steady-state heat conduction equation (Equation 12.6) can be written in the following matrix form:

$$\{D\}([\kappa]\{DT\}) + \dot{q}_{\mathrm{gen}} = 0 \tag{12.20}$$

where

$$\{D\} = \begin{Bmatrix} \partial/\partial x \\ \partial/\partial y \end{Bmatrix} \tag{12.21}$$

$$\{DT\} = \begin{Bmatrix} \partial T/\partial x \\ \partial T/\partial y \end{Bmatrix} \tag{12.22}$$

According to the definition, the heat flux through the boundaries of a 2D orthotropic medium is

$$f_{\mathrm{bound}} = \left(k_x \frac{\partial T}{\partial x} + k_{xy} \frac{\partial T}{\partial y}\right)\cos\vartheta + \left(k_{xy} \frac{\partial T}{\partial x} + k_y \frac{\partial T}{\partial y}\right)\cos\varphi \tag{12.23}$$

where $\cos\vartheta$ and $\cos\varphi$ are the direction cosines of a vector directed normal to the boundary. Therefore, taking into account Equation (12.19), the above equation can be written as:

$$f_{\mathrm{bound}} = \{C\}^{\mathrm{T}}[\kappa]\{DT\} \tag{12.24}$$

where

$$\{C\} = \begin{Bmatrix} \cos\vartheta \\ \cos\varphi \end{Bmatrix} \tag{12.25}$$

Equations (12.20) and (12.24) can be also generalized for the case of 3D steady-state heat conduction in Cartesian coordinates:

$$[\kappa] = \begin{bmatrix} k_x & k_{xy} & k_{xz} \\ k_{yx} & k_y & k_{yz} \\ k_{zx} & k_{zy} & k_z \end{bmatrix} = [\Theta^*]^{\mathrm{T}} \begin{bmatrix} k_X & 0 & 0 \\ 0 & k_Y & 0 \\ 0 & 0 & k_Z \end{bmatrix} [\Theta^*] \tag{12.26}$$

$$\{D\} = \begin{Bmatrix} \partial/\partial x \\ \partial/\partial y \\ \partial/\partial z \end{Bmatrix} \tag{12.27}$$

$$\{DT\} = \begin{Bmatrix} \partial T/\partial x \\ \partial T/\partial y \\ \partial T/\partial z \end{Bmatrix} \tag{12.28}$$

$$\{C\} = \begin{Bmatrix} \cos\vartheta \\ \cos\varphi \\ \cos\rho \end{Bmatrix} \tag{12.29}$$

$$[\Theta^*] = \begin{bmatrix} \partial x/\partial X & \partial y/\partial X & \partial z/\partial X \\ \partial x/\partial Y & \partial y/\partial Y & \partial z/\partial Y \\ \partial x/\partial Z & \partial y/\partial Z & \partial z/\partial Z \end{bmatrix} \tag{12.30}$$

Very often the solids under investigation are cylindrical (Figure 12.5), for example, pipes, made by orthotropic layer. For such cases, Equations (12.20) and (12.24) remain applicable if we replace Equations (12.26)–(12.30) by the following equations:

$$[\kappa] = \begin{bmatrix} \kappa_\vartheta & \kappa_{\vartheta y} & \kappa_{\vartheta r} \\ \kappa_{y\vartheta} & \kappa_y & \kappa_{yr} \\ \kappa_{r\vartheta} & \kappa_{ry} & \kappa_r \end{bmatrix} = [\Theta^*]^{\mathrm{T}} \begin{bmatrix} k_X & 0 & 0 \\ 0 & k_Y & 0 \\ 0 & 0 & k_Z \end{bmatrix} [\Theta^*] \tag{12.31}$$

$$\{D\} = \begin{Bmatrix} \partial/\partial\vartheta \\ \partial/\partial y \\ \partial/\partial r \end{Bmatrix} \tag{12.32}$$

$$\{DT\} = \begin{Bmatrix} \partial T/\partial\vartheta \\ \partial T/\partial y \\ \partial T/\partial r \end{Bmatrix} \tag{12.33}$$

$$\{C\} = \begin{Bmatrix} \cos\alpha \\ \cos\beta \\ \cos 0 \end{Bmatrix} \tag{12.34}$$

$$[\Theta^*] = \begin{bmatrix} \partial\vartheta/\partial X & \partial y/\partial X & \partial r/\partial X \\ \partial\vartheta/\partial Y & \partial y/\partial Y & \partial r/\partial Y \\ \partial\vartheta/\partial Z & \partial y/\partial Z & \partial r/\partial Z \end{bmatrix} \tag{12.35}$$

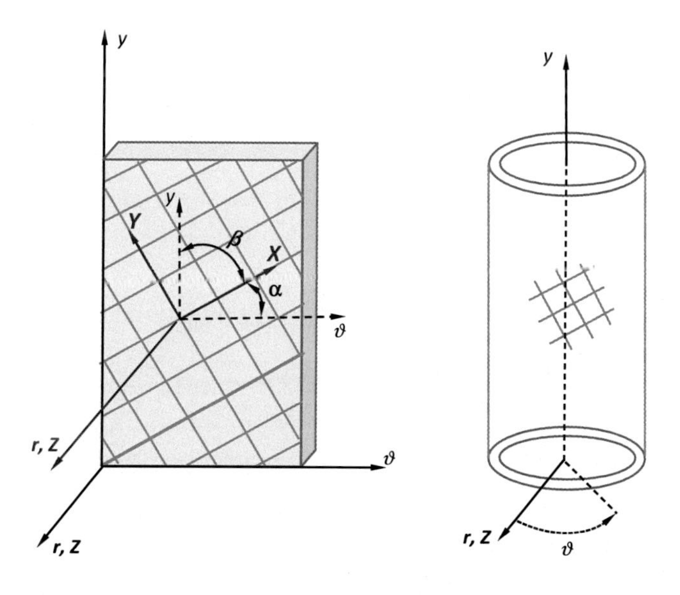

FIGURE 12.5

Pipe made by a layer with orthotropic thermal conductivity coefficients k_X, k_Y, k_Z.

12.2 CONVECTION HEAT TRANSFER

When a surface with temperature T_w is exposed to a stream of a fluid with temperature T_∞ (Figure 12.6), the temperature gradient on the surface is influenced by the velocity profile of the flow, the viscosity, the density, and the thermal properties of the fluid (thermal conductivity and specific heat).

Since the fluid flow is the physical mechanism that carries the heat away, a high fluid velocity yields high value of temperature gradient on the surface. The heat transfer due to the interaction of the fluid flow with the surface temperature is called convection heat transfer. Analytic modeling of the above mechanism is difficult. However, experimental procedures have indicated that the overall effect of the convection heat flux f_{bound} can be described by the following Newton's formula:

$$f_{\text{bound}} = h(T_w - T_\infty) \tag{12.36}$$

The parameter h is the convection heat transfer coefficient, and its units are W/m^2 K.

It should be noticed that heat convection can occur even when a surface is exposed to ambient room air without an external source of air ventilation. In this case, a movement of the air can happen due to its density gradients near the surface. This type of heat convection is called natural (or free) convection.

12.3 FINITE ELEMENT FORMULATION

12.3.1 ONE-DIMENSIONAL HEAT TRANSFER MODELING USING A VARIATIONAL METHOD

1D heat transfer finite element (FE) modeling can be formulated by the same functions with the FE modeling of the spring and bar elements. In order to derive the element equation, it is convenient to use a variational method. To this scope, the following functional, which is analogous to the potential energy, should be minimized:

$$\Pi = \int \left[\frac{1}{2}\{DT\}^{\text{T}}[\kappa]\{DT\} - \dot{q}_{\text{gen}}T\right] \text{d}V - \int \left[f_{\text{bound}}T + hT_\infty T - \frac{1}{2}hT^2\right] \text{d}S \tag{12.37}$$

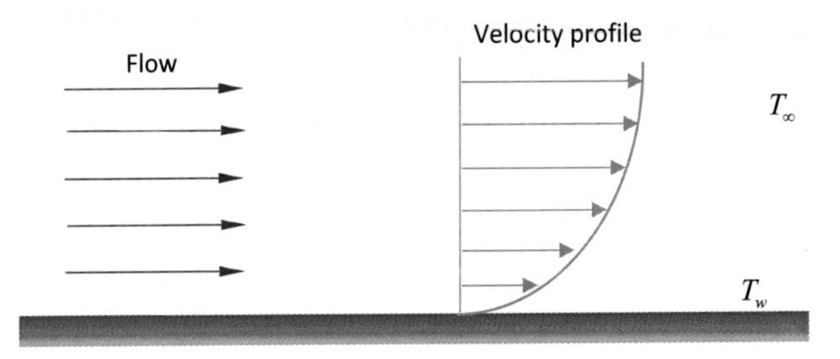

FIGURE 12.6

Convection heat transfer.

For the temperature distribution T, we can use shape functions correlating T to the nodal temperatures. As it has already been shown in Chapters 3 and 4, the temperature distribution can be described by the following formulae:

$$T = [N_1 \ N_2]\begin{Bmatrix} T_1 \\ T_2 \end{Bmatrix} \tag{12.38}$$

or in an abbreviated form:

$$T = [N]\{T_e\} \tag{12.39}$$

For the derivation of the shape functions N_1 and N_2 we recall Equations (4.12) and (4.13):

$$N_1 = 1 - \frac{x}{L} \tag{12.40}$$

$$N_2 = \frac{x}{L} \tag{12.41}$$

Using the matrix $[N]$, the vector $\{DT\}$ can now be correlated to the nodal temperatures $\{T_e\}$:

$$\{DT\} = [B]\{T_e\} \tag{12.42}$$

where

$$[B] = \{D\}[N] \tag{12.43}$$

Taking into account the FE formulation described in Chapter 10, it is obvious that the vector $\{DT\}$ is analogous to the vector of strains $\{\varepsilon\}$, and the vector $\{T_e\}$ is analogous to the vector of nodal displacements $\{d\}$.

Using Equations (12.39) and (12.42), the potential Π given by Equation (12.37) can be correlated to the nodal temperatures $\{T_e\}$:

$$\begin{aligned} \Pi = {} & \frac{1}{2}\{T_e\}^{\mathrm{T}}\left[\int_V [B]^{\mathrm{T}}[\kappa][B]\mathrm{d}V + \int_V [N]^{\mathrm{T}}[N]h\mathrm{d}S\right]\{T_e\} \\ & -\{T_e\}^{\mathrm{T}}\left[\int_S [N]^{\mathrm{T}} f_{\mathrm{bound}}\mathrm{d}S + \int_S [N]^{\mathrm{T}} hT_{\infty}\mathrm{d}S + \int_V [N]^{T}\dot{g}_{\mathrm{gen}}\mathrm{d}V\right] \end{aligned} \tag{12.44}$$

The condition for minimization of the above functional

$$\left\{\frac{\partial \Pi}{\partial T}\right\} = \{0\} \tag{12.45}$$

yields the following element equation:

$$[k]\{T\} = \{R\} \tag{12.46}$$

where

$$[k] = \underbrace{\int_V [B]^{\mathrm{T}}[\kappa][B]\mathrm{d}V}_{\text{conduction part } [k_{\mathrm{CD}}]} + \underbrace{\int_S [N]^{\mathrm{T}}[N]h\,\mathrm{d}S}_{\text{convection part } [k_{\mathrm{CV}}]} \tag{12.47}$$

and

$$\{R\} = \underbrace{\int_S [N]^T f_{bound}\, dS}_{\text{conduction part}} + \underbrace{\int_S [N]^T hT_\infty\, dS}_{\text{convection part}} + \underbrace{\int_V [N]^T \dot{g}_{gen}\, dV}_{\substack{\text{internal heat}\\ \text{generationpart}}} \tag{12.48}$$

Since for 1D heat transfer, the matrix $[\kappa]$ should be replaced by the conductivity coefficient k_x, then, taking into account Equations (12.39)–(12.41) and (12.43), the conduction part of Equation (12.47) can be written:

$$[k_{CD}] = \int_V [B]^T [\kappa][B] dV = \frac{Ak_x}{L}\begin{bmatrix} 1 & -1 \\ -1 & 1 \end{bmatrix} \tag{12.49}$$

The above equation has same functional form with the stiffness matrix of bar elements. If the conduction coefficient k_x is replaced by the modulus of elasticity E, Equation (12.49) yields Equation (4.36).

If we now replace the shape functions N_1 and N_2 to the convection part of Equation (12.47), the following formula can be obtained for the convection from the lateral surface:

$$[k^l_{CV}] = \int_{S_{lateral}} [N]^T [N] h dS = \frac{hPL}{6}\begin{bmatrix} 2 & 1 \\ 1 & 2 \end{bmatrix} \tag{12.50}$$

where P is the perimeter (Figure 12.7) of the 1D element and $dS = Pdx$. The "force" vector for heat convection from the lateral surface can be derived if we replace the shape functions N_1 and N_2 of Equation (12.48).

$$\{R^l\} = \left(\underbrace{\frac{f_{bound} PL}{2}}_{\substack{\text{conduction}\\ \text{part}}} + \underbrace{\frac{hT_\infty PL}{2}}_{\substack{\text{convection}\\ \text{part from}\\ \text{the lateral}\\ \text{surface}}} + \underbrace{\frac{\dot{q}_{gen} AL}{2}}_{\substack{\text{internal}\\ \text{energy}\\ \text{generation}\\ \text{part}}} \right) \begin{Bmatrix} 1 \\ 1 \end{Bmatrix} \tag{12.51}$$

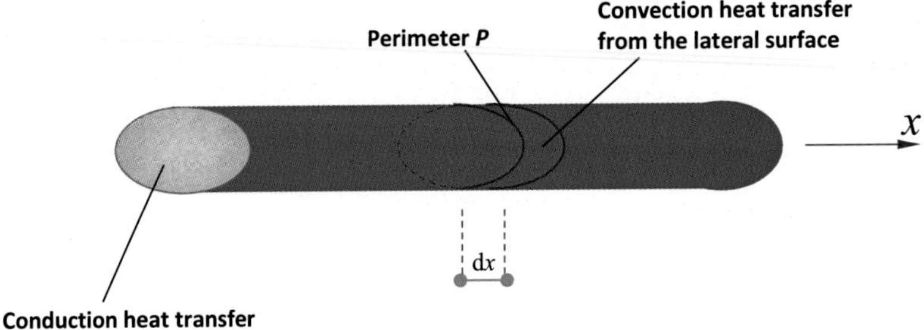

FIGURE 12.7

Perimeter P of a 1D element.

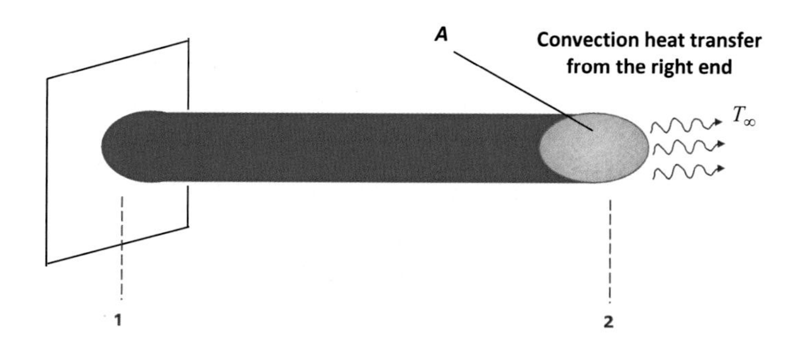

FIGURE 12.8

Convection heat transfer from the right end of the element 1-2.

In case of convection from a free end of an element, let us say from the right end (Figure 12.8), the additional contribution to the stiffness matrix $[K_{CV}]$ can be obtained by setting into convection part of Equation (12.47) the values of N_1 and N_2 corresponding to $x = L$, that is,

$$N_1 = 0 \tag{12.52}$$

$$N_2 = 1 \tag{12.53}$$

Then, for heat convection from the right end the following formula can be derived:

$$\left[k_{CV}^{re}\right] = \int_{S_{end}} [N]^T [N] h_{re} dS = \int_{S_{end}} \begin{Bmatrix} 0 \\ 1 \end{Bmatrix} [0 \;\; 1] h_{re} dS = h_{re} A \begin{bmatrix} 0 & 0 \\ 0 & 1 \end{bmatrix} \tag{12.54}$$

Therefore, the final form for the "stiffness" matrix of an element containing the convection part from the lateral surface and the convection part from the right end is

$$[k_{CV}] = \left[k_{CV}^{l}\right] + \left[k_{CV}^{re}\right] \tag{12.55}$$

In the case of convection from the left free end of an element, that is, $x = 0$, the shape functions are $N_1 = 1$, $N_2 = 0$ yielding

$$\left[k_{CV}^{le}\right] = \int_{S_{end}} [N]^T [N] h_{le} dS = \int_{S_{end}} \begin{Bmatrix} 0 \\ 1 \end{Bmatrix} [0 \;\; 1] h_{le} dS = h_{le} A \begin{bmatrix} 1 & 0 \\ 0 & 0 \end{bmatrix} \tag{12.56}$$

and then, the final form for the "stiffness" matrix of an element containing the convection part from the lateral surface and the convection part from the left end is:

$$[k_{CV}] = \left[k_{CV}^{l}\right] + \left[k_{CV}^{le}\right] \tag{12.57}$$

The "force" vector due to convection from the right end is

$$\{R^{re}\} = hT_\infty A \begin{Bmatrix} 0 \\ 1 \end{Bmatrix} \tag{12.58}$$

and the "force" vector due to convection from the left end is

$$\{R^{le}\} = hT_\infty A \begin{Bmatrix} 1 \\ 0 \end{Bmatrix} \tag{12.59}$$

Therefore, combining Equations (12.51), (12.58), and (12.59) the final form of the force vector due to lateral conduction and convection, as well as convection at the right and left ends is

$$\{R\} = \{R^{l}\} + \{R^{re}\} + \{R^{le}\} \tag{12.60}$$

After the derivation of the element equation, the subsequent steps regarding (a) the expansion of the local element matrices to the degrees of freedom of the whole structure, and (b) the assembly of the local matrices to derive the global one, can be performed by the procedures exhibited in Chapters 3 and 4.

EXAMPLE 1: HEAT TRANSFER THROUGH A MULTILAYERED WALL

The wall of a building is composed from the layers shown in the figure. Determine: (a) the temperatures in the inner and outer surfaces as well as in the interfaces, (b) the temperature distribution within each layer, and (c) the heat flux transferred through each layer.

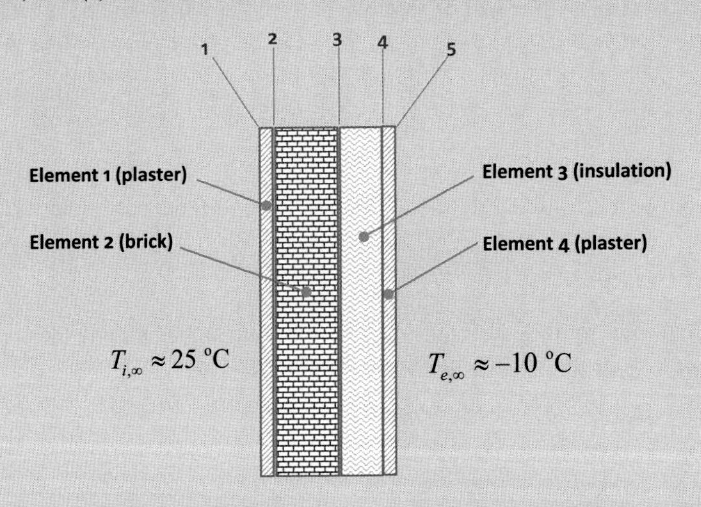

Data

$T_{i,\infty} = 25\,°\mathrm{C}$ (inner ambient temperature)
$T_{e,\infty} = -10\,°\mathrm{C}$ (outer ambient temperature)
$h_i = 8\,\mathrm{W/m^2\,°C}$ (internal surface convection coefficient)
$h_e = 24\,\mathrm{W/m^2\,°C}$ (external surface convection coefficient)
$k_P = 0.82\,\mathrm{W/m\,°C}$ (conduction coefficient for plaster)
$k_B = 0.64\,\mathrm{W/m\,°C}$ (conduction coefficient for brick-wall)
$k_I = 0.034\,\mathrm{W/m\,°C}$ (conduction coefficient for insulation)
$L_P = 3.0\,\mathrm{cm}$ (thickness of the plaster)
$L_B = 15.0\,\mathrm{cm}$ (thickness of the brick-wall)
$L_I = 5.0\,\mathrm{cm}$ (thickness of the insulation)

Solution

Question (a): Temperatures in the Inner and Outer Surfaces and in the Interfaces

Step 1: Derivation of the element matrices. *Element 1, nodes 1 and 2*

Taking into account Equations (12.49) and (12.56), the following element equations can be obtained:

$$\left[\frac{Ak_P}{L_P}\begin{bmatrix}1 & -1\\ -1 & 1\end{bmatrix} + h_iA\begin{bmatrix}1 & 0\\ 0 & 0\end{bmatrix}\right]\begin{Bmatrix}T_1\\ T_2\end{Bmatrix} = \begin{Bmatrix}r_1\\ r_2\end{Bmatrix} \tag{12.e1}$$

or

$$A\left[\frac{0.82}{3\times10^{-2}}\begin{bmatrix}1 & -1\\ -1 & 1\end{bmatrix} + 8\begin{bmatrix}1 & 0\\ 0 & 0\end{bmatrix}\right]\begin{Bmatrix}T_1\\ T_2\end{Bmatrix} = \begin{Bmatrix}r_1\\ r_2\end{Bmatrix} \tag{12.e2}$$

yielding

$$A\begin{bmatrix}35.333 & -27.333\\ -27.333 & 27.333\end{bmatrix}\begin{Bmatrix}T_1\\ T_2\end{Bmatrix} = \begin{Bmatrix}r_1\\ r_2\end{Bmatrix} \tag{12.e3}$$

Element 2, nodes 2 and 3

Taking into account Equation (12.49), the following local "stiffness" matrix can be obtained:

$$\frac{Ak_B}{L_B}\begin{bmatrix}1 & -1\\ -1 & 1\end{bmatrix}\begin{Bmatrix}T_2\\ T_3\end{Bmatrix} = \begin{Bmatrix}r_2\\ r_3\end{Bmatrix} \tag{12.e4}$$

or

$$\frac{0.64A}{15\times10^{-2}}\begin{bmatrix}1 & -1\\ -1 & 1\end{bmatrix}\begin{Bmatrix}T_2\\ T_3\end{Bmatrix} = \begin{Bmatrix}r_2\\ r_3\end{Bmatrix} \tag{12.e5}$$

yielding

$$A\begin{bmatrix}4.266 & -4.266\\ -4.266 & 4.266\end{bmatrix}\begin{Bmatrix}T_2\\ T_3\end{Bmatrix} = \begin{Bmatrix}r_2\\ r_3\end{Bmatrix} \tag{12.e6}$$

Element 3, nodes 3 and 4

Taking into account Equation (12.49), it can be written as:

$$\frac{Ak_I}{L_I}\begin{bmatrix}1 & -1\\ -1 & 1\end{bmatrix}\begin{Bmatrix}T_3\\ T_4\end{Bmatrix} = \begin{Bmatrix}r_3\\ r_4\end{Bmatrix} \tag{12.e7}$$

or

$$\frac{0.034A}{5\times10^{-2}}\begin{bmatrix}1 & -1\\ -1 & 1\end{bmatrix}\begin{Bmatrix}T_3\\ T_4\end{Bmatrix} = \begin{Bmatrix}r_3\\ r_4\end{Bmatrix} \tag{12.e8}$$

yielding

$$A\begin{bmatrix}0.68 & -0.68\\ -0.68 & 0.68\end{bmatrix}\begin{Bmatrix}T_3\\ T_4\end{Bmatrix} = \begin{Bmatrix}r_3\\ r_4\end{Bmatrix} \tag{12.e9}$$

Continued

EXAMPLE 1: HEAT TRANSFER THROUGH A MULTILAYERED WALL—CONT'D

Element 4, nodes 4 and 5

The conduction part of the stiffness matrix is given by Equation (12.49). Since now the convection takes place from the right end, the convection part of the stiffness matrix can be obtained from Equation (12.54).

$$\left[\frac{Ak_P}{L_P}\begin{bmatrix}1 & -1\\ -1 & 1\end{bmatrix} + h_eA\begin{bmatrix}0 & 0\\ 0 & 1\end{bmatrix}\right]\begin{Bmatrix}T_4\\ T_5\end{Bmatrix} = \begin{Bmatrix}r_4\\ r_5\end{Bmatrix} \quad (12.e10)$$

or

$$A\left[\frac{0.82}{3\times 10^{-2}}\begin{bmatrix}1 & -1\\ -1 & 1\end{bmatrix} + 24\begin{bmatrix}0 & 0\\ 0 & 1\end{bmatrix}\right]\begin{Bmatrix}T_4\\ T_5\end{Bmatrix} = \begin{Bmatrix}r_4\\ r_5\end{Bmatrix} \quad (12.e11)$$

yielding

$$A\begin{bmatrix}27.333 & -27.333\\ -27.333 & 51.333\end{bmatrix}\begin{Bmatrix}T_4\\ T_5\end{Bmatrix} = \begin{Bmatrix}r_4\\ r_5\end{Bmatrix} \quad (12.e12)$$

Step 2: Expansion of the element equations to the degrees of freedom of the structure. *Element 1, nodes 1 and 2*

$$A\begin{bmatrix}35.333 & -27.333 & 0 & 0 & 0\\ -27.333 & 27.333 & 0 & 0 & 0\\ 0 & 0 & 0 & 0 & 0\\ 0 & 0 & 0 & 0 & 0\\ 0 & 0 & 0 & 0 & 0\end{bmatrix}\begin{Bmatrix}T_1\\ T_2\\ T_3\\ T_4\\ T_5\end{Bmatrix} = \begin{Bmatrix}r_1\\ r_2\\ r_3\\ r_4\\ r_5\end{Bmatrix} \quad (12.e13)$$

Element 2, nodes 2 and 3

$$A\begin{bmatrix}0 & 0 & 0 & 0 & 0\\ 0 & 4.266 & -4.266 & 0 & 0\\ 0 & -4.266 & 4.266 & 0 & 0\\ 0 & 0 & 0 & 0 & 0\\ 0 & 0 & 0 & 0 & 0\end{bmatrix}\begin{Bmatrix}T_1\\ T_2\\ T_3\\ T_4\\ T_5\end{Bmatrix} = \begin{Bmatrix}r_1\\ r_2\\ r_3\\ r_4\\ r_5\end{Bmatrix} \quad (12.e14)$$

Element 3, nodes 3 and 4

$$A\begin{bmatrix}0 & 0 & 0 & 0 & 0\\ 0 & 0 & 0 & 0 & 0\\ 0 & 0 & 0.68 & -0.68 & 0\\ 0 & 0 & -0.68 & 0.68 & 0\\ 0 & 0 & 0 & 0 & 0\end{bmatrix}\begin{Bmatrix}T_1\\ T_2\\ T_3\\ T_4\\ T_5\end{Bmatrix} = \begin{Bmatrix}r_1\\ r_2\\ r_3\\ r_4\\ r_5\end{Bmatrix} \quad (12.e15)$$

Element 4, nodes 4 and 5

$$A\begin{bmatrix}0 & 0 & 0 & 0 & 0\\ 0 & 0 & 0 & 0 & 0\\ 0 & 0 & 0 & 0 & 0\\ 0 & 0 & 0 & 27.333 & -27.333\\ 0 & 0 & 0 & -27.333 & 51.333\end{bmatrix}\begin{Bmatrix}T_1\\ T_2\\ T_3\\ T_4\\ T_5\end{Bmatrix} = \begin{Bmatrix}r_1\\ r_2\\ r_3\\ r_4\\ r_5\end{Bmatrix} \quad (12.e16)$$

Step 3: Assembly of the element equations. Superposition of the element equations (Equations 12.e13–12.e16) yields

$$A\begin{bmatrix} 35.333 & -27.333 & 0 & 0 & 0 \\ -27.333 & 27.333+4.266 & -4.266 & 0 & 0 \\ 0 & -4.266 & 4.266+0.68 & -0.68 & 0 \\ 0 & 0 & -0.68 & 0.68+27.333 & -27.333 \\ 0 & 0 & 0 & -27.333 & 51.333 \end{bmatrix}\begin{Bmatrix} T_1 \\ T_2 \\ T_3 \\ T_4 \\ T_5 \end{Bmatrix} = \begin{Bmatrix} R_1 \\ R_2 \\ R_3 \\ R_4 \\ R_5 \end{Bmatrix} \tag{12.e17}$$

or

$$A\begin{bmatrix} 35.333 & -27.333 & 0 & 0 & 0 \\ -27.333 & 31.599 & -4.266 & 0 & 0 \\ 0 & -4.266 & 4.946 & -0.68 & 0 \\ 0 & 0 & -0.68 & 28.013 & -27.333 \\ 0 & 0 & 0 & -27.333 & 51.333 \end{bmatrix}\begin{Bmatrix} T_1 \\ T_2 \\ T_3 \\ T_4 \\ T_5 \end{Bmatrix} = \begin{Bmatrix} R_1 \\ R_2 \\ R_3 \\ R_4 \\ R_5 \end{Bmatrix} \tag{12.e18}$$

Step 4: Derivation of the field values. An alternative formulation of Equation (12.e18) is the following:

$$\begin{bmatrix} 35.333 & -27.333 & 0 & 0 & 0 & -1 & 0 & 0 & 0 & 0 \\ -27.333 & 31.599 & -4.266 & 0 & 0 & 0 & -1 & 0 & 0 & 0 \\ 0 & -4.266 & 4.946 & -0.68 & 0 & 0 & 0 & -1 & 0 & 0 \\ 0 & 0 & -0.68 & 28.013 & -27.333 & 0 & 0 & 0 & -1 & 0 \\ 0 & 0 & 0 & -27.333 & 51.333 & 0 & 0 & 0 & 0 & -1 \end{bmatrix}\begin{Bmatrix} T_1 \\ T_2 \\ T_3 \\ T_4 \\ T_5 \\ R_1/A \\ R_2/A \\ R_3/A \\ R_4/A \\ R_5/A \end{Bmatrix} = \begin{Bmatrix} 0 \\ 0 \\ 0 \\ 0 \\ 0 \end{Bmatrix} \tag{12.e19}$$

Boundary conditions. The above matrix equation represents an algebraic system of four equations with eight unknowns. In order the above system to be solvable, four more equations should be added. The missing equations can be specified by the following boundary conditions. The "forces" at the ends 1 and 5 of the wall are the convection fluxes given by the following equations:

$$R_1 = h_i T_{i,\infty} A = 8 \times 25 \times A = 200\text{A} \tag{12.e20}$$

and

$$R_5 = h_e T_{e,\infty} A = 24 \times (-10) \times A = -240\text{A} \tag{12.e21}$$

The "forces" on the interfaces 2, 3, and 4 are

$$R_2 = 0 \tag{12.e22}$$

$$R_3 = 0 \tag{12.e23}$$

$$R_4 = 0 \tag{12.e24}$$

Continued

EXAMPLE 1: HEAT TRANSFER THROUGH A MULTILAYERED WALL—CONT'D

The boundary conditions of Equation (12.e20)–(12.e24) can be expanded in a 5 × 10 matrix as follows:

$$\begin{bmatrix} 0 & 0 & 0 & 0 & 0 & 1 & 0 & 0 & 0 & 0 \\ 0 & 0 & 0 & 0 & 0 & 0 & 1 & 0 & 0 & 0 \\ 0 & 0 & 0 & 0 & 0 & 0 & 0 & 1 & 0 & 0 \\ 0 & 0 & 0 & 0 & 0 & 0 & 0 & 0 & 1 & 0 \\ 0 & 0 & 0 & 0 & 0 & 0 & 0 & 0 & 0 & 1 \end{bmatrix} \begin{Bmatrix} T_1 \\ T_2 \\ T_3 \\ T_4 \\ T_5 \\ R_1/A \\ R_2/A \\ R_3/A \\ R_4/A \\ R_5/A \end{Bmatrix} = \begin{Bmatrix} 200 \\ 0 \\ 0 \\ 0 \\ -240 \end{Bmatrix} \tag{12.e25}$$

Final solution. Superposition of the matrix equations (Equations 12.e19 and 12.e25) yields an algebraic system of 10 equations with 10 unknowns providing the nodal temperatures T_1, T_2, …, T_5 and the fluxes R_2/A, R_3/A, R_4/A at the corresponding interfaces:

$$\begin{bmatrix} 35.333 & -27.333 & 0 & 0 & 0 & -1 & 0 & 0 & 0 & 0 \\ -27.333 & 31.599 & -4.266 & 0 & 0 & 0 & -1 & 0 & 0 & 0 \\ 0 & -4.266 & 4.946 & -0.68 & 0 & 0 & 0 & -1 & 0 & 0 \\ 0 & 0 & -0.68 & 28.013 & -27.333 & 0 & 0 & 0 & -1 & 0 \\ 0 & 0 & 0 & -27.333 & 51.333 & 0 & 0 & 0 & 0 & -1 \\ 0 & 0 & 0 & 0 & 0 & 1 & 0 & 0 & 0 & 0 \\ 0 & 0 & 0 & 0 & 0 & 0 & 1 & 0 & 0 & 0 \\ 0 & 0 & 0 & 0 & 0 & 0 & 0 & 1 & 0 & 0 \\ 0 & 0 & 0 & 0 & 0 & 0 & 0 & 0 & 1 & 0 \\ 0 & 0 & 0 & 0 & 0 & 0 & 0 & 0 & 0 & 1 \end{bmatrix} \begin{Bmatrix} T_1 \\ T_2 \\ T_3 \\ T_4 \\ T_5 \\ R_1/A \\ R_2/A \\ R_3/A \\ R_4/A \\ R_5/A \end{Bmatrix} = \begin{Bmatrix} 0 \\ 0 \\ 0 \\ 0 \\ 0 \\ 200 \\ 0 \\ 0 \\ 0 \\ -240 \end{Bmatrix} \tag{12.e26}$$

Solving the above algebraic system, the following results will be obtained:

$$\begin{Bmatrix} T_1 \\ T_2 \\ T_3 \\ T_4 \\ T_5 \\ R_1/A \\ R_2/A \\ R_3/A \\ R_4/A \\ R_5/A \end{Bmatrix} = \begin{Bmatrix} 22.75 \\ 22.09 \\ 17.87 \\ -8.59 \\ -9.25 \\ 200 \\ 0 \\ 0 \\ 0 \\ -240 \end{Bmatrix} \tag{12.e27}$$

Question (b): Temperature Distribution within Each Layer

The temperature distribution within each layer can be obtained from Equations (12.38), (12.40), and (12.41):

$$T_{12}(x) = \left[\left(1 - \frac{x}{L_P}\right) \quad \left(\frac{x}{L_P}\right) \right] \begin{Bmatrix} 22.75 \\ 22.09 \end{Bmatrix} = 22.7505 - 21.947x$$

$$T_{23}(x) = \left[\left(1 - \frac{x}{L_B}\right) \quad \left(\frac{x}{L_B}\right) \right] \begin{Bmatrix} 22.09 \\ 17.87 \end{Bmatrix} = 22.092 - 28.1237x$$

$$T_{34}(x) = \left[\left(1 - \frac{x}{L_I}\right) \quad \left(\frac{x}{L_I}\right) \right] \begin{Bmatrix} 17.87 \\ -8.59 \end{Bmatrix} = 17.8735 - 529.305x$$

$$T_{45}(x) = \left[\left(1 - \frac{x}{L_P}\right) \quad \left(\frac{x}{L_P}\right) \right] \begin{Bmatrix} -8.59 \\ -9.25 \end{Bmatrix} = -8.59174 - 21.947x$$

Question (c): Heat Flux Through Each Layer

The heat flux through each layer can be derived from the Fourier's law:

$$f_{12} = -k_P \frac{\partial T_{12}(x)}{\partial x} = (-0.82)(-21.947) = 18\,\mathrm{W/m^2}$$

$$f_{23} = -k_B \frac{\partial T_{23}(x)}{\partial x} = (-0.64)(-28.1237) = 18\,\mathrm{W/m^2}$$

$$f_{34} = -k_I \frac{\partial T_{34}(x)}{\partial x} = (-0.034)(-529.305) = 18\,\mathrm{W/m^2}$$

$$f_{45} = -k_P \frac{\partial T_{45}(x)}{\partial x} = (-0.82)(-21.947) = 18\,\mathrm{W/m^2}$$

EXAMPLE 2: HEAT TRANSFER IN A HEAT EXCHANGER

A heat exchanger uses rods to remove heat of a hot plate. Determine (a) the temperature distribution along each rod and (b) the heat flux from the element 1.

Data

Temperature of plate: $T_1 = 300$ °C (constant)

Length of the rod: $3\,L = 300$ mm

Diameter of the rod: $d = 120$ mm

Conductivity coefficient: $k = 62$ W/m °C

Continued

EXAMPLE 2: HEAT TRANSFER IN A HEAT EXCHANGER—CONT'D

Convection coefficient: $h = 730$ W/m °C

Temperature of the air surrounding the rod: $T_\infty = 25^\circ$C

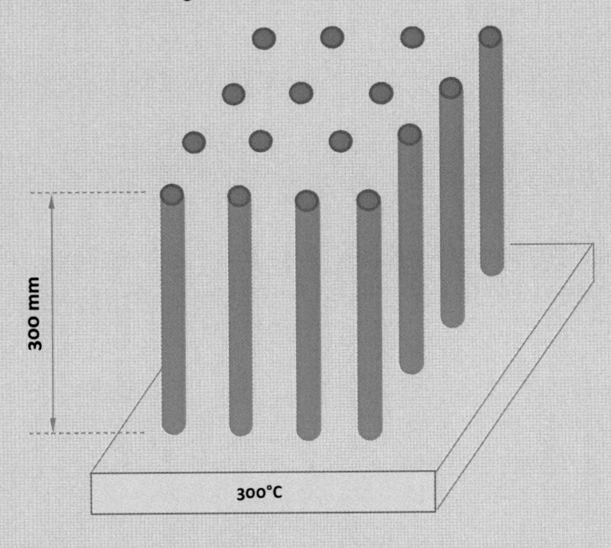

Solution

We shall discretize the rod into three FEs (more elements should increase the accuracy of the results, however, aim of the example is to teach how we can use the finite element method to solve 1D heat transfer problems).

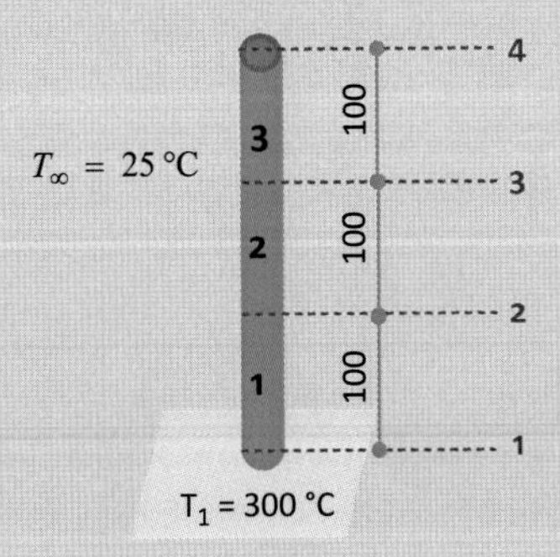

Step 1: Derivation of the element matrices

Element 1, nodes 1 and 2

In the element 1, there is convection only through its lateral surface. Therefore, the convection part of the "stiffness" matrix is given by Equation (12.50):

$$\left[k_{CV}^{1}\right] = \frac{hPL}{6}\begin{bmatrix} 2 & 1 \\ 1 & 2 \end{bmatrix} \tag{12.e1}$$

where P is the rod's perimeter: $P = \pi d = 120\pi \times 10^{-3}$ m and L is the element's length: $L = 100 \times 10^{-3}$ m.

Therefore, the convection part of the stiffness matrix is

$$[k_{CV}^1] = \frac{(730) \times (120\pi \times 10^{-3}) \times (100 \times 10^{-3})}{6} \begin{bmatrix} 2 & 1 \\ 1 & 2 \end{bmatrix} \quad (12.e2)$$

The conduction part of the "stiffness" matrix of this element is given by Equation (12.49):

$$[k_{CD}] = \frac{Ak}{L} \begin{bmatrix} 1 & -1 \\ -1 & 1 \end{bmatrix} = \frac{(\pi d^2/4) \times 62}{100 \times 10^{-3}} \begin{bmatrix} 1 & -1 \\ -1 & 1 \end{bmatrix} \quad (12.e3)$$

Therefore, the "stiffness" matrix of the element 1 is

$$[k_1] = [k_{CV}^1] + [k_{CD}] = \begin{bmatrix} 16.1855 & -2.425 \\ -2.425 & 16.1855 \end{bmatrix} \quad (12.e4)$$

Element 2, nodes 2 and 3

Element 2 has the same properties and boundary conditions as element 1. Therefore:

$$[k_2] = [k_1] = \begin{bmatrix} 16.1855 & -2.425 \\ -2.425 & 16.1855 \end{bmatrix} \quad (12.e5)$$

Element 3, nodes 3 and 4

Element 3 has the same geometry and conduction coefficient as elements 1 and 2. However, in addition to the lateral surface it has convection heat transfer from the right end surface too. Therefore, apart from Equations (12.49) and (12.50), we should take into account and Equation (12.54) as well:

$$[k_3] = \frac{(\pi d^2/4)k}{L} \begin{bmatrix} 1 & -1 \\ -1 & 1 \end{bmatrix} + \frac{h(\pi d)L}{6} \begin{bmatrix} 2 & 1 \\ 1 & 2 \end{bmatrix} + h(\pi d^2/4) \begin{bmatrix} 0 & 0 \\ 0 & 1 \end{bmatrix} \quad (12.e6)$$

or

$$[k_3] = \begin{bmatrix} 16.1855 & -2.425 \\ -2.425 & 24.442 \end{bmatrix} \quad (12.e7)$$

Step 2: Expansion of the element equations to the degrees of freedom of the structure

Element 1, nodes 1 and 2

$$\begin{bmatrix} 16.1855 & -2.425 & 0 & 0 \\ -2.425 & 16.1855 & 0 & 0 \\ 0 & 0 & 0 & 0 \\ 0 & 0 & 0 & 0 \end{bmatrix} \begin{Bmatrix} T_1 \\ T_2 \\ T_3 \\ T_4 \end{Bmatrix} = \begin{Bmatrix} r_1 \\ r_2 \\ r_3 \\ r_4 \end{Bmatrix} \quad (12.e8)$$

Element 2, nodes 2 and 3

$$\begin{bmatrix} 0 & 0 & 0 & 0 \\ 0 & 16.1855 & -2.425 & 0 \\ 0 & -2.425 & 16.1855 & 0 \\ 0 & 0 & 0 & 0 \end{bmatrix} \begin{Bmatrix} T_1 \\ T_2 \\ T_3 \\ T_4 \end{Bmatrix} = \begin{Bmatrix} r_1 \\ r_2 \\ r_3 \\ r_4 \end{Bmatrix} \quad (12.e9)$$

Continued

EXAMPLE 2: HEAT TRANSFER IN A HEAT EXCHANGER—CONT'D

Element 3, nodes 3 and 4

$$\begin{bmatrix} 0 & 0 & 0 & 0 \\ 0 & 0 & 0 & 0 \\ 0 & 0 & 16.1855 & -2.425 \\ 0 & 0 & -2.425 & 24.442 \end{bmatrix} \begin{Bmatrix} T_1 \\ T_2 \\ T_3 \\ T_4 \end{Bmatrix} = \begin{Bmatrix} r_1 \\ r_2 \\ r_3 \\ r_4 \end{Bmatrix} \tag{12.e10}$$

Step 3: Assembly of the element equations

Superposition of the element equations (Equations 12.e8–12.e10) yields

$$\begin{bmatrix} 16.1855 & -2.425 & 0 & 0 \\ -2.425 & 16.1855+16.1855 & -2.425 & 0 \\ 0 & -2.425 & 16.1855+16.1855 & -2.425 \\ 0 & 0 & -2.425 & 24.442 \end{bmatrix} \begin{Bmatrix} T_1 \\ T_2 \\ T_3 \\ T_4 \end{Bmatrix} = \begin{Bmatrix} R_1 \\ R_2 \\ R_3 \\ R_4 \end{Bmatrix} \tag{12.e11}$$

or

$$\begin{bmatrix} 16.1855 & -2.425 & 0 & 0 \\ -2.425 & 32.371 & -2.425 & 0 \\ 0 & -2.425 & 32.371 & -2.425 \\ 0 & 0 & -2.425 & 24.442 \end{bmatrix} \begin{Bmatrix} T_1 \\ T_2 \\ T_3 \\ T_4 \end{Bmatrix} = \begin{Bmatrix} R_1 \\ R_2 \\ R_3 \\ R_4 \end{Bmatrix} \tag{12.e12}$$

Step 4: Derivation of the field values

An alternative formulation of Equation (12.e12) is the following:

$$\begin{bmatrix} 16.1855 & -2.425 & 0 & 0 & -1 & 0 & 0 & 0 \\ -2.425 & 32.371 & -2.425 & 0 & 0 & -1 & 0 & 0 \\ 0 & -2.425 & 32.371 & -2.425 & 0 & 0 & -1 & 0 \\ 0 & 0 & -2.425 & 24.442 & 0 & 0 & 0 & -1 \end{bmatrix} \begin{Bmatrix} T_1 \\ T_2 \\ T_3 \\ T_4 \\ R_1 \\ R_2 \\ R_3 \\ R_4 \end{Bmatrix} = \begin{Bmatrix} 0 \\ 0 \\ 0 \\ 0 \end{Bmatrix} \tag{12.e13}$$

Boundary conditions. We are looking for four boundary conditions in order to complete the above 4×8 algebraic system of equations. One boundary condition is given by the problem;

$$T_1 = 300\,^\circ\mathrm{C} \tag{12.e14}$$

The remaining three boundary conditions are the "forces" at the nodes 2, 3, and 4.

"Forces" at the ends of the element 1

Element 1 has convection from its lateral surface and from the end located on the node 1. Since this end is bonded on a plate with unknown ambient temperature conditions, the heat convection flux at this end is unknown. Therefore, the "force" vector for the element 1 can be obtained from Equation (12.60):

$$\{R_1\} = \{R_1^{l}\} + \{R_1^{le}\} \tag{12.e15}$$

where $\left\{\widetilde{R_1^{l}}\right\}$ is given by Equation (12.51) and $\{R_1^{le}\}$ is given by Equation (12.59).

Since there is not a conduction part or internal energy generation, implementation of Equation (12.51) for element 1 yields

$$\{R_1^{\mathrm{l}}\} = \frac{hT_{\infty}PL}{2}\begin{Bmatrix} 1 \\ 1 \end{Bmatrix} = \begin{Bmatrix} 344 \\ 344 \end{Bmatrix} \quad (12.e16)$$

Since the convection at the node 1 of the element 1 is unknown, the following equation can be written as:

$$\{R_1^{\mathrm{le}}\} = R_1^{\mathrm{le}}\begin{Bmatrix} 1 \\ 0 \end{Bmatrix} \quad (12.e17)$$

Then, taking into account Equations (12.e15)–(12.e17), the "force" vector for the element 1 is

$$\{R_1\} = \begin{Bmatrix} R_1^1 \\ R_1^2 \end{Bmatrix} = \begin{Bmatrix} 344 + R_1^{\mathrm{le}} \\ 344 \end{Bmatrix} \quad (12.e18)$$

The symbol R_i^j refers to the "force" at the node j of the element i.

"Forces" at the ends of the element 2

Element 2 has convection only on its lateral surface. Therefore, taking into account the procedure adopted for element 1, the following "force" vector for element 2 can be obtained:

$$\{R_2\} = \begin{Bmatrix} R_2^2 \\ R_2^3 \end{Bmatrix} = \begin{Bmatrix} 344 \\ 344 \end{Bmatrix} \quad (12.e19)$$

"Forces" at the ends of the element 3

Element 3 has convection from its lateral surface and from its right end. Therefore, the "force" vector for this element can be derived from Equation (12.60):

$$\{R_3\} = \{R_3^{\mathrm{l}}\} + \{R_3^{\mathrm{re}}\} \quad (12.e20)$$

where

$$\{R_3^{\mathrm{l}}\} = \begin{Bmatrix} 344 \\ 344 \end{Bmatrix} \quad (12.e21)$$

and according to Equation (12.58),

$$\{R_3^{\mathrm{re}}\} = hT_{\infty}A\begin{Bmatrix} 0 \\ 1 \end{Bmatrix} = 730 \times 25 \times \left[\pi x \left(120 \times 10^{-3}\right)^2/4\right]\begin{Bmatrix} 0 \\ 1 \end{Bmatrix} \quad (12.e22)$$

Then, Equation (12.e20) yields

$$\{R_3\} = \begin{Bmatrix} R_3^3 \\ R_3^4 \end{Bmatrix} = \begin{Bmatrix} 344 \\ 550.4 \end{Bmatrix} \quad (12.e23)$$

Performing superposition of the results given in Equations (12.e18), (12.e19), and (12.e23), the following "forces" at nodes 1, 2, 3, and 4 can be obtained:

$$R_1 = R_1^1 = 344 + R_1^{\mathrm{le}} \ (\mathrm{W}) \quad (12.e24)$$

Continued

EXAMPLE 2: HEAT TRANSFER IN A HEAT EXCHANGER—CONT'D

$$R_2 = R_1^2 + R_2^2 = 344 + 344 = 688\,\text{W} \tag{12.e25}$$

$$R_3 = R_2^3 + R_3^3 = 344 + 344 = 688\,\text{W} \tag{12.e26}$$

$$R_4 = R_3^4 = 550.4\,\text{W} \tag{12.e27}$$

Taking into account Equations (12.e14) and (12.e25)–(12.e27), the boundary conditions can be written in the following matrix form:

$$\begin{bmatrix} 1 & 0 & 0 & 0 & 0 & 0 & 0 & 0 \\ 0 & 0 & 0 & 0 & 0 & 1 & 0 & 0 \\ 0 & 0 & 0 & 0 & 0 & 0 & 1 & 0 \\ 0 & 0 & 0 & 0 & 0 & 0 & 0 & 1 \end{bmatrix} \begin{Bmatrix} T_1 \\ T_2 \\ T_3 \\ T_4 \\ R_1 \\ R_2 \\ R_3 \\ R_4 \end{Bmatrix} = \begin{Bmatrix} 300 \\ 688 \\ 688 \\ 550.4 \end{Bmatrix} \tag{12.e28}$$

It should be noticed that since the value of the "force" R_1 is unknown, we have used the boundary conditions given by Equations (12.e14) and (12.e25)–(12.e27). The value of R_1 and the values of T_2, T_3, T_4 will be obtained by the solution of the following algebraic system, which is the superposition of the matrix equations (Equations 12.e13 and 12.e28):

$$\begin{bmatrix} 16.1855 & -2.425 & 0 & 0 & -1 & 0 & 0 & 0 \\ -2.425 & 32.371 & -2.425 & 0 & 0 & -1 & 0 & 0 \\ 0 & -2.425 & 32.371 & -2.425 & 0 & 0 & -1 & 0 \\ 0 & 0 & -2.425 & 24.442 & 0 & 0 & 0 & -1 \\ 1 & 0 & 0 & 0 & 0 & 0 & 0 & 0 \\ 0 & 0 & 0 & 0 & 0 & 1 & 0 & 0 \\ 0 & 0 & 0 & 0 & 0 & 0 & 1 & 0 \\ 0 & 0 & 0 & 0 & 0 & 0 & 0 & 1 \end{bmatrix} \begin{Bmatrix} T_1 \\ T_2 \\ T_3 \\ T_4 \\ R_1 \\ R_2 \\ R_3 \\ R_4 \end{Bmatrix} = \begin{Bmatrix} 0 \\ 0 \\ 0 \\ 0 \\ 300 \\ 688 \\ 688 \\ 550.4 \end{Bmatrix} \tag{12.e29}$$

The solution of the above algebraic equation yields

$$\begin{Bmatrix} T_1 \\ T_2 \\ T_3 \\ T_4 \\ R_1 \\ R_2 \\ R_3 \\ R_4 \end{Bmatrix} = \begin{Bmatrix} 300 \\ 45.72 \\ 26.56 \\ 25.15 \\ 4744.76 \\ 688.00 \\ 688.00 \\ 550.40 \end{Bmatrix} \tag{12.e30}$$

The rate of the heat flow through element 1 can be obtained by Equation (12.1):

$$f_x = -k\frac{\partial T}{\partial x} \tag{12.e31}$$

The temperature gradient $\{DT\}$ (see Equation 12.22) is correlated to the nodal temperatures $\{T_e\} = [T_1\ \ T_2]^{\mathrm{T}}$ by Equation (12.42):

$$\{DT\} = [B]\{T_e\} \tag{12.e32}$$

Where, according to Equation (12.43)

$$[B] = \{D\}[N] \tag{12.e33}$$

In the above equation, $\{D\}$ is the operator given by Equation (12.21), and

$$[N] = \left[1 - \frac{x}{L}\ \ \frac{x}{L}\right] \tag{12.e34}$$

Therefore, taking into account Equations (12.e31)–(12.e34) it can be written:

$$f_x = -k\frac{\partial}{\partial x}\left[1 - \frac{x}{L}\ \ \frac{x}{L}\right]\begin{Bmatrix} T_1 \\ T_2 \end{Bmatrix} \tag{12.e35}$$

or

$$f_x = -k\left[-\frac{1}{L}\ \ \frac{1}{L}\right]\begin{Bmatrix} T_1 \\ T_2 \end{Bmatrix} \tag{12.e36}$$

yielding

$$f_x = -\left[-\frac{1}{100 \times 10^{-3}}\ \ \frac{1}{100 \times 10^{-3}}\right]\begin{Bmatrix} 300 \\ 45.72 \end{Bmatrix} = 157{,}653\,\mathrm{W/m^2} \tag{12.e37}$$

12.3.2 TWO-DIMENSIONAL AND THREE-DIMENSIONAL HEAT TRANSFER MODELING USING A VARIATIONAL METHOD

We recall from the previous section that the minimization of the functional Π yields the following "stiffness" matrix

$$[k] = \underbrace{\int_V [B]^{\mathrm{T}}[\kappa][B]\,\mathrm{d}V}_{\substack{\text{conduction part} \\ [k_{\mathrm{CD}}]}} + \underbrace{\int_{S_{\mathrm{CV}}} [N]^{\mathrm{T}}[N]h\,\mathrm{d}S}_{\substack{\text{convection part} \\ [k_{\mathrm{CV}}]}} \tag{12.61}$$

where

$$[B] = \begin{Bmatrix} \partial/\partial x \\ \partial/\partial y \\ \partial/\partial z \end{Bmatrix}[N] \tag{12.62}$$

and $[\kappa]$ is a matrix containing the conduction coefficients given by Equation (12.19). It should be noted that S_{CV} denotes the area where heat convection takes place.

Therefore, the formulation of the "stiffness" matrix depends on the functional form of the matrix $[N]$ containing the shape functions. For common types of 2D and 3D element, for example, linear triangular elements, quadratic triangular elements, bilinear rectangular elements, tetrahedral solid elements, eight node rectangular solid elements, and isoparametric bilinear quadrilateral elements, analytical expressions of the shape functions $[N]$ can be obtained from Chapter 10. However, it should be noted that unlike the 2D or 3D structural elements of Chapter 10, where each node contains two (or three) degrees of freedom on x-y (or x-y-z) directions, in 2D or 3D heat transfer elements only a single scalar value (nodal temperature) is the unknown variable at each node. In this chapter, the required formulae for "stiffness" matrix derivation of common 2D and 3D heat transfer elements will be summarized. For the solution of 2D or 3D heat transfer problems, apart from the "stiffness" matrix, the "force" vector is also required. As it has already been stated, the "force" vector is given by the following equation:

$$\{R\} = \underbrace{\int_{S_f} [N]^T f_{\text{bound}}\, dS}_{\substack{\text{conduction part} \\ \{R_f\}}} + \underbrace{\int_{S_{CV}} [N]^T h T_\infty\, dS}_{\substack{\text{convection part} \\ \{R_{CV}\}}} + \underbrace{\int_V [N]^T \dot{g}_{\text{gen}}\, dV}_{\substack{\text{internal heat} \\ \text{generation part} \\ \{R_{\text{gen}}\}}} \tag{12.63}$$

It should be noted that S_f denotes the area where heat flux f_{bound} is applied, and S_{CV} denotes the area subjected to convection losses $h(T - T_\infty)$. We cannot specify applied heat flux and convection losses on the same surface because they cannot occur simultaneously.

Implementation of Equation (12.63) on the several common types of 2D or 3D elements is based on the form of the matrix of shape functions $[N]$.

Linear triangular heat transfer element

Taking into account the procedure described in Chapter 10, the following steps should be followed in order to derive the "stiffness" matrix and the "force" vector for the linear triangular element (Figure 12.9). In the subsequent formulae, t denotes the thickness of the element, and $\dot{q}_{\text{gen}}$ denotes the thermal power per unit volume (W/m^3).

Step 1: Derivation of the coordinates vector

$$[\Xi] = \begin{bmatrix} 1 & x_i & y_i \\ 1 & x_j & y_j \\ 1 & x_m & y_m \end{bmatrix} \tag{12.64}$$

Step 2: Derivation of the thermal conductivity matrix

We recall that the thermal conductivity matrix of an orthotropic material is given by the following equation

$$[\kappa] = \begin{bmatrix} k_x & k_{xy} \\ k_{xy} & k_y \end{bmatrix} \tag{12.65}$$

When conductivity coefficients are expressed with respect to the principal coordinate system X-Y (which is different from the coordinate system x-y), then, the matrix $[\kappa]$ can be obtained from

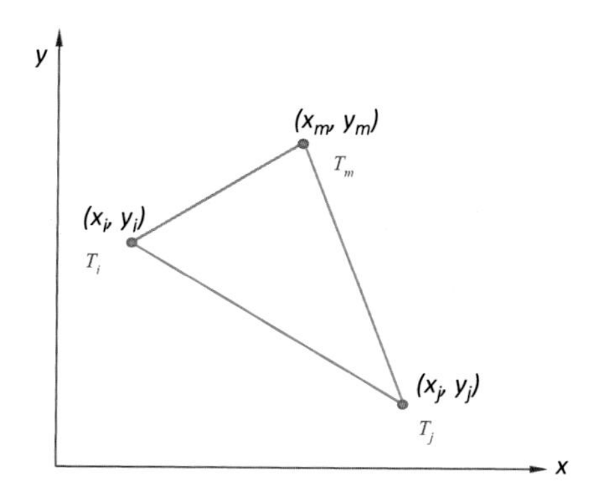

FIGURE 12.9

Coordinate system of a linear triangular heat transfer element.

Equation (12.19). When the material is isotropic, the assumptions $k_{xy} = 0$ and $k_x = k_y = k$ should be applied in Equation (12.65).

Step 3: Derivation of the vector of the nodal temperatures

$$\{T_e\} = \begin{Bmatrix} T_i \\ T_j \\ T_m \end{Bmatrix} \tag{12.66}$$

Step 4: Derivation of the matrix [B]

The matrix $[B]$ is given in Equation (12.62). The shape functions $[N] = [N_i \ N_j \ N_m]$ can be obtained from Equations (10.54)–(10.56), where $a_i, a_j, a_m, \beta_i, \beta_j, \beta_m, \gamma_i, \gamma_j, \gamma_m$ are constant parameters derived from the nodal coordinates (see Equations 10.40–10.48). After algebraic operations, the matrix $[B]$ can be derived as:

$$[B] = \begin{bmatrix} 0 & 1 & 0 \\ 0 & 0 & 1 \end{bmatrix} [\Xi]^{-1} \tag{12.67}$$

Step 5: Derivation of the elements area

$$A = \frac{1}{2} \det [\Xi] \tag{12.68}$$

Step 6: Derivation of the "stiffness" matrix

Since the thickness t of 2D elements is usually constant and according to Equation (12.67) the matrix $[B]$ is also constant, then, taking into account Equation (12.61) the conduction part of the "stiffness" matrix can be derived by the following equation:

$$[k_{CD}] = [B]^{\mathrm{T}} [\kappa] [B] A t \tag{12.69}$$

Apart from the conduction part, the convection part should also be derived. Unlike the conduction part, the convection part of Equation (12.61) contains the nonconstant matrices $[N]^{\mathrm{T}}$ and $[N]$. The integration over the triangles area of the matrix product $[N]^{\mathrm{T}}[N]$ is more complicated. The integral

$$[k_{\mathrm{CV}}]=\int_S [N]^{\mathrm{T}}[N]\mathrm{d}S \tag{12.70}$$

should be carried out by the following formula:

$$[k_{\mathrm{CV}}]=h\int_{x=x_i}^{x_m}\left(\int_{y=y_{ij}}^{y_{im}}[N]^{\mathrm{T}}[N]\mathrm{d}y\right)\mathrm{d}x+h\int_{x=x_m}^{x_j}\left(\int_{y=y_{ij}}^{y_{mj}}[N]^{\mathrm{T}}[N]\mathrm{d}y\right)\mathrm{d}x \tag{12.71}$$

where the integration limits y_{ij}, y_{im}, y_{mj} represent (see Figure 12.10) the equations of the lines $i-j, i-m, m-j$, respectively:

$$y_{ij}=y_i+\frac{(x-x_i)(y_j-y_i)}{x_j-x_i} \tag{12.72}$$

$$y_{im}=y_i+\frac{(x-x_i)(y_m-y_i)}{x_m-x_i} \tag{12.73}$$

$$y_{mj}=y_m+\frac{(x-x_m)(y_j-y_m)}{x_j-x_m} \tag{12.74}$$

Making the required algebraic operations, the following formula can be obtained:

$$[k_{\mathrm{CV}}]=\frac{hA}{12}\begin{bmatrix}2 & 1 & 1\\ 1 & 2 & 1\\ 1 & 1 & 2\end{bmatrix} \tag{12.75}$$

The above formula is valid in the case that we have heat convection in the three sides of the triangle. When heat convection occurs in individual sides of the triangle, the following formula for the convection part of the "stiffness" matrix should be adopted:

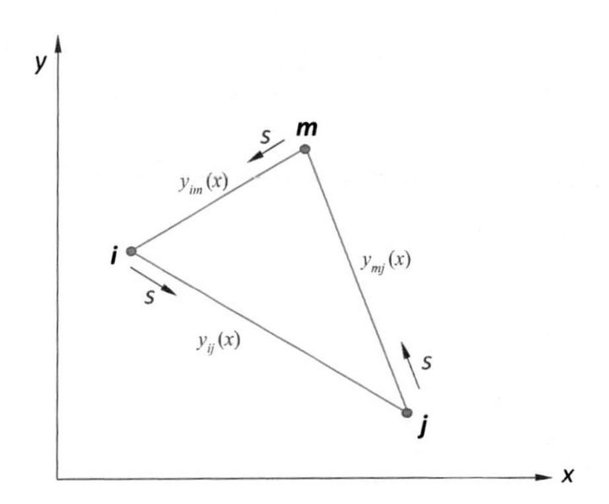

FIGURE 12.10

Representation of the sides of the triangle by the functions $y_{ij}(x), y_{mj}(x), y_{im}(x)$.

Convection of the sides i-j, j-m, *and* m-i

Let the symbol s denote the independent variable along the sides of the triangle (Figure 12.10). The matrix containing the shape functions along the side i-j of the triangle should be linear, that is,

$$[N^s] = \left[1 - \frac{s}{L_{ij}} \ \frac{s}{L_{ij}} \ 0\right] \tag{12.76}$$

Therefore,

$$\left[k_{CV}^{ij}\right] = \int_{s=0}^{L_{ij}} [N^s]^{\mathrm{T}}[N^s] h \mathrm{d}s = \frac{hL_{ij}}{6}\begin{bmatrix} 2 & 1 & 0 \\ 1 & 2 & 0 \\ 0 & 0 & 0 \end{bmatrix} \tag{12.77}$$

where L_{ij} is the length of the side i-j.

Following the same concept, the convection part of the "stiffness" matrix for the sides j-m and m-i is

$$\left[k_{CV}^{jm}\right] = \frac{hL_{jm}}{6}\begin{bmatrix} 0 & 0 & 0 \\ 0 & 2 & 1 \\ 0 & 1 & 2 \end{bmatrix} \tag{12.78}$$

and

$$\left[k_{CV}^{mi}\right] = \frac{hL_{mi}}{6}\begin{bmatrix} 2 & 0 & 1 \\ 0 & 0 & 0 \\ 1 & 0 & 2 \end{bmatrix} \tag{12.79}$$

Step 7: Derivation of the "load" vectors

Taking into account the expressions for the matrix $[N]$, the "load" vectors can be derived from the integrals of Equation (12.63). Using the following integration

$$\int_S [T]^{\mathrm{T}} \mathrm{d}S = \int_{x=x_i}^{x_m} \left(\int_{y=y_{ij}}^{y_{im}} [N]^{\mathrm{T}} \mathrm{d}y\right) \mathrm{d}x + \int_{x=x_m}^{x_j} \left(\int_{y=y_{ij}}^{y_{mj}} [N]^{\mathrm{T}} \mathrm{d}y\right) \mathrm{d}x \tag{12.80}$$

or

$$\int_S [T]^{\mathrm{T}} \mathrm{d}S = \frac{A}{3}\begin{Bmatrix} 1 \\ 1 \\ 1 \end{Bmatrix} \tag{12.81}$$

the conduction, convection, and heat transfer generation parts of the "force" vector $\{R\}$ have the following form:

$$\{R_{\mathrm{f}}\} = \frac{f_{\mathrm{bound}} A}{3}\begin{Bmatrix} 1 \\ 1 \\ 1 \end{Bmatrix} \tag{12.82}$$

$$\{R_{\mathrm{CV}}\} = \frac{hT_{\infty} A}{3}\begin{Bmatrix} 1 \\ 1 \\ 1 \end{Bmatrix} \tag{12.83}$$

$$\{R_{\text{gen}}\} = \frac{\dot{g}_{\text{gen}} V}{3} \begin{Bmatrix} 1 \\ 1 \\ 1 \end{Bmatrix} \tag{12.84}$$

The above equation means that heat is generated in three equal parts to the nodes.

Equations (12.82) and (12.83) assume that the applied heat flux f_{bound} or the heat convection $h(T - T_{\infty})$ take place on the three sides of the triangle. It should be remembered that the heat flux and the heat convection losses cannot occur simultaneously on the same surface. For the derivation of the load matrices for the heat convection losses or for the applied thermal flux on individual sides, we should use the shape functions $[N^s]$ given in Equation (12.76). Thus, the vectors $\{R_{\text{f}}\}$ and $\{R_{\text{CV}}\}$ can be obtained by the following formulae:

Applied heat flux on the side i-j

$$\left\{R_f^{ij}\right\} = \frac{f_{\text{bound}} L_{ij} t}{2} \begin{Bmatrix} 1 \\ 1 \\ 0 \end{Bmatrix} \tag{12.85}$$

Applied heat flux on the side j-m

$$\left\{R_f^{jm}\right\} = \frac{f_{\text{bound}} L_{jm} t}{2} \begin{Bmatrix} 0 \\ 1 \\ 1 \end{Bmatrix} \tag{12.86}$$

Applied heat flux on the side m-i

$$\left\{R_f^{mi}\right\} = \frac{f_{\text{bound}} L_{mi} t}{2} \begin{Bmatrix} 1 \\ 0 \\ 1 \end{Bmatrix} \tag{12.87}$$

Heat convection losses on the side i-j

$$\left\{R_{\text{CV}}^{ij}\right\} = \frac{h T_{\infty} t}{2} \begin{Bmatrix} 1 \\ 1 \\ 0 \end{Bmatrix} \tag{12.88}$$

Heat convection losses on the side j-m

$$\left\{R_{\text{CV}}^{jm}\right\} = \frac{h T_{\infty} t}{2} \begin{Bmatrix} 0 \\ 1 \\ 1 \end{Bmatrix} \tag{12.89}$$

Heat convection losses on the side m-i

$$\left\{R_{\text{CV}}^{mi}\right\} = \frac{h T_{\infty} t}{2} \begin{Bmatrix} 1 \\ 0 \\ 1 \end{Bmatrix} \tag{12.90}$$

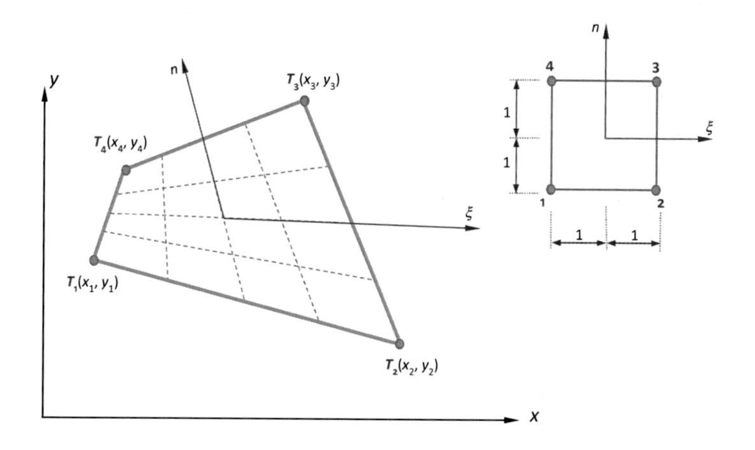

FIGURE 12.11

Coordinates of a bilinear quadrilateral heat transfer isoparametric element.

Bilinear quadrilateral heat transfer isoparametric element

Taking into account Equations (12.60)–(12.63), as well as the procedure reported in Chapter 10, the derivation of the "stiffness" matrix and the "force" matrix for the element shown in Figure 12.11 can be achieved by the following steps.

Step 1: Correlation of the variables x,y to the variables ξ,n and to the nodal coordinates (x_i, y_i)

Taking into account Equations (10.166)–(10.171), the variables x,y can be correlated to the variables ξ, n and to the nodal coordinates (x_1, y_1), (x_2, y_2), (x_3, y_3), (x_4, y_4) as follows:

$$\begin{Bmatrix} x \\ y \end{Bmatrix} = \begin{bmatrix} N_1 & 0 & N_2 & 0 & N_3 & 0 & N_4 & 0 \\ 0 & N_1 & 0 & N_2 & 0 & N_3 & 0 & N_4 \end{bmatrix} \begin{Bmatrix} x_1 \\ y_1 \\ x_2 \\ y_2 \\ x_3 \\ y_3 \\ x_4 \\ y_4 \end{Bmatrix} \tag{12.91}$$

where

$$N_1 = \frac{(1-\xi)(1-\eta)}{4} \tag{12.92}$$

$$N_2 = \frac{(1+\xi)(1-\eta)}{4} \tag{12.93}$$

$$N_3 = \frac{(1+\xi)(1+\eta)}{4} \tag{12.94}$$

$$N_4 = \frac{(1-\xi)(1+\eta)}{4} \tag{12.95}$$

Step 2: Derivation of the Jacobian matrix [J]

For the derivation of the Jacobian matrix $[J]$ we recall Equation (10.176):

$$[J] = \begin{bmatrix} \partial x/\partial \xi & \partial y/\partial \xi \\ \partial x/\partial n & \partial y/\partial n \end{bmatrix} \tag{12.96}$$

The functions for x and y can be obtained by Equation (12.91).

Step 3: Derivation of the matrix [B]

According to Chapter 10, the matrix $[B]$ is given by the following equation:

$$[B] = [J]^{-1} \begin{bmatrix} \partial/\partial \xi \\ \partial/\partial n \end{bmatrix} [N_1 \ \ N_2 \ \ N_3 \ \ N_4] \tag{12.97}$$

Step 4: Derivation of the conduction "stiffness" and "force" matrices

The "stiffness" matrix can be derived from the following (already known from Chapter 10) equation:

$$[k_{\mathrm{CD}}] = \int_{-1}^{1}\int_{-1}^{1} [B]^{\mathrm{T}}[\kappa][B] tJ \mathrm{d}\xi \mathrm{d}n \tag{12.98}$$

In the above integral the parameter J (Jacobian) is

$$J = \det[J] \tag{12.99}$$

The "force" matrix can be obtained by the following equation:

$$[R_{\mathrm{gen}}] = \int_{-1}^{1} [N]^{\mathrm{T}} \dot{g}_{gen} tJ \mathrm{d}\xi \mathrm{d}n \tag{12.100}$$

where

$$[N] = [N_1 \ \ N_2 \ \ N_3 \ \ N_4] \tag{12.101}$$

Eight-node isoparametric heat transfer 2D element

The "stiffness" matrix and the "force" matrix for the element shown in Figure 12.12 will be derived following a similar procedure to the bilinear quadrilateral heat transfer isoparametric element. However, we have now eight nodes and the shape functions N_1, N_2, ..., N_8 are given from different formulae.

Step 1: Correlation of the variables x, y to the variables ξ, n and to the nodal coordinates (x_i, y_i)

Following a similar procedure to that of the previous element, the variables x, y can be correlated to the variables ξ, n and to the nodal coordinates (x_i, y_i), $i=1,2,\ldots,8$ by the following formulae:

$$x = \sum_{i=1}^{8} N_i x_i \tag{12.102}$$

$$y = \sum_{i=1}^{8} N_i y_i \tag{12.103}$$

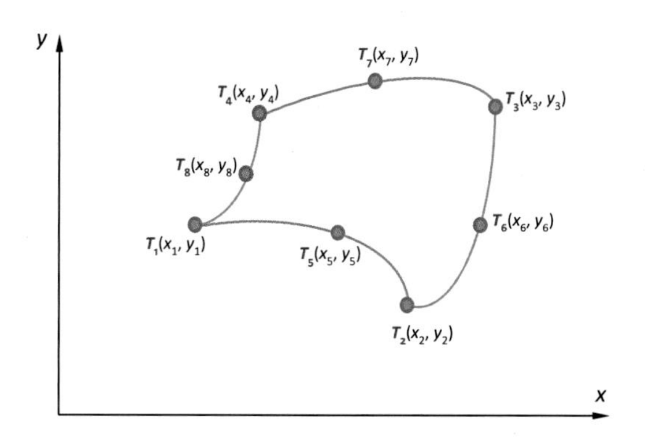

FIGURE 12.12

Coordinates of an eight-node isoparametric heat transfer 2D element.

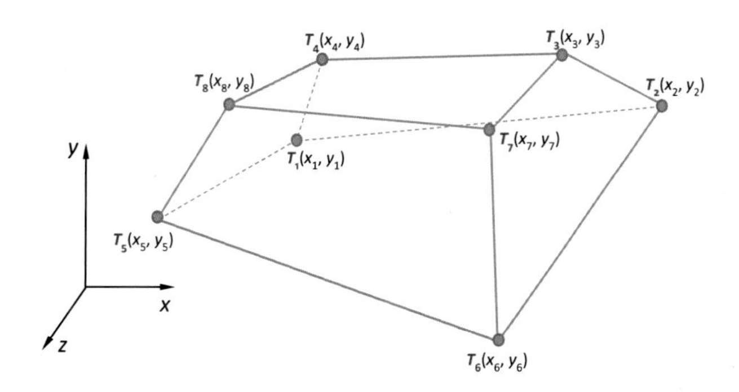

FIGURE 12.13

Coordinates of an eight-node isoparametric heat transfer 3D element.

where

$$N_1 = -\frac{1}{4}(1-\xi)(1-n)(1+\xi+n) \tag{12.104}$$

$$N_2 = -\frac{1}{4}(1+\xi)(1-n)(1-\xi+n) \tag{12.105}$$

$$N_3 = -\frac{1}{4}(1+\xi)(1+n)(1-\xi-n) \tag{12.106}$$

$$N_4 = -\frac{1}{4}(1-\xi)(1+n)(1+\xi-n) \tag{12.107}$$

$$N_5 = \frac{1}{2}(1-\xi^2)(1-n) \tag{12.108}$$

$$N_6 = \frac{1}{2}(1+\xi)(1-n^2) \tag{12.109}$$

$$N_7 = \frac{1}{2}(1-\xi^2)(1+n) \quad (12.110)$$

$$N_8 = \frac{1}{2}(1-\xi)(1-n^2) \quad (12.111)$$

Step 2: Derivation of the Jacobian matrix [J]

Taking into account the Equations (12.102) and (12.103), the Jacobian matrix can be derived by Equation (12.96).

Step 3: Derivation of the matrix [B]

The matrix [B] should be derived by the following equation:

$$[B] = [J]^{-1} \begin{bmatrix} \partial/\partial\xi \\ \partial/\partial n \end{bmatrix} [N_1 \;\; N_2 \;\; N_3 \;\; N_4 \;\; N_5 \;\; N_6 \;\; N_7 \;\; N_8] \quad (12.112)$$

Step 4: Derivation of the conduction "stiffness" and "force" vector $\{R_{gen}\}$

The conduction "stiffness" matrix and the "force" vector $\{R_{\text{gen}}\}$ will be derived from Equations (12.98) and (12.100), respectively. The matrices [N] and [B] are given by Equations (12.104)–(12.111) and (12.112).

Eight-node isoparametric heat transfer 3D element

Adopting the same concept with the previous 2D elements, the "stiffness" and "force" matrices for the element shown in Figure 12.13 can be derived as follows:

Step 1: Correlation of the variables x, y, z to the variables ξ, n, ζ and to the nodal coordinates (x_i, y_i, z_i), i = 1,2,…,8

$$x = \sum_{i=1}^{8} N_i x_i \quad (12.113)$$

$$y = \sum_{i=1}^{8} N_i y_i \quad (12.114)$$

$$z = \sum_{i=1}^{8} N_i z_i \quad (12.115)$$

where

$$N_1 = \frac{1}{8}(1-\xi)(1-n)(1-\zeta) \quad (12.116)$$

$$N_2 = \frac{1}{8}(1+\xi)(1-n)(1-\zeta) \quad (12.117)$$

$$N_3 = \frac{1}{8}(1+\xi)(1+n)(1-\zeta) \quad (12.118)$$

$$N_4 = \frac{1}{8}(1-\xi)(1+n)(1-\zeta) \tag{12.119}$$

$$N_5 = \frac{1}{8}(1-\xi)(1-n)(1+\zeta) \tag{12.120}$$

$$N_6 = \frac{1}{8}(1+\xi)(1-n)(1+\zeta) \tag{12.121}$$

$$N_7 = \frac{1}{8}(1+\xi)(1+n)(1+\zeta) \tag{12.122}$$

$$N_8 = \frac{1}{8}(1-\xi)(1+n)(1+\zeta) \tag{12.123}$$

Step 2: Derivation of the Jacobian matrix [J]

For 3D elements, the Jacobian matrix $[B]$ can be derived by the following formula:

$$[J] = \begin{bmatrix} \partial x/\partial \xi & \partial y/\partial \xi & \partial z/\partial \xi \\ \partial x/\partial n & \partial y/\partial n & \partial z/\partial n \\ \partial x/\partial \zeta & \partial y/\partial \zeta & \partial z/\partial \zeta \end{bmatrix} \tag{12.124}$$

where x, y, and z are given in Equations (12.113)–(12.115).

Step 3: Derivation of the matrix [B]

The matrix $[B]$ for the eight-node 3D element can be derived by the following equation:

$$[B] = [J]^{-1} \begin{Bmatrix} \partial/\partial \xi \\ \partial/\partial n \\ \partial/\partial \zeta \end{Bmatrix} [N_1 \; N_2 \; N_3 \; N_4 \; N_5 \; N_6 \; N_7 \; N_8] \tag{12.125}$$

Step 4: Derivation of the conduction "stiffness" and "force" vector {R_{gen}}

Since now the element is 3D, before the derivation of the conduction "stiffness" matrix, we should derive the thermal conductivity matrix $[\kappa]$. For an orthotropic material, the matrix $[\kappa]$ has the form:

$$[\kappa] = \begin{bmatrix} k_x & k_{xy} & k_{xz} \\ k_{yx} & k_y & k_{yz} \\ k_{zx} & k_{zy} & k_{zz} \end{bmatrix} \tag{12.126}$$

If the conductivity coefficients are expressed with respect to the principal coordinate system X, Y, Z which is different than the coordinate system x, y, z, the matrix $[\kappa]$ can be obtained from Equation (12.26). For isotropic materials the substitutions $k_{xy} = k_{xz} = k_{yx} = k_{yz} = k_{zx} = k_{zy} = 0$ and $k_x = k_y = k_z = k$ should be applied in Equation (12.126).

For axisymmetric orthotropic solids, the matrix $[\kappa]$ can be derived by Equation (12.31). Then, using Equations (12.124)–(12.126), the conduction "stiffness" matrix and the "force" vector $\{R_{\text{gen}}\}$ can be derived by the following equations:

$$[\kappa] = \int_{-1}^{1}\int_{-1}^{1}\int_{-1}^{1} [B]^T [\kappa][B] J \, d\xi \, dn \, d\zeta \tag{12.127}$$

where $J = \det[J]$.

EXAMPLE 3: TEMPERATURE DISTRIBUTION IN A 2D HEAT TRANSFER PROBLEM

In the 2D body shown in the following figure, the sides AB and ED are insulated, the side AE is maintained at 60 °C and the sides BC and CD are exposed in heat convection with $h=108$ W/m^2 C and ambient temperature 12 °C. The thermal conductivity coefficients are $k_x = k_y = 45\,\mathrm{W/m^\circ C}$, the thickness is $t=1.0$ m and the dimensions of the sides are demonstrated in the figure. Determine the temperature distribution.

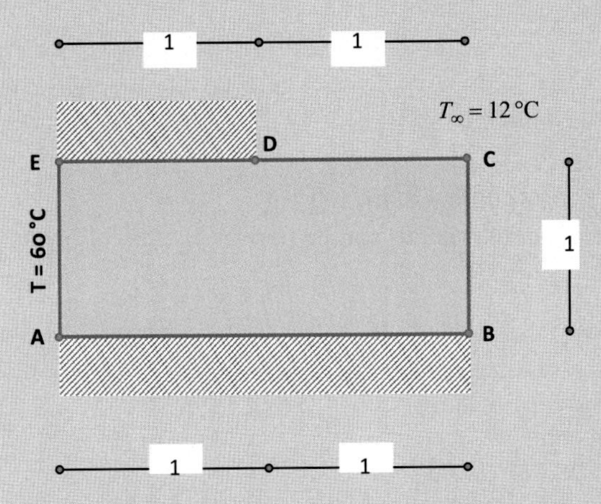

Solution

The FE discretization of the physical model is shown in the following figure. It should be noted that during derivation of stiffness matrices, the order of the nodes of each element should be in a counterclockwise manner.

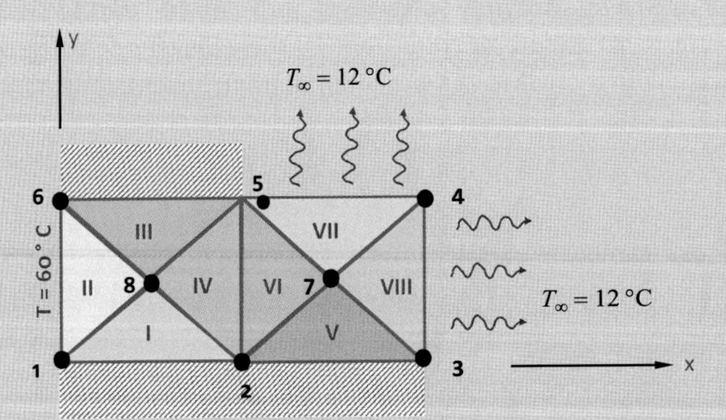

For the solution of the above FE model, the following steps should be followed.

Step 1: Derivation of the "stiffness" matrices

The FE model consists of eight elements. The "stiffness" matrix of the elements I, II, III, IV, V, VI consists from only the conduction part, while the "stiffness" matrix of the elements VII and VIII consists of both conduction and convection part.

Element I, nodes 1-2-8

$$[\Xi]=\begin{bmatrix}1&0&0\\1&1&0\\1&0.5&0.5\end{bmatrix}$$

$$[B]=\begin{bmatrix}0&1&0\\0&0&1\end{bmatrix}[\Xi]^{-1}=\begin{bmatrix}-1&1&0\\-1&-1&2\end{bmatrix}$$

$$[\kappa]=\begin{bmatrix}45&0\\0&45\end{bmatrix}$$

$$A=\frac{1}{2}(1)(0.5)=0.25m^2$$

$$t=1$$

Then, the local "stiffness" matrix due to conduction heat transfer is

$$\left[k_{\mathrm{CD}}^{(\mathrm{I})}\right]=[B]^{\mathrm{T}}[\kappa][B]At=\begin{bmatrix}22.5&0&-22.5\\0&22.5&-22.5\\-22.5&-22.5&45\end{bmatrix}$$

The following matrix is called "topology" matrix and contains the position of the members of the local matrix within the global matrix:

$$\begin{bmatrix}[1,1]&[1,2]&[1,8]\\ [2,1]&[2,2]&[2,8]\\ [8,1]&[8,2]&[8,8]\end{bmatrix}$$

The derivation of the above "topology" matrix is based on the node numbers 1-2-8 of the element I.

Taking into account the "topology" matrix, the "stiffness" matrix $[k_{\mathrm{CD}}^{(\mathrm{I})}]$ is expanded to the degrees of freedom of the whole structure as follows:

$$[K_I]=\begin{bmatrix}22.5&0&0&0&0&0&0&-22.5\\0&22.5&0&0&0&0&0&-22.5\\0&0&0&0&0&0&0&0\\0&0&0&0&0&0&0&0\\0&0&0&0&0&0&0&0\\0&0&0&0&0&0&0&0\\0&0&0&0&0&0&0&0\\-22.5&-22.5&0&0&0&0&0&45\end{bmatrix}$$

Continued

EXAMPLE 3: TEMPERATURE DISTRIBUTION IN A 2D HEAT TRANSFER PROBLEM—CONT'D

Element II, nodes 1-8-6

$$[\Xi] = \begin{bmatrix} 1 & 0 & 0 \\ 1 & 0.5 & 0.5 \\ 1 & 0 & 1 \end{bmatrix}$$

$$[B] = \begin{bmatrix} 0 & 1 & 0 \\ 0 & 0 & 1 \end{bmatrix} [\Xi]^{-1} = \begin{bmatrix} -1 & 2 & -1 \\ -1 & 0 & 1 \end{bmatrix}$$

$$[\kappa] = \begin{bmatrix} 45 & 0 \\ 0 & 45 \end{bmatrix}$$

$$A = \frac{1}{2}(1)(0.5) = 0.25\text{m}^2$$

$$t = 1$$

Then, the local "stiffness" matrix due to conduction heat transfer is

$$\left[k_{\text{CD}}^{(\text{II})}\right] = [B]^{\text{T}}[\kappa][B]At = \begin{bmatrix} 22.5 & -22.5 & 0 \\ -22.5 & 45 & -22.5 \\ 0 & -22.5 & 22.5 \end{bmatrix}$$

Based on the node numbers 1-8-6 of the element II, the "topology" matrix containing the position of the members of the above local matrix within the global matrix is

$$\begin{bmatrix} [1,1] & [1,8] & [1,6] \\ [8,1] & [8,8] & [8,6] \\ [6,1] & [6,8] & [6,6] \end{bmatrix}$$

Then, the corresponding expansion of the "stiffness" matrix $[k_{\text{CD}}^{(\text{II})}]$ to the degrees of freedom of the whole structure is the following:

$$[K_{II}] = \begin{bmatrix} 22.5 & 0 & 0 & 0 & 0 & 0 & 0 & -22.5 \\ 0 & 0 & 0 & 0 & 0 & 0 & 0 & 0 \\ 0 & 0 & 0 & 0 & 0 & 0 & 0 & 0 \\ 0 & 0 & 0 & 0 & 0 & 0 & 0 & 0 \\ 0 & 0 & 0 & 0 & 0 & 0 & 0 & 0 \\ 0 & 0 & 0 & 0 & 0 & 22.5 & 0 & -22.5 \\ 0 & 0 & 0 & 0 & 0 & 0 & 0 & 0 \\ -22.5 & 0 & 0 & 0 & 0 & -22.5 & 0 & 45 \end{bmatrix}$$

Element III, nodes 8-5-6

$$[\Xi] = \begin{bmatrix} 1 & 0.5 & 0.5 \\ 1 & 1 & 1 \\ 1 & 0 & 1 \end{bmatrix}$$

$$[B] = \begin{bmatrix} 0 & 1 & 0 \\ 0 & 0 & 1 \end{bmatrix}[\Xi]^{-1} = \begin{bmatrix} 0 & 1 & -1 \\ -2 & 1 & 1 \end{bmatrix}$$

$$[\kappa] = \begin{bmatrix} 45 & 0 \\ 0 & 45 \end{bmatrix}$$

$$A = \frac{1}{2}(1)(0.5) = 0.25\text{m}^2$$

$$t = 1$$

Then, the local "stiffness" matrix due to conduction heat transfer is

$$\left[k_{\text{CD}}^{(\text{III})}\right] = [B]^{\text{T}}[\kappa][B]At = \begin{bmatrix} 45 & -22.5 & -22.5 \\ -22.5 & 22.5 & 0 \\ -22.5 & 0 & 22.5 \end{bmatrix}$$

"Topology" matrix:

$$\begin{bmatrix} [8,8] & [8,5] & [8,6] \\ [5,8] & [5,5] & [5,6] \\ [6,8] & [6,5] & [6,6] \end{bmatrix}$$

Expanded "stiffness" matrix:

$$[K_{III}] = \begin{bmatrix} 0 & 0 & 0 & 0 & 0 & 0 & 0 & 0 \\ 0 & 0 & 0 & 0 & 0 & 0 & 0 & 0 \\ 0 & 0 & 0 & 0 & 0 & 0 & 0 & 0 \\ 0 & 0 & 0 & 0 & 0 & 0 & 0 & 0 \\ 0 & 0 & 0 & 0 & 22.5 & 0 & 0 & -22.5 \\ 0 & 0 & 0 & 0 & 0 & 22.5 & 0 & -22.5 \\ 0 & 0 & 0 & 0 & 0 & 0 & 0 & 0 \\ 0 & 0 & 0 & 0 & -22.5 & -22.5 & 0 & 45 \end{bmatrix}$$

Element IV, nodes 8-2-5

$$[\Xi] = \begin{bmatrix} 1 & 0.5 & 0.5 \\ 1 & 1 & 0 \\ 1 & 1 & 1 \end{bmatrix}$$

Continued

EXAMPLE 3: TEMPERATURE DISTRIBUTION IN A 2D HEAT TRANSFER PROBLEM—CONT'D

$$[B]=\begin{bmatrix}0 & 1 & 0\\ 0 & 0 & 1\end{bmatrix}[\Xi]^{-1}=\begin{bmatrix}-2 & 1 & 1\\ 0 & -1 & 1\end{bmatrix}$$

$$[\kappa]=\begin{bmatrix}45 & 0\\ 0 & 45\end{bmatrix}$$

$$A=\frac{1}{2}(1)(0.5)=0.25\mathrm{m}^2$$

$$t=1$$

Local "stiffness" matrix due to conduction heat transfer:

$$\left[k_{\mathrm{CD}}^{(\mathrm{IV})}\right]=[B]^{\mathrm{T}}[\kappa][B]At=\begin{bmatrix}45 & -22.5 & -22.5\\ -22.5 & 22.5 & 0\\ -22.5 & 0 & 22.5\end{bmatrix}$$

"Topology" matrix:

$$\begin{bmatrix}[8,8] & [8,2] & [8,5]\\ [2,8] & [2,2] & [2,5]\\ [5,8] & [5,2] & [5,5]\end{bmatrix}$$

Expanded "stiffness" matrix:

$$[K_{IV}]=\begin{bmatrix}0 & 0 & 0 & 0 & 0 & 0 & 0 & 0\\ 0 & 22.5 & 0 & 0 & 0 & 0 & 0 & -22.5\\ 0 & 0 & 0 & 0 & 0 & 0 & 0 & 0\\ 0 & 0 & 0 & 0 & 0 & 0 & 0 & 0\\ 0 & 0 & 0 & 0 & 22.5 & 0 & 0 & -22.5\\ 0 & 0 & 0 & 0 & 0 & 0 & 0 & 0\\ 0 & 0 & 0 & 0 & 0 & 0 & 0 & 0\\ 0 & -22.5 & 0 & 0 & -22.5 & 0 & 0 & 45\end{bmatrix}$$

Element V, nodes 2-3-7

$$[\Xi]=\begin{bmatrix}1 & 1 & 0\\ 1 & 2 & 0\\ 1 & 1.5 & 0.5\end{bmatrix}$$

$$[B]=\begin{bmatrix}0 & 1 & 0\\ 0 & 0 & 1\end{bmatrix}[\Xi]^{-1}=\begin{bmatrix}-1 & 1 & 0\\ -1 & -1 & 2\end{bmatrix}$$

$$[\kappa]=\begin{bmatrix}45 & 0\\ 0 & 45\end{bmatrix}$$

$$A=\frac{1}{2}(1)(0.5)=0.25\text{m}^2$$

$$t=1$$

Local "stiffness" matrix due to conduction heat transfer:

$$\left[k_{\text{CD}}^{(\text{V})}\right]=[B]^{\text{T}}[\kappa][B]At=\begin{bmatrix}22.5 & 0 & -22.5\\ 0 & 22.5 & -22.5\\ -22.5 & -22.5 & 45\end{bmatrix}$$

"Topology" matrix:

$$\begin{bmatrix}[2,2] & [2,3] & [2,7]\\ [3,2] & [3,3] & [3,7]\\ [7,2] & [7,3] & [7,7]\end{bmatrix}$$

Expanded "stiffness" matrix:

$$[K_V]=\begin{bmatrix}0 & 0 & 0 & 0 & 0 & 0 & 0 & 0\\ 0 & 22.5 & 0 & 0 & 0 & 0 & -22.5 & 0\\ 0 & 0 & 22.5 & 0 & 0 & 0 & -22.5 & 0\\ 0 & 0 & 0 & 0 & 0 & 0 & 0 & 0\\ 0 & 0 & 0 & 0 & 0 & 0 & 0 & 0\\ 0 & 0 & 0 & 0 & 0 & 0 & 0 & 0\\ 0 & -22.5 & -22.5 & 0 & 0 & 0 & 45 & 0\\ 0 & 0 & 0 & 0 & 0 & 0 & 0 & 0\end{bmatrix}$$

Element VI, nodes 2-7-5

$$[\Xi]=\begin{bmatrix}1 & 1 & 0\\ 1 & 1.5 & 0.5\\ 1 & 1 & 1\end{bmatrix}$$

$$[B]=\begin{bmatrix}0 & 1 & 0\\ 0 & 0 & 1\end{bmatrix}[\Xi]^{-1}=\begin{bmatrix}-1 & 2 & -1\\ -1 & 0 & 1\end{bmatrix}$$

$$[\kappa]=\begin{bmatrix}45 & 0\\ 0 & 45\end{bmatrix}$$

$$A=\frac{1}{2}(1)(0.5)=0.25\text{m}^2$$

$$t=1$$

Continued

EXAMPLE 3: TEMPERATURE DISTRIBUTION IN A 2D HEAT TRANSFER PROBLEM—CONT'D

Local "stiffness" matrix due to conduction heat transfer:

$$\left[k_{CD}^{(VI)}\right]=[B]^{T}[\kappa][B]At=\begin{bmatrix}22.5 & -22.5 & 0\\ -22.5 & 45 & -22.5\\ 0 & -22.5 & 22.5\end{bmatrix}$$

"Topology" matrix:

$$\begin{bmatrix}[2,2] & [2,7] & [2,5]\\ [7,2] & [7,7] & [7,5]\\ [5,2] & [5,7] & [5,5]\end{bmatrix}$$

Expanded "stiffness" matrix:

$$[K_{VI}]=\begin{bmatrix}0 & 0 & 0 & 0 & 0 & 0 & 0 & 0\\ 0 & 22.5 & 0 & 0 & 0 & 0 & -22.5 & 0\\ 0 & 0 & 0 & 0 & 0 & 0 & 0 & 0\\ 0 & 0 & 0 & 0 & 0 & 0 & 0 & 0\\ 0 & 0 & 0 & 0 & 22.5 & 0 & -22.5 & 0\\ 0 & 0 & 0 & 0 & 0 & 0 & 0 & 0\\ 0 & -22.5 & 0 & 0 & -22.5 & 0 & 45 & 0\\ 0 & 0 & 0 & 0 & 0 & 0 & 0 & 0\end{bmatrix}$$

Element VII, nodes 7-4-5

It is reminded that the element VII, apart from the conduction part of the "stiffness" matrix, contains and a convection part as well.

Conduction part of the "stiffness" matrix

$$[\Xi]=\begin{bmatrix}1 & 1.5 & 0.5\\ 1 & 2 & 1\\ 1 & 1 & 1\end{bmatrix}$$

$$[B]=\begin{bmatrix}0 & 1 & 0\\ 0 & 0 & 1\end{bmatrix}[\Xi]^{-1}=\begin{bmatrix}0 & 1 & -1\\ -2 & 1 & 1\end{bmatrix}$$

$$[\kappa]=\begin{bmatrix}45 & 0\\ 0 & 45\end{bmatrix}$$

$$A=\frac{1}{2}(1)(0.5)=0.25\text{m}^2$$

$$t=1$$

Local "stiffness" matrix due to conduction heat transfer:

$$\left[k_{CD}^{(VII)}\right] = [B]^T[\kappa][B]At = \begin{bmatrix} 45 & -22.5 & -22.5 \\ -22.5 & 22.5 & 0 \\ -22.5 & 0 & 22.5 \end{bmatrix}$$

Convection part of the "stiffness" matrix. Since one side of this element is exposed to heat losses due to convection, the convection part of its "stiffness" matrix can be derived if we correspond the indexes i, j, m to the element's nodes as $i=7, j=4, m=5$. Then, since $4=j$ and $5=m$, the element VII has convection on the side j-m. Therefore, implementation of Equation (12.78) for this case yields

$$\left[k_{CV}^{4-5}\right] = \frac{hL_{4-5}}{6}\begin{bmatrix} 0 & 0 & 0 \\ 0 & 2 & 1 \\ 0 & 1 & 2 \end{bmatrix}$$

or

$$\left[k_{CV}^{4-5}\right] = \frac{(108)(1)}{6}\begin{bmatrix} 0 & 0 & 0 \\ 0 & 2 & 1 \\ 0 & 1 & 2 \end{bmatrix}$$

Element's "stiffness" matrix. Taking into account the conduction and the convection part of heat transfer, the local "stiffness" matrix of the element VII is

$$[k_{VII}] = \left[k_{CD}^{VII}\right] + \left[k_{CV}^{4-5}\right] = \begin{bmatrix} 45 & -22.5 & -22.5 \\ -22.5 & 58.5 & 18 \\ -22.5 & 18 & 58.5 \end{bmatrix}$$

"Topology" matrix:

$$\begin{bmatrix} [7,7] & [7,4] & [7,5] \\ [4,7] & [4,4] & [4,5] \\ [5,7] & [5,4] & [5,5] \end{bmatrix}$$

Expanded "stiffness" matrix:

$$[K_{VII}] = \begin{bmatrix} 0 & 0 & 0 & 0 & 0 & 0 & 0 & 0 \\ 0 & 0 & 0 & 0 & 0 & 0 & 0 & 0 \\ 0 & 0 & 0 & 0 & 0 & 0 & 0 & 0 \\ 0 & 0 & 0 & 58.5 & 18 & 0 & -22.5 & 0 \\ 0 & 0 & 0 & 18 & 58.5 & 0 & -22.5 & 0 \\ 0 & 0 & 0 & 0 & 0 & 0 & 0 & 0 \\ 0 & 0 & 0 & -22.5 & -22.5 & 0 & 45 & 0 \\ 0 & 0 & 0 & 0 & 0 & 0 & 0 & 0 \end{bmatrix}$$

Element VIII, nodes 3-4-7

Like element VII, element VIII in addition to the conduction heat transfer also has convection heat losses. Following similar procedure, the conduction and the convection parts of this element can be derived as follows:

Continued

EXAMPLE 3: TEMPERATURE DISTRIBUTION IN A 2D HEAT TRANSFER PROBLEM—CONT'D

Conduction part of the "stiffness" matrix

$$[\Xi]=\begin{bmatrix}1 & 2 & 0\\ 1 & 2 & 1\\ 1 & 1.5 & 0.5\end{bmatrix}$$

$$[B]=\begin{bmatrix}0 & 1 & 0\\ 0 & 0 & 1\end{bmatrix}[\Xi]^{-1}=\begin{bmatrix}1 & 1 & -2\\ -1 & 1 & 0\end{bmatrix}$$

$$[\kappa]=\begin{bmatrix}45 & 0\\ 0 & 45\end{bmatrix}$$

$$A=\frac{1}{2}(1)(0.5)=0.25\mathrm{m}^2$$

$$t=1$$

Local "stiffness" matrix due to conduction heat transfer:

$$\left[k_{CD}^{(VIII)}\right]=[B]^{T}[\kappa][B]At=\begin{bmatrix}22.5 & 0 & -22.5\\ 0 & 22.5 & -22.5\\ -22.5 & -22.5 & 45\end{bmatrix}$$

Convection part of the "stiffness" matrix. For the derivation of the convection part of the "stiffness" matrix the nomenclature $i=3$, $j=4$, $m=7$ should be used. Therefore, since the convection losses take place on the side i-j, Equation (12.77) should be used:

$$\left[k_{CV}^{3-4}\right]=\frac{hL_{3-4}}{6}\begin{bmatrix}2 & 1 & 0\\ 1 & 2 & 0\\ 0 & 0 & 0\end{bmatrix}$$

or

$$\left[k_{CV}^{3-4}\right]=\frac{(108)(1)}{6}\begin{bmatrix}2 & 1 & 0\\ 1 & 2 & 0\\ 0 & 0 & 0\end{bmatrix}$$

Element's "stiffness" matrix. Taking into account the conduction and the convection part of heat transfer, the local "stiffness" matrix of the element VIII is

$$[k_{VIII}]=\left[k_{CD}^{VIII}\right]+\left[k_{CV}^{3-4}\right]=\begin{bmatrix}58.5 & 18 & -22.5\\ 18 & 58.5 & -22.5\\ -22.5 & -22.5 & 45\end{bmatrix}$$

"Topology" matrix:

$$\begin{bmatrix}[3,3] & [3,4] & [3,7]\\ [4,3] & [4,4] & [4,7]\\ [7,3] & [7,4] & [7,7]\end{bmatrix}$$

Expanded "stiffness" matrix:

$$[K_{VIII}]=\begin{bmatrix} 0 & 0 & 0 & 0 & 0 & 0 & 0 & 0 \\ 0 & 0 & 0 & 0 & 0 & 0 & 0 & 0 \\ 0 & 0 & 58.5 & 18 & 0 & 0 & -22.5 & 0 \\ 0 & 0 & 18 & 58.5 & 0 & 0 & -22.5 & 0 \\ 0 & 0 & 0 & 0 & 0 & 0 & 0 & 0 \\ 0 & 0 & 0 & 0 & 0 & 0 & 0 & 0 \\ 0 & 0 & -22.5 & -22.5 & 0 & 0 & 45 & 0 \\ 0 & 0 & 0 & 0 & 0 & 0 & 0 & 0 \end{bmatrix}$$

Global "stiffness" matrix of the structure. Taking into account the expressions of the expanded "stiffness" matrices of all elements, the global "stiffness" matrix of the structure is:

$$[K]=[K_{\mathrm{I}}]+[K_{\mathrm{II}}]+[K_{\mathrm{III}}]+[K_{\mathrm{IV}}]+[K_{\mathrm{V}}]+[K_{\mathrm{VI}}]+[K_{\mathrm{VII}}]+[K_{\mathrm{VIII}}]$$

or

$$[K]=\begin{bmatrix} 45 & 0 & 0 & 0 & 0 & 0 & 0 & -45 \\ 0 & 90 & 0 & 0 & 0 & 0 & -45 & -45 \\ 0 & 0 & 81 & 18 & 0 & 0 & -45 & 0 \\ 0 & 0 & 18 & 117 & 18 & 0 & -45 & 0 \\ 0 & 0 & 0 & 18 & 126 & 0 & -45 & -45 \\ 0 & 0 & 0 & 0 & 0 & 45 & 0 & -45 \\ 0 & -45 & -45 & -45 & -45 & 0 & 180 & 0 \\ -45 & -45 & 0 & 0 & -45 & -45 & 0 & 180 \end{bmatrix}$$

Using the above "stiffness" matrix, the structure equation can be written in the following abbreviated notation:

$$[K]\{T_e\}=\{R\}$$

where

$$\{T_e\}=[T_1\ \ T_2\ \ T_3\ \ T_4\ \ T_5\ \ T_6\ \ T_7\ \ T_8]^{\mathrm{T}}$$

$$\{R\}=[R_1\ \ R_2\ \ R_3\ \ R_4\ \ R_5\ \ R_6\ \ R_7\ \ R_8]^{\mathrm{T}}$$

The above structure equation can be also expressed by the following alternative format:

$$[K]\{T_e\}-[I]\{R\}=\{O_{8\times1}\}$$

where $[I]$ is the unit matrix, and $\{O_{8\times1}\}$ is an vector with size 8×1 containing zeros.

It is more convenient for the subsequent steps to express the last equation as follows:

$$[[K]\ \ [-I]]\begin{Bmatrix}\{T_e\}\\ \{R\}\end{Bmatrix}=\{O_{8\times1}\} \tag{12.a1}$$

Continued

EXAMPLE 3: TEMPERATURE DISTRIBUTION IN A 2D HEAT TRANSFER PROBLEM—CONT'D

Step 2: Derivation of the "load" vectors

In the structure under consideration we do not have applied heat flux on the boundaries ($f_{bound}=0$) or internal heat generation ($\dot{q}_{gen}=0$). There is only convective heat transfer on side 4-5 of the element VII and on side 3-4 of the element VIII. The "loads" on the nodes 3, 4, and 5 will be calculated, the "loads" on the nodes 2, 7, and 8 are zero, (i.e., $R_2=0$, $R_7=0$, $R_8=0$), and the "loads" on the nodes 1 and 6 are unknown.

"Load" vector for the element VII (nodes 7-4-5). Since $i=7, j=4, m=5$, this element has convective heat transfer on the side j-m. Therefore, its local "load" vector can be derived from Equation (12.89):

$$\left\{R_{CV}^{j-m}\right\}=\frac{hT_{\infty}t}{2}\begin{Bmatrix}0\\1\\1\end{Bmatrix}$$

or

$$\left\{R_{CV}^{4-5}\right\}=\frac{(108)(12)(1)}{2}\begin{Bmatrix}0\\1\\1\end{Bmatrix}=\begin{Bmatrix}0\\648\\648\end{Bmatrix}$$

In the above result, the first row corresponds to node 7, the second row corresponds to node 4, and the third row corresponds to node 5. Therefore, the participation of the element VII on the "loads" of nodes 4 and 5 is:

$$R_4^{VII}=648$$

$$R_5^{VII}=648$$

"Load" vector for the element VIII (nodes 3-4-7). For the element VIII, the notations we should use for the nodes 3, 4, and 7 are: $i=3$, $j=4$, and $m=7$. This element has convective heat transfer on the side i-j. Therefore, its local "load" vector can be derived from Equation (12.88):

$$\left\{R_{CV}^{i-j}\right\}=\frac{hT_{\infty}t}{2}\begin{Bmatrix}1\\1\\0\end{Bmatrix}$$

or

$$\left\{R_{CV}^{3-4}\right\}=\frac{(108)(12)(1)}{2}\begin{Bmatrix}1\\1\\0\end{Bmatrix}=\begin{Bmatrix}648\\648\\0\end{Bmatrix}$$

In the above result, the first row corresponds to node 3, the second row corresponds to node 4, and the third row corresponds to node 7. Therefore, the participation of the element VIII on the "loads" of the nodes 3 and 4 is

$$R_3^{VIII}=648$$

$$R_4^{VIII}=648$$

Step 3: Derivation of the matrix containing the boundary conditions

Finally, the known values of the nodal "loads" are

$$R_2 = 0$$

$$R_7 = 0$$

$$R_8 = 0$$

$$R_3 = R_3^{\text{VIII}} = 648$$

$$R_4 = R_4^{\text{VII}} + R_4^{\text{VIII}} = 648 + 648 = 1296$$

$$R_5 = R_5^{\text{VIII}} = 648$$

Moreover, the known values of the nodal temperatures are

$$T_1 = 60$$

$$T_6 = 60$$

The above two known values for the temperatures T_1 and T_6 along with the six known values for the nodal loads are the boundary conditions of the problem. These boundary conditions can be written in a matrix form as follows:

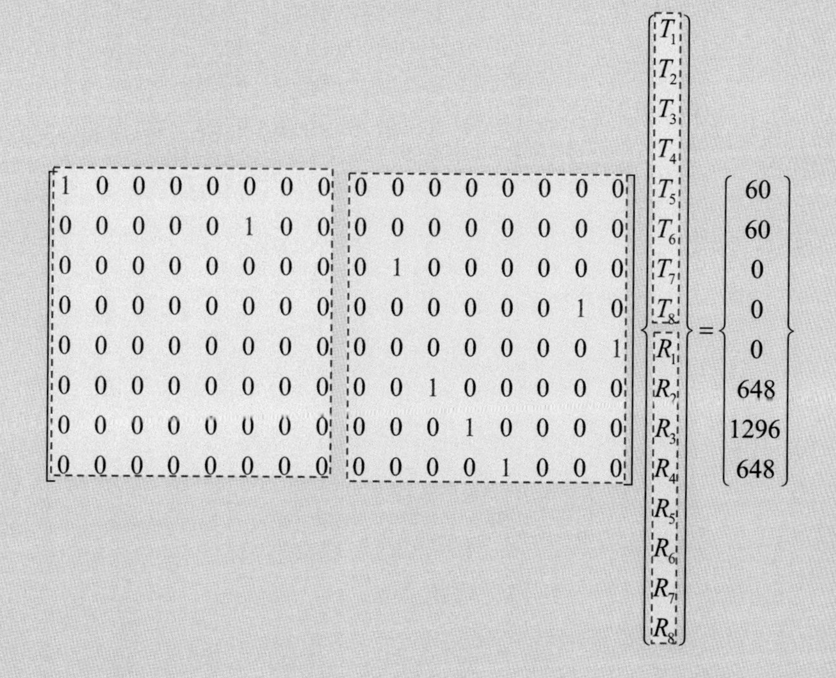

$$\left[\begin{array}{cccccccc|cccccccc} 1&0&0&0&0&0&0&0&0&0&0&0&0&0&0&0 \\ 0&0&0&0&0&1&0&0&0&0&0&0&0&0&0&0 \\ 0&0&0&0&0&0&0&0&0&1&0&0&0&0&0&0 \\ 0&0&0&0&0&0&0&0&0&0&0&0&0&0&1&0 \\ 0&0&0&0&0&0&0&0&0&0&0&0&0&0&0&1 \\ 0&0&0&0&0&0&0&0&0&0&1&0&0&0&0&0 \\ 0&0&0&0&0&0&0&0&0&0&0&1&0&0&0&0 \\ 0&0&0&0&0&0&0&0&0&0&0&0&1&0&0&0 \end{array}\right] \begin{Bmatrix} T_1 \\ T_2 \\ T_3 \\ T_4 \\ T_5 \\ T_6 \\ T_7 \\ T_8 \\ R_1 \\ R_2 \\ R_3 \\ R_4 \\ R_5 \\ R_6 \\ R_7 \\ R_8 \end{Bmatrix} = \begin{Bmatrix} 60 \\ 60 \\ 0 \\ 0 \\ 0 \\ 648 \\ 1296 \\ 648 \end{Bmatrix}$$

Continued

EXAMPLE 3: TEMPERATURE DISTRIBUTION IN A 2D HEAT TRANSFER PROBLEM—CONT'D

or in an abbreviated form:

$$[[H_1]\ [H_2]]\begin{Bmatrix} \{T_e\} \\ \{R\} \end{Bmatrix} = \{\Phi\} \tag{12.b1}$$

where

$$[H_1] = \begin{bmatrix} 1 & 0 & 0 & 0 & 0 & 0 & 0 & 0 \\ 0 & 0 & 0 & 0 & 0 & 1 & 0 & 0 \\ 0 & 0 & 0 & 0 & 0 & 0 & 0 & 0 \\ 0 & 0 & 0 & 0 & 0 & 0 & 0 & 0 \\ 0 & 0 & 0 & 0 & 0 & 0 & 0 & 0 \\ 0 & 0 & 0 & 0 & 0 & 0 & 0 & 0 \\ 0 & 0 & 0 & 0 & 0 & 0 & 0 & 0 \\ 0 & 0 & 0 & 0 & 0 & 0 & 0 & 0 \end{bmatrix}$$

$$[H_2] = \begin{bmatrix} 0 & 0 & 0 & 0 & 0 & 0 & 0 & 0 \\ 0 & 0 & 0 & 0 & 0 & 0 & 0 & 0 \\ 0 & 1 & 0 & 0 & 0 & 0 & 0 & 0 \\ 0 & 0 & 0 & 0 & 0 & 0 & 1 & 0 \\ 0 & 0 & 0 & 0 & 0 & 0 & 0 & 1 \\ 0 & 0 & 1 & 0 & 0 & 0 & 0 & 0 \\ 0 & 0 & 0 & 1 & 0 & 0 & 0 & 0 \\ 0 & 0 & 0 & 0 & 1 & 0 & 0 & 0 \end{bmatrix}$$

$$\{\Phi\} = [60\ \ 60\ \ 0\ \ 0\ \ 0\ \ 648\ \ 1296\ \ 648]^{\mathrm{T}}$$

Equations (12.a1) and (12.b1) can now be combined, yielding

$$\begin{bmatrix} [K] & [-I] \\ [H_1] & [H_2] \end{bmatrix}\begin{Bmatrix} \{T_e\} \\ \{R\} \end{Bmatrix} = \begin{Bmatrix} \{O_{8\times 1}\} \\ \{\Phi\} \end{Bmatrix}$$

or in the expanded form:

$$\left[\begin{array}{cccccccc|cccccccc}
45 & 0 & 0 & 0 & 0 & 0 & 0 & -45 & -1 & 0 & 0 & 0 & 0 & 0 & 0 & 0 \\
0 & 90 & 0 & 0 & 0 & 0 & -45 & -45 & 0 & -1 & 0 & 0 & 0 & 0 & 0 & 0 \\
0 & 0 & 81 & 18 & 0 & 0 & -45 & 0 & 0 & 0 & -1 & 0 & 0 & 0 & 0 & 0 \\
0 & 0 & 18 & 117 & 18 & 0 & -45 & 0 & 0 & 0 & 0 & -1 & 0 & 0 & 0 & 0 \\
0 & 0 & 0 & 18 & 126 & 0 & -45 & -45 & 0 & 0 & 0 & 0 & -1 & 0 & 0 & 0 \\
0 & 0 & 0 & 0 & 0 & 45 & 0 & -45 & 0 & 0 & 0 & 0 & 0 & -1 & 0 & 0 \\
0 & -45 & -45 & -45 & -45 & 0 & 180 & 0 & 0 & 0 & 0 & 0 & 0 & 0 & -1 & 0 \\
-45 & -45 & 0 & 0 & 0 & 0 & 0 & 0 & 0 & 0 & 0 & 0 & 0 & 0 & 0 & -1 \\
\hline
1 & 0 & 0 & 0 & 0 & 0 & 0 & 0 & 0 & 0 & 0 & 0 & 0 & 0 & 0 & 0 \\
0 & 0 & 0 & 0 & 0 & 1 & 0 & 0 & 0 & 0 & 0 & 0 & 0 & 0 & 0 & 0 \\
0 & 0 & 0 & 0 & 0 & 0 & 0 & 0 & 0 & 1 & 0 & 0 & 0 & 0 & 0 & 0 \\
0 & 0 & 0 & 0 & 0 & 0 & 0 & 0 & 0 & 0 & 0 & 0 & 0 & 0 & 1 & 0 \\
0 & 0 & 0 & 0 & 0 & 0 & 0 & 0 & 0 & 0 & 0 & 0 & 0 & 0 & 0 & 1 \\
0 & 0 & 0 & 0 & 0 & 0 & 0 & 0 & 0 & 0 & 1 & 0 & 0 & 0 & 0 & 0 \\
0 & 0 & 0 & 0 & 0 & 0 & 0 & 0 & 0 & 0 & 0 & 1 & 0 & 0 & 0 & 0 \\
0 & 0 & 0 & 0 & 0 & 0 & 0 & 0 & 0 & 0 & 0 & 0 & 1 & 0 & 0 & 0
\end{array}\right]
\begin{Bmatrix} T_1 \\ T_2 \\ T_3 \\ T_4 \\ T_5 \\ T_6 \\ T_7 \\ T_8 \\ R_1 \\ R_2 \\ R_3 \\ R_4 \\ R_5 \\ R_6 \\ R_7 \\ R_8 \end{Bmatrix}
=
\begin{Bmatrix} 0 \\ 0 \\ 0 \\ 0 \\ 0 \\ 0 \\ 0 \\ 0 \\ 60 \\ 60 \\ 0 \\ 0 \\ 0 \\ 648 \\ 1296 \\ 648 \end{Bmatrix}$$

The solution of the above system yields

$$\begin{Bmatrix} T_1 \\ T_2 \\ T_3 \\ T_4 \\ T_5 \\ T_6 \\ T_7 \\ T_8 \end{Bmatrix} = \begin{Bmatrix} 60 \\ 34.48 \\ 18.10 \\ 12.99 \\ 27.92 \\ 60 \\ 23.37 \\ 45.6 \end{Bmatrix}$$

and

$$\begin{Bmatrix} R_1 \\ R_2 \\ R_3 \\ R_4 \\ R_5 \\ R_6 \\ R_7 \\ R_8 \end{Bmatrix} = \begin{Bmatrix} 648 \\ 0 \\ 648 \\ 1296 \\ 648 \\ 648 \\ 0 \\ 0 \end{Bmatrix}$$

EXAMPLE 4: ANSYS IMPLEMENTATION TO SOLVE A CHIP-COOLING PROBLEM

The 2D heat transfer problem in the accompanying figure consists of an electronic chip (made of epoxy) and an aluminum fin used for the chip's cooling. The bottom of the structure is insulated. The thermal power of 22 Watt generated in the chip is convected to the environment through the fin surface. The ambient temperature is $T=30$ °C and the convection coefficient is $h=6$ W/m^2 °C. Taking into account that the thermal conductivity of the chip's and fin's material is $k_{epoxy}=0.22$ W/m^2 °C and $k_{AL}=250$ W/m^2 °C respectively, determine: (a) the temperature distribution, and (b) the direction of the heat flux.

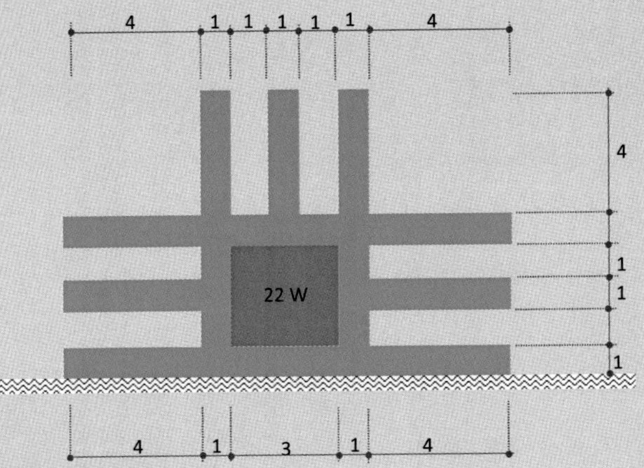

Main Menu>Preferences
Select "**Thermal**" and press "**OK**"

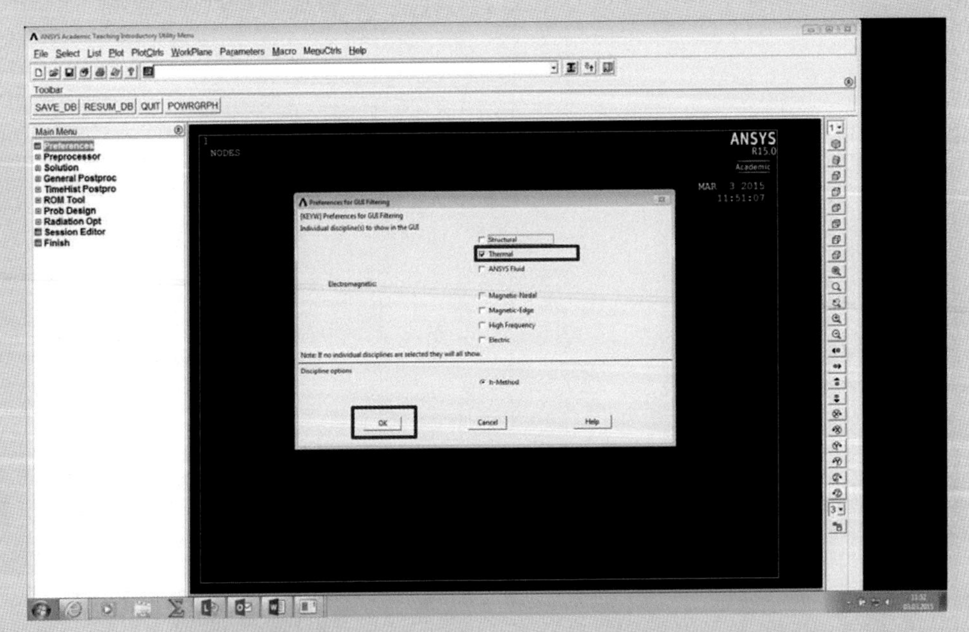

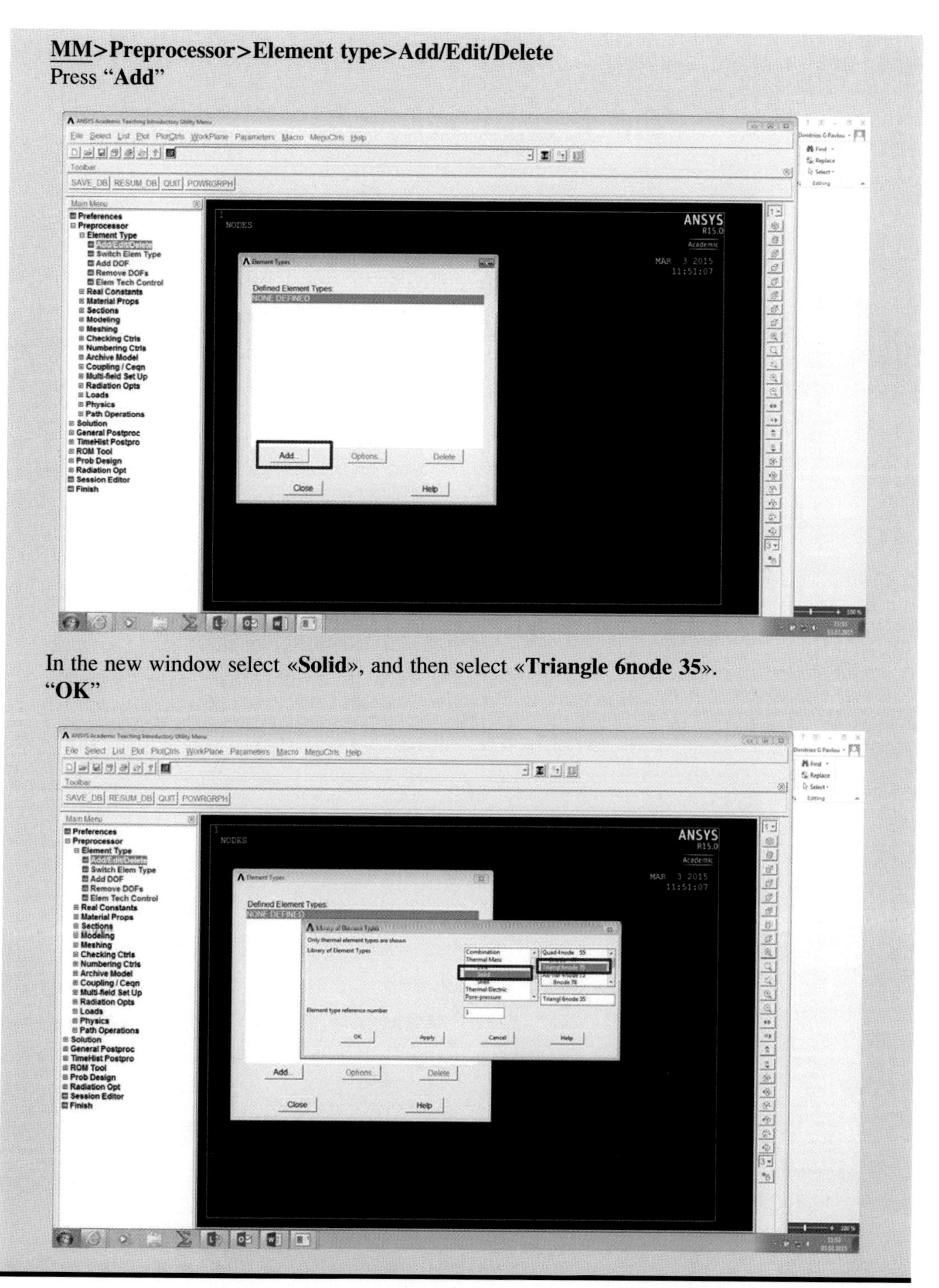

MM>Preprocessor>Element type>Add/Edit/Delete
Press "**Add**"

In the new window select «**Solid**», and then select «**Triangle 6node 35**».
"**OK**"

Continued

EXAMPLE 4: ANSYS IMPLEMENTATION TO SOLVE A CHIP-COOLING PROBLEM—CONT'D

«Close»

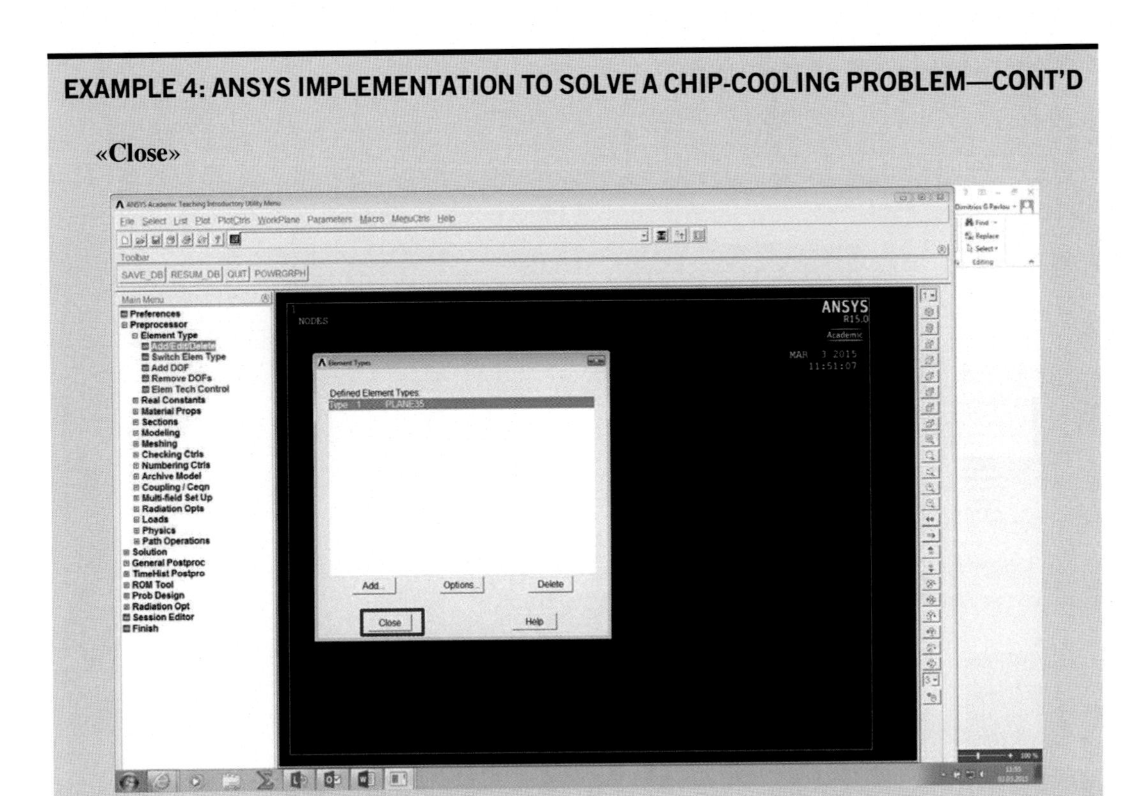

MM>**Preprocessor>Material Props>Material Models**
Then press: **Thermal>Conductivity>Isotropic**

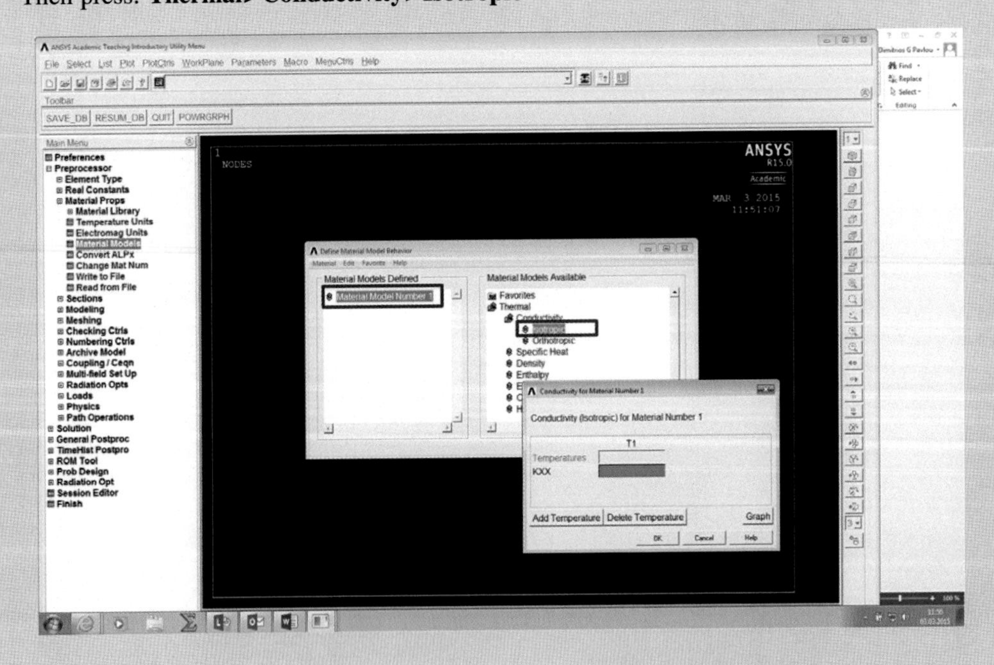

In the new window type the conductivity coefficient of the fin (material 1): **KXX=250,** then press **"OK".**

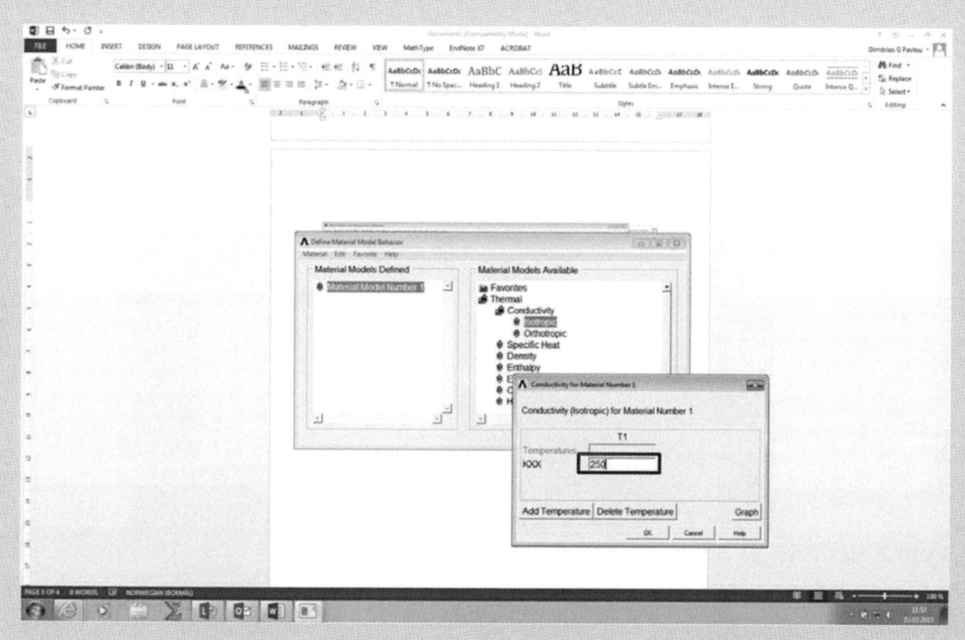

Now is time to define the material 2 (Chip). Therefore select: **Material>New Model,** and type "2" in the box "**Define Material ID**". Then press "**OK**":

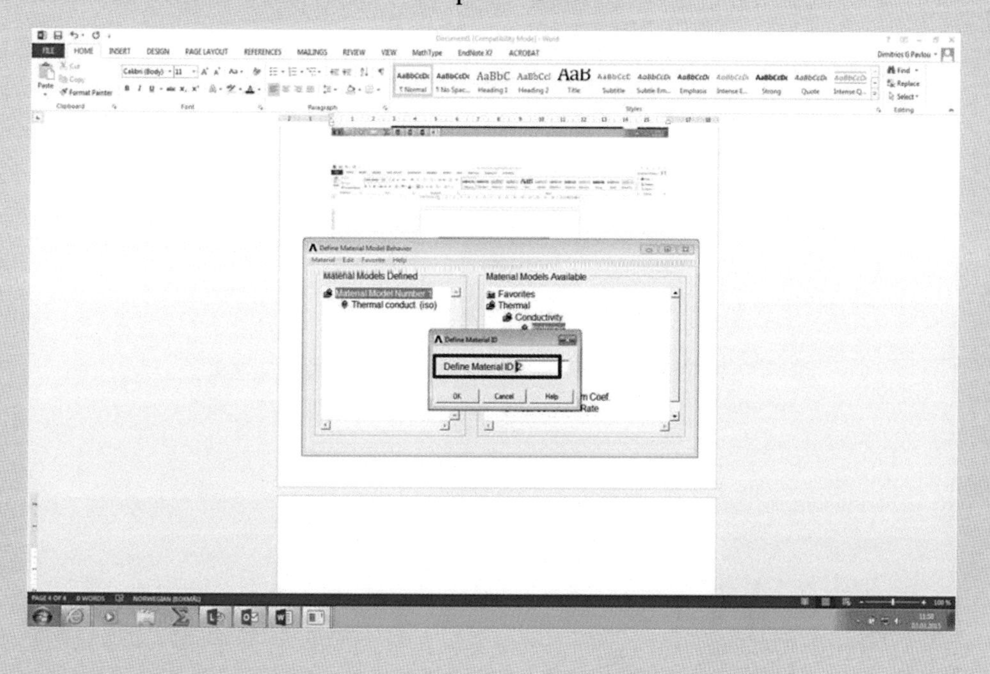

Continued

In the new window, select «**Material Model 2**», and then select "**Thermal>Conductivity>Isotropic**".

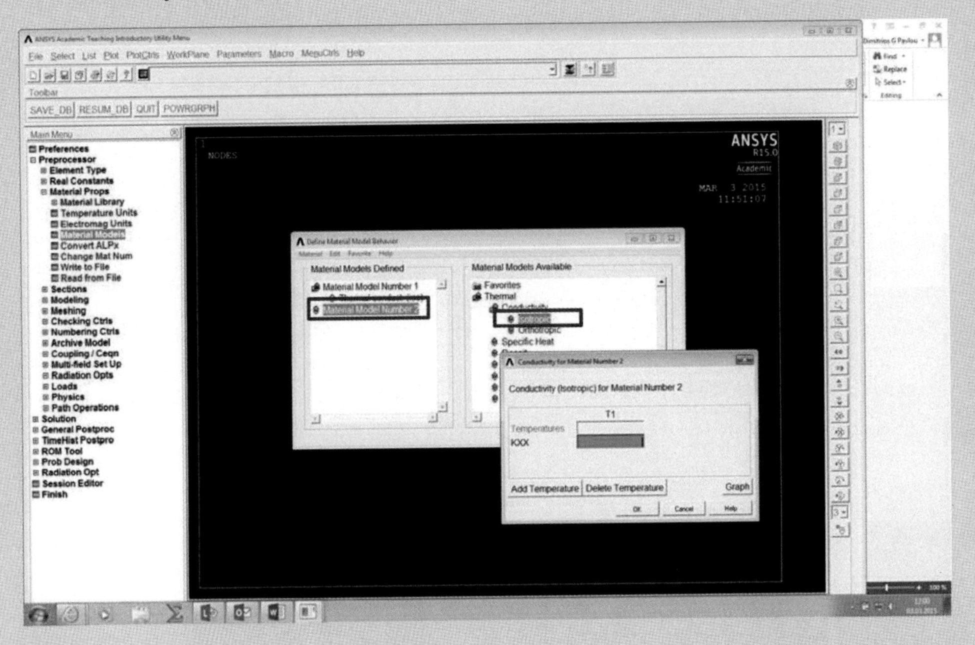

In the next window, type the conductivity coefficient of the chip (material 2): **KXX = 0.22,** then press **"OK",** and close the material model behavior window.

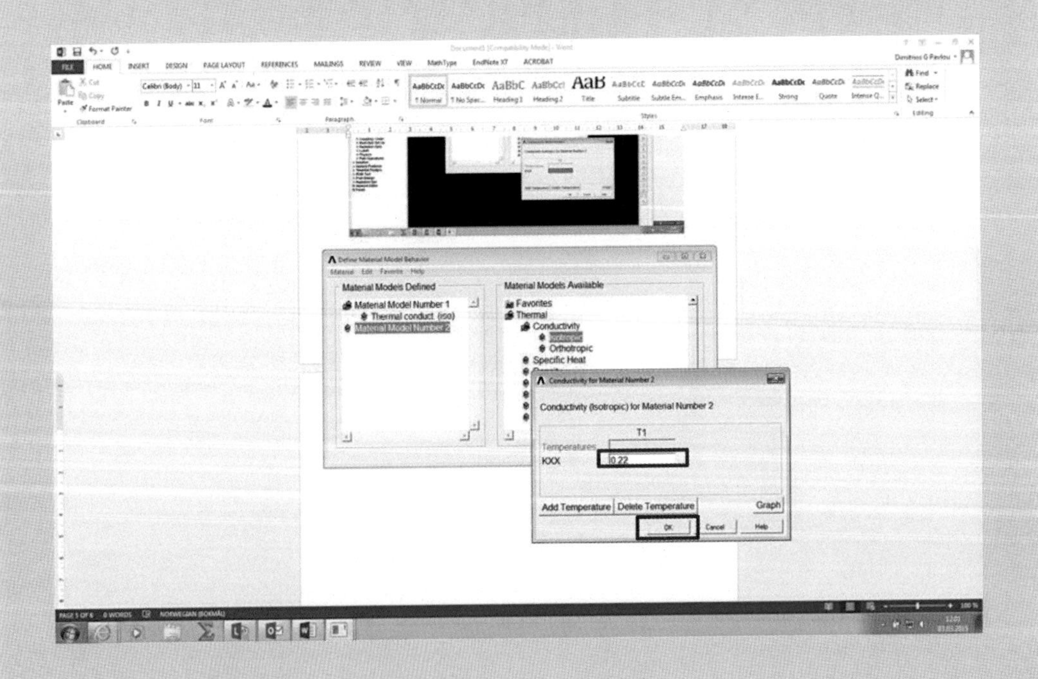

Now is time to create the geometry of the structure. It is very convenient to create a universal grid on the work space, and specify on this grid the keypoints, lines, and areas. To this scope, the space is going to be divided into squares with side length of the smallest distance, that is, 0.01 m. Therefore, the following settings should be used. First, in the next window we should select "**Grid only**". Then, the value "**0.01**" should by typed in "**Snap Incr**" and in "**Spacing**". Since the maximum size of the structure is 0.13 m, the values "**0**" and "**0.13**" should be typed in "**Minimum**" and in "**Maximum**" respectively. Then, press "**OK**".

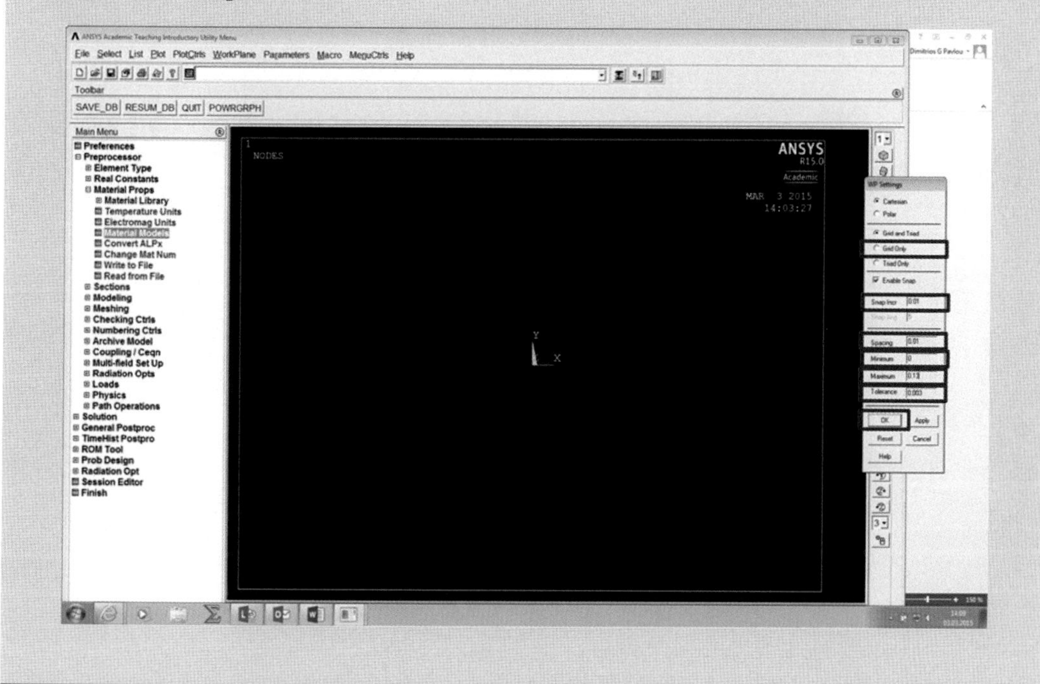

Continued

EXAMPLE 4: ANSYS IMPLEMENTATION TO SOLVE A CHIP-COOLING PROBLEM—CONT'D

Then, select "**ANSYS Utility Meny>WorkPlane>Display Working Plane**", and "**ANSYS Utility Menu>PlotCtrls>Pan Zoom Rotate**". The consequence of the above settings is the following grid:

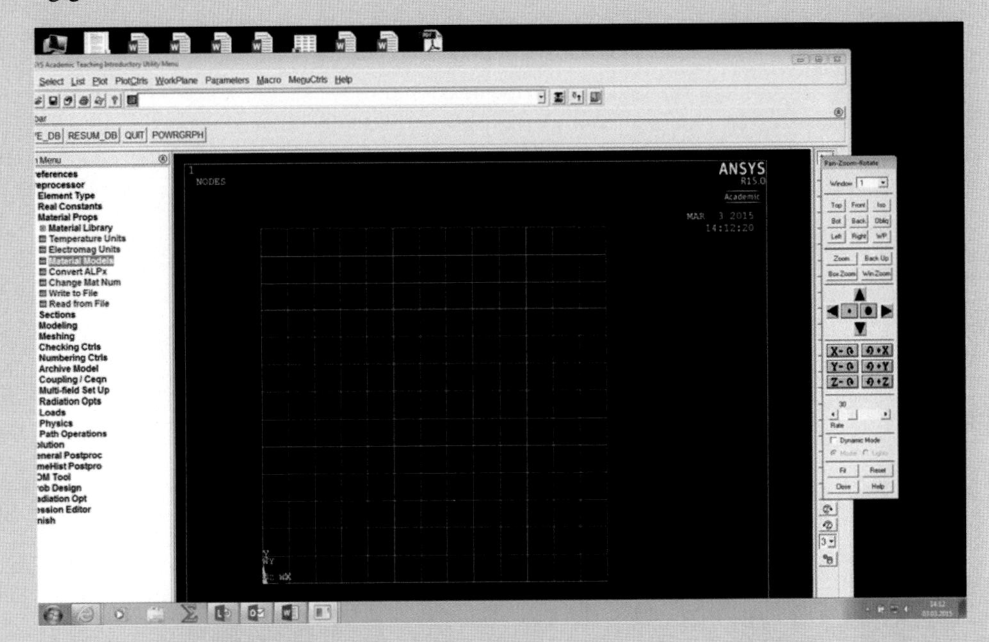

On the above grid, you can specify the keypoints.

Select "**ANSYS Main Menu>Preprocessor>Modeling>Create>Key points>On Working Plane**" and use the mouse to click on the grid the locations of the key point. Then, the following figure will be composed. Press "**OK**" to continue.

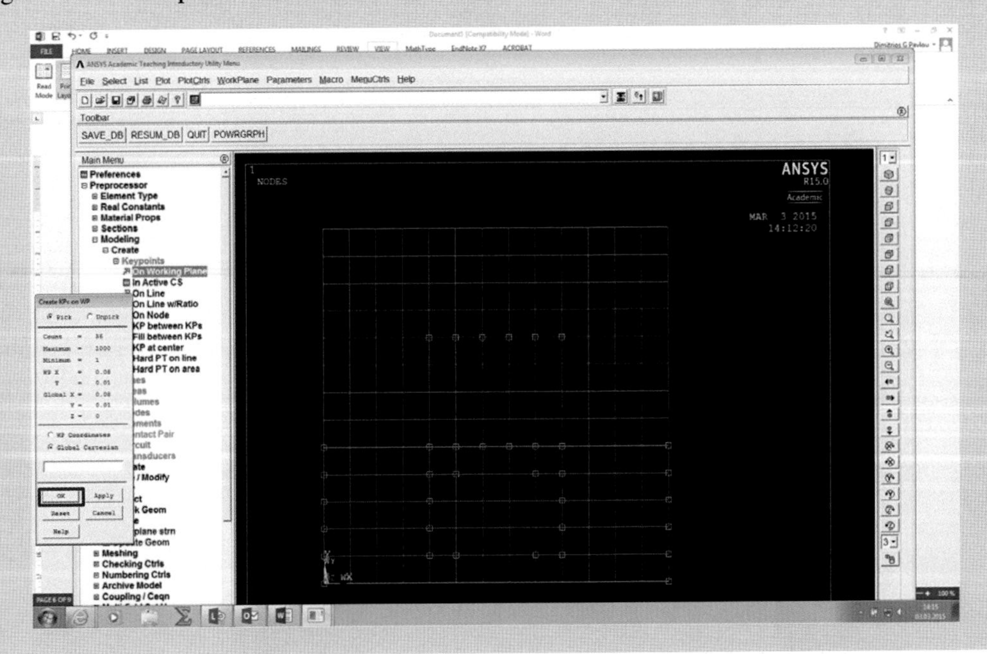

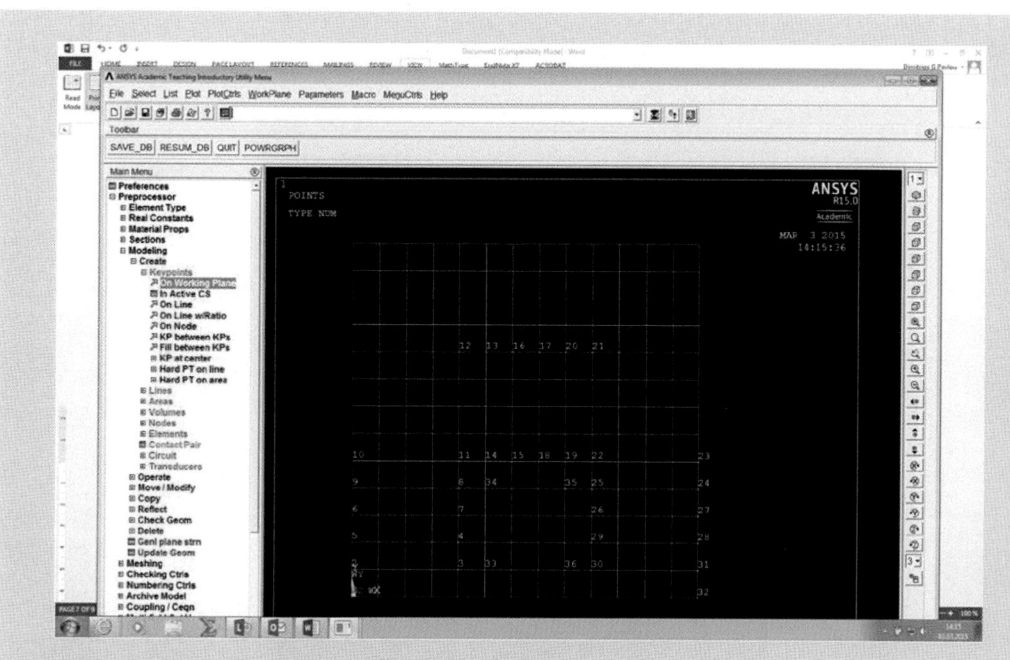

To create the lines bounding the structure select:

"**ANSYS Main Menu>Preprocessor>Modeling>Create>Lines>Lines>Straight Line**". Then click on the pairs of key points to create the corresponding lines. Do it for all lines, and press "**OK**". Then the created lines are shown as follows:

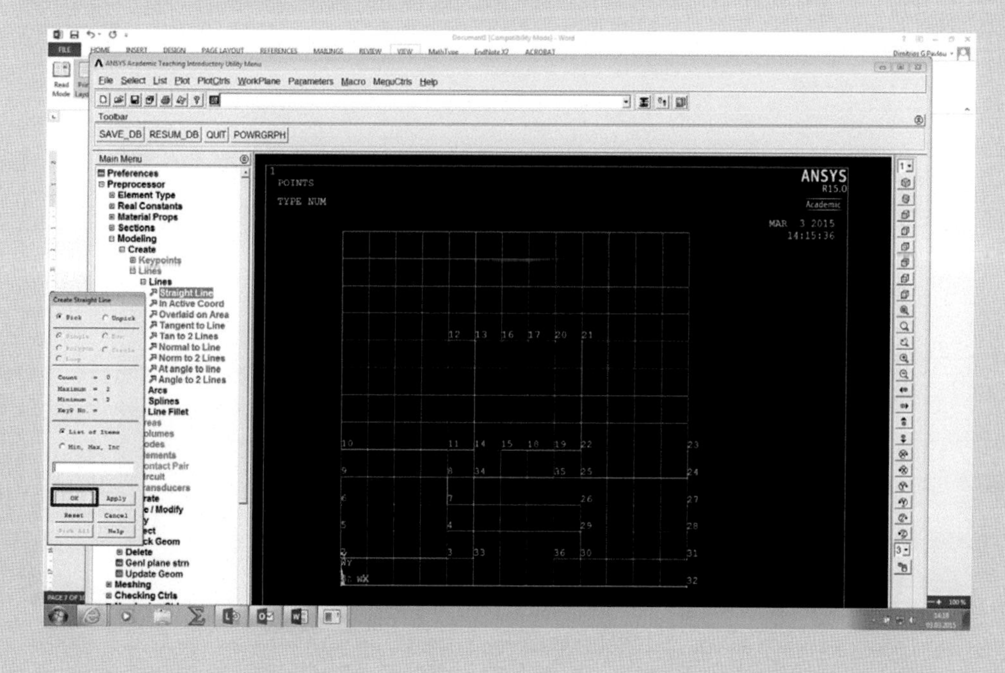

Continued

Now is time to specify the areas. Select:

"ANSYS Main Menu>Preprocessor>Modeling>Create>Areas>Arbitrary>By Lines". Then, click on the lines surrounding the chip, and press "**Apply**".

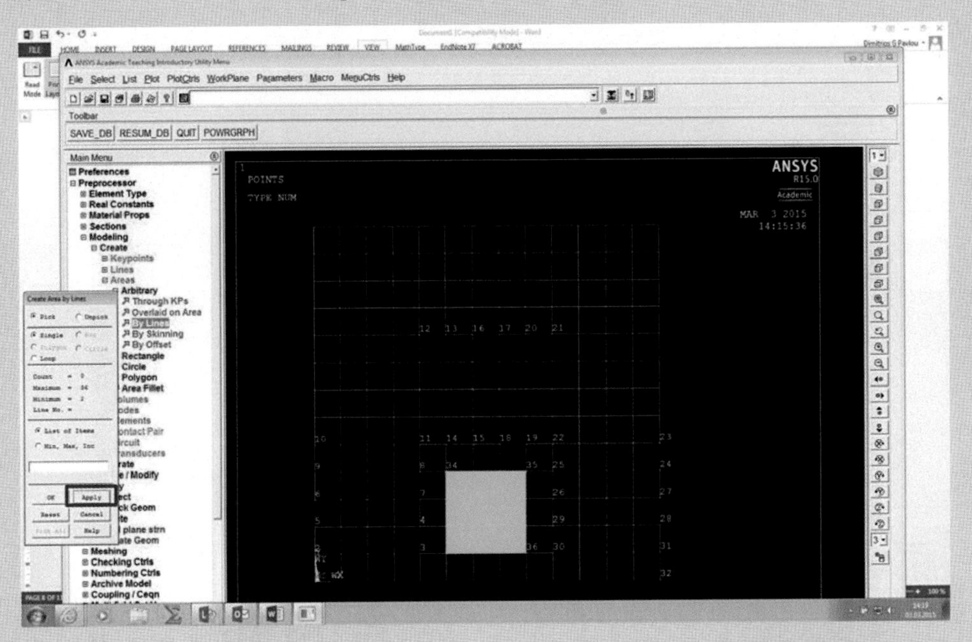

Next step is to click on the lines composing the exterior and interior perimeters of the fin and then, click "**OK**". The geometry of the structure is shown below:

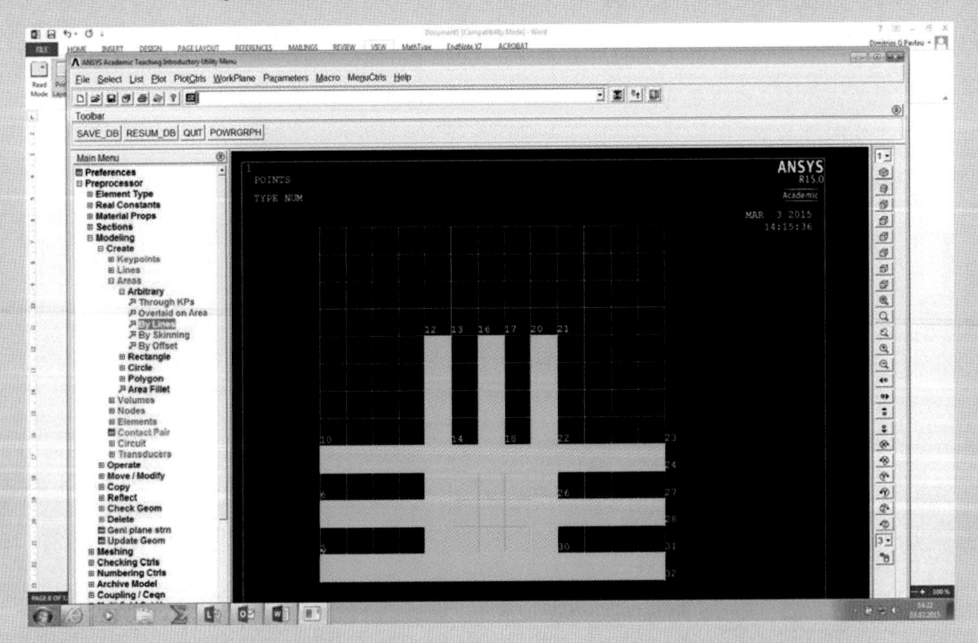

The next step is to create the mesh. Select: "**Main Menu>Preprocessor>Meshing>Mesh Tool**". In the new window select "**Areas**" and click on "**Set**"

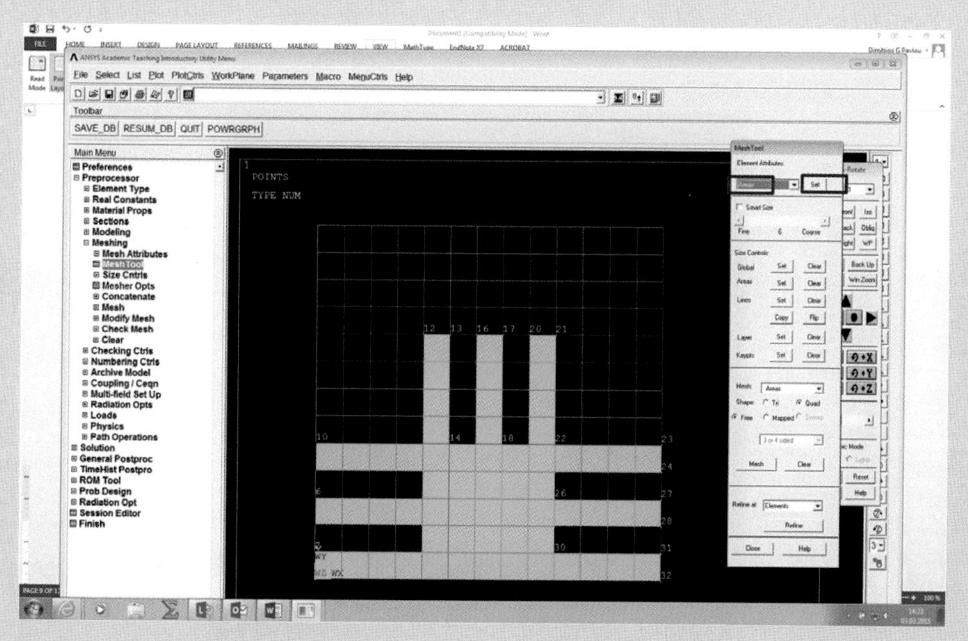

Use the mouse to select the chip area and press "**OK**".

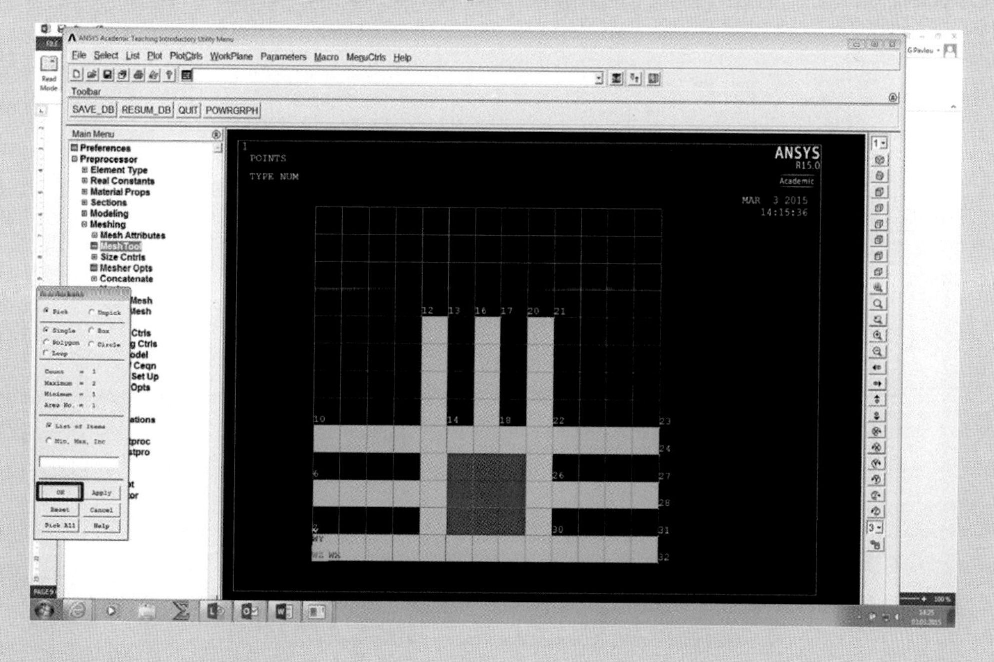

EXAMPLE 4: ANSYS IMPLEMENTATION TO SOLVE A CHIP-COOLING PROBLEM—CONT'D

In the new window, go to the line "**MAT Material number**", select "**2**" and press "**OK**". Then the set of material properties "**2**" will be assigned to the chip. The fin will take by default the set of material properties "**1**".

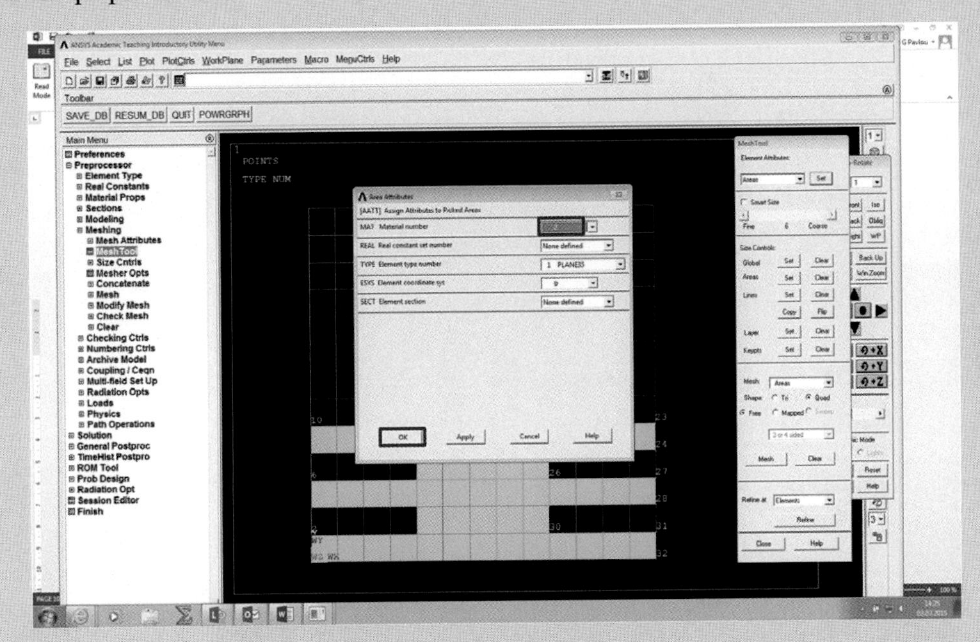

Now we should go again to the Mesh Tool. Therefore, select:

"**Main Menu">Preprocessor>Meshing>Mesh Tool**". In the Mesh Tool window select: "**Smart Size**" and set the level to "**1**". Then, press "**Mesh**", click on "**Pick All**", and "**Close**".

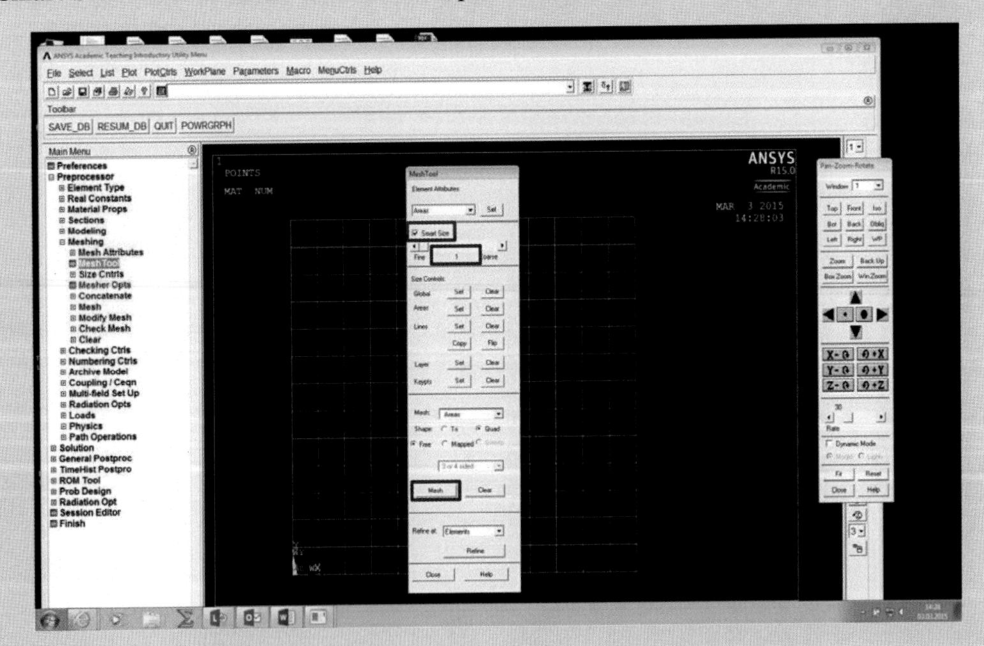

The above setting yields the following mesh:

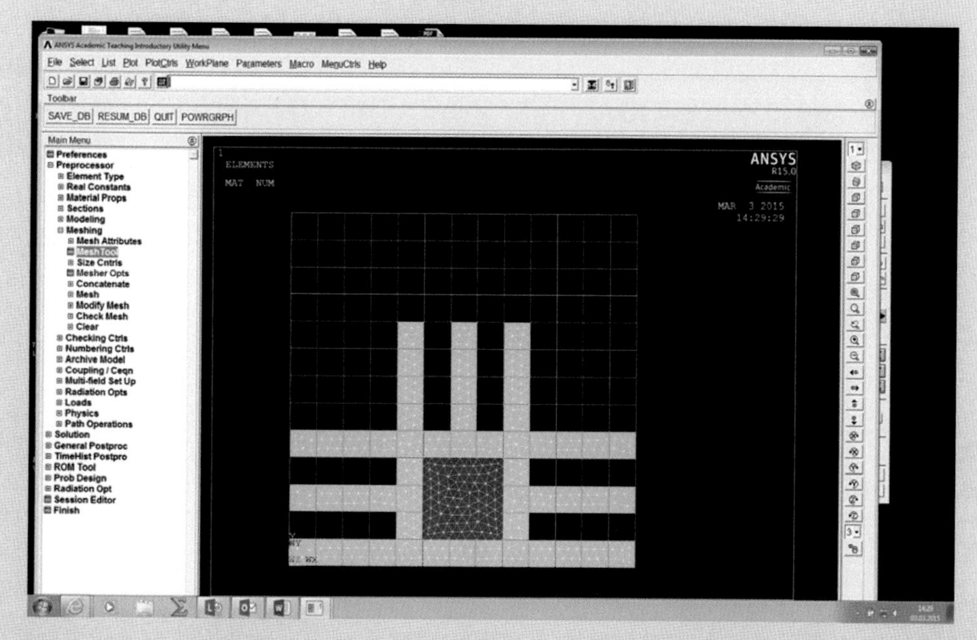

The next step is to specify the boundary conditions. If no boundary conditions are specified on a surface, the software will consider by default that this is an insulated surface (bottom of the fin). To specify the heat convection through the external surfaces of the fin we should select "**Main Menu>Solution>Define Load>Apply>Thermal>Convection>On Lines**". Then, click on the convective surfaces of the fin and press "**OK**". Insert the value "**6**" in the "**Film Coefficient**", and the value "**30**" in "**Bulk temperature**", and press "**OK**".

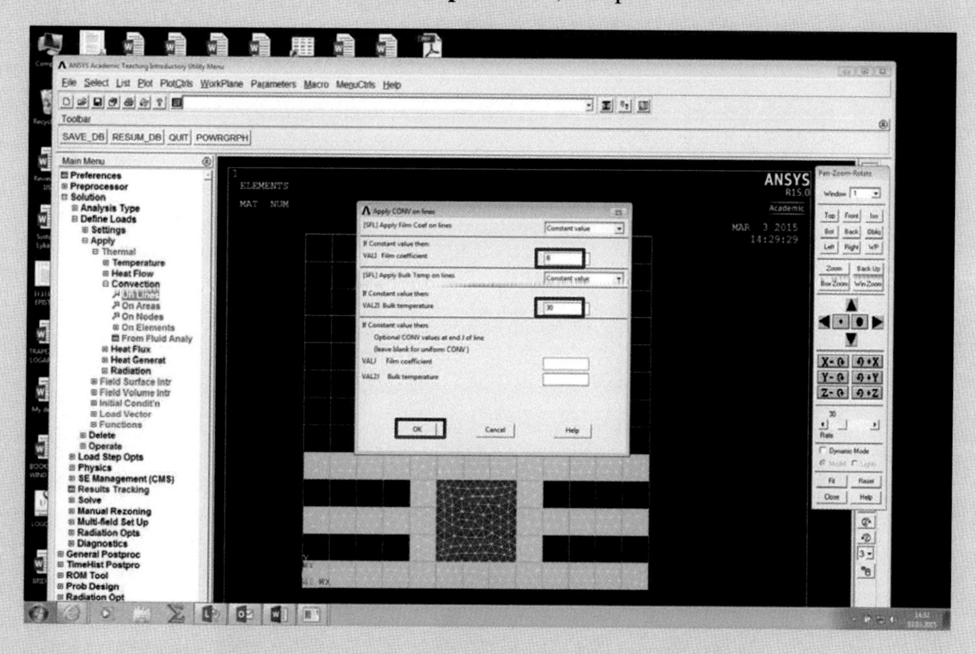

Continued

EXAMPLE 4: ANSYS IMPLEMENTATION TO SOLVE A CHIP-COOLING PROBLEM—CONT'D

The following graphical representation of convective surfaces will be appear.

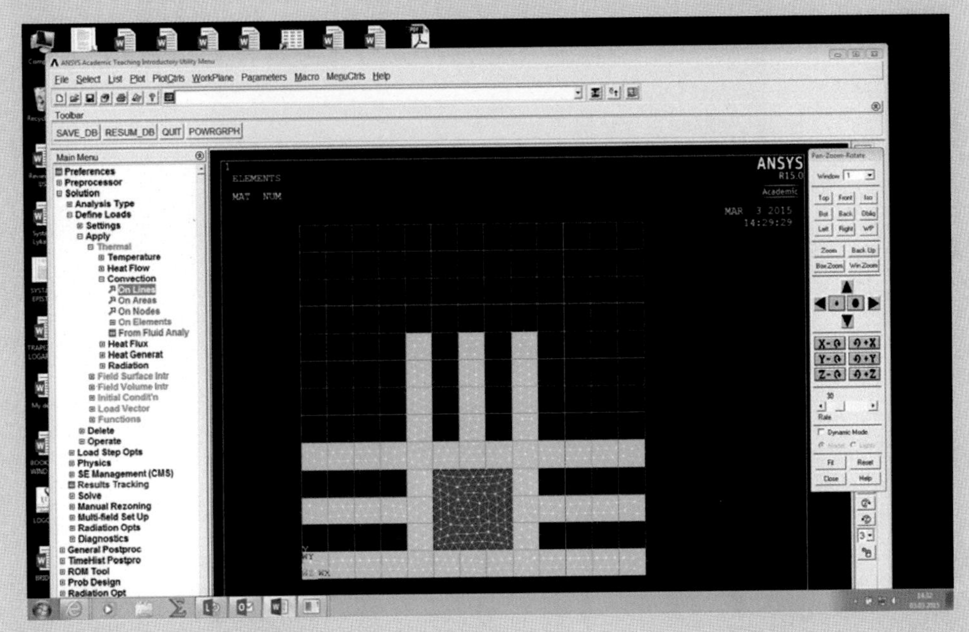

The heat generation per unit volume is $22\,\mathrm{W}/(0.03 \times 0.03)\mathrm{m}^2 = 24{,}444\,\mathrm{W}/\mathrm{m}^2$. To insert this value select: "**Main Menu>Solution>Define Load>Apply>Thermal>Heat Generation>On Area**", click on the chip area, press "**OK**" and type "**24444**" in "**Load HGEN**". Then press "**OK**".

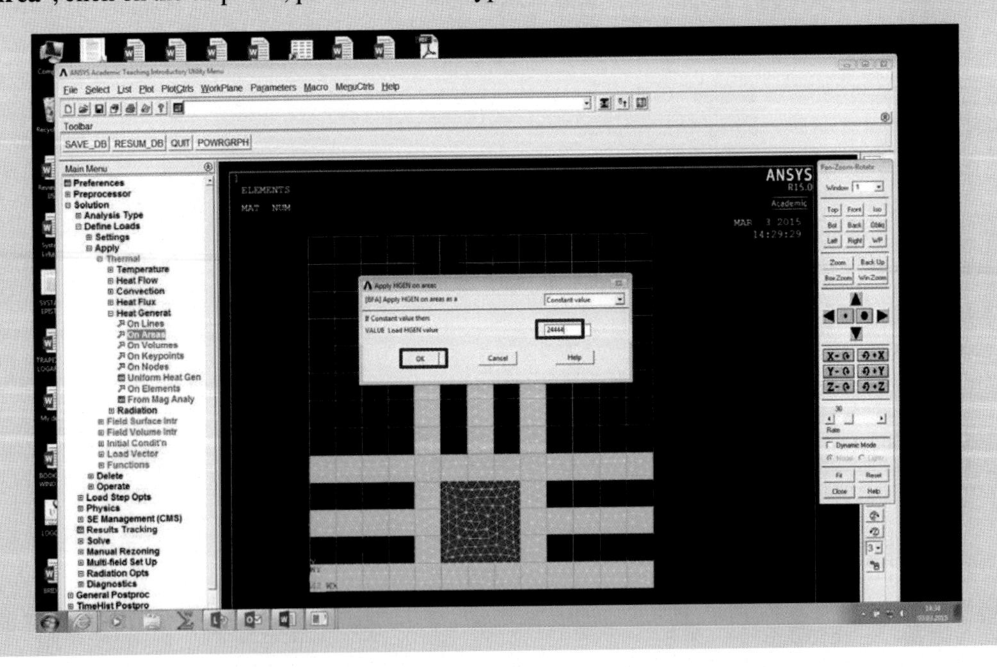

The data entry is completed. Next step is to solve the problem. Select: "**Main Menu>Solution>Solve>Current LS**". Then press "**OK**" to start the solution.

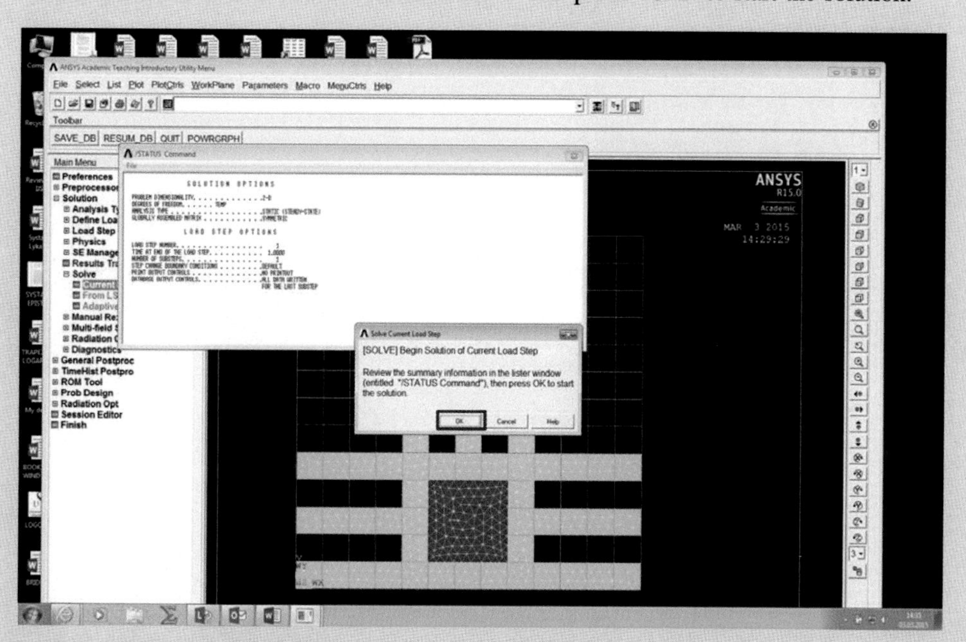

The new window indicates that the solution is done. Then press "**Close**".

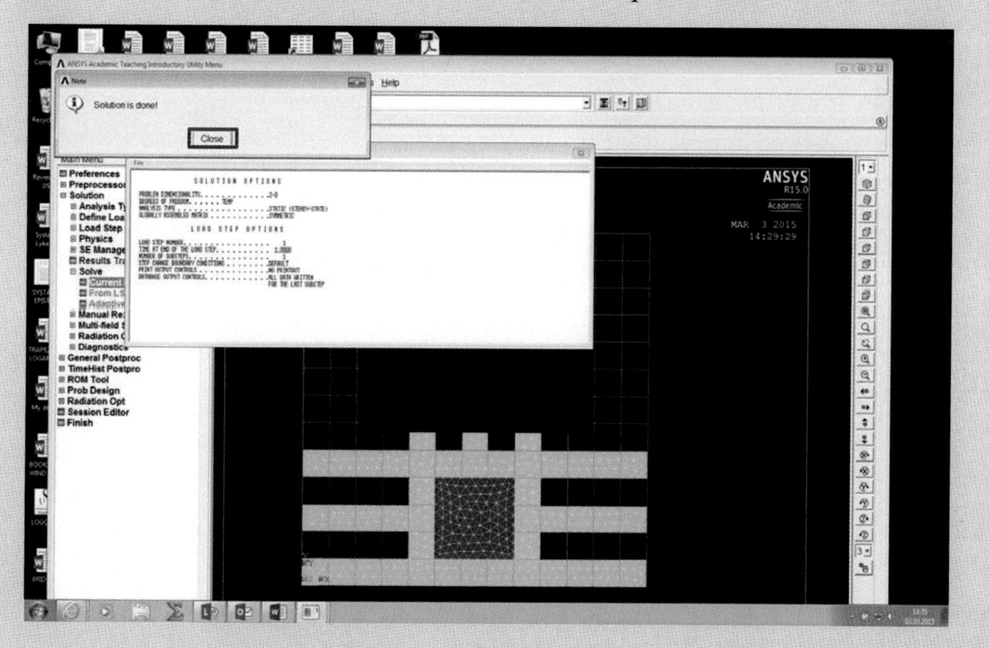

Continued

EXAMPLE 4: ANSYS IMPLEMENTATION TO SOLVE A CHIP-COOLING PROBLEM—CONT'D

Now it is time to demonstrate the results. The map of the temperature distribution can be demonstrated by selecting:

"**Main Menu>General Postproc>Plot Results>Contour Plot>Nodal Solution**". In the new window click on "**Nodal Solution>DOF Solution>Nodal Temperature**", and then "**OK**".

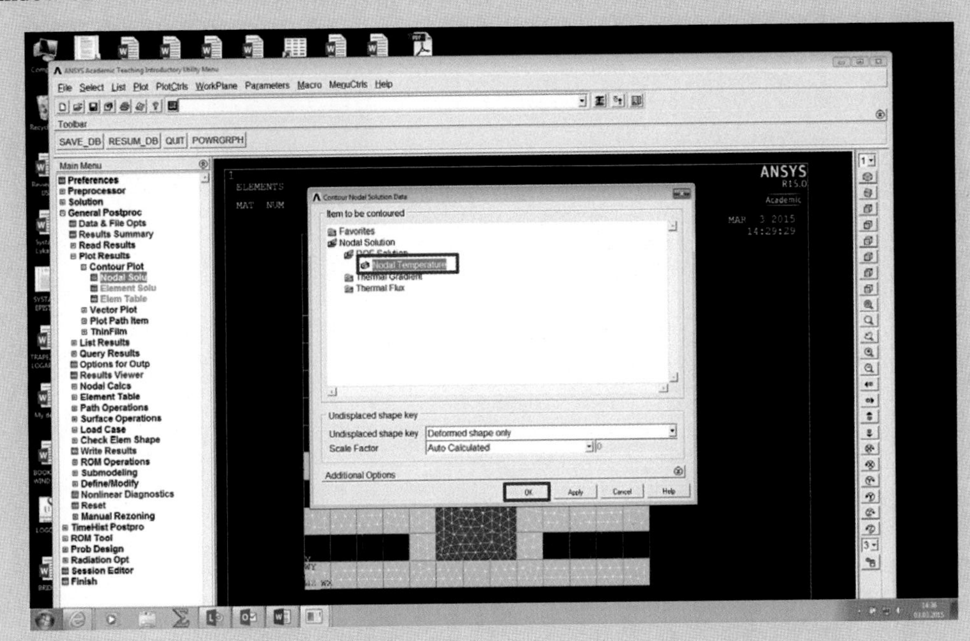

The above selections yield the following temperature distribution map:

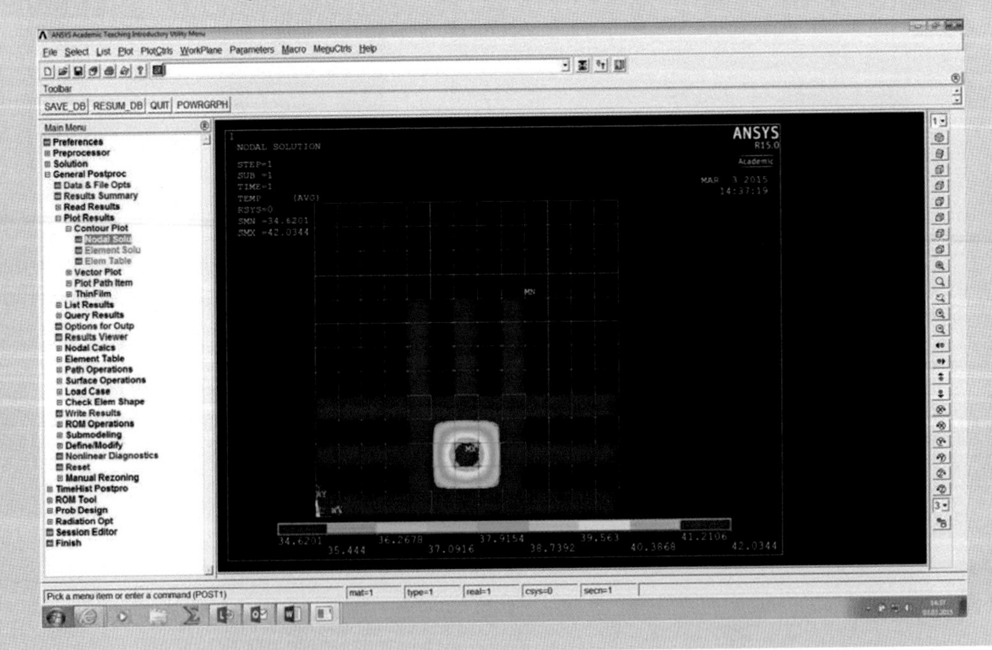

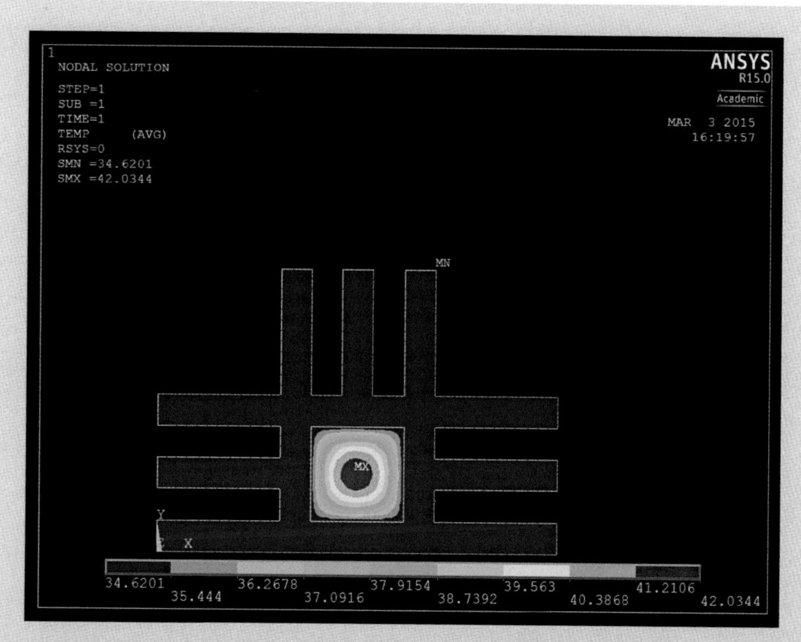

The above distribution indicates that the temperature on the fin surfaces is 34.62 °C (lowest temperature), while in the center of the chip is 41.21 °C (highest temperature). Now, we can create the map of the heat flux distribution by selecting:

"**Main Menu>General Postproc>Plot Results>Vector Plot>Predefined**"

In the new window, "**Thermal flux TF**" should be selected,

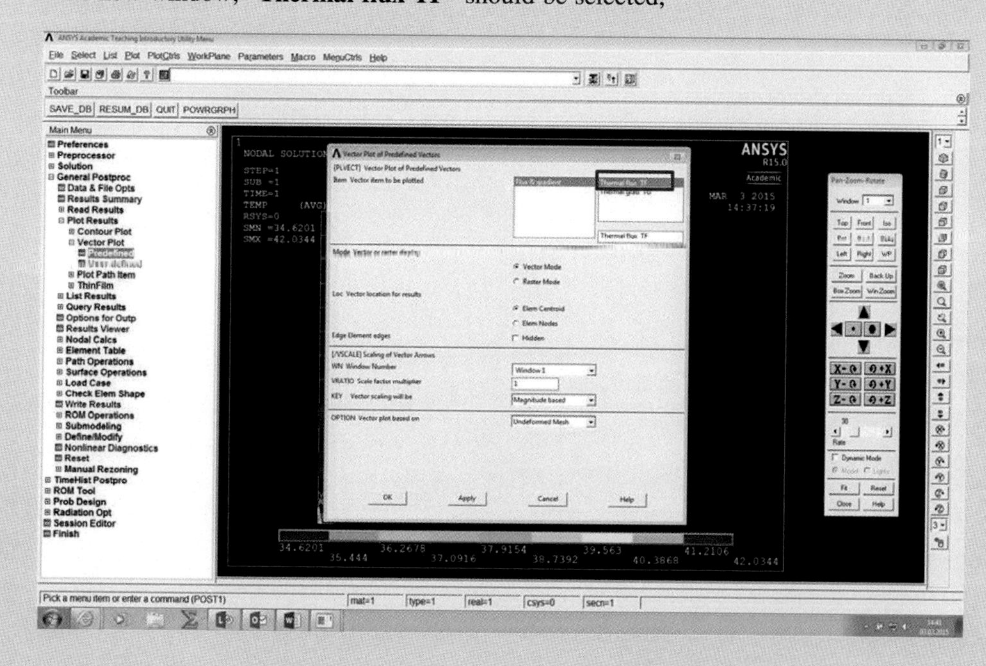

Continued

EXAMPLE 4: ANSYS IMPLEMENTATION TO SOLVE A CHIP-COOLING PROBLEM—CONT'D

yielding the following map:

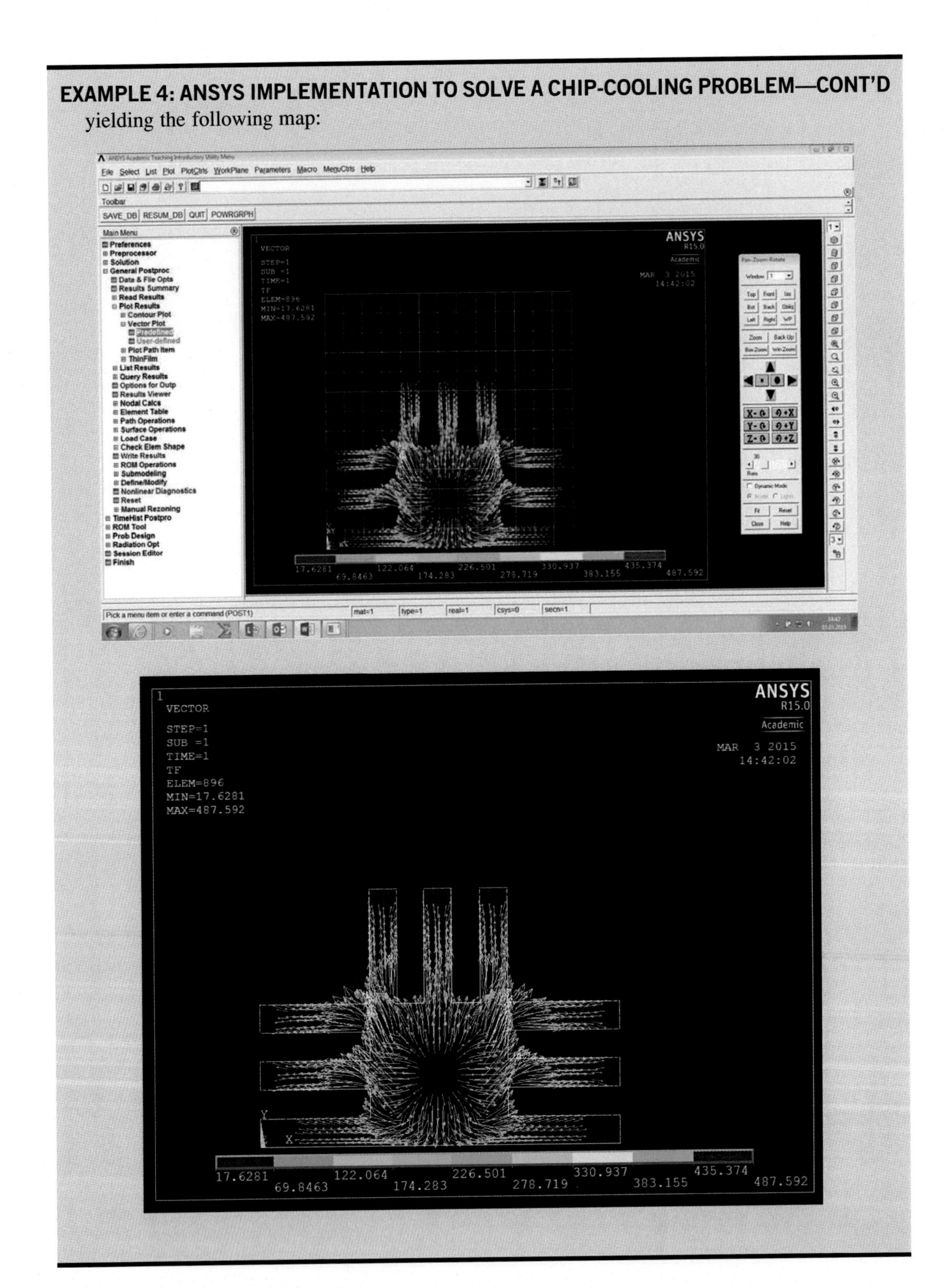

REFERENCES

[1] Holman J. Heat transfer. New York: McGraw-Hill Science/Engineering/Math; 2009.
[2] Bergman TL, Lavine AS. Introduction to heat transfer. Hoboken: Wiley; 2011.
[3] Alawadhi EM. Finite element simulations using ANSYS. Boca Raton: CRC Press; 2010.
[4] Logan DL. A first course in the Finite Element Method. Boston, MA: Cengage Learning; 2012.
[5] Cook RD, Malkus DS, Plesha ME, Witt RJ. Concepts and applications of finite element analysis. Hoboken: John Wiley & Sons; 2002.
[6] Fish J, Belytschko T. A first course in finite elements. New York: John Wiley & Sons; 2007.
[7] ANSYS, User e-manual, Version 13.

Index

Note: Page numbers followed by *b* indicate boxes, *f* indicate figures and *t* indicate tables.

D

E

P

Q

R

S

T